Differentiation of Normal and Neoplastic Hematopoietic Cells

BOOK A

First row: J. Adamson/D. Baltimore/G. Johnson/H. Kaplan
Second row: V. Ingram/G. Price, P. Fialkow
Third row: J. D. Rowley, M. Andreef, R. O'Reilly/L. Sachs, J. D. Watson
Fourth row: B. Clarkson, P. A. Marks/J. Till

Differentiation of Normal and Neoplastic Hematopoietic Cells

BOOK A

edited by

Bayard Clarkson
Memorial Sloan-Kettering Cancer Center

Paul A. Marks
Columbia University

James E. Till
Ontario Cancer Institute

**COLD SPRING HARBOR CONFERENCES ON CELL PROLIFERATION
VOLUME 5**

**Cold Spring Harbor Laboratory
1978**

**Cold Spring Harbor Conferences
on Cell Proliferation**

Volume 1   Control of Proliferation in Animal Cells
Volume 2   Proteases and Biological Control
Volume 3   Cell Motility
Volume 4   Origins of Human Cancer
Volume 5   Differentiation of Normal and
           Neoplastic Hematopoietic Cells

**Differentiation of
Normal and Neoplastic
Hematopoietic Cells**

© 1978 by Cold Spring Harbor Laboratory
All rights reserved

Printed in the United States of America

Cover and book design by Emily Harste

Library of Congress Cataloging in Publication Data

Main entry under title:

Differentiation of normal and neoplastic hematopoietic cells.

  (Cold Spring Harbor conferences on cell proliferation ; v. 5)
    Includes index.
    1. Carcinogenesis—Congresses. 2. Hematopoietic system—Congresses. 3. Cell differentiation—Congresses. 4. Myeloproliferative disorders—Congresses. I. Clarkson, Bayard. II. Marks, Paul A. III. Till, James E. IV. Series.
RC268.5.D54   616.9'94   78-60391
ISBN 0-87969-121-2

Contents

Preface   xiii

BOOK A

Section 1. Ontogeny of Hematopoietic Development and Stem Cells

Introduction   3
J. E. Till

Ontogeny of Hematopoietic Organs Studied in Avian
Embryo Interspecific Chimeras   5
N. M. Le Douarin

Fetal Hematopoietic Origins of the Adult
Hematolymphoid System   33
I. Weissman, V. Papaioannou, and R. Gardner

Clonal Analysis in Vitro of Fetal Hepatic Hematopoiesis   49
G. R. Johnson and D. Metcalf

In Vitro Analysis of Self-Renewal and Commitment of
Hematopoietic Stem Cells   63
**T. M. Dexter, T. D. Allen, L. G. Lajtha, F. Krizsa, N. G. Testa, and
M. A. S. Moore**

Approaches to the Evaluation of Human
Hematopoietic Stem-Cell Function   81
**J. E. Till, S. Lan, R. N. Buick, P. Sousan, J. E. Curtis, and
E. A. McCulloch**

The Dynamic Two-State Model of the Kinetic Behavior
of Cell Populations   93
S. I. Rubinow

Stem-Cell Heterogeneity: Pluripotent and Committed
Stem Cells of the Myeloid and Lymphoid Systems   109
R. A. Phillips

Histamine H_2-Receptor and the Hematopoietic Stem
Cell (CFU-S)   121
J. W. Byron

Human Myeloproliferative Disorders: Clonal Origin in
Pluripotent Stem Cells   131
P. J. Fialkow, A. M. Denman, J. Singer, R. J. Jacobson, and
M. N. Lowenthal

Section 2. Erythrocyte Differentiation and Regulation

Introduction   147
P. A. Marks

Regulation of the Population Size of Erythropoietic
Progenitor Cells   155
A. A. Axelrad, D. L. McLeod, S. Suzuki, and M. M. Shreeve

Competitive Effects of Erythropoietin and Colony-
Stimulating Factor   165
G. Van Zant and E. Goldwasser

In Vitro Studies of Erythropoietic Progenitor Cell
Differentiation   179
C. J. Gregory and A. C. Eaves

Differentiation of Murine Erythroleukemia Cells: The
Central Role of the Commitment Event   193
D. Housman, J. Gusella, R. Geller, R. Levenson, and S. Weil

Erythroleukemia Cells: Commitment to Differentiate
and the Role of the Cell Surface   209
R. A. Rifkind, E. Fibach, R. C. Reuben, Y. Gazitt, H. Yamasaki,
I. B. Weinstein, U. Nudel, I. Sumida, M. Terada, and P. A. Marks

Induction of Murine Erythroleukemia Cells to Differentiate: Cell-Cycle-Related Events in Expression
of Erythroid Differentiation   221
P. A. Marks, M. Terada, E. Fibach, Y. Gazitt, U. Nudel, R. Reuben,
A. Bank, and R. A. Rifkind

Modulation of in Vitro Erythropoiesis: Hormonal Interactions and Erythroid Colony Growth   235
J. W. Adamson, W. J. Popovic, and J. E. Brown

The Cell Surface and the Control of Friend-Cell
Differentiation   249
A. Bernstein, D. L. Mager, M. MacDonald, M. Letarte, F. Loritz,
M. McCutcheon, R. G. Miller, and T. W. Mak

Chromatin Changes and DNA Synthesis in Friend
Erythroleukemia and HeLa Cells during Treatment
with DMSO and n-Butyrate   261
J. R. Neumann, M. G. Riggs, H. K. Hagopian, R. G. Whittaker,
and V. M. Ingram

Biochemical and Genetic Analysis of Erythroid Differentiation in Friend Virus-Transformed Murine Erythroleukemia Cells   277
H. Eisen, F. Keppel-Ballivet, C. P. Georgopoulos, S. Sassa,
J. Granick, I. Pragnell, and W. Ostertag

Regulation of the Individual Globin Genes   295
A. W. Nienhuis, D. Axelrod, J. E. Barker, E. J. Benz, Jr.,
R. Croissant, D. Miller, and N. Young

The State of the Globin Gene in Friend Erythroleukemia Cells   311
S. K. Dube, R. B. Wallace, and J. Bonner

Sequential Gene Expression during Induced Differentiation of Cultured Friend Erythroleukemia Cells   319
M. Obinata, R. Kameji, Y. Uchiyama, and Y. Ikawa

Section 3. Granulocyte and Monocyte Differentiation and Regulation

Introduction   337
M. A. S. Moore

Regulation of Hematopoietic Differentiation and
Proliferation by Colony-Stimulating Factors   339
A. W. Burgess, D. Metcalf, and S. Russell

Inhibitors of Hematopoiesis Obtained from Mature
Neutrophils, Established Cell Lines, and Patients
with Bone-Marrow-Failure Syndromes   359
M. J. Cline, T. Olofsson, M. Frölich, S. Herman, and D. W. Golde

Use of Molecular Probes for Detection of Human
Hematopoietic Progenitors   371
G. B. Price and R. L. Krogsrud

The Maturation and Activation of Mononuclear
Phagocytes   383
Z. A. Cohn

Regulatory Interactions in Normal and Leukemic
Myelopoiesis   393
M. A. S. Moore, J. Kurland, and H. E. Broxmeyer

Microenvironmental Influences on Granulopoiesis in
Acute Myeloid Leukemia   405
P. Greenberg and B. Mara

Regulation of Normal Cell Differentiation and
Malignancy in Myeloid Leukemia 411
L. Sachs

Section 4. Lymphocyte Differentiation and Regulation

Introduction 427
H. Cantor and E. A. Boyse

Correlating Terminal Deoxynucleotidyl Transferase
and Cell-Surface Markers in the Pathway of Lymphocyte Ontogeny 433
A. E. Silverstone, N. Rosenberg, D. Baltimore, V. L. Sato,
M. P. Scheid, and E. A. Boyse

Polypeptides Regulating Lymphocyte Differentiation 455
G. Goldstein

Isolation and Structure of HLA Antigens 467
J. L. Strominger, W. Ferguson, A. Fuks, J. Kaufman, H. Orr,
P. Parham, R. Robb, C. Terhorst, M. Giphart, and D. Mann

Utility of B- and T-Cell-Specific Antisera in the Classification of Human Leukemias 479
S. E. Sallan, L. Chess, E. Frei III, C. O'Brien, D. G. Nathan,
J. L. Strominger, and S. F. Schlossman

B-Lymphocyte Development and Growth Regulation 485
F. Melchers

Murine IgD and B-Lymphocyte Differentiation 505
J. W. Uhr, J. C. Cambier, F. S. Ligler, J. Kettman, E. S. Vitetta,
I. Zan-Bar, and S. Strober

Differentiation of Precursor Cytotoxic T Lymphocytes
following Alloantigenic Stimulation 519
G. Sundharadas, M. L. Sopori, C. E. Hayes, P. R. Narayanan,
B. J. Alter, M. L. Bach, and F. H. Bach

BOOK B

Section 5. Viruses—Transformation and Differentiation

Introduction 531
D. Baltimore

The Origin, Physical Map, and Coding Properties of
the Kirsten Murine Sarcoma Virus and the Friend
Strain of Spleen Focus-Forming Virus 535
E. M. Scolnick, T. Y. Shih, D. H. Troxler, R. S. Howk, and
H. A. Young

Generation of Diversity among Murine C-Type Viruses
via Envelope Gene Recombination 553
J. H. Elder, J. W. Gautsch, F. C. Jensen, and R. A. Lerner

Analysis of the Genomes of Leukemia Viruses of the
AKR Mouse   561
J. Rommelaere, D. V. Faller, and N. Hopkins

A Receptor-Mediated Model of Viral Leukemogenesis:
Hypothesis and Experiments   577
M. S. McGrath and I. L. Weissman

The Effect of the $Fv\text{-}2^r$ Gene on Spleen Focus-Forming
Virus and on Embryonic Development   591
R. A. Steeves, F. Lilly, G. Steinheider, and K. J. Blank

Feline Oncornavirus-Associated Cell-Membrane
Antigen: A FeLV- and FeSV-Induced Tumor-Specific
Antigen   601
W. D. Hardy, Jr., E. E. Zuckerman, M. Essex, E. G. MacEwen,
and A. A. Hayes

In Vitro Transformation of Hematopoietic Cells by
Avian Erythroid and Myeloid Leukemia Viruses: A
Model System for the Differentiation of Normal and
Neoplastic Cells   625
T. Graf, H. Beug, B. Royer-Pokora, and W. Meyer-Glauner

Differentiation and Leukemic Transformation of
Thymus-Derived Lymphocytes   641
S. D. Waksal

Effects of Murine Leukemia Virus Infection on Differentiation of Hematopoietic Cells in Vitro   657
N. M. Teich and T. M. Dexter

Human Hematopoietic Cells: A Search for Retrovirus-
Related Information in Primary Tissues and Development of Liquid Suspension Culture Systems for
Studies of Cell Growth and Differentiation and of
Virus-Cell Interactions   671
R. C. Gallo, R. E. Gallagher, and F. Ruscetti

Studies of an RNA Virus Isolated from a Human
Histiocytic Lymphoma Cell Line   695
H. S. Kaplan

Section 6. Cytogenetics and Expression of Cell-Surface Antigens

Nonrandom Involvement of Chromosomal Segments in
Human Hematologic Malignancies   709
J. D. Rowley

Genetic Control of Hemoglobin Production by Murine
Erythroleukemic Cells   723
A. I. Skoultchi, S. Benoff, S. A. Bruce, P. F. Lin, H. Lonial, and
J. Pyati

Analysis of the Biosynthesis of T25 (Thy-1) in Mutant
Lymphoma Cells: A Model for Plasma Membrane
Glycoprotein Biosynthesis   741
R. Hyman and I. Trowbridge

Section 7. Marrow Architecture and Microenvironment

Hematopoietic Stem-Cell Seeding of a Cellular Matrix:
A Principle of Initiation and Regeneration of
Hematopoiesis   757
T. M. Fliedner and W. Calvo

Cellular and Architectural Factors Influencing the
Proliferation of Hematopoietic Stem Cells   775
B. I. Lord

The Orderliness of Cell Proliferation and Cell Differentiation in Relation to Kinetic Heterogeneity in Mouse
Bone Marrow   789
S. E. Shackney

Section 8. Clinical-Pathological Relationships and Differentiation of Human Hematopoietic Tumors

Introduction   807
K. Lennert

Cell Differentiation in B- and T-Lymphoproliferative
Disorders   811
D. Catovsky, M. Cherchi, D. A. G. Galton, A. V. Hoffbrand, and
K. Ganeshaguru

Membrane Phenotypes of Human Leukemic Cells and
Leukemic Cell Lines: Clinical Correlates and Biological
Implications   823
M. Greaves, G. Janossy, G. Francis, and J. Minowada

Adult Lymphoid Neoplasias of T- and Null-Cell Types   843
B. Koziner, R. Mertelsmann, D. A. Filippa, R. A. Good, and
B. D. Clarkson

Membrane Markers in Human Lymphoid Malignancies: Clinicopathological Correlations and Insight
into the Differentiation of Normal and Neoplastic Cells   859
M. Seligmann, J.-L. Preud'homme, and J.-C. Brouet

The Evolution of Immunoblastic Lymphomas in
Morphologically Nonneoplastic Immunoproliferative
Diseases   877
H. Rappaport, G. A. Pangalis, and B. N. Nathwani

Malignant Lymphomas: Models of Differentiation and
Cooperation of Lymphoreticular Cells   897
K. Lennert, E. Kaiserling, and H. K. Müller-Hermelink

Biology of the Human Malignant Lymphomas. V. Differentiation and Neoplastic Properties of Established Lymphoma Cell Lines — 915
A. L. Epstein and H. S. Kaplan

An Immunologic Approach to Classification of Malignant Lymphomas: A Cytokinetic Model of Lymphoid Neoplasia — 935
R. J. Lukes, T. L. Lincoln, J. W. Parker, and M. J. Alavaikko

Differentiation of Normal and Neoplastic Hematopoietic Cells: A Summary — 953
B. Clarkson

Name Index — 969

Subject Index — 989

Preface

The first Cold Spring Harbor meeting in this series, held in 1973, was on the subject of control of proliferation in animal cells. At that meeting, investigators from various disciplines, working with many different cell systems, presented their latest findings on basic aspects of the control of proliferation. This first meeting generally was held to be successful, and the idea behind the present meeting was to adopt a similar multidisciplinary approach in reviewing recent advances in a closely related area, cellular differentiation. Since this topic is so broad, it was decided to focus on the hematopoietic system. This system was chosen because the major pathways of differentiation are relatively well defined and considerable knowledge already exists of the regulatory factors and of the interactions occurring between different types of cells. Moreover, there have been many recent advances in methodology for examining precursor cells and the factors controlling their growth and differentiation and for identifying specific markers to more sharply define the specific steps in differentiation within the various lineages. Instead of concentrating in depth on a single lineage or limiting consideration to a few well-defined in vitro systems, we chose to consider most of the principal hematopoietic differentiation pathways and to include a broad range of topics dealing with various aspects of hematopoietic differentiation in both normal and diseased states.

Modern research methods have become so complicated and require such a high degree of specialized knowledge that nowadays it is difficult for investigators to become competent in more than one area of research. Moreover, when one is working intensively in one restricted area, it is hard to keep abreast of new developments in related areas, to sort out the good research from the bad, and to appreciate its relevance to one's own work and to the overall field. We organized this meeting to help remedy this state of affairs by bringing together prominent investigators, working on different problems related to hematopoietic differentiation, in order to discuss the latest developments in their fields, exchange information, and promote interdisciplinary research.

From the comments of those who attended the meeting, it appears that the main objectives of the meeting were satisfied and that most of us came away with a much better appreciation of what is known about hematopoietic differentiation and what remains to be learned.

We wish to express our thanks to Jim Watson for encouraging us to arrange this meeting and for his many helpful suggestions, to the session chairpersons for presenting critical overviews of the topics of their sessions and for regulating the presentations, and to all of the speakers for coming to Cold Spring Harbor and sharing their latest findings and thoughts with us. We also wish to thank Mrs. Joan Cook Carpenter and Mrs. Gladys Kist for their invaluable help in organizing the meeting and Mrs. Nancy Ford, Mrs. Annette Zaninovic, and Ms. Virginia Cleary in the Editorial Office for editing and preparing the manuscripts for publication.

Bayard Clarkson
Paul A. Marks
James E. Till

Differentiation of Normal and Neoplastic Hematopoietic Cells
BOOK A

Section 1

ONTOGENY OF HEMATOPOIETIC
DEVELOPMENT AND STEM CELLS

Introduction

J. E. Till

Ontario Cancer Institute, University of Toronto
Toronto, Ontario M4X 1K9, Canada

Central problems in studies of hematopoiesis have been, first, the identification of stem cells and, second, the elucidation of their properties. The stem cells, capable of extensive proliferation and self-renewal, able to give rise to differentiated descendants, and responsive to regulatory mechanisms, are considered to be the cells ultimately responsible for the integrity of the blood-forming system. Historical controversies about the origins of hematopoietic cells involved two opposing viewpoints: the monophyletic concept of pluripotent stem cells able to give rise to all the differentiated cells of the blood-forming system and the polyphyletic concept of several classes of stem cells, each with a restricted potential for differentiation. Work carried out over the past 15 years has demonstrated that adult hematopoietic tissue contains pluripotent stem cells. One line of evidence came from studies in which the spleen-colony technique was used in mice. This technique allows hematopoietic stem cells to be detected by their ability to give rise to macroscopic colonies of hematopoietic cells when transplanted into appropriate, lethally irradiated or genetically anemic mice. The single colony-forming units that give rise to colonies in the spleen (CFU-S) possess the capacity for differentiation into more than one cell type. Cytogenetic studies of cell populations derived from heavily irradiated cell populations have also provided evidence for pluripotent hematopoietic stem cells; the progeny of individual surviving stem cells can often be recognized by the presence of unique radiation-induced marker chromosomes. Such studies have indicated that marked stem cells can give rise to various differentiated cells found in myeloid and lymphoid tissues. For at least some of the marked stem cells, the recognizably differentiated progeny include lymphocytes as well as granulocytes, erythrocytes, and megakaryocytes.

A third line of evidence for the existence of pluripotent hematopoietic stem cells is based upon studies of human myeloproliferative disorders, especially chronic myelocytic leukemia (CML). Such evidence has come both from studies of the Philadelphia chromosome (Ph^1), an abnormality characteristic of CML, and from studies of female CML patients heterozygous for two different types of glucose-6-phosphate dehydrogenase separable by electrophoresis. These studies show that the clinically recognizable phase of CML has a clonal origin in a pluripotent stem cell that is an ancestor of granulocytes, erythrocytes, platelets, monocytes, and macrophages.

These and other studies have established the existence of pluripotent stem cells. In this regard, the historical controversy about the presence or absence of pluripotent stem cells has been resolved. However, many questions remain to be answered. What factors regulate the migration of pluripotent stem cells and their ability to colonize tissues, especially during the development of hematopoietic organs? How is the proliferation and differentiation of pluripotent stem cells regulated, and can these regulatory mechanisms be successfully transferred to cell-culture systems? How heterogeneous are populations of stem cells; for example, do cells exist that possess the self-replicative ability of stem cells but are restricted only to one or two lines of differentiation? Can assays for stem cells provide useful information about the pathogenesis of human hematopoietic disorders? These and related questions will be considered in this section.

Ontogeny of Hematopoietic Organs Studied in Avian Embryo Interspecific Chimeras

N. M. Le Douarin

Institut d'Embryologie
Centre National de la Recherche Scientifique et Collège de France
94130 Nogent-sur-Marne, France

There are a number of questions remaining regarding the development of the hematopoietic system, i.e., the origin and developmental potentialities of blood-forming cells, the differentiating and regulatory signals they receive from the hematopoietic organs, and the interrelationships of the different components of the system.

Histological observation of developing hematopoietic organs has produced a considerable amount of information related to these questions. On many points, however, these studies have resulted in controversies due to the ambiguous nature of most histological data when dynamic processes are involved. Through the introduction, during the last decades, of radioisotope labels and chromosome markers, it has become possible to trace cells in hematopoietic systems and to reevaluate some of the conclusions reached previously. Sex-chromosome markers have been especially useful in that they have allowed the blood-cell migration pattern to be followed in avian embryo parabionts or in twin embryos (Moore and Owen 1965; Metcalf and Moore 1971). Extensive traffic of cells between various anatomical compartments was demonstrated and appeared primordial for the ontogenesis of the hematopoietic system.

This presentation reports further insights into hematopoiesis during avian embryo development; these insights were derived from studies in which a stable and natural cell-marking technique was used. The latter, which has been described in detail elsewhere (Le Douarin 1969, 1973, 1976), is based on differences in the nuclear structure of two species of birds, the Japanese quail (*Coturnix coturnix japonica*) and the chick (*Gallus gallus*), which are closely related taxonomically. In the quail, a large amount of heterochromatin is associated with the nucleolus, thus making this organelle strongly Feulgen-positive. In chick cells, in contrast, the chromatin is evenly dispersed in the nucleoplasm with only some scattered chromocenters and a small amount of nucleolus-associated chromatin. As a result, when quail and chick cells have been combined and stained with Feulgen, they can be easily identified in the chimeras by their nuclear structure. At the electron microscope level, quail nucleoli are also distinguishable from their chick counterparts by the presence in the quail of a large mass of densely coiled fibrillar material — the heterochromatic DNA — associated with the usual fibrillar and granular components (Le Douarin 1971).

The developmental fate of the embryological components of the primary lymphoid organs was followed by observation of various types of chimeras whose epithelial, mesenchymal, or lymphoid cells were either of quail or chick type, according to the experimental design. A similar approach was used to trace the cell types in bone-marrow differentiation during endochondral ossification. These experiments will be described briefly in the first section of this article; the second part will deal with the hematopoietic stem cells: their embryological origin, the mechanisms that ensure their homing in the hematopoietic organs, and their developmental potentialities.

DEVELOPMENTAL POTENTIALITIES OF THE EARLY THYMIC AND BURSAL RUDIMENTS

Extrinsic Origin of the Lymphoid Stem Cells

The developmental potentialities of the early rudiments of the primary lymphoid organs were tested by heterospecific grafting between quail and chick either on the chorioallantoic membrane (CAM) of 6-day-old hosts or into the somatopleure of 3-day-old recipients. The thymic anlage was taken from the 15-somite stage to 10 days of incubation and the bursa from 5 to 15 days (Fig. 1). The grafts were retrieved when the organs had reached the total age of 14 days for the thymus and of 19 and 20 days for the bursa (age at grafting time + duration of grafting).

Normal development of the lymphoid organs occurs in all types of grafts, but the nature of the lymphoid population that develops in the explants varies according to the age of the rudiment at the time of transplantation.

In quail transplants, the lymphocytes in thymuses excised from the donor before the end of the 5th day of incubation are entirely of host origin (chick); they are a mixture of host and donor cells in explants from the 6th day; and they are all of donor type if the thymuses are taken after the end of the 6th day.

In the reverse type of transplantation, i.e., in chick thymuses grown in quail embryos, identical results can be obtained with only a slight difference in timing. The lymphocytes are of host type in rudiments taken before 6.5 days of incubation and of donor type in thymuses from 8-day-old embryos. A mixture of quail and chick lymphocytes populates the explants taken in the intermediate period, from 6.5 to 8 days. Thus, it can be assumed that the lymphoid population developing in the thymus has an extrinsic origin and derives from stem cells that invade the rudiment during a precise and limited period of time lasting about 24 hours in the quail and 36 hours in the chick (Figs. 2, 3, and 4).

The first inflow of stem cells is followed by a short period during which very few or no cells home in the organ. If "postcolonization" quail thymuses (i.e., from quail embryos from 7 to 8 days of incubation onward) are grafted

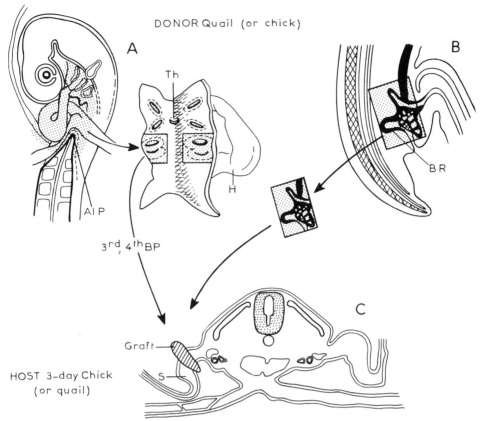

Figure 1
Interspecific graft of the thymic and bursal rudiments into the somatopleure of a 3-day-old host. (*A*) Removal of the thymic rudiment from a 30-somite embryo from which the ventral side of the pharynx comprising the branchial arches and the cardiac primordium has been removed. Bilateral areas corresponding to the 3rd and 4th pharyngeal pouches (endoderm + mesenchyme) have been selectively taken and grafted into the somatopleure of a 3-day-old host. (*B*) Removal of the bursal rudiment from a 5-day-old embryo. (*C*) Graft of the thymic or of the bursal rudiments into the somatopleure of the host embryo at 3 days' incubation. The donor and the graft were always of different species, i.e., quail or chick.

ABBREVIATIONS: AIP, anterior intestinal portal; BP, branchial pouches; BR, bursal rudiment; H, heart; S, somatopleure; Th, thyroid rudiment.

into chick embryos, chick hemocytoblasts appear in the external cortex only 4 to 5 days after the implantation. After 9 days in graft, chick lymphocytes progressively replace the initial quail lymphoid population. The replacement proceeds from the external to the internal cortex and reaches the medulla after 12 to 15 days in graft (Le Douarin and Jotereau 1975).

Similar experiments performed on the bursa of Fabricius have shown that the bursal rudiment also is not able to produce its own lymphoid stem cells but, like the thymus, is invaded by lymphocyte precursors. The invasion period of the bursal rudiment lasts longer than that of the thymus and proceeds for several days (Fig. 4) (Houssaint et al. 1976).

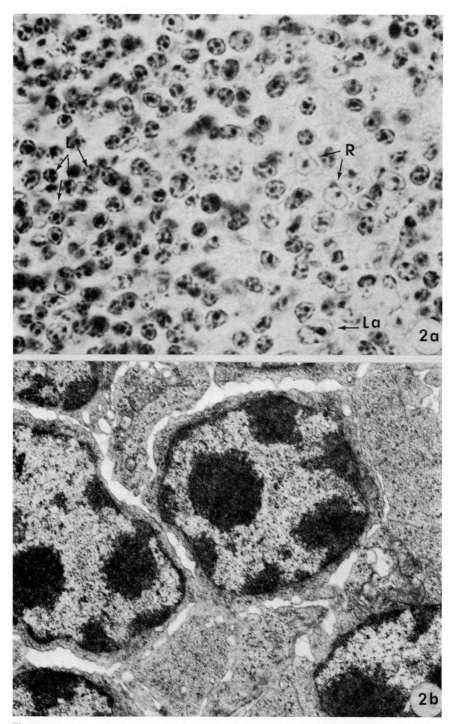

Figure 2
Quail and chick thymuses stained according to the Feulgen-Rossenbeck technique. In the quail (a,b), the lymphocytes' nuclei contain one large heterochromatic mass and several smaller ones attached to the nuclear membrane. In the chick (c,d), several small chromocenters are dispersed in the nuclear membrane. The lymphoblast shows a large nucleus with a Feulgen-positive

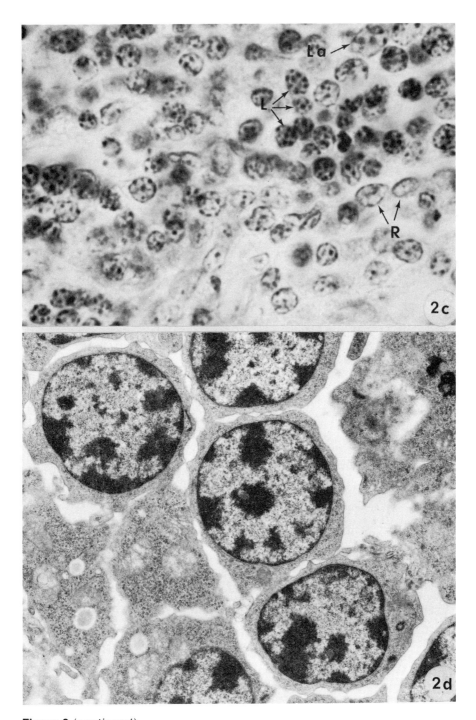

Figure 2 (*continued*)
centronuclear condensation in the quail (*a*) whereas the nucleus in the chick (*c*) is clearly stained. Reticular cells and connective cells of quail (*a*) usually have one single heterochromatic mass. In the same cell type of the chick (*c*), the nucleus contains a fine network of evenly distributed chromatin. (*a,c*) 1300×; (*b,d*) 10,500×.

ABBREVIATIONS: L, lymphocytes; La, lymphoblast; R, reticular cells.

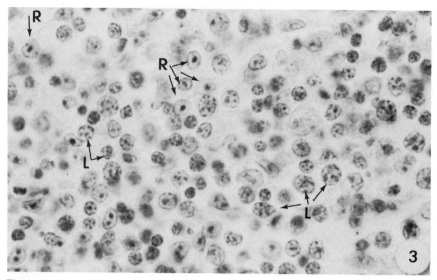

Figure 3
Ectopic thymus developed in the chick embryo body wall following the graft of the 3rd and 4th branchial pouches of a 30-somite quail embryo into the somatopleure (cf. Fig. 1). The reticular and connective cells show the typical quail nucleus, whereas the lymphocytes are of the chick host type. Feulgen-Rossenbeck staining. 1000×.

ABBREVIATIONS: L, lymphocytes; R, reticular cells.

Stem cells characterized by their highly basophilic cytoplasm home first in the bursal mesenchyme and begin to penetrate the epithelium at the apex of the bursal pouch from 10 days of incubation in the chick. From the 12th

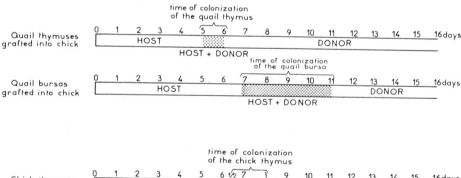

Figure 4
Time of colonization of the thymus and the bursa of Fabricius by lymphoid stem cells in chick and quail embryos.

day, the number of these cells increases rapidly, first in the mesenchyme and later (13th day) in the epithelium. Some basophilic cells remain in the mesenchyme where they differentiate into granulocytes. The most conspicuous granulopoietic activity in the bursal mesenchyme takes place from 8 to 12 days both in the quail and chick. From 11 to 12 days of incubation, follicle primordia start to develop from the epithelium. Later, lymphocyte differentiation and follicular growth proceed rapidly while, conversely, granulopoiesis decreases in the mesenchyme.

Developmental Relationships between Endoderm and Mesenchyme in Thymic and Bursal Ontogeny

The primary lymphoid organs arise as appendages of the digestive tract and, like all other components of this system, are composed primarily of an epithelial sheet surrounded by mesenchymal cells. The experiments reported above show that neither the endodermal epithelium nor the mesenchyme of the lymphoid organ rudiments can give rise to lymphocyte precursors. Further investigations have been performed on the differentiating capabilities of each of the tissue components of the bursa and thymus and on their developmental relationships during ontogeny.

A considerable body of data has shown that tissue interactions occur during development between the two components of the epitheliomesenchymal rudiments and are of decisive significance in their differentiation (for reviews, see Le Douarin 1975; Saxen et al. 1976). Similar observations were made by Auerbach (1960) for the thymus in the mouse and by me for the thymus in the avian embryo.

The pharyngeal endoderm of the 3rd and 4th pharyngeal pouches was isolated by trypsinization from 15- to 30-somite stages according to a technique that has been described previously (Le Douarin 1967; Le Douarin and Jotereau 1975). It was then implanted interspecifically into the somatopleure of a 3-day-old host, where it differentiated and formed an heterotopic thymic lobe (Fig. 5). Analyses of cell components of the thymic tissue showed that the grafted endoderm had differentiated into reticular epithelial cells, while the other cell types (connective cells, endothelium, and lymphocytes) were contributed by the host. This result indicates that the thymic mesenchyme does not exert a specific effect on the branchial pouch endoderm but can be replaced by lateral plate mesoderm. The latter is able to promote thymic differentiation in the epithelium.

To identify the differentiated cell types derived from the thymic mesenchyme, the latter was selectively labeled in a second step (Le Douarin and Jotereau 1973, 1975).

In previous studies the quail-chick marker system was used by our group to follow the migration of the neural crest cells. It was shown that the mesenchyme that develops in the branchial arches derives from the mesencephalic and rhombencephalic parts of the neural crest: it is the so-called mesectoderm. Therefore, the interspecific graft of a fragment of the quail neural primordium into a chick at the adequate level results in the colonization of the host branchial arches by labeled cells, and it is possible to follow the participation of the mesectodermal cells in thymic ontogeny. These mesecto-

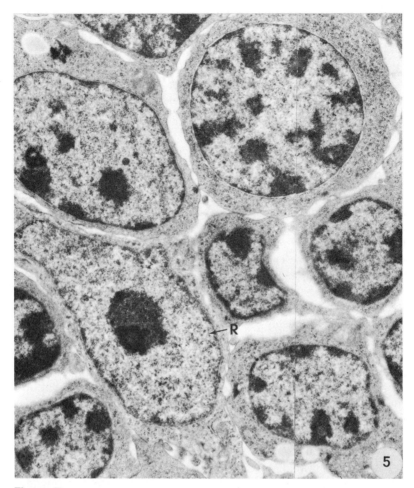

Figure 5
Ectopic thymic lobe resulting from the graft of the isolated thymic endoderm taken from a 30-somite quail embryo into the somatopleure of a chick host. The reticular cells have the large DNA-rich nucleolus of quail cells, whereas the lymphocytes belong to the chick species. Connective cells of the thymic tissue in this experiment were derived from the chick somatopleural mesenchyme. Uranyl acetate-lead citrate staining. 9360×.

ABBREVIATION: R, reticular cells.

dermal cells appear to give rise only to the connective tissue lining the blood vessels in the cortex and the medulla. They were never found to contribute to the lymphoid population or to constitute the endothelial wall of the blood vessels, the latter always arising entirely from the host mesoderm (Le Douarin and Jotereau 1975; Le Lièvre and Le Douarin 1975).

The developmental relationships between the epithelial and mesenchymal components of the bursa of Fabricius appeared to obey stricter requirements than was the case for the thymus. Endoderm can be separated from mesoderm in the bursa by enzymatic digestion from 5 to 12 days of incubation in both quail and chick species (Houssaint et al. 1976). Homo- or heterospecific recombinations of dissociated endoderm and mesenchyme can be performed and lead to a normal development of the bursa when the recombined organs are grafted into a host embryo.

On the contrary, if the endoderm is associated with the somatopleural mesenchyme of a 3-day-old host, it survives but does not differentiate into follicles. The basophilic cells which the endoderm contained at the time of grafting leave the epithelium. Some can be found from 2 to 3 days after being grafted in the surrounding mesenchyme of the host.

Several other sources of mesenchyme, such as the somites and intestine were found to be associated with the bursal endoderm. None were able to promote the development of the epithelium, which neither retained lymphoid stem cells nor formed follicular buds (E. Houssaint and N. M. Le Douarin, unpubl.).

The bursal mesenchyme from 5- to 7-day embryos was also associated with the thymic endoderm taken from the pharynx at 25- to 30-somite stages. No lymphoid organs resulted from these chimeras; the thymic epithelium, identified by its nuclei, developed poorly and did not give rise to reticular cells.

Inasmuch as the bursal epithelium arises from the posterior gut, it seemed interesting to investigate the ability of the mesoderm of the bursa to induce bursal differentiation in the digestive endoderm. Associations of the endoderm isolated from the rectum of 5-day-old embryos with the mesenchyme of the bursa at the same developmental stage did not result in bursal histogenesis. The gut epithelium did not develop follicles and was not colonized by lymphoid stem cells.

The developmental potentialities of the bursal mesenchyme, when separated from the endoderm, were also studied. This was done by isolating the mesenchyme of 10- to 12-day-old quail bursas and grafting it into the 3-day-old chick somatopleure; the latter tissue being considered as a culture medium for the graft. The lymphoid stem cells present in the explant at the time of grafting began to disperse into the surrounding mesenchyme soon after the implantation and later disappeared. Basophilic cells were detectable in the quail mesenchyme for about 3 days after implantation; whether they died or penetrated the host blood vessels could not be determined due to their small number. When granulopoiesis was in progress in the bursas at the time of grafting, granulocytes could be seen for several days inside of and in the vicinity of the explant.

Thus, tissue interactions between the mesenchymal and the endodermal components of the bursal primordium are of decisive importance for the lymphoid differentiation of the organ. The differentiating signals exchanged between the epithelium and the mesenchyme appear to be very specific, since none of the designed heterologous combinations resulted in bursal development. Such a situation is exceptional in epitheliomesenchymal (resulting from the cooperation of endoderm + mesoderm or of ectoderm + mesoderm) organ histogenesis in which the effect exerted by the mesenchyme is usually "permissive" in nature and shows a high degree of nonspecificity. In most cases, the mesenchyme of the organ can actually be replaced by mesenchymal tissues taken from nearly any other rudiment at a similar developmental stage (for review, see Saxen et al. 1976). The morphogenetic signal(s) coming from the mesenchyme provides conditions that enable the target epithelial cells to express preexisting differentiative bias. In the case of the bursa of Fabricius, one of the striking consequences of the separation of the epithelium from its mesenchyme was that neither one nor the other remained capable of retaining lymphoid stem cells when morphogenetic tissue interactions were prevented. Therefore, the cellular mechanisms that control stem-cell homing depend upon reciprocal tissue interactions occurring between the two components of the rudiment.

It is important to stress that thymic histogenesis does not involve the same strictly specific tissue interactions between the epithelial and the mesenchymal parts of the rudiment. Any stroma originating from the lateral plate mesoderm is able to promote the development of thymic endoderm. However, neither bursal mesoderm nor somite-derived mesenchymal tissues support the differentiation of the thymic epithelium (Le Douarin et al. 1968).

The effect of testosterone on bursal development has been described frequently (Kirkpatrick and Andrews 1944; Glick 1957, 1964; Meyer et al. 1959; Glick and Sadler 1961; Warner and Burnet 1961; Papermaster et al. 1962). Treatment of embryos with testosterone at any developmental stage stops the development of the bursa and inhibits B-cell differentiation. When 2.5 mg of testosterone propionate are injected into the egg, the lymphoid population of the bursa disappears completely within a few days; only granulocytes remain abundant in the mesenchyme when the injection is done from day 9 of incubation onward. The follicles either do not grow or they regress, depending upon the age at which the hormone was administered.

In a series of grafting experiments, Moore and Owen (1966) showed that normal bursas transplanted into testosterone-treated host embryos underwent lymphoid differentiation, although the host bursa completely failed to develop. This indicates that bursal lymphoid precursor cells are not injured by testosterone treatment but that stem-cell homing is suppressed. Inasmuch as this latter process depends on tissue interactions between the epithelium and the mesenchyme, it seemed interesting to learn whether testosterone prevents only one or both of the bursal tissue components from playing a role in bursal ontogeny. To answer this question, the following experiment was devised (N. M. Le Douarin, in prep.). After being incubated for 7 days, quail and chick eggs were injected with 2.5 mg testosterone propionate. Two days later, the bursas were removed and dissociation of the endoderm from the mesenchyme was carried out by using a solution of 1% trypsin in Ca^{++}-Mg^{++}-free Tyrode solution. Normal bursas from 9-day-old quail and chick embryos were treated in the same way. Thereafter, the following combinations were made:

Group 1 – Normal chick endoderm + testosterone-treated chick mesenchyme; normal quail endoderm + testosterone-treated quail mesenchyme; normal chick endoderm + testosterone-treated quail mesenchyme; and normal quail endoderm + testosterone-treated chick mesenchyme.

Group 2 – Normal chick mesenchyme + testosterone-treated chick endoderm; normal quail mesenchyme + testosterone-treated quail endoderm; normal chick mesenchyme + testosterone-treated quail endoderm; and normal quail mesenchyme + testosterone-treated chick endoderm.

The associated tissues were cultured for 12 hours on a semisolid culture medium according to the technique of Wolff and Haffen (1952) and were grafted for 10 days into the somatopleure of 3-day-old chick embryos.

In 12 of the 17 explants of group 1, normal bursal tissue with lymphoid follicles developed. In contrast, bursal histogenesis failed to occur in all the explants of group 2. In these cases, the epithelium remained undifferentiated in a way similar to that of bursas in testosterone-treated whole embryos (Fig. 6).

In one case, the endoderm was only partly removed from a normal quail bursa and replaced by testosterone-treated chick endoderm. Lymphoid follicle formation normally occurred in the untreated quail endoderm, whereas the neighboring treated chick endoderm, although in contact with the same mesenchyme, did not differentiate. One can conclude from these results that the endoderm definitely loses its developmental capabilities when treated with testosterone and cannot respond to the morphogenetic signals originating from the mesenchyme. In contrast, in these experiments, the mesenchyme was only transitorily affected by the androgen and remained capable of promoting the differentiation of an intact endoderm. It was also able to retain the hematopoietic stem cells if it was associated with a normal bursal epithelium.

ORIGIN OF BONE MARROW CELLS IN ENDOCHONDRAL OSSIFICATION

In a series of experiments carried out by our group, the ontogeny of bone marrow was studied during endochondral ossification. Interspecific CAM transplantations of limb buds (taken from 3 to 4 days of incubation) and of cartilaginous femur rudiments were performed between quail and chick embryos. The graft endured from 6 to 15 days. Endochondral ossification occurred and, owing to the nuclear labeling, the contribution of host and donor cells to the differentiated bone marrow could be observed. Chondrocytes, osteoblasts, osteocytes, and reticular cells of the bone marrow appeared to be derived from the limb bud mesenchyme which gave rise to the cartilaginous primordium of the bone. In contrast, the endothelium of the blood vessels that penetrated the cartilage and all the hematopoietic cells that colonized the bone marrow were of host origin. The osteoclasts, always of the same nuclear type as the hematopoietic cells, apparently arose by fusion of blood-borne mononucleate cells.

From parabiosis experiments on chick embryos, Moore and Owen (1967) have suggested an extraosseous origin of the hematopoietic population of the marrow. However, in that study, sex chromosomes used as markers did not provide identification of the entire cell population of the marrow. Therefore, a participation of limb-bud mesenchymal cells to hematopoiesis could not be ruled out. With the quail-chick marker system, we have shown that

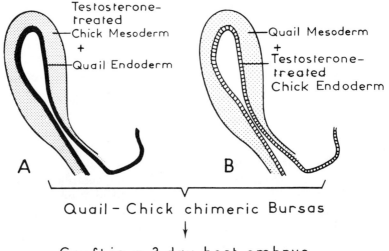

Figure 6

(*a*) Heterospecific combinations of bursal endoderm and mesenchyme between testosterone-treated chick bursa and normal quail bursa at 7 days' incubation. (*b*) Development of normal bursa shown in part a-A; the connective cells are of chick type. Lymphoid follicle contains quail reticular cells and chick lymphocytes. 1000×. (*c*) In part a-B, the testosterone-treated endoderm does not differentiate lymphoid follicles. Feulgen staining. 1000×.

ABBREVIATIONS: BC, bursal cavity; C, connective cells; Ce, testosterone-treated endoderm; L, chick lymphocytes; Qe, quail epithelium; Qm, quail bursal mesenchyme; R, quail reticular cells.

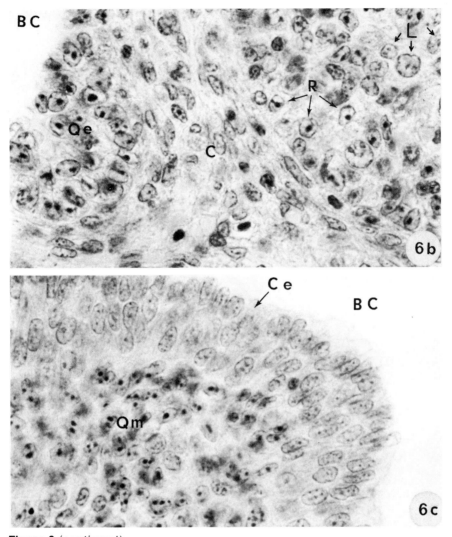

Figure 6 (continued)

the limb-bud mesenchyme and, later, the perichondral cells give rise to the osteogenic cell line; therefore, this line is distinct from the hematogenic line that is extraosseous in origin and includes endothelial cells, preosteoclasts, and hematopoietic stem cells (Le Douarin et al. 1975; Jotereau and Le Douarin 1978).

Inasmuch as all the intraembryonic organs fail to become hematopoietic if they are not colonized by extrinsic stem cells, it follows that the yolk sac is the only active blood-forming tissue provided with its own stem cells. The developmental capabilities of the embryonic stem cells, the mechanisms that control their homing to the primary lymphoid organs, and their embryological origin are discussed in the next section.

INVESTIGATIONS OF THE DEVELOPMENTAL POTENTIALITIES, HOMING, AND EMBRYONIC ORIGIN OF THE LYMPHOID STEM CELLS

Differentiation of Lymphocytes in Interspecific Combinations

During the experiments described in the first part of this review, the chick lymphoid stem cells were able to colonize quail thymic and bursal rudiments (and vice versa) and to undergo cell divisions and morphogenetic changes characteristic of lymphocyte differentiation. However, it remained dubious whether the lymphocytes developing in the microenvironment of a foreign species acquired a normal differentiation pattern. To settle this problem, I searched for some surface antigens that could be detected on normal lymphocytes.

Six-day-old chick precolonization thymic rudiments were grafted into 3-day-old quail embryos for 8 days and 4-day-old quail thymuses were grafted for 10 days into 3-day-old chicks. The lymphocytes were suspended and treated with an antithymic serum. (For details on antibody preparation, see Wick et al. 1973.) The lymphoid cells that showed the nuclear marker of the host embryo were found to bear surface antigens as in control thymuses of the same developmental stage (Fig. 7). The same experiment was done for the bursa of Fabricius. Double labeling for surface immunoglobulin (Ig) was performed using a rabbit antibody raised against chicken Ig and a goat reagent raised against rabbit Ig and labeled with fluorescein isothiocyanate. It was applied first to normal bursal lymphocytes of 19-day-old chick and 16-day-old quail embryos and appeared to label quail as well as chick surface Ig. In a second step, the antisera were applied to chick lymphocytes that

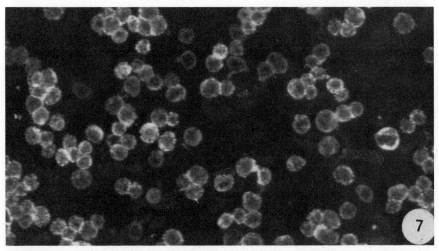

Figure 7
Application of an antithymic serum to 14-day thymocytes developed in a thymus resulting from the graft of a 30-somite quail thymic rudiment into a chick embryo. The thymocytes originating from the host differentiate surface antigens in heterospecific tissular environment as they do in a normal thymus of the same age. 1000×.

had differentiated in a quail bursa grafted in a chick host and to quail lymphocytes that had developed in a chick bursa grafted into a quail host. As in the experiments described above for the thymus, the bursal rudiments were implanted before their colonization by lymphoid stem cells and the grafts were collected when the total age of the explant was 19 days.

Under these conditions, the lymphocytes in the graft developed surface Ig with a time course paralleling that observed in control quail and chick (N. M. Le Douarin, unpubl.). Thus, lymphoid stem cells of the quail were able to respond to the differentiating signal(s) originating from the bursal and thymic chick epithelium, and vice versa. In our experimental conditions, no barrier existed between these two closely related species that would hamper differentiating processes leading to lymphocyte functional development.

Mechanisms Controlling the Time of Homing of Lymphoid Stem Cells

Studies reported above have shown that stem-cell colonization of the embryonic thymus is limited to a brief period and begins at a precise stage of development (6.5 days of incubation in the chick and 5 days in the quail). The next step was to investigate whether the onset of stem-cell inflow depends on the availability of hemocytoblasts in the blood or on the intrinsic capability of the thymic rudiment to retain them. The following experiment was devised toward answering this question. The thymic rudiment was removed from quail embryos prior to the time of its invasion by lymphoid stem cells (after 4 days incubation). It was transplanted first into a 3-day-old chick host for 2 days and then moved into a 3-day-old quail acting as a second host for 8 days. When the total age of the grafted thymus had reached 14 days, the organ was observed after being fixed and stained with Feulgen. The lymphoid population of the organ was found to be entirely of chick type, whereas reticular and connective cells belonged to the quail species. This result shows that stem cells able to colonize the thymus are available in the blood as early as 3 to 4 days of incubation in the chick; that is, at least 2 days earlier than the time of thymic colonization in this species. Therefore, the onset of stem-cell homing depends upon the maturation of the thymic anlage and not upon the availability of hemocytoblasts in the blood (Le Douarin and Jotereau 1975).

The various transplantation experiments that were carried out clearly indicate that, after 24 hours in the quail and 36 hours in the chick, stem cells enter the thymus at a significantly slower rate or even stop completely for a while. Consequently, I wondered whether the thymic rudiment loses its capacity to retain stem cells at the same stage as in normal development (8 days of incubation in the chick, 6 days in the quail) after the homing process has been blocked experimentally. Thymic rudiments, taken before their colonization had begun, were cultured either in vitro or in a diffusion chamber on the CAM of 8-day-old chick embryos for 4 to 6 days (Fig. 8). No stem cells could penetrate the primordium, and the normal colonization period was over at the end of the culture time. The explant was then transplanted heterospecifically into the somatopleure of 3-day-old hosts (chick for quail thymus and conversely) and observed 10 days after grafting. In all cases,

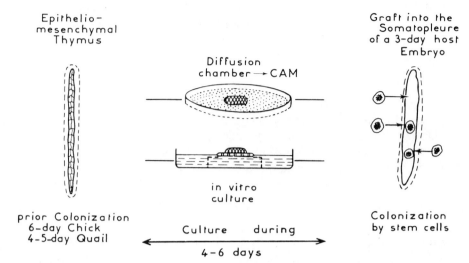

Figure 8
Experiment designed to test the maintenance of the receptive capacity of the thymic rudiment for lymphoid stem cells in either diffusion chamber or in vitro cultures.
ABBREVIATION: CAM, chorioallantoic membrane of the chick.

normal typical histogenesis occurred and the lymphoid population that developed in the explants was of the host type (Le Douarin et al. 1976).

This result demonstrates that the capacity to attract stem cells is maintained in the epitheliomesenchymal thymic rudiment isolated in culture, as long as it is not colonized. Arrest of the homing process seems to depend on the amount of stem cells present in the explant.

The next point of interest concerned the mechanism by which blood-borne hemocytoblasts halt near the receptive thymic anlage and subsequently enter it. Two series of experiments were devised to see whether the thymic and bursal rudiments attract the circulating stem cells through the production of a chemotactic substance.

Is the Homing of Stem Cells Regulated by a Chemotactic Mechanism?

Transplantation of a Chick Lymphoid Rudiment in the Receptive State in Contact with a Quail Lymphoid Organ at the Postreceptive Period

Quail and chick organs were inserted into a slit made in a 3-day-old chick somatopleure (Fig. 9). Six days after being grafted, the explants were observed in serial sections that had been stained according to the Feulgen-Rossenbeck technique. The various types of combinations performed are indicated in Table 1. The chimerism found in certain of the organs showed that stem cells that have already homed to the thymus or to the bursa can migrate out of the rudiment after being grafted and home for a second time to another lymphoid organ, provided the latter is in a receptive state. In these conditions, stem cells that have already penetrated the bursa of Fabricius are able to differentiate along the lymphoid line in the thymic environment.

In another experiment, lymphoid rudiments of both quail and chick, already colonized by their hemocytoblasts, were cografted into the chick somatopleure for 6 days. No traffic of stem cells between the grafts was observed in these cases. It was noticeable in these experiments that the two confronted organs remained separate and that no fusion occurred even if they were in close contact at the time of transplantation (Le Douarin et al. 1977).

In another series of experiments, the ability of cells that have already homed to the bursal epithelium to colonize the thymus was tested. Bursas from 11- and 12-day-old quail embryos were dissociated by trypsinization, and the epithelial part of the rudiment was grafted into the somatopleure of a 3-day-old chick in contact with 6-day-old chick thymuses. In 4 out of 10 grafts, about 75% of the lymphoid cells in the host thymus were quail and, thus, had migrated from the grafted bursal epithelial component.

These observations suggest that stem cells that spread from the "postcolonization" bursal graft are attracted by the "receptive transplant," which probably produces a chemotactic substance. They also show that the same stem cells can colonize indifferently the thymus and the bursa primordium. On the other hand, these experiments raise the question of whether the bursal stem cells that invade the thymus differentiate into B or T lymphocytes. It is generally accepted that the thymic and bursal environment are responsible for the orientation of lymphoid stem cells along the T- or B-cell differentiation pathways.

If the cells that have homed for some time to the bursa have already acquired the B-cell developmental program, they may express the B-cell phenotype even when located in the thymic environment. Alternatively, they may have remained pluripotent and may adjust their developmental fate to the new milieu in which they have become localized. Experiments are now in progress to study this question and to investigate the stage at which the cells become irreversibly committed to differentiate along the B- and T-cell lines.

Transfilter Cultures of Quail and Chick Lymphoid Organ Rudiments

The chemotactic hypothesis was tested in transfilter cultures. Six-day-old chick thymuses and 11-day-old quail bursas were placed on each side of a nucleopore filter with a pore diameter of 5 μm and cultured for 2 to 4 days. At that time, the thymus either was fixed for histology or was grafted into the somatopleure of a 3-day-old chick host to allow subsequent lymphoid differentiation. After 2 to 4 days in culture, a number of quail cells were found in the chick thymus (Fig. 9; Table 2). If transplanted into a chick host for a further 6 days, the chick thymus contained a mixture of lymphocytes from both species. This means that the lymphoid stem cells that had first homed into the quail bursa crossed the filter and completed their lymphoid differentiation in the thymic environment.

The organ specificity of thymus "attractivity" was tested by the replacement of the thymic primordium by other organs taken from chick embryos at the same developmental stage: 6-day-old liver and mesonephric tissues were associated in a *cis-trans* position with bursas from 11-day-old quails.

The number of quail cells originating from the bursa was counted after the organs had been 48 and 72 hours in culture. The results reported in Table 2 show that practically no cells migrated into the kidney tissue, a certain number of cells migrated into the liver, and that many more cells migrated to the thymic rudiment. In addition, quail cells in thymic transfilter cultures penetrated deep into the explant, whereas in liver and kidney cultures, they remained predominantly in contact with the filter.

These experiments suggest a chemotactic mechanism for the invasion of the developing lymphoid organs by stem cells. However, since the transfilter cultures lasted only 48 hours, an active stimulation of stem-cell division in the thymus may be partly responsible for the large number of quail cells found in this rudiment. Although further investigations are needed to clarify this point, I propose, as a working hypothesis, that the epithelium of the thymic rudiment, from 6.5 days' incubation in the chick, produces a chemotactic substance; and that this substance diffuses through the organ and arrests the stem cells, causing them to cross the wall of the blood

I TRANSFILTER CULTURE

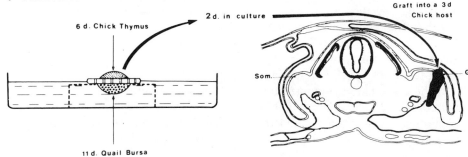

II DOUBLE TRANSPLANTATION

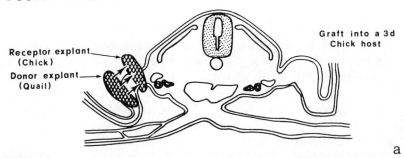

a

Figure 9
(*a*) Experimental procedures to induce lymphoid stem cell migration from a postcolonization quail lymphoid rudiment to a precolonization chick lymphoid organ. (*b*) Result of the transfilter culture; after 2 days in culture, numerous quail cells have colonized the rudiment. 1000×. (*c*) The same explant as shown in *b* has been transplanted into a chick embryo. Numerous quail lymphocytes have developed in the thymus. These arose from the cells originating from the quail bursa. 1000×.

ABBREVIATIONS: d, day; G, graft; Som., somatopleure.

vessel and to migrate actively into the epithelium through the basement membrane along an increasing gradient of concentration. When settled in the thymus, the stem cells make close membrane contacts with the epithelial cells (Le Douarin et al. 1977). When a certain number of stem cells have seeded the thymic epithelium, the latter becomes unattractive, probably because the production of the substance responsible for the attraction has stopped. At that time, the stem cells can leave the organ for a period of time if they are experimentally subjected to the attraction of another lymphoid rudiment.

Embryonic Origin of Lymphoid Stem Cells

According to the hypothesis of Moore and Owen (1965, 1967), intraembryonic hematopoiesis is seeded by blood-borne stem cells originating

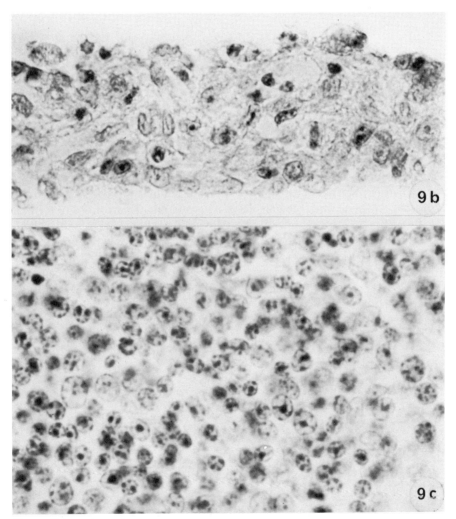

Figure 9 (continued)

Table 1
Results of Combinations of Chick and Quail Lymphoid Rudiments Grafted at Two Developmental Stages

Developmental stages and combinations	Number of cultures	Number of chimeras	
		chick organ	quail organ
Receptive period × postreceptive period			
6-day-old chick thymus × 11-day-old quail bursa	13	10	0
6-day-old chick thymus × 8-day-old quail thymus	6	4	0
7-day-old chick bursa × 11-day-old quail bursa	8	1	0
Postreceptive period × postreceptive period			
10-day-old chick thymus × 8-day-old quail thymus	10	0	0
10-day-old chick thymus × 11-day-old quail bursa	28	0	0

in the yolk sac. In cultures of presomite mouse embryos deprived of yolk sac, normal development of the hematopoietic organs occurred but no evidence of hematopoiesis or hematopoietic stem-cell development was obtained. In contrast, cultures of presomite yolk sac showed marked erythropoiesis with increased numbers of colony-forming units—culture (CFU-C) and colony-forming units—spleen (CFU-S). On the other hand, in intact presomite embryos maintained in vitro, normal development of hematopoiesis occurred in both the intraembryonic organs and in the yolk sac (Moore and Metcalf 1970; Moore and Johnson 1976). In the avian embryo, suspensions of cells from the yolk sac were shown to be able to reconstitute the blood-cell population when they were injected into sublethally irradiated embryos (Moore and Owen 1967).

An exclusive yolk-sac origin of stem cells was challenged recently by Dieterlen-Lièvre (1975), and by Dieterlen-Lièvre et al. (1976), who used another experimental approach based on the quail-chick marker system. Quail embryos, separated from their own yolk sac, were grafted prior to

Table 2
Migration of Quail Cells in Transfilter Cultures of 11-Day-Old Quail Bursa and 6-Day-Old Chick Organ Rudiments

Chick organs	Numbers of quail cells after a 48-hour culture (mean per culture)
Thymus	291
Liver	85
Mesonephros	3

Six cultures were observed per organ. Numbers of quail cells were counted in 10 sections of each culture.

the onset of circulation onto a chick embryo yolk sac in the chick egg (Fig. 10). In successful grafts, vascular anastomosis rapidly developed between the yolk sac and the embryo, and free circulation of blood cells occurred between the two components of the chimera according to a normal pattern. However, the lymphoid organs and the spleen of the quail embryos did not become colonized by chick yolk-sac stem cells but predominantly—or even exclusively in most cases—by quail cells. On the other hand, Dieterlen-Lièvre's experiments also provide indications that erythrocytes are of the quail type as soon as bone-marrow erythropoiesis starts in the embryo and takes over in blood production. This indicates that hematopoietic foci able to provide hematopoietic stem cells exist inside the embryo and probably are responsible for the colonization of the intraembryonic blood-forming organs in normal conditions. In Moore's and Owen's experiments with injections of yolk-sac cells into an irradiated host, the cells were taken from 7-day-old embryos. At that stage, cellular traffic between intra- and extraembryonic stem-cell compartments has occurred and, in view of Dieterlen-Lièvre's results, one wonders whether stem cells of intraem-

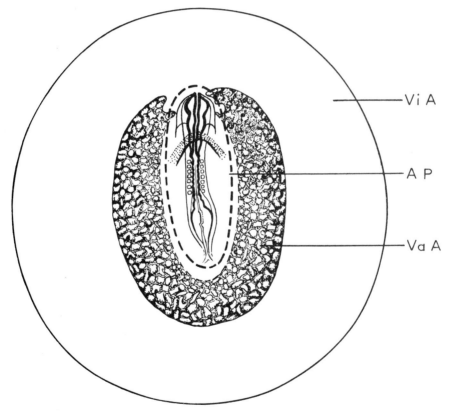

Figure 10
Quail-chick yolk sac chimeras obtained by grafting a quail area pellucida onto a chick yolk sac in a chick egg before the beginning of blood circulation. Drawing courtesy of Dr. F. Dieterlen-Lièvre.
ABBREVIATIONS: AP, area pellucida; VaA, vascular area; ViA, vitelline area.

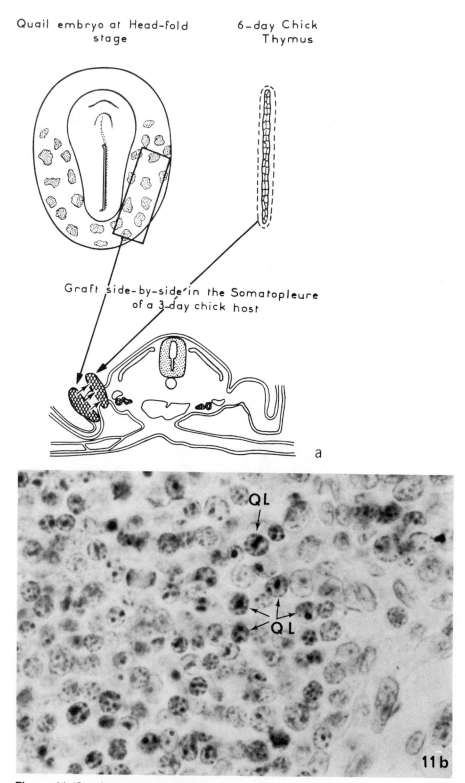

Figure 11 (See facing page for legend.)

bryonic origin were not solely responsible for the repopulation of the intraembryonic hematopoietic organ. It was intriguing, therefore, both to determine whether stem cells of definite yolk-sac origin have the potentiality to home to the thymus and bursa of Fabricius and to follow the lymphocyte differentiation pathway. Adequate experimental conditions were designed by associating a fragment of quail yolk sac, taken from the head-process stage to the two-somite stage with a 6-day-old chick embryonic thymus. The two tissues were grafted together into the somatopleure of a 3-day-old chick host. In 6 out of 10 explants observed 9 days after grafting, numerous quail thymocytes were seen in the chick thymus (Fig. 11). These cells were morphologically similar to control quail lymphocytes at the same age. Whether they bear thymus-specific surface antigens has not been determined, but it does seem likely that they do (Jotereau and Le Douarin 1978).

CONCLUSIONS AND DISCUSSION

The quail-chick cell-marker technique can be used profitably to investigate some aspects of the development of the hematopoietic system in the avian embryo. For example, it has clarified some problems regarding the ability of the rudiments of the intraembryonic blood-forming organs to produce hematopoietic stem cells. The early steps of development of the hemocytoblasts and the tissular environment in which they differentiate appear unequivocally to be independent.

After a transitory phase where hemocytoblasts differentiate in situ in the yolk-sac blood islands, further evolution of hematopoiesis is conditioned by stem-cell seeding in the specific milieu provided by the developing hematopoietic tissues.

Cell-to-cell interactions between the hemocytoblasts and the hematopoietic rudiments are of decisive importance in the differentiation of both components: the differentiating pathways followed by the hemocytoblasts depend on the nature of the microenvironment in which they have become localized, whereas growth and differentiation of the blood-forming organs are stimulated, at least in some cases, by the presence of hematopoietic cells. One example of this reciprocal interaction is the fact that lymphoid follicles appear in the bursal epithelium at the precise sites of stem-cell invasion (Grossi et al. 1976), whereas follicle resorption always follows the departure of lymphoid cells induced by testosterone treatment.

The primary embryological origin of the hemocytoblasts that colonize the intraembryonic blood-forming organs is controversial. It is important

Figure 11
(a) Association of yolk-sac blood islands of a presomitic quail embryo with a 6-day-old chick thymus. (b) Results of the experiment shown in 11a. After 8 days in graft, the quail lymphoid cells have differentiated into the chick thymus. They originated from yolk-sac stem cells. 1000×.
ABBREVIATION: QL, quail lymphoid cells.

to learn their origin, since hematopoiesis that occurs throughout the entire life is conditioned by the evolution of the primitive hematogenic cell lines. The quail-chick yolk-sac chimeras devised by Dieterlen-Lièvre (1975) clearly demonstrate the developmental significance of blood islands located in the embryonic body in amniote embryos. Although they were described a long time ago following careful histological studies of early embryogenesis (for review, see Le Douarin 1966), their evolution has not drawn as much attention as has their conspicuous extraembryonic counterparts. It is very likely, however, that their role in the hematopoietic system development is decisive, since they seem to provide most—if not all—of the perennial stem-cell pool of the adult. In contrast, the yolk-sac stem cells probably become exhausted during the embryonic life. This different life span may be related to the fact that the yolk-sac stem cells in birds are strongly stimulated during the first half of the embryonic period, whereas relatively little hematopoiesis occurs simultaneously in the embryo. As a result, the hemocytoblasts mobilized via the blood stream at the time when the hematopoietic rudiments are ready to receive them are of intraembryonic origin.

Because yolk-sac hematopoiesis is more discrete and shorter in mammal embryos than it is in bird embryos, the contribution of vitelline stem cells to definitive hematopoiesis may be quite different in the two vertebrate classes.

Although it is normally the erythroid line that develops in the yolk sacs of chicks and quails, other experiments demonstrate that the vitelline stem cells have other developmental potentialities (F. V. Jotereau and N. M. Le Douarin, unpubl.). If adequate experimental conditions are devised, these cells can invade the thymic rudiment and differentiate into lymphocytes. Experiments are in progress to see whether yolk-sac stem cells can also colonize the bursa of Fabricius and, eventually, differentiate along the B-cell line.

The investigations reported in this article about the early events of the histogenesis of the lymphoid organs have shown that, even though competent stem cells are present in the blood from an early developmental stage, they can seed the bursal and thymic rudiments only when the latter have reached a certain state of maturation. On the other hand, the seeding process is followed by a "postreceptive" period during which no additional stem cells can penetrate the organs. The mechanism by which the immigration stops does not depend on the age of the thymic or bursal rudiment but on the number of stem cells that have homed to the organ.

Various types of interspecific associations of thymic and bursal anlagen at the pre- and postreceptive periods have shown that a traffic of hemocytoblasts can be induced from a postcolonization to a precolonization rudiment. In contrast, such an exchange of cells did not occur between two postcolonization organs.

On the other hand, cocultures in a filter *cis-trans* position, of a 6-day-old chick thymus or liver or mesonephros with an 11-day-old quail bursa have shown a significant difference in the migration of cells from the bursa to the thymus as compared with other organs. The quail bursal cells colonize the thymic anlage and, if adequate conditions are provided, can differentiate into lymphocytes in the thymic environment.

As a working hypothesis, I suggest that the colonization of hemopoietic organs by circulating stem cells obeys a chemotactic mechanism due to an attractive substance produced by the developing blood-forming organs. If this hypothesis is true, the hemocytoblasts would be similarly sensitive to the attraction exerted by the bursa or the thymus, since bursal stem cells can be "attracted" by a thymic primordium.

The transfer of lymphoid stem cells from the bursa to the thymus at various developmental stages can be used as a tool to investigate the period at which the cells become committed toward the B- or T-differentiation pathway.

Acknowledgments

Antithymic serum used to treat lymphocytes was kindly provided by Dr. G. Wick, Institut für Allgemeine und Experimentelle Pathologie der Universität Innsbrück; antiimmunoglobin serum was provided by Dr. M. Feldman, University College, London. This work has been supported by the Centre National de la Recherche Scientifique and a grant from the Delégation Générale à la Recherche Scientifique et Technique 76-7-0972.

REFERENCES

Auerbach, R. 1960. Morphogenetic interactions in the development of the mouse thymus gland. *Dev. Biol.* **2**: 271.

Dieterlen-Lièvre, F. 1975. On the origin of haemopoietic stem cells in the avian embryo: An experimental approach. *J. Embryol. Exp. Morphol.* **33**: 607.

Dieterlen-Lièvre, F., D. Beaupain, and C. Martin. 1976. Origin of erythropoietic stem cells in avian development: Shift from the yolk sac to an intraembryonic site. *Ann. Immunol.* (Paris) **127 C**: 857.

Glick, B. 1957. Experimental modification of the growth of the bursa of Fabricius. *Poult. Sci.* **36**: 18.

———. 1964. The bursa of Fabricius and the development of immunologic competence. In *The thymus in immunobiology* (eds. R. A. Good and A. E. Gabrielsen), p. 348. Hoeber, New York.

Glick, B., and C. R. Sadler. 1961. The elimination of the bursa of Fabricius: Reduction of antibody production in birds from eggs dipped in hormone solutions. *Poult. Sci.* **40**: 185.

Grossi, C. E., P. M. Lydyard, and M. D. Cooper. 1976. B-cell ontogeny in the chicken. *Ann. Immunol.* (Paris) **127 C**: 931.

Houssaint, E., M. Belo, and N. M. Le Douarin. 1976. Investigations on cell lineage and tissue interactions in the developing bursa of Fabricius through interspecific chimeras. *Dev. Biol.* **53**: 250.

Jotereau, F. V., and N. M. Le Douarin. 1978. The developmental relationship between osteocytes and osteoclasts: A study using the quail-chick nuclear marker in endochondral ossification. *Dev. Biol.* **63**. (In press.)

Kirkpatrick, C. M., and F. N. Andrews. 1944. The influence of the sex hormones on the bursa of Fabricius and the pelvis in the ring-necked pheasant. *Endocrinology* **34**: 340.

Le Douarin, N. 1966. L'hématopoïèse dans les formes embryonnaires et jeunes des vertébrés. *Annee Biol.*, ser 4 **5**: 105.

———. 1967. Détermination précoce des ébauches de la thyroïde et du thymus chez l'embryon de Poulet. *C. R. Acad. Sci.*, ser D **264**: 940.

———. 1969. Particularités du noyau interphasique chez la caille japonaise (*Coturnix coturnix japonica*): Utilisation de ces particularités comme "marquage biologique" dans les recherches sur les interactions tissulaires et les migrations cellulaires au cours de l'ontogenèse. *Bull. Biol. Fr. Belg.* **103**: 435.

———. 1971. Caractéristiques ultrastructurales du noyau interphasique chez la Caille et chez le Poulet et utilisation de cellules de Caille comme "marqueurs biologiques" en embryologie expérimentale. *Ann. Embryol. Morphog.* **4**: 125.

———. 1973. A biological cell labeling technique and its use in experimental embryology. *Dev. Biol.* **30**: 217.

———. 1975. An experimental analysis of liver development. *Med. Biol.* **53**: 427.

———. 1976. Cell migration in early vertebrate development studied in interspecific chimaeras. In *Embryogenesis in mammals, Ciba Foundation Symposium*, p. 71. Elsevier-Excerpta Medica-North-Holland, Amsterdam.

Le Douarin, N. M., and F. Jotereau. 1973. Origin and renewal of lymphocytes in avian embryo thymuses. *Nat. New Biol.* **246**: 25.

———. 1975. Tracing of cells of the avian thymus through embryonic life in interspecific chimaeras. *J. Exp. Med.* **142**: 17.

Le Douarin, N., C. Bussonnet, and F. Chaumont. 1968. Etude des capacités de différenciation et du rôle morphogène de l'endoderme pharyngien chez l'embryon d'oiseau. *Ann. Embryol. Morphog.* **1**: 29.

Le Douarin, N. M., E. Houssaint, and F. Jotereau. 1977. Differentiation of the primary lymphoid organs in avian embryos: Origin and homing of the lymphoid stem cells. In *Avian immunology* (ed. A. A. Benedict), p. 29. Plenum Press, New York.

Le Douarin, N. M., E. Houssaint, F. V. Jotereau, and M. Belo. 1975. Origin of hemopoietic stem cells in embryonic bursa of Fabricius and bone marrow studied through interspecific chimeras. *Proc. Natl. Acad. Sci. U.S.A.* **72**: 2701.

Le Douarin, N. M., F. V. Jotereau, E. Houssaint, and M. Belo. 1976. Ontogeny of the avian thymus and bursa of Fabricius studied in interspecific chimeras. *Ann. Immunol.* (Paris) **127C**: 849.

Le Lièvre, C. S., and N. M. Le Douarin. 1975. Mesenchymal derivatives of the neural crest: Analysis of chimaeric quail and chick embryos. *J. Embryol. Exp. Morphol.* **34**: 125.

Metcalf, D., and M. A. S. Moore. 1971. *Haemopoietic cells*. North Holland Research Monographs. *Front. Biol.* **24**.

Meyer, R. K., M. A. Rao, and R. L. Aspinall. 1959. Inhibition on the development of the bursa of Fabricius in the embryos of the common fowl by 19-nortestosterone. *Endocrinology* **64**: 890.

Moore, M. A. S., and G. R. Johnson. 1976. Hemopoietic stem cells during embryonic development and growth. In *Stem cells of renewing cell populations* (eds. A. B. Cairnie, P. K. Lala, and D. G. Osmond), p. 323. Academic Press, New York.

Moore, M. A. S., and D. Metcalf. 1970. Ontogeny of the haemopoietic system: Yolk sac origin of in vivo and in vitro colony forming cells in the developing mouse embryo. *Br. J. Haematol.* **18**: 279.

Moore, M. A. S., and J. J. T. Owen. 1965. Chromosome marker studies on the development of the haemopoietic system in the chick embryo. *Nature* **208**: 956, 989.

———. 1966. Experimental studies on the development of the bursa of Fabricius. *Dev. Biol.* **14**: 40.

———. 1967. Stem cell migration in developing myeloid and lymphoid systems. *Lancet* **2**: 658.

Papermaster, B. W., D. J. Friedman, and R. A. Good. 1962. Relationship of the bursa of Fabricius to immunologic responsiveness and homograft immunity in the chicken. *Proc. Soc. Exp. Biol. Med.* **110**: 62.

Saxen, L., M. Karkinen-Jaaskelainen, E. Lehtonen, S. Nordling, and J. Wartiovaara. 1976. Inductive tissue interactions. In *The cell surface in animal embryogenesis and development* (eds. G. Poste and G. L. Nicolson), p. 331. North-Holland, Amsterdam.

Warner, N. L., and F. M. Burnet. 1961. The influence of testosterone treatment on the development of the bursa of Fabricius in the chick embryo. *Autr. J. Biol. Sic.* **14**: 580.

Wick, G., B. Albini, and F. Milgrom. 1973. Antigenic surface determinant of chicken lymphoid cells. I. Serologic properties of anti-bursa and anti-thymus sera. *Clin. Exp. Immunol.* **15**: 237.

Wolff, E., and K. Haffen. 1952. Sur une méthode de culture d'organes embryonnaires in vitro. *Tex. Rep. Biol. Med.* **10**: 463.

Fetal Hematopoietic Origins of the Adult Hematolymphoid System

I. Weissman

Laboratory of Experimental Oncology, Department of Pathology
Stanford University Medical Center, Stanford, California 94305
and Medical Research Council Cellular Immunology Unit
Oxford OX1 3RE, United Kingdom

V. Papaioannou and R. Gardner

Department of Zoology
Medical Research Council Cellular Immunology Unit
Oxford OX1 3PS, United Kingdom

Many aspects of the embryonic origins of the adult hematolymphoid system are still matters of controversy. In the developing embryo, the yolk-sac blood islands are the first site of appearance of cells bearing unequivocal markers of hematopoietic differentiation (such as hemoglobin synthesis—Metcalf and Moore 1971). As blood islands mature, their walls become lined with endothelial cells while their cavities fill with cells that apparently represent several stages of hematopoietic (primarily erythroid) differentiation. These blood islands eventually connect with vitelline vessels and become vessels themselves. In developing mouse and chicken embryos, this maturation process immediately precedes the movement of hematopoiesis from the yolk sac to the fetal liver (Moore and Metcalf 1970). Moore and Metcalf (1970) have suggested that all fetal hepatic cells and adult hematolymphoid cells eventually are derived from nests of hematopoietic stem cells developing in these yolk-sac blood islands.

In this paper we test the postulate that yolk-sac blood-island cells can leave the yolk sac and give rise to adult bone-marrow hematopoietic stem cells as well as to thymic lymphoid cells. To do this, we developed a technique of orthotopic, synchronic transplantation of yolk-sac blood-island cells into the yolk sacs of hosts differing in cell-surface-membrane antigens H-2 or Thy-1. Because these markers are expressed on individual cells and can be detected unequivocally by immunofluorescence microscopy, we were able to confirm the hypothesis of Metcalf and Moore that yolk-sac blood-island cells could, in fact, populate the adult hematolymphoid cell system.

MATERIALS AND METHODS

Mice

All mice were obtained from the Medical Research Council Laboratory Animals Center, Carshalton, Surrey, and maintained at the Medical Research Council Cellular Immunology Unit in Oxford. Embryos were dated from the time of plugging, the day of the plug being designated as day 0.

Method of Obtaining Yolk-Sac Blood-Island Cells

Embryo donors, 8, 9, or 10 days old, were isolated from the uterine horns of hosts that had been perfused intravenously with Dulbecco's buffered saline (DBS) preterminally under anesthesia with Avertin. The femoral arteries of these hosts were incised following the initiation of tail-vein infusion, and perfusion continued until 10 cc of clear fluid had emerged from an incised femoral artery. The uterine horns were removed aseptically and rinsed in 10 ml DBS; each decidua was then isolated and washed in a second petri dish containing 15 ml of DBS. Following removal of the trophoblast, the embryo and extra embryonic membranes were removed from their attachment to the ectoplacental cone via the vitelline vessels and were washed in yet another vessel with DBS. Final dissection of the yolk sac from the embryo was carried out in another petri dish containing DBS. The yolk sacs were collected in cold DBS until sufficient numbers for isolation of yolk-sac blood-island cells had been obtained.

Two techniques for isolation of yolk-sac blood-island cells were utilized, each giving essentially the same results. The first technique involved a 15-minute incubation in a solution of trypsin 0.25%-pancreatin 0.5% in Tris-buffered, citrated saline; two volumes of neat fetal calf serum (FCS) were then added to halt the proteolytic process. The digested yolk sacs were flushed vigorously with a Pasteur pipet several times until dissociated. The second technique of dissociation involved no proteolytic digestion. In this method, the yolk sacs were minced to small pieces, and the cells were expressed gently with a blunt object onto the surface of a fine-mesh stainless-steel screen in DBS to which 20% FCS had been added. In both techniques, the resulting suspension contained free-floating, individual, mononuclear blood-island cells as well as clumps of epithelial cells. The epithelial cells were removed largely by velocity sedimentation, and the resulting cell suspension in either case comprised 92% to 98% mononuclear round cells of greater than 90% viability as tested by trypan-blue exclusion and 2% to 8% contaminating epithelial cells of less than 10% viability. All cell suspensions were washed in DBS with 20% FCS at least three times prior to being adjusted to the concentration for microinjection.

Injection of Embryos in Utero

Embryos were injected in utero at 8 to 10 days gestation; this procedure was done under transillumination, using an Oxford micromanipulator. Microneedles were prepared (using the de Fonbrane microforge), each with a side hole of 15–20 μm inside diameter adjoining the closed end. Test injections of pontamine sky-blue dye, India ink, and acridine orange-stained bone-marrow cells revealed that insertion of the microneedle at a point approximately three-fourths the distance between the mesenteric attachment to the uterine wall and the antimesenteric side of the uterus almost invariably resulted in deposition of the injected material into the yolk-sac cavity of 8–11-day-gestation hosts. Yolk-sac mononuclear cells, 1×10^4 to 1×10^5, were injected into the yolk-sac cavity of each conceptus in a litter.

Assay for Hematopoietic Stem Cells

Hematopoietic stem cells were tested by injection of limiting doses of bone-marrow or yolk-sac cells intravenously into lethally irradiated mice. Spleens were harvested from these hosts 8–10 days following injection; spleen colonies were counted by direct inspection of unfixed spleens under a dissecting microscope; and individual spleen colonies were dissected free and suspended for cellular immunofluorescence assays. In some cases, prospective hosts were immunized prior to irradiation with $1–5 \times 10^7$ viable spleen cells from an H-2-different donor so that spleen colonies would not form from cells bearing H-2 antigens shared by that donor. Such hosts never gave a donor-type spleen colony, even when injected with up to 10^7 bone-marrow cells.

Antisera and Immunofluorescence

The expression of H-2 or Thy-1 alloantigens on cells was assessed by immunofluorescence of cell suspensions. In a standard assay, 1×10^6 cells in a total volume of 20 μl were incubated for 1 hour on ice with 5 μl neat, aggregate-free, normal mouse serum or anti-H-2-haplotype serum, or with 2 μl of anti-Thy-1 reagent; this mixture was then layered over 100 μl of FCS and centrifuged at 1000 rpm for 10 minutes. The cells were resuspended in 20 μl of DBS, and then an equal volume of reconstituted, aggregate-free, fluoresceinated rabbit antimouse immunoglobulin (Cappel lot no. 6571) was added and the mixture incubated from 20 to 60 minutes on ice prior to being rewashed through FCS as before. The cells were resuspended in 50 μl of DBS and then were brought to 1% final concentration of formaldehyde buffered to pH 7.4. These cell suspensions were stored at room temperature in the dark, and they maintained a constant percentage of fluorescence-positive cells for at least 1 year. Alloantisera defining the H-2D haplotype (BALB/c) were SAS-347K, a (B10.AKM $\times$ C3H)F$_1$ anti-B10.A ([H-2^m $\times$ H-2^k] anti-H-2^a), and SAS-329H, a B10.M anti-B10.D2 (H-2^f anti-H-2^d). These two sera, which give equivalent specificity when tested on a number of relevant mouse strains by immunofluorescence, shall hereafter be referred to as anti-D. Antisera defining the H-2^k haplotype were SAS-316J, a (DBA/2 $\times$ B10.F)F$_1$ anti-B10.A antiserum ([H-2^d $\times$ H-2^n] anti-H-2^a), and SAS-340C, a B10.D2 anti-B10.A (H-2^d anti-H-2^a) antiserum. These two sera shall hereafter be called anti-K. The anti-Thy-1.1 antiserum was a C3H anti-AKR thymocyte serum which contained 0.5 mg/ml immunoglobulin eluted from Thy-1.1-purified glycoproteins. In addition, a second system was used for detecting H-2D-haplotype cells in organs containing high proportions of Ig-positive (B) cells. This system employed a dinitrophenylated C57BL/Ka anti-BALB/c serum (15.1 dinitrophenyl [Dnp] moles per mole IgG), which was detected with the second-stage reagent fluorescein isothiocyanate (FITC)-modified myeloma protein

Table 1
Antiserum Specificity

Antiserum	BALB	AKR	A/J	C57BL
Anti-D	+	−	+	−
Anti-K	−	+	+	−
BL anti-BALB	+	NT	+	−
Dnp BL anti-BALB	+	NT	+	−
Anti-Thy 1.1 (thymocytes)	−	+	−	−

NT, not tested.

MOPC-315 (F/P ratio, 4.7), an IgA myeloma which binds Dnp groups with a high affinity (R. Coffman and I. Weissman, in prep.). Table 1 demonstrated the immunofluorescence specificity of these antisera on the mouse strains used in this study.

In most experiments, tubes were given letter codes by one person, and cell suspensions were read and scored by another person before the code was revealed.

RESULTS

Direct Examination of Mice Injected during Embryogenesis with Yolk-Sac Cells for Persistence of Progeny of These Cells in Bone Marrow, Lymphoid Tissues, and Thymus

Tables 2 through 5 demonstrate the results of immunofluorescence analysis of several litters for the presence of original-donor yolk-sac cells in the

Table 2
Direct Analysis for Chimerism in an Injected Litter

Mouse no.	Cells	Identifying sera	Positive/total	Percentage	χ^2	p[a,b]
[(B10-AKR)F_1 × B10](8 Days) → B10 → Sacrifice at 43 Days after Birth						
15	TH	NMS	5/2800	0.2		
		anti-Thy 1.1	50/3400	1.5	27.7	<0.01
					1.3	NS
		anti-K	150/1300	11.5	23.0	<0.01
16	TH	NMS	2/1600	0.1		
		anti-Thy 1.1	43/1900	2.3	29.6	<0.01
					4.6	<0.05
	BM	NMS	79/1600	4.9		
		anti-K	129/1300	9.9	26.0	<0.01
17	TH	NMS	4/3560	0.1		
		anti-Thy 1.1	8/2000	0.4	3.7	NS
					0.8	NS

Abbreviations: TH, thymocytes; BM, bone marrow cells; NMS, normal mouse spleen.
[a] In this and subsequent tables, for anti-Thy 1.1 comparisons the first χ^2/p compares NMS and anti-Thy 1.1, and the second compares anti-Thy 1.1 on the experimental thymus with the same serum on a B10, BALB/c, or A/J thymus.
[b] NS, not significant; $p > 0.05$.

Table 3
Two Litters with Insignificant Thymic Chimerism

Mouse no.	Cells	Identifying sera	Positive/ total	Percentage	χ^2	p
(A) $(AKR \times A)F_1 \rightarrow CBA \rightarrow$ Sacrifice at 13 Days after Birth							
31	TH	NMS	2/610	0.3		
		anti-Thy 1.1	7/647	1.0	1.6	NS
					0.05	NS
		anti-D	7/622	1.1	1.7	NS
	BM	NMS	5/358	1.4		
		anti-D	13/420	3.1	1.8	NS
32	TH	NMS	0/1100	0		
		anti-Thy 1.1	14/943	1.5	14.3	<0.01
					0.9	NS
		anti-D	8/779	1.0	9.1	<0.01
	BM	NMS	2/254	0.8		
		anti-D	16/360	4.4	5.8	<0.025
A/J	TH	NMS	1/145	0.7		
		anti-Thy 1.1	1/82	1.2	0.1	NS
		anti-D	4/121	3.3	1.2	NS
		anti-K	1/118	0–1	0.3	NS
	BM	NMS	4/522	0.8		
		anti-D	111/123	90.2	537.9	<0.01
		anti-K	100/116	86.2	501.6	<0.01
(B) $[(B10\text{-}AKR)F_1 \times B10](8\ Days) \rightarrow BALB/c\ (8\ Days) \rightarrow$ Sacrifice at 43 Days after Birth							
6	TH	NMS	0/2000	0		
		anti-Thy 1.1	16/2000	0.8	14.2	<0.01
					0.06	NS
7	TH	NMS	3/3000	0.1		
		anti-Thy 1.1	14/3000	0.5	5.9	<0.05
					0.6	NS
8	TH	NMS	11/6000	0.2		
		anti-Thy 1.1	15/6000	0.3	0.4	<0.6
					0.05	NS

hematolymphoid tissues of injected hosts. In Table 2, two of three hosts apparently were bone-marrow chimeras, with insignificant or low levels of thymic chimerism. In Table 3, two separate litters failed to show significant thymic chimerism. In Table 4, however, two of three hosts were clear thymic chimeras. The litter in Table 5 showed bone-marrow chimerism in both hosts and thymic chimerism in neither.

Analysis of Hematopoietic Stem-Cell Chimerism by Retransfer of Cells into Preimmunized, Lethally Irradiated Hosts

The results of our first attempt to demonstrate chimerism of hematopoietic stem cells are shown in Table 6, where spleen and blood cells from newborn putative chimeras enabled irradiated BALB/c hosts that had been pre-

Table 4
A Litter with Significant Thymic Chimerism

Mouse no.	Cells	Identifying sera	Positive/ total	Percentage	χ^2	p
[(B10-AKR)F_1 × B10](8 Days) → B10 (8 Days) → Sacrifice at 36 Days after Birth						
9	TH	NMS	3/1800	0.2		
		anti-Thy 1.1	266/1700	15.7	293.1	<0.01
					95.8	<0.01
10	TH	NMS	0/2100	0		
		anti-Thy 1.1	37/2100	1.8	35.3	<0.01
					2.3	NS
11	TH	NMS	4/1300	0.3		
		anti-Thy 1.1	80/1300	6.2	69.2	<0.01
					27.2	<0.01

immunized prior to irradiation with primary host cells to survive. The first assay of spleen colonies from yolk-sac-injected hosts is shown in Table 7, where cells from 10-day-gestation C57BL mice, injected 2 days earlier with A-strain yolk-sac cells, gave rise to spleen colonies in A hosts that had been preimmunized with C57BL tissues prior to irradiation. However, when all of the tissues in a single litter were pooled, the number of cells

Table 5
A Litter with Bone-Marrow Chimerism in Both Hosts and Thymic Chimerism in Neither

Mouse no.	Cells	Identifying sera	Positive/ total	Percentage	χ^2	p
BALB → [(B10 × AKR)F_1 × B10 → Sacrifice at 17 Days after Birth						
29	TH	NMS	2/797	0.3		
		anti-D	31/858	3.6		NS
	BM	NMS	28/669	4.2		
		anti-D	52/487	10.7	17.4	<0.01
		anti-K	147/277	53.1	307.2	<0.01
30	TH	NMS	8/903	0.9		
		anti-D	7/680	1.0	<0.1	NS
	BM	NMS	9/589	1.5		
		anti-D	22/627	3.5	4.0	<0.05
		anti-K	121/211	57.4	351.6	<0.01
A/J	TH	NMS	1/145	0.7		
		anti-Thy 1.1	1/82	1.2	0.1	NS
		anti-D	4/121	3.3	1.2	NS
		anti-K	1/118	0.9	0.3	NS
	BM	NMS	4/522	0.8		
		anti-D	111/123	90.2	537.9	<0.01
		anti-K	100/116	86.2	501.6	<0.01

Table 6
A Pilot Retransfer Experiment wherein Primary Host Cells Are Immunologically Excluded

Mouse no.	No. and type of cells	Results
BALB (9 Days) $\to$ *B10 (8 Days)* $\to$ *BALB anti-BL*		
A-45	4.8×10^6 spleen cells	survived
A-46	4.0×10^6 spleen cells	survived
	2.8×10^5 blood cells	survived
C57BL	5×10^6 bone marrow	died

available did not give convincing evidence that these spleen-colony numbers were greater than background.

In contrast, even low levels of bone-marrow chimerism can result in significant numbers of hematopoietic stem cells, as measured by spleen-colony formation (Tables 8–10); in these cases, all spleen colonies were also typed by immunofluorescence to determine whether they had arisen from primary donors or primary hosts. In Table 8A, B10 hosts that had been injected with BALB yolk-sac blood-island cells were sacrificed 51 days after birth. None of these hosts had significant thymic chimerism, but two out of three showed significant bone-marrow chimerism. These hosts were also tested for chimerism in tissues containing high numbers of endogenous B cells, and, therefore, were tested with the Dnp C57BL anti-BALB serum. The two hosts that were bone-marrow chimeras were also lymphoid-tissue chimeras. The third host, which had insignificant levels of bone-marrow and thymic chimerism, was not significantly chimeric in other lymphoid tissues. However, all three hosts contained cells in their bone marrow that were capable of causing the appearance of spleen colonies in irradiated BALB hosts that had been preimmunized with C57BL tissues. In each case, the spleen colonies carried BALB cell-surface antigens.

Because it was theoretically possible that the cells present in these chimeric bone marrows could somehow cause regeneration of irradiated-

Table 7
Do Injected Yolk-Sac Cells Get into Blood Islands?

No. and type of cells injected	Host	No. of spleen colonies
A-Strain 8-Day Yolk-Sac Cells $\to$ *8-Day C57BL* $\to$ *Sacrifice 2 Days Later* $\to$ *A(αBL)*		
None	A anti-BL	0
Yolk-sac cavity (6×10^5)	A anti-BL	3
Yolk-sac blood islands (7×10^5)	A anti-BL	2
Embryo blood (5×10^5)	A anti-BL	1

Table 8
Direct and Spleen-Colony Analysis of a Litter for Hematopoietic Pluripotential Stem-Cell Chimerism

Mouse no.	Cells	Identifying sera	Positive/total	Percentage	χ^2	p
(A) Direct Analysis BALB (9 Days) → B10 (8 Days) → Sacrifice at 51 Days after Birth						
12	TH	NMS	6/1100	0.5		
		anti-D	16/1100	1.5	3.7	NS
	LN	NMS	1/1000	0.1		
		Dnp BL anti-BALB	31/1100	2.8	24.0	<0.01
	BM	NMS	15/660	2.3		
		anti-D	55/762	7.2	17.4	<0.01
13	TH	NMS	2/1000	0.2		
		anti-D	21/2438	0.9	3.7	NS
	BM	NMS	26/698	3.7		
		anti-D	140/969	14.5	50.8	<0.01
	SP	NMS	5/505	1.0		
		Dnp BL anti-BALB	62/562	11.0	43.9	<0.01
	LN	NMS	6/329	1.8		
		Dnp BL anti-BALB	23/496	4.6	3.8	NS
14	TH	NMS	3/603	0.5		
		anti-D	16/916	1.8	3.6	NS
	BM	NMS	32/1032	3.1		
		anti-D	23/580	4.0	0.6	NS
	SP	NMS	7/807	0.9		
		Dnp BL anti-BALB	7/266	2.63	3.6	NS
	LN	NMS	11/811	1.36		
		Dnp BL anti-BALB	17/817	2.1	0.9	NS

Additional abbreviations: LN, lymph-node cells; SP, spleen nucleated cell

Mouse no.	Chimera (%)	No. of cells	Spleen colonies (spleen weight)	Identifying sera	Positive/ total	Percentage
(B) Spleen-Colony Analysis						
BALB (H-$2^{d/d}$) (9 Days) $\rightarrow$ B10 (H-$2^{b/b}$) (8 Days) $\rightarrow$ Sacrifice at 51 Days after Birth $\rightarrow$ BALB (H-$2^{d/d}$) (anti-BL)						
12	5.0	1.1×10^7	TMTC[a] } pool	NMS	6/311	1.9
			TMTC[a]	BL anti-BALB	105/167	62.9
13	10.7	2.2×10^7	5	NMS	8/124	6.5
				BL anti-BALB	75/100	75.0
14	0.9	1.4×10^7	30, 24} pool	NMS	3/107	2.8
				BL anti-BALB	55/100	55.0
				anti-D	70/103	68.0
				anti-K	18/82	22.0
B10 control			BM	NMS	20/320	6.3
				anti-D	9/202	4.5
				anti-K	3/81	3.7
				BL anti-BALB	2/144	1.4
BALB control			BM	NMS	1/250	0.4
				anti-D	51/102	50.0
				anti-K	5/133	3.7
				BL anti-BALB	59/136	43.4

[a] TMTC, too many spleen colonies to count accurately.

Table 9
Direct and Hematopoietic Stem Cell Spleen-Colony Analysis of a Litter for Chimerism

Mouse no.	Cells	Identifying sera	Positive/total	Percentage	χ^2	p
		(A) Direct Analysis				
		BALB (9 Days) → [(B10 × AKR)F_1 × B10] (9 Days) → Sacrifice at 1 Day after Birth				
1	TH	NMS	1/491	0.2		
		anti-D	11/524	2.1	6.3	<0.05
		anti-K	2/175	1.1	0.9	NS
	BM	NMS	10/278	3.6		
		anti-D	43/546	7.9	5.4	<0.05
		anti-K	4/171	2.3	0.2	NS
2	TH	NMS	1/1500	0.1		
		anti-D	3/682	0.5	1.8	NS

Mouse no.	Chimera (%)	No. of cells	Spleen colonies	Identifying sera	Positive/ total	Percentage

(B) Spleen-Colony Analysis

BALB ($H-2^{d/d}$) (9 Days) → B10 AKR × B10 ($H-2^{b/k}$ or $H-2^{b/b}$) (9 Days) → Sacrifice 1 Day after Birth → BALB ($H-2^{d/d}$) (anti-BL)

Mouse no.	Chimera (%)	No. of cells	Spleen colonies	Identifying sera	Positive/ total	Percentage
1	4.3	3.2×10^6	28, 30	NMS	0/145	0
				BL anti-BALB	114/202	56.4
2	NT	6×10^6	20, TMTC	NMS	1/212	0.5
				BL anti-BALB	138/164	84.2
3	NT	1.4×10^7	18/22	NMS	9/223	4.0
				BL anti-BALB	101/171	59.1
5	NT	7×10^6	12/14	NMS	1/99	1.0
				BL anti-BALB	74/118	62.7

BALB ($H-2^{d/d}$) (9 Days) → B10 AKR × B10 ($H-2^{b/k}$ or $H-2^{b/b}$) (9 Days) → Sacrifice 1 Day after Birth → A ($H-2^{a/a}$) (anti BL)

Mouse no.	Chimera (%)	No. of cells	Spleen colonies	Identifying sera	Positive/ total	Percentage
Pool 1, 2, 3, 4, 5,			3, 4	NMS	6/77	7.8
				anti-D	25/33	75.8
				anti-K	3/25	12.0
				BL anti-BALB	66/71	93.0

Table 10
Direct and Hematopoietic Stem Cell Spleen-Colony Analysis of a Litter for Chimerism

Mouse no.	Cells	Identifying sera	Positive/total	Percentage	χ^2	p
		(A) Direct Analysis				
		$BALB \times AKR$ *(10 Days)* $\rightarrow$ $BL \times AKR$ *(9 Days)* $\rightarrow$ *Sacrifice at 34 Days after Birth*				
20	TH	NMS	2/502	0.4		
		anti-D	18/1300	1.4	2.4	NS
		Dnp BL anti-BALB	16/1100	1.5	2.6	NS
	BM	NMS	2/280	0.7		
		anti-D	20/354	5.7	9.9	<0.01
		Dnp BL anti-BALB	34/600	5.7	10.7	<0.01
21	TH	NMS	2/1000	0.2		
		anti-D	4/1000	0.4	0.2	NS
		Dnp BL anti-BALB	8/1000	0.8	2.5	NS
	BM	NMS	9/415	2.2		
		anti-D	19/439	4.3	2.5	NS
		Dnp BL anti-BALB	66/1100	6.0	8.6	<0.01
BL	BM	NMS	12/610	2.0		
		anti-D	7/212	3.3	0.7	NS
		Dnp BL anti-BALB	10/613	1.6	0.1	NS
BALB	BM	NMS	1/201	0.5		
		anti-D	74/101	73.3	186.8	<0.01
		Dnp BL anti-BALB	88/157	56.1	142.7	<0.01

Mouse no.	Chimera (%)	Spleen colonies (spleen weight)	Identifying sera	Positive/total	Percentage
(B) Spleen-Colony Analysis					
BALB × AKR ($H-2^{d/k}$) (10 Days) → BL × AKR ($H-2^{b/k}$) (9 Days) → Sacrifice 34 Days after Birth → BALB ($H-2^{d/d}$) (anti-BL)					
20	4.9	TMTC (150 mg)	NMS	3/123	2.4
			anti-D	100/140	71.4
			anti-K	184/209	88.0
21	2.2	TMTC (149 mg)	NMS	13/168	7.7
			anti-D	83/150	53.3
			anti-K	107/122	87.7
No cells injected		0 (34 mg)	NMS	3/18	16.7
			anti-D	42/48	87.5
			anti-K	12/57	21.1
BALB		BM	NMS	1/163	0.6
			anti-D	122/136	89.7
			anti-K	8/304	2.6
AKR × BL		BM	NMS	7/258	2.7
			anti-D	5/207	2.4
			anti-K	78/126	61.9

host hematopoietic stem cells, we assayed two litters for chimerism using strain combinations and antisera that could distinguish primary donors, primary hosts, and secondary hosts. In Table 9A, for example, [(B10 × AKR)F$_1$ × B10] backcross offspring that had been injected in utero with BALB/c cells were tested for original-donor (H-2D) and AKR-determined host (H-2K) cell-surface antigens. Neither of the two hosts tested was H-2K, and only mouse no. 1 appeared to be chimeric by direct assay. However, these 1-day-old hosts contained bone-marrow cells that could give rise to spleen colonies in irradiated secondary hosts, and these colonies were demonstrated to be of original-donor origin by the discriminant antiserum assay shown at the bottom of Table 9B.

The final litter assayed is shown in Table 10, where two (B10 × AKR)F$_1$ 34-day-old hosts that had been injected with (BALB × AKR)F$_1$ yolk-sac cells showed low-level bone-marrow chimerism and no thymic chimerism, but their bone-marrow hematopoietic stem cells did have a significant original-donor component (Table 10B). In this case, original-donor cells would have been positive with anti-D and anti-K sera, original host cells would have been positive only with anti-K sera, and secondary host cells would have been positive only with anti-D sera. Spleen colonies all contained a majority population of cells bearing both H-2^d and H-2^k markers, thus formally establishing their origin from (BALB × AKR)F$_1$ yolk-sac blood-island cells.

DISCUSSION

These experiments demonstrate that orthotopic, synchronic injection of yolk-sac blood-island cells can result in the appearance of these cells in bone marrow, thymus, and lymphoid tissues. A careful analysis of all of the litters so injected showed that such chimerism was always at low levels and was not always established. We tested simultaneously all original hosts for chimerism by assaying for donor-specific glucose phosphate dehydrogenase isoenzymes, and in no instance did we find sufficient levels of original-donor isoenzymes in any tissue to confirm chimerism. Also, in collaboration with Dr. Charles Ford, we examined a few litters for chromosomal (T-6) chimerism in which T-6/T-6 donor yolk-sac cells had been injected into CBA host embryos (and even host blastocysts), and, again, no chimerism was detectable. We have not yet assayed for isoenzyme or T-6 chimerism of the hematopoietic stem cells by the filtration technique used for the last spleen-colony experiments discussed above. We plan to carry out these types of experiments using cell-surface, chromosomal, and isoenzyme markers.

This type of experiment presents one major problem in terms of interpretation. Injection of blood-island cells into the yolk-sac cavity of a same-aged embryo in utero does not result in homing of all such cells from the yolk-sac cavity into the yolk-sac blood islands. In fact, short-term analysis by autoradiography and injection of stained cells, as well as spleen-colony analysis, revealed that only a small percentage of such cells ended up colonizing the blood islands. Therefore, we cannot state conclusively whether adult hematolymphoid cells bearing original-donor cell-surface markers derived from ancestors in yolk-sac blood islands or from ancestors in the yolk-sac cavity itself. It is currently beyond our technical expertise to inject cells reliably and directly into the yolk-sac blood islands.

Nevertheless, these experiments strongly support the contention by Moore and Metcalf (1971) that hematopoiesis in the yolk sac results in cells that can colonize future sites of hematopoiesis in the embryo. The experiments presented here did not test the hypothesis that *all* adult hematopoiesis derives from ancestors in the embryonic yolk sac. Nor, of course, did these experiments test the hypothesis of Moore and Metcalf that hematopoietic stem cells within the yolk sac are generated from endogenous yolk-sac precursors. In fact, Le Douarin (this volume) has convincingly demonstrated that even yolk-sac blood-island cells are derived from embryonic sites. Thus, the yolk sac appears to be a site of extraembryonic location of hematopoietic stem cells as well as of primitive germinal cells (Mintz and Russell 1957), both cell types immigrating from the early developing embryo and migrating back into the developing fetus. Unfortunately, we did not carry out breeding experiments to test if our hematopoietic chimeras were germ-cell chimeras. These experiments and experiments defining the role of endogenous yolk-sac epithelial cells in hematopoietic maturation and homing are to be the subject of future investigations.

Acknowledgments

We are grateful to L. Hackfath for excellent technical assistance. The work was carried out in the laboratory of Professor J. L. Gowans (MRC Cellular Immunology Research Unit), and we are grateful for many helpful discussions with him, as well as for his support of the project. We would also like to thank Dr. Jack Stimpfling, McLaughlin Research Institute, Great Falls, Montana, for his gift of the antisera that define the H-2D and the H-2^k haplotypes, and Dr. Allen Williams, Department of Biochemistry, University of Oxford, for the anti-Thy-1.1 antiserum.

This research was supported by American Cancer Society Grant IM-56, U.S. Public Health Service Grant AI-09072, and by the Medical Research Council. I. L. Weissman was a Josiah Macy, Jr., Foundation Faculty Scholar and an American Cancer Society Faculty Research Awardee during these projects.

REFERENCES

Metcalf, D., and M. A. S. Moore. 1971. *Haemopoietic cells.* North-Holland Research Monographs. *Front. Biol.* **24.**

Mintz, B., and E. S. Russell. 1957. Gene-induced embryological modification of primordial germ cells in the mouse. *J. Exp. Zool.* **134:** 207.

Moore, M. A. S., and D. Metcalf. 1970. Ontogeny of the haemopoietic system: Yolk sac origin of *in vivo* and *in vitro* colony forming cells in the developing mouse embryo. *Brit. J. Haematol.* **18:** 279.

Clonal Analysis in Vitro of Fetal Hepatic Hematopoiesis

G. R. Johnson and D. Metcalf

Cancer Research Unit, The Walter and Eliza Hall Institute of Medical Research
Melbourne 3050, Australia

During murine ontogeny, the development of the hematopoietic system proceeds through several distinct phases (for review, see Metcalf and Moore 1971). From the 8th to the 10th day of gestation, the yolk sac is the only site of hematopoiesis, but this function is superseded by the developing fetal liver on the 10th day of gestation (Rifkind et al. 1969; Moore and Metcalf 1970). The development of hepatic parenchymal cells also occurs at this time and, because of the intimate structural relationships that exist between this diversity of cell types, the origin of hepatic hematopoietic cells has long been a point of controversy. Two major theories have arisen. One hypothesis suggests that multipotential hematopoietic stem cells migrate from the yolk sac and colonize the developing hepatic primordium. Hematopoietic cells failed to develop in the livers of 9–10-day-old fetuses that had been grown intact in vitro from day 7 in the absence of the yolk sac (Moore and Metcalf 1970). According to the other view, hepatic hematopoietic tissue arises from the transformation of liver mesenchymal cells (Rifkind et al. 1969) and, thus, has no direct relationship with vitelline hematopoiesis.

Recent work favors the yolk-sac-derived hematopoietic cell migration theory. It has been established that hepatic hematopoiesis commences precisely at the 28th somite stage of development, hematopoiesis occurring in organ transplants of this age or older when they are placed in vitro in Rose chambers or under the kidney capsules of irradiated hosts (Johnson and Jones 1973; Johnson and Moore 1975). Pre-28th somite hepatic organ grafts placed under the kidney capsules of adult hosts only became hematopoietic if supplied with an exogenous source of hematopoietic cells (Johnson and Moore 1975; Moore and Johnson 1976).

Although there have been many morphological studies of fetal hepatic hematopoiesis, it has been possible to study the differentiation of the immigrant yolk-sac-derived cells into morphologically recognizable cells only since the introduction of the in vivo spleen-colony assay for multipotential cells, CFU-S (colony-forming units – spleen) (Till and McCulloch 1961) and the in vitro agar cloning of granulocyte-macrophage colony-forming cells (GM-CFC) (Pluznik and Sachs 1965; Bradley and Metcalf 1966).

The present work describes some of the physical properties of the progenitor cells in the fetal liver that form in vitro hematopoietic colonies and the factors regulating this colony formation in comparison with comparable cells in the adult bone marrow.

EXPERIMENTAL PROCEDURES

CBA/CaH Wehi fetal and adult tissues were obtained as described previously (Johnson and Metcalf 1977a). The basic techniques for in vitro culturing of hematopoietic cells in agar have been detailed elsewhere (Metcalf 1977). The procedure for growing erythroid colonies is identical to that used for growing granulocytes and macrophages except that the fetal calf serum (FCS) and/or horse serum is replaced by 20% human plasma.

Growth of neutrophil and macrophage colonies is dependent upon the addition to the culture of a source of granulocyte-macrophage colony-stimulating factor (GM-CSF), and the following sources have been utilized for the work reported here: endotoxin serum (ES) from C57BL mice that had been injected with 5 μg endotoxin 3 hours previously (Metcalf 1971); semipurified human urinary CSF (Stanley and Metcalf 1969); semipurified GM-CSF obtained from mouse-lung conditioned media (MLCM) and prepared according to the method of Sheridan and Metcalf (1973); concentrated yolk-sac conditioned media (YSCM) obtained from 14-day CBA yolk sacs (Johnson and Metcalf 1977b).

Erythroid colonies were stimulated by pokeweed-mitogen-stimulated spleen conditioned media (2×10^6 C57BL spleen cells per milliliter of RPMI-1640 containing 5% human plasma and 5% pokeweed mitogen previously diluted 1:15 in distilled water). Human urinary erythropoietin (fraction E6-3-15 LSL, fractionated by gel filtration and affinity chromatography on concanavalin A-Sepharose, 80 units/mg) was used for the studies reported here and was supplied by Dr. P. P. Dukes on behalf of the Division of Blood Diseases and Resources of the National Heart, Lung, and Blood Institute, Bethesda, Maryland.

All colonies (> 50 cells) were scored unstained at 35× magnification or removed and stained with acetoorcein (Metcalf et al. 1967) or with benzidine and Giemsa (Johnson and Metcalf 1977c).

Separation of cells by velocity sedimentation was performed as described by Metcalf and MacDonald (1975). Recovery of viable nucleated cells from all velocity sedimentation experiments exceeded 60% of the starting cell number.

RESULTS

In Vitro Clonal Analysis of Hepatic Granulocyte-Macrophage Progenitor Cells

Although the fetal liver is primarily an erythropoietic organ (Silini et al. 1967), the immediate precursors of neutrophils and macrophages, GM-

CFC, have been detected at all stages of hepatic development (Moore and Metcalf 1970; Moore and Williams 1973). Since the GM-CFC from early fetal liver formed colonies that differentiated mainly into macrophages, Moore and Williams (1973) proposed that the low level of granulopoiesis but high frequency of GM-CFC in the early fetal liver was due to the fact that GM-CFC were intrinsically biased toward macrophage differentiation.

Recent studies have confirmed that colonies derived from fetal liver GM-CFC are composed primarily of macrophages (Johnson and Metcalf 1977a). However, the proportion of macrophage colonies varies according to the source of GM-CSF used to stimulate the cultures. In all cases, the proportion of macrophage colonies in fetal liver cultures stimulated by plateau concentrations of GM-CSF (determined by dose-response curves) exceeded the proportion of macrophage colonies in the equivalent bone-marrow cultures (Table 1).

Mouse-lung conditioned medium, serum from endotoxin-injected mice, or human urinary GM-CSF all indicated a frequency of GM-CFC in the 12-day fetal liver of at least 2.5 times that of GM-CFC in adult bone marrow. With optimal concentrations of MLCM, up to 30% of the colonies formed by GM-CFC from 12-day fetal liver were granulocytic (Johnson and Metcalf 1977a). Whereas in cultures of bone-marrow cells the proportion of macrophage colonies decreased as MLCM concentration increased, the proportions of fetal-liver-derived colony types remained relatively constant regardless of the MLCM concentration used (Johnson and Metcalf 1977a). A similar result was obtained with cultures stimulated with ES. With adult marrow cells, as the concentration of ES was increased, the proportion of macrophage colonies decreased from 100% to 52% over the range of dilutions 1:8 to 1:1 (Table 1). However, in parallel cultures of 12-day fetal

Table 1
Frequency and Type of GM-CFC in Fetal Mouse Liver

Tissue	Dilution of ES	Total colonies[a]	Percentage of colonies[b]		
			granulocyte	mixed	macrophage
Adult bone marrow	1:1	72 ± 4	26	22	52
	1:2	57 ± 4	17	18	65
	1:4	25 ± 2	21	9	70
	1:8	7 ± 2	0	0	100
	1:16	0	—	—	—
12-Day fetal liver	1:1	242 ± 28	10	7	83
	1:2	196 ± 12	20	20	60
	1:4	140 ± 8	10	10	80
	1:8	56 ± 6	14	—	86
	1:16	10 ± 2	10	—	90

Cultures were stimulated by 0.1 ml of serum from endotoxin-injected C57BL mice at varying dilutions. Adult CBA bone-marrow cultures contained 40,000 viable nucleated cells; CBA fetal liver cultures, 20,000.

[a] Colony numbers, mean ± standard deviations from four replicate cultures scored after 7 days of incubation.

[b] Colony morphology determined from 45 sequential colonies sampled at each dilution of stimulus.

liver cells stimulated over the same range of ES dilutions, the porportion of macrophage colonies remained relatively constant in the range 80–90% (Table 1). In fetal liver cultures maximally stimulated by the GM-CSF in human urine, all colonies were macrophagic in type (Johnson and Metcalf 1977a).

The GM-CSF both in MLCM and in ES is of adult origin and normally might not impinge upon fetal GM-CFC. A study was initiated, therefore, to determine which 12-day fetal tissues were capable of producing material with GM colony-stimulating activity (CSA). Only fetal peripheral blood and yolk sacs were able to stimulate colony formation in cultures of adult marrow or fetal liver cells (Johnson and Metcalf 1977b). At no gestational age was colony-stimulating activity detected in fetal liver cells, as assessed by their ability to produce active conditioned media or when they were placed directly into underlayers.

Since fetal yolk sacs consistently produced active conditioned media (YSCM), further studies were conducted on concentrated preparations of this material obtained from 14-day fetal yolk sacs. Preparations of this conditioned media, after absorption onto DEAE-cellulose, were able to stimulate as many GM colonies to form in bone-marrow cultures as did optimal concentrations of MLCM (Johnson and Metcalf 1977b). Concentrated YSCM prepared in this manner could be diluted at least 1:8 and still stimulate the same number of colonies as were stimulated by optimal concentrations of MLCM (Fig. 1). Note in Figure 1 that, with bone-marrow cells, dilution of YSCM produced a typical dose-response curve for GM-CSF (a twofold dilution producing a twofold reduction in colonies), unlike

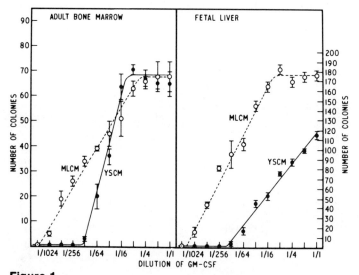

Figure 1

Titration of (o----o) semipurified MLCM and (●———●) concentrated YSCM in cultures of adult CBA bone-marrow cells and 12-day CBA fetal liver cells. Fetal liver cultures contained 20,000 viable nucleated cells; adult marrow cultures, 40,000 viable nucleated cells. Mean colony counts ± standard deviations of four replicate cultures were scored after 7 days of incubation.

the characteristic flat titration curve previously described for MLCM (Sheridan and Metcalf 1973).

Although both YSCM and MLCM, when used at optimal concentrations, stimulated the formation of identical numbers of colonies, the distribution of colony types after 7 days of culture differed considerably. Thus, bone-marrow cultures stimulated with YSCM usually contained no pure granulocytic colonies but did contain more than 95% macrophage colonies (Table 2). In contrast, cultures maximally stimulated by MLCM consistently contained, after 7 days of incubation, 20–40% pure granulocytic colonies (Table 2). When bone-marrow cultures were maximally stimulated by mixtures of both stimuli, the proportions of different colonies were similar to the proportions found in cultures stimulated by MLCM alone (Table 2), although total colony numbers in all three types of culture were similar.

The same preparations of MLCM and YSCM were also tested in parallel cultures of 12-day fetal liver cells. As expected from previous work (Johnson and Metcalf 1977a), the frequency of colonies with fetal liver cells was higher than with bone-marrow cells (Fig. 1). However, when YSCM was used at a concentration supramaximal for bone-marrow cultures, it was found to be capable of stimulating at most only 60% of the fetal liver colonies that had been stimulated by maximal MLCM concentrations (Fig. 1).

Whether or not this relative unresponsiveness of fetal liver GM-CFC to YSCM was due to a specific inhibitor was tested with mixing experiments, performed with optimal concentrations of MLCM. In no experiments was any inhibition detected in fetal liver cultures containing both stimuli, the number of colonies developing being identical to that found in cultures maximally stimulated by MLCM alone (Table 2). As with bone-marrow cultures, the proportion of different colonies in fetal liver cultures containing both stimuli were similar to those in cultures stimulated by MLCM alone (Table 2). With all concentrations of YSCM, cultures of fetal liver cells in-

Table 2
Effect of Mixing MLCM and YSCM upon Colony Frequency and Morphology in Cultures of Adult Marrow Cells or 12-Day Fetal Liver Cells

Cells	Stimulus	Mean colony no.	Percentage of colonies		
			granulocytic	mixed	macrophage
Adult bone marrow	YSCM	102 ± 7	0	5	95
	MLCM	99 ± 7	40	29	31
	YSCM + MLCM	130 ± 6	52	23	25
Fetal liver	YSCM	75 ± 7	0	0	100
	MLCM	225 ± 7	38	10	52
	YSCM + MLCM	270 ± 1	31	0	69

Data are from four replicate cultures ± standard deviations. Fetal liver cultures contained 20,000 cells; adult bone-marrow cultures, 40,000 cells. All cultures were stimulated by 0.1 ml of appropriate stimulus at optimal concentrations for adult bone marrow. Thirty sequential colonies were sampled from each type of culture.

variably contained no pure granulocytic colonies, and more than 95% of the colonies were composed of macrophages.

In fractionation studies, 12-day fetal liver GM-CFC were found to sediment with two major peaks, at 7–8 mm/hour and at 9–10 mm/hour, when fractionated cells were maximally stimulated by MLCM. By comparison, adult bone-marrow GM-CFC sedimented with a single peak of these cells at 4.5–5 mm/hour (Fig. 2). These data indicate that 12-day fetal liver cells are larger cells (i.e., sediment with a higher sedimentation velocity) than the adult bone-marrow GM-CFC. However, a slower sedimentation rate was observed with fetal liver GM-CFC from older fetuses, and GM-CFC from 18-day fetal liver cells sedimented at a rate similar to adult bone-marrow GM-CFC (Fig. 2).

When submaximal concentrations of MLCM were used to stimulate fractionated 12-day fetal liver populations, the more rapidly sedimenting GM-CFC were found to be more responsive than the more slowly sedimenting GM-CFC (Johnson and Metcalf 1977a). Similar results were obtained with 18-day fetal liver cells (Fig. 3). In this aspect, fetal liver GM-CFC resembled GM-CFC in adult marrow populations (Metcalf and MacDonald 1975).

With GM-CFC in marrow, the most rapidly sedimenting cells tend to form macrophage colonies, whereas the most slowly sedimenting cells form granulocyte colonies (Metcalf and MacDonald 1975). However, with GM-CFC in fetal liver, no evidence of segregation was exhibited, and cells of all sedimentation velocities produced comparable proportions of granulocyte, mixed, and macrophage colonies as found in control cultures of unfractionated cells (Fig. 4).

The morphology of fetal-liver-derived colonies is distinctive and readily

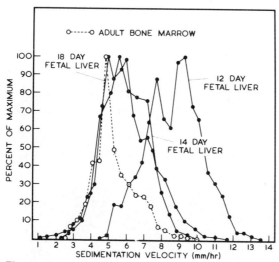

Figure 2

Velocity sedimentation separation of CBA fetal liver cells and adult bone-marrow GM-CFC. Data are expressed as percentage of peak colony number. Each point represents mean number of colonies from four replicate cultures maximally stimulated by MLCM and scored after 7 days of incubation.

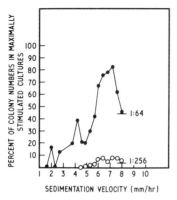

Figure 3
Responsiveness of colony-forming cells from 18-day CBA fetal liver in various velocity sedimentation fractions to stimulation by GM-CSF. Replicate cultures contained 0.1 ml of MLCM undiluted or diluted 1:64 or 1:256. Colony counts in submaximally stimulated cultures are expressed as a percentage of colony counts in maximally stimulated cultures. Colony counts in submaximally stimulated cultures of unfractionated fetal liver cells are indicated by horizontal bars.

permits a culture of fetal liver cells to be distinguished from a culture of adult marrow cells. Colonies tend to be smaller and more uniform in size and shape than those in adult marrow cultures. Furthermore, macrophage colonies have a characteristically fragmented appearance, colonies appearing as multiple, tight fragments instead of as a loose cloud of cells.

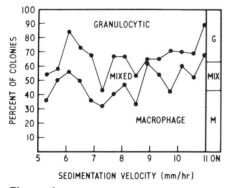

Figure 4
Segregation of granulocyte, mixed granulocyte and macrophage, and macrophage colonies by velocity sedimentation fractionation of 12-day CBA fetal liver cells. Data are from colonies in unfractionated sample shown in right-hand column (ON). Proportions of various colonies produced by cells from each fraction are expressed as percentages of the total colonies present in the culture dish. All cultures maximally stimulated by 0.1 ml of MLCM and incubated for 7 days.

Fetal Liver Erythroid-Colony Formation

Pokeweed-mitogen-stimulated spleen conditioned medium (SCM) has been shown to be able to stimulate adult marrow cells to form colonies not only of neutrophils and macrophages but also of eosinophils and megakaryocytes (Metcalf et al. 1974, 1975). When used in fetal liver cultures, SCM can stimulate the formation of numerous GM colonies and occasional eosinophil and megakaryocyte colonies from the few progenitors present in this population (G. R. Johnson and D. Metcalf, unpubl.). In addition, however, after 7 days of incubation, cultures of CBA fetal liver cells also develop large red colonies (Johnson and Metcalf 1977c). These colonies are composed of erythroid cells at all stages of differentiation that stain positively for hemoglobin with benzidine and, in the electron microscope, exhibit the typical morphology of erythroid cells (D. Metcalf et al., unpubl.).

Experiments were conducted to determine the optimal conditions for erythroid-colony development. Although FCS added to Dulbecco's modified Eagle's medium (DMEM) could support erythroid-colony growth, the size and frequency of these colonies was increased when FCS was replaced by human plasma. Titration of the human plasma showed that, for a given concentration of SCM, optimal erythroid-colony formation occurred at human plasma concentrations between 16% and 20%, and these conditions were maintained for all further experiments. Of eight different sources with known GM-CSF activity—monkey peripheral blood cells (Moore et al. 1973), human-placenta conditioned medium (Burgess et al. 1977), MLCM (Sheridan and Metcalf 1973), serum from ES-injected mice (Metcalf 1971), partially purified human urinary GM-CSF (Stanley and Metcalf 1969), mouse YSCM (Johnson and Metcalf 1977b), mouse-heart conditioned medium (Metcalf et al. 1975), and mouse-kidney conditioned medium (Metcalf et al. 1975)—only SCM was found to be capable of stimulating erythroid-colony formation in fetal liver cultures. Pokeweed mitogen alone at a concentration equivalent to that contained in SCM was unable to produce any proliferation in fetal liver cultures.

To determine the optimal time at which to score fetal liver cultures for erythroid colonies, cultures of 13-day CBA fetal liver cells were established and then scored on days 2 through 14 of incubation. These data are summarized in Table 3. Aggregates of more than 50 cells (i.e., colonies) first appeared after 4 days of incubation. In addition to the obvious red erythroid colonies, pale, degenerating colonies were also observed and were scored separately. These latter colonies had a characteristic ragged appearance and, when picked off and stained, contained benzidine-positive erythrocytes as well as numerous degenerating cells. Maximum numbers of red erythroid colonies were observed between days 4 and 7 (Table 3). The number of degenerating erythroid colonies increased with increasing time in culture, up to day 13. When intact cultures were stained directly with benzidine, all of the red colonies and a varied proportion of degenerating erythroid colonies stained positive (Table 3).

Erythropoietin assays of SCM and of human plasma were performed in hypertransfused, polycythemic C57BL mice (Johnson and Metcalf 1977a). Reticulocyte responses or 24-hour uptake of ^{59}Fe indicated no detectable

Table 3
Sequential Analysis of Erythroid-Colony Development

	Mean no. of aggregates		
Day of incubation	benzidine-positive	degenerating erythroid	erythroid
2[a]	80	0	0
4	25	23	11
5	18	20	12
6	20	21	14
7	20	20	11
8	21	23	8
9	11	25	8
10	8	21	4
13	11	36	0
14	1	16	0

All cultures contained 20,000 13-day CBA fetal liver cells, 1 ml DMEM with 20% human plasma, and 0.2 ml SCM. Mean data are from two cultures at each timepoint.

[a] At day 2 of incubation, only clusters were present; from day 4 onward, all aggregates contained more than 50 cells.

erythropoietin, either in SCM or in the human plasma used in cultures. Spleen conditioned medium did not inhibit the ability of injected erythropoietin to stimulate erythropoiesis in test animals. It was calculated from the response obtained from titrations of erythropoietin that, even if the SCM or human plasma contained erythropoietin, with the volumes used, less than 0.05 units of erythropoietin would have been added to each culture.

The effects of the addition of erythropoietin to fetal liver cultures were also studied. With cultures of 12-day CBA fetal liver cells in medium containing either FCS or human plasma, the addition of up to 4 units of erythropoietin did not result in any erythroid-colony formation after 7 days of culture, although occasional macrophage colonies were observed (Table 4). Spleen conditioned medium alone stimulated the formation of erythroid colonies in cultures containing human plasma and in cultures containing FCS (Table 4). In the latter case, however, the erythroid colonies were not as large or as red as those in the cultures containing human plasma. With cultures containing both erythropoietin and SCM, the number of erythroid colonies was almost twice as high in those cultures that also contained either FCS or human plasma as in cultures containing SCM alone (Table 4). In cultures containing human plasma, the increase in erythroid colonies associated with the addition of erythropoietin was usually accompanied by a decrease in the number of degenerating erythroid colonies. The addition of erythropoietin also appeared to increase the degree of hemoglobinization of colony cells and/or colony size in cultures containing FCS, since such colonies exhibited a deeper red color than did those in cultures containing SCM alone.

A survey of different mouse fetal and adult hematopoietic tissues cultured in 20% human plasma and 0.2 ml of SCM has shown that the highest

Table 4
Effect of Addition of Erythropoietin on Erythroid-Colony Formation in Cultures of CBA Fetal Liver Cells

Medium	Stimulus added[a]		No. of colonies[b]		
	erythropoietin	SCM	degenerating erythroid	erythroid	nonerythroid
FCS	−	−	0	0	0
FCS	−	+	35 ± 7	17 ± 4	70
FCS	+	−	0	0	3
FCS	+	+	35 ± 7	41 ± 6	50 ± 14
HP	−	−	0	0	0
HP	−	+	35 ± 7	28 ± 2	70 ± 14
HP	+	−	0	0	0
HP	+	+	25 ± 7	58 ± 13	50 ± 7

All cultures contained 20,000 12-day CBA fetal liver cells, with either 20% FCS or 20% human plasma (HP).

[a] Cultures (1 ml) stimulated by either 4 units of erythropoietin and/or 0.2 ml of pokeweed-mitogen-stimulated spleen conditioned medium.

[b] Number of colonies ± standard deviations from four replicate cultures were scored on day 7 of incubation.

frequency of erythroid colony-forming cells occurred in the fetal liver between days 10 and 12 of gestation (G. R. Johnson and D. Metcalf, in prep.). At day 11, approximately one erythroid colony-forming cell appeared per 100 fetal liver cells. Erythroid colony-forming cells were found in every fetal hematopoietic organ, including the peripheral blood, but excluding the thymus, but these cells sharply declined in frequency with increasing fetal age. However, in cultures of adult spleen cells, no erythroid colonies were observed, and a mean of only two erythroid colonies per 10^5 adult marrow cells was obtained from 10 separate experiments. The addition of either 1 or 2 units of erythropoietin to cultures of adult bone-marrow cells stimulated by SCM did not increase the size or number of the few erythroid colonies that were obtained.

Mixing experiments were performed with 12-day CBA fetal liver cells and adult marrow cells stimulated by SCM. In no instance was the number of erythroid colonies in mixed cultures different from that obtained in cultures of fetal liver cells alone.

The cells in 12-day fetal liver capable of producing erythroid colonies when stimulated by SCM have been shown to be radiosensitive (D_0, 110–125 rads), mainly in cycle, and to be nonadherent, light-density cells (G. R. Johnson and D. Metcalf, in prep.). When fractionated by velocity sedimentation, the erythroid colony-forming cells in 12-day fetal liver were found in fractions sedimenting between 4 and 13 mm/hour, with a peak sedimenting between 7 and 9 mm/hour (Fig. 4). As gestational age increased, the peak sedimentation velocity of erythroid colony-forming cells in fetal liver decreased. Thus, by day 18 of gestation, erythroid colony-forming cells sedimented at rates from 4 to 9 mm/hour, with the peak of cells occurring between 5 and 7 mm/hour (Fig. 5). When fractionated fetal liver cells were cultured in parallel to determine the distribution of GM-CFC and erythroid

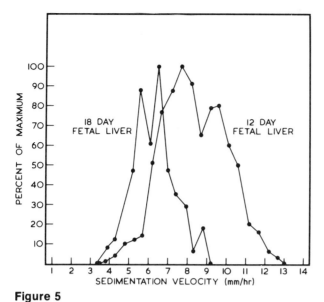

Figure 5
Velocity sedimentation fractionation of 12- and 18-day CBA fetal liver erythroid colony-forming cells. Data are expressed as percentage of fraction with peak number of colony-forming cells for each gestational age. Each point represents mean data from replicate 1-ml cultures stimulated by 0.2 ml of pokeweed-mitogen-stimulated SCM. Colony numbers were determined after 7 days of incubation.

colony-forming cells, the resultant profiles showed no segregation of the two colony-forming cell types (cf. Figs. 2 and 4). Similarly, after buoyant density fractionation of 12-day CBA fetal liver cells, no segregation was obtained of cells capable of forming neutrophil and macrophage colonies from the cells capable of forming erythroid colonies (G. R. Johnson and D. Metcalf, in prep.).

The most spectacular feature of the present type of erythroid colonies derived from fetal liver populations was the fact that approximately 50% contained other hematopoietic cells, the most common being, in order of frequency, macrophages, neutrophils, eosinophils, and megakaryocytes. Although the nature of these mixed erythroid colonies requires further investigation, they appear to differ sharply from the pure erythroid colonies described as forming from progenitor cells in cultures of adult marrow cells (Heath et al. 1976).

DISCUSSION

Although morphological studies have characterized the mature cell types—in particular the erythroid cells (Kovach et al. 1967; Fantoni et al. 1968)—present during fetal liver development, the factors regulating the differentiation of these fetal cells are poorly understood. Analysis of granulocyte and macrophage colonies in agar cultures has considerably increased our knowledge of the regulatory factors controlling the differentiation of these cell types from morphologically unrecognizable precursor cells (Metcalf 1974). This is due in part to the fact that, with properly defined conditions, it is possible to follow the development and progeny of single cells.

The GM precursor cells in the early fetal liver are light-density cells segregating as a relatively homogeneous population compared with adult bone-marrow GM-CFC (Moore et al. 1970). The present velocity sedimentation data indicate that 12-day fetal GM-CFC are relatively large cells compared with adult bone-marrow GM-CFC. Adult GM-CFC sediment as a heterogeneous population, with the most rapidly sedimenting cells tending to form macrophage colonies and being more responsive to low GM-CSF concentrations (Metcalf and MacDonald 1975). The most rapidly sedimenting GM-CFC from 12-day fetal liver are also more responsive to low GM-CSF concentrations but, unlike GM-CFC in bone marrow, they differentiate to produce similar proportions of granulocyte, mixed, and macrophage colonies to the more slowly sedimenting fetal GM-CFC. The relative homogeneity of fetal GM-CFC is also supported by the fact that granulocyte differentiation (when GM-CSF derived from ES or MLCM is used) is not as dependent upon GM-CSF concentration as is adult marrow GM-CFC.

The large erythroid colonies that develop in 12-day fetal liver cultures stimulated by pokeweed-mitogen-stimulated SCM develop from light-density cells that sediment between 7 and 9 mm/hour. The fetal erythroid colony-forming cells that produce pure erythroid colonies are likely to be equivalent to the adult marrow cells (burst-forming unit—erythroid, BFU-E) that produce similar large erythroid colonies. Since these BFU-E have been demonstrated to be slowly sedimenting cells (Heath et al. 1976), this suggests that fetal erythroid progenitor cells also possess physical characteristics different from those present in the equivalent cells in the adult.

As well as physical differences from adult marrow cells, fetal GM and erythroid progenitor cells appear to require different regulatory factors for the initiation and control of their subsequent differentiation. Thus, fetal GM-CFC are relatively unresponsive to the GM-CSF obtained from YSCM. Similarly, the fetal erythropoietic cells stimulated by SCM apparently do not require erythropoietin or, at most, require very low concentrations of this molecule. Relative unresponsiveness of fetal-liver-derived cells to erythropoietin has been proposed to explain the appearance of erythroid colonies in irradiated and reconstituted polycythemic mice (Bleiberg and Feldman 1969; Latsinik et al. 1971).

The findings presented here raise a number of questions that as yet remain unresolved. The data indicate that a distinct fetal hematopoietic population and regulatory system exists, at least for erythropoiesis and granulopoiesis. Since the fetal situation is one of active expansion of all cell populations, do similar regulatory procedures operate following perturbations to normal adult hematopoiesis? The fact that up to 50% of fetal erythroid colonies contain other hematopoietic cells and the fact that such colonies are clones (Johnson and Metcalf 1977c) suggest that these represent the progeny of a cell type very closely related to the multipotential hematopoietic stem cell. Although the cells producing mixed hematopoietic colonies are seldom found in adult marrow, they may represent a typically fetal population. Experiments to test this supposition are being performed with adult irradiated mice reconstituted with either fetal liver or adult marrow cells.

While the present observations provide evidence for the existence of distinct fetal populations of erythropoietic and GM populations, they leave unresolved the question as to whether two distinct types of hematopoietic stem cells exist—one forming fetal populations and the other, adult populations. If some method could be devised for identifying individual spleen colonies as being of fetal or adult type, the nature of stem cells in the fetal animal could be explored in irradiated adult recipients. It is likely, however, that many questions regarding stem-cell populations will remain unresolved until the development of satisfactory in vitro cloning systems for this cell type.

Acknowledgments

The authors are indebted to Mrs. Sue Webb, Ms. Kerry Haynes, Ms. Lynette Wadeson, and Mr. Roderick Mitchell for excellent technical assistance throughout and to Dr. A. W. Burgess for the preparation of concentrated yolk-sac conditioned medium. This work was supported by the Anti-Cancer Council of Victoria and by the National Cancer Institute Contract NOI-CB-33854.

REFERENCES

Bleiberg, I., and M. Feldman. 1969. On the regulation of hemopoietic spleen colonies produced by embryonic and adult cells. *Dev. Biol.* **19:** 566.

Bradley, T. R., and D. Metcalf. 1966. The growth of mouse bone marrow cells in vitro. *Aust. J. Exp. Biol. Med. Sci.* **44:** 287.

Burgess, A. W., E. M. A. Wilson, and D. Metcalf. 1977. Stimulation by human placental conditioned medium of hemopoietic colony formation by human marrow cells. *Blood* **49:** 573.

Fantoni, A., A. De la Chapelle, R. A. Rifkind, and P. A. Marks. 1968. Erythroid cell development in fetal mice: Synthetic capacity for different proteins. *J. Mol. Biol.* **33:** 79.

Heath, D. S., A. A. Axelrad, D. L. McLeod, and M. M. Shreeve. 1976. Separation of the erythropoietin-responsive progenitors BFU-E and CFU-E in mouse bone marrow by unit gravity sedimentation. *Blood* **47:** 777.

Johnson, G. R., and R. O. Jones. 1973. Differentiation of the mammalian hepatic primordium in vitro. *J. Embryol. Exp. Morphol.* **30:** 83.

Johnson, G. R., and D. Metcalf. 1977a. Characterization of mouse fetal liver granulocyte-macrophage colony-forming cells (GM-CFC) using velocity sedimentation. *Exp. Hematol. (Copenh.).* (In press.)

———. 1977b. Sources and nature of granulocyte-macrophage colony stimulating factor in fetal mice. *Exp. Hematol. (Copenh.).* (In press.)

———. 1977c. Pure and mixed erythroid colony formation in vitro stimulation by spleen conditioned medium with no detectable erythropoietin. *Proc. Natl. Acad. Sci. U.S.A.* **74:** 3879.

Johnson, G. R., and M. A. S. Moore. 1975. Role of stem cell migration in initiation of mouse foetal liver haemopoiesis. *Nature* **258:** 726.

Kovach J. S., P. A. Marks, E. S. Russell, and H. Epler. 1967. Erythroid cell development in fetal mice: Ultrastructural characteristics and hemoglobin synthesis. *J. Mol. Biol.* **25:** 131.

Latsinik, N. V., N. L. Samoylina, and J. L. Chertkov. 1971. Susceptibility to polycythemia of hemopoietic spleen colonies produced by cultured embryonal liver cells. *J. Cell. Physiol.* **78:** 405.

Metcalf, D. 1971. Acute antigen-induced elevation of serum colony stimulating factor (CSF) levels. *Immunology* **21:** 427.

―――. 1974. Regulation by colony stimulating factor (CSF) of granulocyte and macrophage colony formation in vitro by normal and leukemic cells. In *Control of proliferation in animal cells* (ed. B. Clarkson and R. Baserga), p. 887. Cold Spring Harbor Laboratory, Cold Spring Harbor, New York.

―――. 1977. *Hemopoietic colonies.* Springer-Verlag, Heidelberg.

Metcalf, D., and H. R. MacDonald. 1975. Heterogeneity of in vitro colony- and cluster-forming cells in the mouse marrow: Segregation by velocity sedimentation. *J. Cell. Physiol.* **85:** 643.

Metcalf, D., and M. A. S. Moore. 1971. *Haemopoietic cells.* North-Holland Research Monographs. *Front. Biol.* **24.**

Metcalf, D., T. R. Bradley, and W. Robinson. 1967. Analysis of colonies developing in vitro from mouse bone marrow cells stimulated by kidney feeder layers or leukemic serum. *J. Cell. Physiol.* **69:** 93.

Metcalf, D., H. R. MacDonald, N. Odartchenki, and B. Sordat. 1975. Growth of mouse megakaryocyte colonies in vitro. *Proc. Natl. Acad. Sci. U.S.A.* **72:** 1744.

Metcalf, D., J. Parker, H. M. Chester, and P. W. Kincade. 1974. Formation of eosinophilic-like granulocytic colonies by mouse bone marrow cells in vitro. *J. Cell. Physiol.* **84:** 275.

Moore, M. A. S., and G. R. Johnson. 1976. Stem cells during embryonic development and growth. In *Stem cells of renewing cell populations* (eds. A. B. Cairnie et al.), p. 323. Academic Press, New York.

Moore, M. A. S., and D. Metcalf. 1970. Ontogeny of the haematopoietic system: Yolk sac origin of in vivo and in vitro colony forming cells in the developing mouse embryo. *Br. J. Haematol.* **18:** 279.

Moore, M. A. S., and N. Williams. 1973. Analysis of proliferation and differentiation of foetal granulocyte-macrophage progenitor cells in haemopoietic tissue. *Cell Tissue Kinet.* **6:** 459.

Moore, M. A. S., T. A. McNeill, and J. S. Haskill. 1970. Density distribution analysis of in vivo and in vitro colony forming cells in developing fetal liver. *J. Cell. Physiol.* **75:** 181.

Moore, M. A. S., N. Williams, and D. Metcalf. 1973. In vitro colony formation by normal and leukemic human hematopoietic cells: Interaction between colony-forming and colony-stimulating cells. *J. Natl. Cancer Inst.* **50:** 591.

Pluznik, D. H., and L. Sachs. 1965. The cloning of normal "mast" cells in tissue culture. *J. Cell. Physiol.* **66:** 319.

Rifkind, R. A., D. Chui, and H. Epler. 1969. An ultrastructural study of early morphogenetic events during the establishment of fetal hepatic erythropoiesis. *J. Cell Biol.* **40:** 343.

Sheridan, J. W., and D. Metcalf. 1973. A low molecular weight factor in lung conditioned medium stimulating granulocyte and monocyte colony formation in vitro. *J. Cell. Physiol.* **81:** 11.

Silini, G., L. V. Pozzi, and S. Pons. 1967. Studies on the haematopoietic stem cells of mouse foetal liver. *J. Embryol. Exp. Morphol.* **17:** 303.

Stanley, E. R., and D. Metcalf. 1969. Partial purification and some properties of the factor in normal and leukemic human urine stimulating mouse bone marrow colony growth in vitro. *Aust. J. Exp. Biol. Med. Sci.* **47:** 467.

Till, J. E., and E. A. McCulloch. 1961. A direct measurement of the radiation sensitivity of normal mouse bone marrow cells. *Radiat. Res.* **14:** 213.

In Vitro Analysis of Self-Renewal and Commitment of Hematopoietic Stem Cells

T. M. Dexter, T. D. Allen, L. G. Lajtha, F. Krizsa, and N. G. Testa

Paterson Laboratories, Christie Hospital and Holt Radium Institute
Withington, Manchester M20 9BX, England

M. A. S. Moore

Sloan-Kettering Institute for Cancer Research
New York, New York 10021

Since the development of in vivo and in vitro techniques for clonal assay of hematopoietic stem cells and their committed progenitors, there has been a rapid expansion in our knowledge and understanding of the hematopoietic system. Till and McCulloch (1961) described the ability of certain murine bone-marrow cells to form spleen colonies when grafted into potentially lethally irradiated mice. Subsequently, unique chromosome markers were used to show that each spleen colony arises from a single cell, the colony-forming unit—spleen (CFU-S), and, furthermore, that such cells can regenerate the different compartments of the hematopoietic system, including lymphocytes (Wu et al. 1968; Edwards et al. 1970). Many studies have also demonstrated that the spleen colonies produced contain retransplantable CFU-S. The features possessed by these spleen colony-forming cells (capacity for self-renewal and pluripotentiality) fully justify their putative role as the hematopoietic stem cell.

Differentiation of the pluripotent stem cells produces the variety of committed or precursor cells (i.e., those that retain some proliferative ability but are committed to a particular maturation pathway), and in vitro systems are available for cloning many of these populations. The granulocyte precursor cells (colony-forming unit—culture; CFU-C), first described by Bradley and Metcalf (1966) and by Pluznik and Sachs (1966), can be recognized by their ability to form granulocyte and macrophage colonies in soft agar, in the presence of appropriate stimulating factors. Erythroid precursor cells are detectable by their capacity to form erythroid colonies in plasma clot or methylcellulose. Erythropoietin is required for the development of these colonies, and colony-forming cells are of two main types: burst-forming unit—erythroid (BFU-E), which may be derived from the CFU-S; and colony-forming unit—erythroid (CFU-E), which are more differentiated cells with only limited proliferative potential (Axelrad et al. 1974; Iscove and Sieber 1975; Gregory 1976). With suitable stimulating factors, the development of megakaryocyte colonies in vitro has been reported (Metcalf et al. 1975a; Williams et al. 1977), and the in vitro growth of mitogen-stimulated T lymphocytes (Sredni et al. 1976) and B lymphocytes grown in the presence of mercaptoethanol (Metcalf et al. 1975b) has also been documented. These in vitro systems permit the elucidation of the nature of the colony-forming cells and the characteristics of the various stimulating factors necessary for their development. However, as Lajtha et al. (1977) have pointed out, hematopoietic control processes in vivo operate essentially on two levels: the control of a steady-state maintenance of stem cells and the control of amplifying divisions of the differentiated and maturing precursor cells, or transit populations, the former being more critical in the long term. Since in vitro clonal assay systems for myeloid and lymphoid cells probably measure committed, transit populations, and since these systems represent essentially short-term cultures in which stem cells are rapidly lost, they are obviously unsuitable for the study of factors controlling stem-cell proliferation and differentiation.

Recently, however, we developed an in vitro system in which the processes of stem-cell proliferation and differentiation into various precursor cells were maintained for several months (Dexter and Lajtha 1976; Dexter and Testa 1976; Dexter et al. 1977a); this system may provide a suitable model for the study of stem-cell control mechanisms. The milieu for stem-cell proliferation was provided by a bone-marrow-derived adherent cell population, which was made up of a heterogeneous collection of different cell types that were probably representative of cells present in bone-marrow stroma (Allen 1977). This adherent layer developed after bone-marrow cells had been cultured from 2 to 3 weeks and, if reinoculated with fresh bone-marrow cells, sustained stem-cell production for many weeks. Since extensive cellular interactions, including fusion of cell membranes, are known to occur between the adherent cells and the maturing granulocytes (Allen and Dexter 1976), the adherent layer probably functions as an in vitro inductive microenvironment. This paper examines the role of the adherent layer in stem-cell proliferation and production of differentiated progeny.

MATERIALS AND METHODS
Mice

Unless otherwise stated, female BDF_1 (C57BL/6 × DBA/2) mice from 6 to 10 weeks of age were used.

Long-Term Bone-Marrow Cultures

The contents of a single femur were flushed into screw-capped, glass culture bottles (approx. 4 oz, United Glass) in 10 ml of Fischer's medium supplemented with 25% horse serum (Flow Laboratories) and with antibiotics as described previously (Dexter and Testa 1976). No attempt was made to obtain a single-cell suspension, as bone-marrow "fragments" are more effective at producing an adherent layer (Dexter et al. 1977b). The cultures were gassed with a mixture of 5% CO_2 in air and incubated at 33°C, a temperature shown previously to facilitate the development of the adherent layer (Dexter et al. 1977a). One-half of the growth medium was removed each week and an equal volume of fresh medium was added. After 3 weeks, the cultures again were fed, but with a growth medium containing 10^7 freshly isolated syngeneic (occasionally semiallogeneic) bone-marrow cells. The cultures were subsequently maintained at 33°C (or, occasionally, at 37°C) and were fed weekly by removal of one-half (5 ml) or all of the growth medium and addition of fresh medium. The nonadherent cells that are removed can be assayed for CFU-S (Till and McCulloch 1961) or CFU-C (Bradley and Metcalf 1966) using methods described previously (Dexter and Testa 1976).

To assay for the production of stem cells and committed precursor cells from the adherent layer, we sacrificed three to four cultures at various time intervals. After all the growth medium (containing nonadherent cells which were pooled to assay for CFU-S and CFU-C) had been removed, the adherent layer was washed twice with 5 ml of Fischer's medium, and these washed layers were assayed for CFU-S. The adherent cells were removed mechanically with a rubber policeman and were similarly assayed.

[^{3}H]TdR Suicide

The technique of assay of [^{3}H]TdR suicide was based on the method of Becker et al. (1965) and Iscove et al. (1970). Nonadherent or adherent cells were incubated in 2-ml aliquots, $1-3 \times 10^6$ cells/ml, at 37°C. [^{3}H]TdR (200 μCi/ml; sp. act. 15 Ci/mmole) or equivalent amounts of thymidine were added to the prewarmed tubes; the cultures were then incubated at 37°C for 30 minutes. Following incubation, the culture tubes were placed on ice. Cells were removed for CFU-S assays and diluted appropriately prior to injection. The remaining cells were washed three times in Fischer's medium supplemented with 10% horse serum and 100 μg/ml thymidine, resuspended in Fischer's medium containing 10 μg/ml thymidine, and assayed for CFU-C.

Erythroid Precursor Cells

Assay for BFU-E was performed according to the technique developed by Iscove and Sieber (1975). The cultured cells were washed twice in alpha medium (Gibco) supplemented with 2% fetal calf serum (FCS) to remove traces of horse serum, which is toxic for BFU-E development (N. G. Testa, unpubl.); 2×10^5 cells were then plated in 35-mm petri dishes (Corning) in 1 ml of alpha medium supplemented with 30% FCS, 2 units of erythropoietin (Step III, Connaught Labs), 10^{-4} M 2-mercaptoethanol, 1% bovine serum albumin (Sigma), 3.4×10^{-6} M human transferrin (Sigma), and 10^{-7} M sodium selenite (Guilbert and Iscove 1976) in 0.8% methylcellulose (Dow Chemical). Cultures were incubated for 9 days in a fully humidified atmosphere of 5% CO_2 in air. The erythroid colonies were recognized by their morphology (Iscove and Sieber 1975) and scored directly at 75× magnification. Colonies were also smeared on slides and stained with benzidine (Gregory 1976) for hemoglobin. The same technique was used to assay CFU-E, but the plates were scored after 2 days of incubation (Axelrad et al. 1974; Iscove and Sieber 1975).

Lymphoid Cells

Cultured cells were assayed for surface immunoglobulin (Ig), characteristic of B lymphocytes, and θ antigen, characteristic of T lymphocytes (Raff and Wortis 1970; Raff et al. 1970).

Direct Labeling for Surface Ig

Cultured cells were washed in McCoy's medium + 15% FCS. After the supernate had been removed, the cell pellet was resuspended in 50 μl fluorescein-labeled rabbit antimouse Kappa antiserum (Bionetics), containing 0.5 mg/ml protein, and placed on ice for 15 minutes. Two ml of McCoy's medium + 10% FCS were added, and the cells were centrifuged. The pellet was layered two times through FCS, washed with 10% FCS, and, finally, resuspended in serum-free McCoy's medium. Cells were kept on ice throughout this time. They were examined directly.

Detection of θ Antigen

Cultured cells were washed in buffered saline solution (BSS) alone, rabbit complement alone, or rabbit complement plus mouse antitheta serum (Bionetics) diluted 1:10, 1:50; or 1:100. Cells were observed at 16× at intervals of 15 minutes, and the percentage of lysed cells was estimated at each point.

Electron Microscopy

The methodology for scanning and transmission electron microscopy has been described previously (Allen and Dexter 1976).

RESULTS

Cellular Interactions

When bone-marrow fragments were cultured for 3 weeks at 33°C, an adherent layer became established that contained several cell types. These cell types often formed a multilayered adherent film which comprised the following types of cells. (1) E cells—these are cells with a smooth surface that form a flattened pavement (Fig. 1). They contain no phagocytic inclusions and are characterized further by a flattened nuclear profile and a small amount of condensed peripheral chromatin (Fig. 2). (2) R cells—these are larger, single, flattened cells with a smooth surface (Fig. 3). They may be derived from reticular cells in the bone marrow (Allen 1977). (3) F cells—these are large, fat-containing cells (Figs. 4-7), which are often grossly enlarged by lipid accumulation. The presence of these cells is a reliable indication that the adherent layer can support stem-cell proliferation (Allen and Dexter 1976). Allen (1977) has suggested that the R cells may act as precursors of the F cells, since he observed early stages of lipid ac-

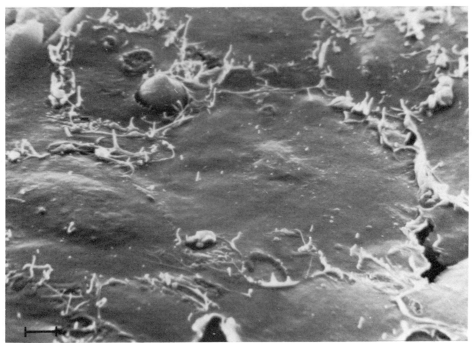

Figure 1
Scanning electron microscope (SEM) micrograph of an area of the pavement epithelial layer formed by the E cells. Individual cell surfaces are smooth, although there are microvilli and ruffles where the cells are in contact. Bar = 5 μm.

cumulation in such cells—although lipid accumulation has also been seen in cells having a fibroblastoid appearance. (4) M cells—these are phagocytic mononuclear cells (Figs. 2 and 8) that are characterized by phagocytic inclusions. Unlike the E, R, and F cells—which, because they tend to aggregate, become localized in certain regions of the cultures—M cells occur at all levels of the adherent multilayer and all over the substratum (Fig. 8).

When such adherent layers were seeded with fresh bone-marrow cells, extensive granulopoiesis occurred for several months. Granulocyte proliferation seemed to take place preferentially in regions containing F cells, particularly on the periphery of such areas where cells were in an early stage of lipid accumulation (Fig. 6) (Allen 1977). The more mature granulocytes tended to be found on the surface of the adherent layer, whereas the immature granulocytes and blast cells were often located within the adherent layer (Figs. 2 and 9) between E and M cells. Junctional complexes often occurred between the developing granulocytes and the surrounding "environmental" cells (Fig. 10), possibly indicating metabolic cooperation between such cells.

Production of CFU-S

To assay for the production of stem cells by the adherent layer, we established multiple cultures and maintained them at 33°C. One week after

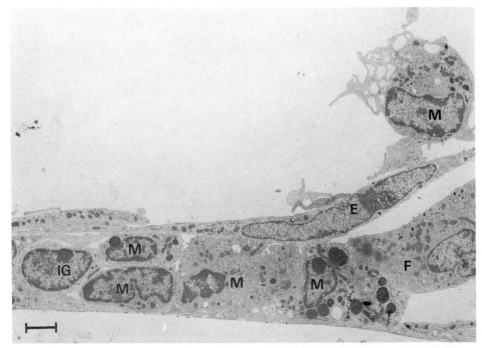

Figure 2
Transmission electron microscope (TEM) micrograph of a vertical section through an area similar to that shown in Fig. 1. The substratum is located just above the bottom of the micrograph. The multilayered nature of the adherent layer is clearly visible. The majority of basal cells are monocytes (M), with characteristic phagocytic inclusions. A fat cell (F) in early stages of lipid accumulation is also visible. A flattened E cell (E) runs across the entire area, and itself acts as a substratum for a monocyte (M). An immature granulocyte (IG) is also visible in the multilayer. Bar = 2 μm.

inoculation of the second marrow population, all the growth medium (containing nonadherent cells) was removed from six cultures, the adherent layer was washed twice gently with Fischer's medium, and the adherent cells were removed mechanically. Both the adherent and nonadherent cell populations and the cells present in the two washes were assayed for CFU-S and CFU-C.

All the growth medium was removed from the remaining cultures and was replaced with fresh medium; the cultures were maintained subsequently at 33°C. Half of the growth medium was removed weekly and fresh medium was added, thus allowing subsequent stem-cell production from the adherent layer to be monitored. At selected time intervals, further cultures were sacrificed and the adherent and nonadherent cells were assayed for CFU-S and CFU-C.

The results (Table 1) show that when the nonadherent cells were removed, their numbers recovered by the following week and were subsequently maintained for several months. These new cells presumably were generated from the adherent layer, the numbers of which were maintained

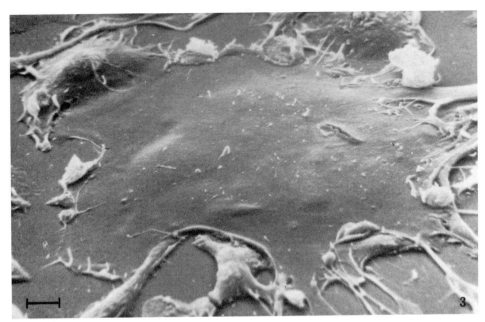

Figure 3
SEM micrograph of a typical R cell showing an epithelioid morphology and smooth surface. Bar = 10 μm.

throughout the culture, and consisted mainly (>80%) of granulocytes in all stages of maturation. Within the adherent layer, mainly immature granulocytes were found. The CFU-S remaining in the adherent layer after all the growth medium had been removed could regenerate themselves and, furthermore, could give rise to the population of nonadherent CFU-S. The nonadherent CFU-S subsequently were maintained at high levels for several months. During this time, CFU-C production was occurring, presumably representing differentiation from the stem-cell compartment.

Figure 4
SEM micrograph of an aggregation of fat cells viewed from above, showing the overall spherical nature of the cells produced by coalescence of individual fat droplets. This preparation was made at the end of the period of establishment of the adherent layer, prior to recharging the cultures with a fresh inoculum. Bar = 20 μm.

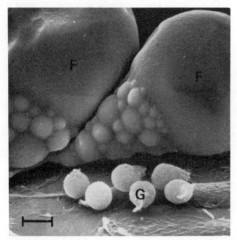

Figure 5
SEM micrograph showing detail of two fat cells (F) and a group of granulocytes (G) illustrating the difference in size. Bar = 5 μm.

Stem-Cell Kinetics in Long-Term Cultures

The adherent layer was obviously a major site of stem-cell production and presumably was the source of the proliferative stimulus. The cycling characteristics of stem cells were measured by determining [^{3}H]TdR suicide at various times after feeding. Cultures were initiated and maintained at

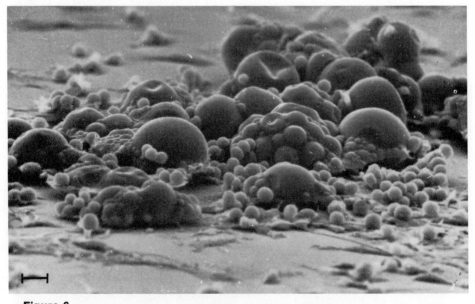

Figure 6
SEM micrograph of an area of active granulopoiesis viewed at high tilt, illustrating the clustering of granulocytes around the periphery of the fat cells. Bar = 10 μm.

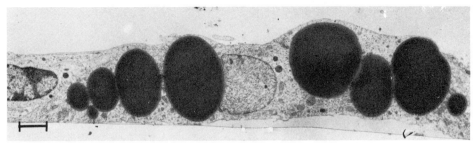

Figure 7
TEM micrograph of a section through a fat cell in an early stage of fat accumulation, showing individual lipid vacuoles. Bar = 2 μm.

33°C. One week after inoculation of the second marrow population, three to four bottles were sacrificed and [^{3}H]TdR suicide rates measured on the adherent and nonadherent CFU-S and CFU-C. The remaining cultures were fed (by demidepopulation) and [^{3}H]TdR suicide assays performed at various time intervals. The results are shown in Figure 11. One week after initiation of the cultures, both the adherent and the nonadherent CFU-S were in a low cycling state (<10% [^{3}H]TdR suicide). One day after these

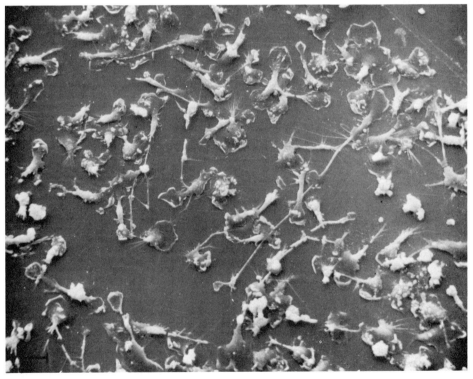

Figure 8
SEM micrograph of an area of the adherent layer showing numerous monocytes. Bar = 20 μm.

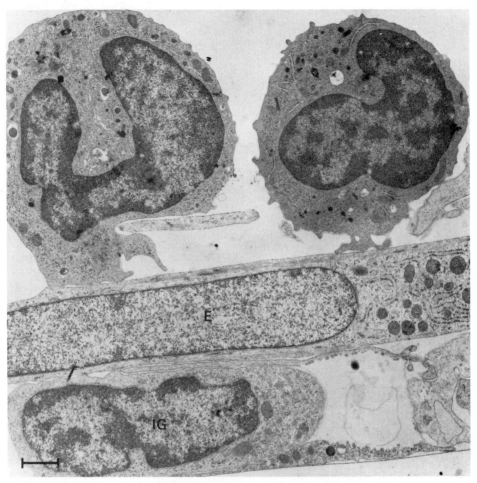

Figure 9
TEM micrograph of a vertical section through the multilayer showing an immature granulocyte (IG) beneath the E cell layer (E) and two maturing granulocytes above. Bar = 1 μm.

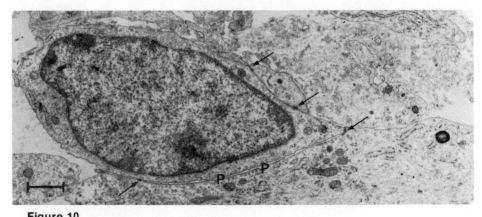

Figure 10
TEM micrograph of an oblique section through an immature cell showing junctional complexes (arrowed) with the neighboring cells. Pinocytotic vesicles (P) are also present at adjacent regions. Bar = 1 μm.

Table 1
The Role of the Adherent Layer in the Maintenance of Stem Cells

Weeks cultured	Total cells × 10⁵				Total CFU-S				Total CFU-C			
	NA	wash I	wash II	Ad	NA	wash I	wash II	Ad	NA	wash I	wash II	Ad
1	39	7	3	30	1300[a]	280	70	1200	48,800	ND	ND	30,000
2	35	7	4	40	1903	410	ND	800	32,000	ND	ND	10,000
3	51	ND	ND	ND	1200	ND	ND	ND	12,600	ND	ND	ND
5	73	9	ND	46	1640	140	40	430	48,100	ND	ND	14,000
7	23	ND	ND	ND	720	ND	ND	ND	12,000	ND	ND	ND
10	11	3	2	28	810	ND	ND	560	10,000	2000	800	7000

NA, nonadherent cells; Ad, adherent cells; ND, not detectable.
[a] All nonadherent cells were removed after 1 week and cells were subsequently fed weekly.

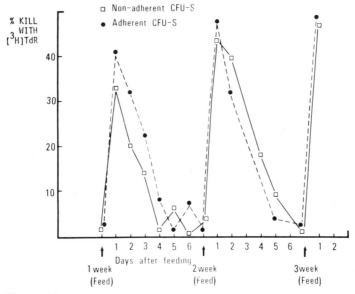

Figure 11
Cycling characteristics of stem cells.

cultures had been refed, CFU-S were in a high cycling state, but this proportion progressively decreased with time after feeding and a low cycling state occurred from 4 to 7 days after the cultures had been refed. A further refeeding then restimulated the CFU-S into a high cycling state. During this time, the CFU-C were maintained in a high cycling state with no apparent decrease in [^{3}H]TdR kill with time after feeding.

Maintenance of Erythroid Precursor Cells

Although the cultures supported granulocyte maturation, we consistently found that no morphologically recognizable (hemoglobin-producing) erythroid cells were present. This suggests either that erythroid precursor cells were not present or that their maturation into hemoglobin-producing cells were inhibited in the long-term cultures. To investigate these possibilities, we assayed cultured cells, established and maintained at 33°C, for BFU-E and CFU-E. The results (Table 2) show that BFU-E were indeed being

Table 2
Production of Erythroid Precursor Cells in Long-Term Bone-Marrow Cultures

Weeks cultured	Cell count × 10^5	Total CFU-S	Total BFU-E	Total CFU-E
5	25	700	100	0
6	24	ND	180	0
8	12	ND	11	0
9	20	200	10	0

ND, not detectable

maintained in the cultures. At no stage, however, were the more differentiated CFU-E detected.

Megakaryopoiesis in Long-Term Cultures

Allen (1977) has shown that mature megakaryocytes are present in the cultures. This finding was confirmed and extended by Williams et al. (1977), who showed further that megakaryocyte precursor cells, detected by their ability to form colonies of mature platelet-producing megakaryocytes in soft agar, are sustained in culture for several weeks.

Lymphopoiesis in Bone-Marrow Cultures

Using the techniques described, we were unable to detect cells possessing surface Ig or θ antigen after 1 week of culture; therefore, no mature B or T lymphocytes were being produced. Also, in preliminary studies using the clonal assay developed by Metcalf et al. (1975) we detected no B-lymphocyte colony-forming cells. However, the possibility remains that pre-B or pre-T cells were present in the cultures, since treatment in vitro with chemical leukemogens (Dexter and Lajtha 1976) or lymphoid leukemia-inducing virus (N. Teich and T. M. Dexter, in prep.) will induce leukemias that show significant N-alkaline phosphatase activity (P. Baines et al., in prep.), a general enzyme marker for lymphoid malignancies (Neumann et al. 1974). Furthermore, some leukemias have been induced in vitro that show terminal deoxynucleotidyl transferase (T. M. Dexter, in prep.), an enzyme thought to be restricted to prethymocytes and thymocytes (Coleman et al. 1974). Whether these leukemias represent transformation of an already committed lymphoid precursor cell or, alternatively, transformation of a stem cell with subsequent expression of lymphoid enzymes remains to be determined.

In Vitro Microenvironments

Preliminary experiments in which bone marrow from genetically anemic mice was used (Dexter and Moore 1977) support our hypothesis that the adherent layer acts as an inductive microenvironment. Mice possessing the Steel (Sl/Sl^d) and W/W^v mutations were used in these studies. Genetically determined macrocytic anemia of Sl/Sl^d mice cannot be cured by injection of stem cells from a normal littermate but can be alleviated by transplantation of normal spleen. It is generally thought that such mice have an environment defective for stem-cell proliferation and/or differentiation. On the other hand, the anemia of W/W^v mice can be cured by injection of stem cells from a normal littermate (or from Sl/Sl^d mice!), and the lesions in W/W^v mice presumably are due to a defect(s) in the pluripotential stem cells (Russell and Bernstein 1966; Bennett et al. 1968; Sutherland et al. 1970).

When adherent layers were established from Sl/Sl^d cells, only low numbers of CFU-C were produced (Table 3). Also, when W/W^v adherent layers were fed with W/W^v marrow cells, CFU-C were barely detectable after 3 weeks in culture. Bone-marrow-derived adherent layers from normal

Table 3
Production of CFU-C on Bone-Marrow-Derived Adherent Layers from W/W^v Mice, Sl/Sl^d Mice, and Their Normal — W(+/+); Sl(+/+) — Littermates

Adherent layer	Cells added	CFU-C production
W(+/+)	W(+/+)	>5 weeks (normal rate)
W/W^v	W/W^v	barely detectable after 1 week
Sl(+/+)	Sl(+/+)	>5 weeks (normal rate)
Sl/Sl^d	Sl/Sl^d	low production, 5 weeks
Sl/Sl^d	W/W^v	not detectable after 1 week
W/W^v	Sl/Sl^d	>5 weeks (normal rate)

littermates could sustain a high level of CFU-C production. Of great interest was our finding that Sl/Sl^d adherent layers (defective environment) inoculated with W/W^v bone marrow (defective stem cells) showed no production of CFU-C, whereas the reciprocal combination of W/W^v adherent layers sustained high levels of CFU-C production when inoculated with Sl/Sl^d bone marrow; i.e., the defects seen in vivo were successfully reproduced and compensated for in vitro.

DISCUSSION

The role of environmental influences in the proliferation and differentiation of stem cells has been the subject of prolonged interest, but studies have been hampered by the lack of suitable in vitro systems for studying interactions between stem cells and the hemopoietic inductive environment. That the soft-gel systems used for the growth of stem-cell progenitor populations (CFU-C, BFU-E) are unsuitable for prolonged survival of stem cells has been well established and probably indicates the importance of a suitable milieu (cell-to-cell contact) for stem-cell proliferation and differentiation. Chang and Anderson (1971) described a liquid culture system where granulocyte *maturation* was maintained for several weeks. Although their studies did not incorporate assay for CFU-S or CFU-C, they were of importance in that they described an adherent feeder cell that appeared to be essential for granulopoiesis and upon which clusters of developing granulocytes regularly were found. We have learned in our studies that the prior establishment of a suitable bone-marrow-derived adherent layer is essential for stem-cell proliferation and granulopoiesis (Dexter, Allen, and Lajtha 1977). In siliconized culture flasks and in cultures showing mediocre development of the adherent layer, stem-cell survival is limited to a few days.

The adherent layer contains several cell types, E, R, M, and giant fat cells, all of which appear to have counterparts in vivo (Allen 1977). Granulopoiesis is largely associated with areas showing fat-cell development, and the possibility must be considered that these fat cells represent those described by Chang and Anderson (1971). This interesting association between fat cells and granulopoiesis has also been observed by G. Brecher (pers. comm.), who implanted bone-marrow cells and irradiated spleen stroma into the omentum of irradiated mice and found that granulopoiesis was usually present in the interstices of the fat cells of the omentum, whereas erythropoiesis was associated with the implanted splenic trabecular stroma (Haley et al. 1975).

The extensive granulopoiesis observed in the cultures for several months presumably is dependent upon the continued production and maturation of the CFU-C, which, in turn, are derived from the CFU-S. A major site of stem-cell production is the adherent layer, where the CFU-S probably are lodged within the cellular architecture in much the same way as the developing granulocytes (Allen and Dexter 1976). We have shown that such adherent CFU-S are capable of extensive proliferation. Feeding the cultures (either by depopulation of all or of one-half the growth medium) seems to provide the stimulus for stem-cell proliferation, since, within 1 day postfeeding, CFU-S are in a high-cycling state. From 4 to 7 days postrefeeding, however, the stem cells have returned to low-cycling, steady-state conditions. A further refeeding will then restimulate these into a high-cycling condition. We have suggested previously (Dexter et al. 1977c) that the CFU-S depopulation (consequent upon feeding) is "recognized," leading to proliferation of adherent CFU-S and to their subsequent release into the surrounding medium. We have suggested further that the number of stem cells is limited by the availability of the cellular architecture necessary for their proliferation. In this hypothesis, the eventual depletion of stem cells, after several months in culture, may be associated with changes in the cellular composition of the adherent layer.

Our studies with genetically anemic Sl/Sl^d and W/W^v mice further emphasize the importance of the adherent layer in stem-cell maintenance. The ability to reproduce and "cure" the environmental defect in vitro using appropriate bone-marrow combinations should be a valuable tool for investigating the importance of cellular interactions in the maintenance of hemopoiesis. Studies along these lines are now in progress.

Concomitant with stem-cell proliferation is production of granulocyte (CFU-C), erythroid (BFU-E), and megakaryocyte (colony-forming unit-megakaryocyte, CFU-M) precurser cells. The extent to which interactions between the stem cell (CFU-S) and the cellular environment determine differentiation into these various compartments is not known. Further work also is needed to examine the roles of the various factors required for the maturation of these precursor cells. Certain anomalous situations have already arisen. For example, the presence of molecules with colony-stimulating activity (CSA) is obligatory for the development of CFU-C in soft-agar cultures (Metcalf 1970). However, in the long-term bone-marrow cultures (where there is extensive granulopoiesis for many months), CSA is not detectable (Dexter et al. 1977a); this casts doubt on the proposed physiological role of CSA as a regulator of granulopoiesis (Moore et al. 1974). Although BFU-E were detectable in the cultures (which did not contain erythropoietin!), their more mature progeny, CFU-E, were not present. This could have been due to the sera used (horse serum generally being inhibitory for CFU-E development) or, alternatively, to a maturation block in the progression from BFU-E to CFU-E. This block could not be overcome and erythropoiesis was not stimulated when the medium from long-term cultures was replaced with medium, serum, and erythropoietin as used in the methylcellulose BFU-E assay (Testa and Dexter 1977). This suggests that the adherent cells may modify the response of the BFU-E to erythropoietin. The culture system described will allow us to examine many of the aforementioned questions.

Acknowledgments

This work was supported by the Medical Research Council and the Cancer Research Campaign.

REFERENCES

Allen, T. D. 1977. Ultrastructural aspects of *in vitro* haemopoiesis. In *The Second Symposium of the British Society for Cell Biology on Stem Cells and Tissue Homeostasis* (eds. B. I. Lord et al.). Cambridge University Press, Cambridge. (In press.)

Allen, T. D., and T. M. Dexter. 1976. Cellular interrelationships during *in vitro* granulopoiesis. *Differentiation* **6:** 191.

Axelrad, A. A., D. L. McLeod, M. M. Shreeve, and D. S. Heath. 1974. Properties of cells that produce erythrocytic colonies in vitro. In *Hemopoiesis in Culture, Second International Workshop* (ed. W. A. Robinson), p. 226. U.S. Government Printing Office, Washington, D.C.

Becker, A. J., E. A. McCulloch, L. Siminovitch, and J. E. Till. 1965. The effect of differing demands for blood cell production on DNA synthesis by hemopoietic colony forming cells of mice. *Blood* **26:** 296.

Bennett, M., G. Cudkowicz, R. S. Foster, Jr., and D. Metcalf. 1968. Hemopoietic progenitor cells of *W* anemic mice studied *in vivo* and *in vitro*. *J. Cell. Physiol.* **71:** 211.

Bradley, T. R., and D. Metcalf. 1966. The growth of mouse bone marrow cells *in vitro*. *Aust. J. Exp. Biol. Med. Sci.* **44:** 287.

Chang, Y. T., and R. N. Andersen. 1971. Cultivation of mouse bone marrow cells. I. Growth of granulocytes. *J. Reticuloendothel. Soc.* **9:** 568.

Coleman, M. S., J. J. Hutton, P. De Simone, and F. J. Bollum. 1974. Terminal deoxyribonucleotidyl transferase in human leukemia. *Proc. Natl. Acad. Sci. U.S.A.* **71:** 4404.

Dexter, T. M., and L. G. Lajtha. 1976. Proliferation of haemopoietic stem cells and development of potentially leukaemic cells in vitro. In *Comparative leukaemia research 1975* (eds. J. Clemmesen and D. S. Yohn). *Bibl. Haematol.* **43:** 1.

Dexter, T. M., and M. A. S. Moore. 1977. *In vitro* duplication and 'cure' of haemopoietic defects in genetically anaemic mice. *Nature* **269:** 412.

Dexter, T. M., and N. G. Testa. 1976. Differentiation and proliferation of hemopoietic cells in culture. In *Methods in cell biology* (ed. D. M. Prescott), vol. 14, p. 387. Academic Press, New York.

Dexter, T. M., T. D. Allen, and L. G. Lajtha. 1977a. Conditions controlling the proliferation of haemopoietic stem cells in vitro. *J. Cell. Physiol.* **91:** 335.

Dexter, T. M., D. Scott, and N. M. Teich. 1977b. Infection of bone marrow cells *in vitro* with FLV: Effects on stem cell proliferation, differentiation and leukemogenic capacity. *Cell* **12:** 355.

Dexter, T. M., E. G. Wright, F. Krizsa, and L. G. Lajtha. 1977c. The regulation of haemopoietic stem cell proliferation in long term bone marrow cultures. *Biomedicine* (Paris). (In press.)

Edwards, G. E., R. G. Miller, and R. A. Phillips. 1970. Differentiation of rosette-forming cells from myeloid stem cells. *J. Immunol.* **105**: 719.

Guilbert, L. J., and N. N. Iscove. 1976. Partial replacement of serum by selenite, transferrin, albumin and lecithin in haemopoietic cell cultures. *Nature* **263**: 594.

Gregory, C. J. 1976. Erythropoietin sensitivity as a differentiation marker in the haemopoietic system: Studies of three erythropoietic colony responses in culture. *J. Cell. Physiol.* **89**: 289.

Haley, J. E., J. H. Tjio, W. W. Smith, and G. Brecher. 1975. Hematopoietic differentiative properties of murine spleen implanted in the omenta of irradiated and non-irradiated hosts. *Exp. Hematol. (Copenh.)* **3**: 187.

Iscove, N. N., and F. Sieber. 1975. Erythroid progenitors in mouse bone marrow detected by macroscopic colony formation in culture. *Exp. Hematol. (Copenh.)* **3**: 32.

Iscove, N. N., J. E. Till, and E. A. McCulloch. 1970. The proliferative states of mouse granulopoietic progenitor cells. *Proc. Soc. Exp. Biol. Med.* **134**: 33.

Lajtha, L. G., B. I. Lord, T. M. Dexter, E. G. Wright, and T. D. Allen. 1977. Inter-relationships of differentiation and proliferation control in haemopoietic stem cells. In *30th Annual Symposium on Fundamental Cancer Research*. University of Texas and M. D. Anderson Tumor Institute, Houston, Texas. (In press.)

Metcalf, D. 1970. Studies on colony formation *in vitro* by mouse bone marrow cells. II. Action of colony stimulating factor. *J. Cell. Physiol.* **76**: 89.

Metcalf, D., H. R. MacDonald, N. Odartchenko, and B. Sordat. 1975a. Growth of mouse megakaryocyte colonies *in vitro*. *Proc. Natl. Acad. Sci. U.S.A.* **72**: 1744.

Metcalf, D., N. L. Warner, G. J. V. Nossal, J. F. A. P. Miller, K. Shortman, and E. Rabellino. 1975b. Growth of B-lymphocyte colonies *in vitro* from mouse lymphoid organs. *Nature* **255**: 630.

Moore, M. A. S., G. Spitzer, D. Metcalf, and D. G. Pennington. 1974. Monocyte production of colony stimulating factor in familial neutropenia. *Brit. J. Haematol.* **27**: 47.

Neumann, H., E. M. Moran, R. M. Russell, and I. H. Rosenberg. 1974. Distinct alkaline phosphatase in serum of patients with lymphatic leukemia and infectious mononucleosis. *Science* **186**: 151.

Pluznik, D. H., and L. Sachs. 1966. The induction of clones of normal 'mast' cells by a substance from conditioned medium. *Exp. Cell Res.* **43**: 553.

Raff, M. C., and H. H. Wortis. 1970. Thymus dependence of θ-bearing cells in the peripheral lymphoid tissues of mice. *Immunology* **18**: 931.

Raff, M. C., M. Sternberg, and R. B. Taylor. 1970. Immunoglobulin determinants on the surface of mouse lymphoid cells. *Nature* **225**: 553.

Russell, E. S., and S. E. Bernstein. 1966. Blood and blood formation. In *Biology of the laboratory mouse,* 2nd ed. (ed. E. L. Green), p. 351. McGraw-Hill, New York.

Sredni, B., Y. Kalechman, H. Michlin, and L. A. Rozenszajn. 1976. Development of colonies *in vitro* of mitogen stimulated mouse T-lymphocytes. *Nature* **259**: 130.

Sutherland, D. J. A., J. E. Till, and E. A. McCulloch. 1970. A kinetic study of the genetic control of hemopoietic progenitor cells assayed in culture and in vivo. *J. Cell. Physiol.* **75**: 267.

Testa, N. G., and T. M. Dexter. 1977. Long term production of erythroid precursor cells (BFU) in bone marrow cultures. *Differentiation* **9**: 193.

Till, J. E., and E. A. McCulloch. 1961. A direct measurement of the radiation sensitivity of normal mouse bone marrow cells. *Radiat. Res.* **14**: 213.

Williams, N., H. Jackson, A. P. C. Sneridan, M. J. Murphy, A. Elste, and M. A. S.

Moore. 1977. Regulation of megakaryopoiesis in long term bone marrow cultures. *Blood.* (In press.)

Wu, A. M., J. E. Till, L. Siminovitch, and E. A. McCulloch. 1968. Cytological evidence for a relationship between normal hematopoietic colony forming cells and cells of the lymphoid system. *J. Exp. Med.* **127:** 455.

Approaches to the Evaluation of Human Hematopoietic Stem-Cell Function

J. E. Till, S. Lan, R. N. Buick, P. Sousan, J. E. Curtis, and E. A. McCulloch

Ontario Cancer Institute, Toronto, Ontario M4X 1K9 Canada

The progenitor cells of the blood-forming system exist as a minority subpopulation, dominated numerically by their more differentiated descendants. Selective assays, such as those based on colony formation, have been developed for their study. A major application of these assays for colony-forming progenitor cells has been the analysis of lineage relationships between the various classes of progenitors, and a hierarchy of such relationships has been described (see, for example, Chervenick et al. 1975). These cell-lineage relationships have been investigated primarily in the mouse, and the major limitation on extension of these studies to human cells has been the lack of an assay for human pluripotent stem cells analogous to the spleen-colony assay for rodent stem cells (Till and McCulloch 1961). In its absence, cell-culture methods are being used to explore various approaches (Dexter; Johnson and Metcalf; both this volume).

Even though an assay for pluripotent human hematopoietic stem cells is not yet available, indirect evidence about these stem cells can still be obtained, because of their close relationship to progenitor cells already committed to a single major pathway of differentiation. Studies of individual murine spleen colonies have shown that a shared relationship to a common ancestral cell class (the pluripotent stem cells) can lead to significant correlations between the results of assays for two different classes of committed progenitor cells. An example is provided by the granulopoietic progenitor cells (colony-forming units—culture [CFU-C]) (Pluznik and Sachs 1965; Bradley and Metcalf 1966) and the erythropoietic progenitors (colony-forming units—erythropoietin-responsive [CFU-E] and burst-forming units —erythroid [BFU-E]) (Stephenson et al. 1971; Axelrad et al. 1974). The frequency of each of these progenitors varies greatly from one spleen colony to another, but the results of all three assays correlate significantly. However, when their shared relationship to pluripotent stem cells (colony-forming units—spleen [CFU-S]) is taken into account, the correlation between CFU-C and BFU-E (or CFU-E) largely disappears (Gregory and Henkelman 1977). In contrast, the correlation observed between the parent-progeny pair of erythropoietic progenitors (BFU-E and CFU-E) does not depend on a "shared-ancestor" relationship and is not eliminated even when their relationship to CFU-S is taken into account.

This demonstration of "shared-ancestor" correlations provides a basis for an approach to the investigation of cell-lineage relationships in normal and abnormal human hematopoiesis. The purpose of this paper is to summarize studies of correlations between numbers of committed progenitor cells (BFU-E, CFU-E, and CFU-C) in marrow samples both from patients with leukemia and from patients with nonhematological malignancies. The prognostic value of these assays as predictors of remission after combination chemotherapy has also been investigated. The results indicate that correlation analysis provides a useful way to study cell-lineage relationships in human leukemia, and that more information is needed about the blast cells, the major cell population in leukemia.

COMMITTED PROGENITOR CELLS IN ACUTE MYELOBLASTIC LEUKEMIA

Assays of progenitor cells committed to granulopoiesis (CFU-C) in marrow samples from patients with leukemia have revealed considerable patient-to-patient variation (Curtis et al. 1975). This variation is somewhat analogous to the variation seen when assays for CFU-C are carried out on individual splenic colonies. In the latter situation, the variation does not appear to be a result primarily of heterogeneity in the population of pluripotent stem cells from which the splenic colonies arise. Instead, random fluctuations in the timing of events involved in colony formation (Till et al. 1964) and heterogeneity in the microenvironments in which the colonies are formed (Curry and Trentin 1967) have been proposed as the major sources of variation.

The patient-to-patient variation in frequency of CFU-C in marrow samples does not appear to be a reflection of subclasses of acute myeloblastic leukemia (AML), not readily defined on the basis of usual clinical criteria. If it were, one might expect the level of CFU-C in marrow to be characteristic of the subclass and to persist even after perturbations such as those introduced by chemotherapy. This is not the case, as indicated by the results of assays for CFU-C carried out on marrow samples obtained from the same patients prior to treatment and 8 weeks after initiation of treatment. The results are shown in Figure 1. The data obtained prior to treatment and those obtained after treatment both show extensive patient-to-patient variation. However, the patients yielding high levels of marrow CFU-C prior to treatment did not necessarily yield high CFU-C levels after treatment; indeed, the results before and after treatment showed no significant correlation (Spearman's rank correlation coefficient $R_s = 0.11$). This result makes it more likely that the patient-to-patient variation in the level of CFU-C after treatment reflects fluctuations in factors influencing the repopulation of the granulopoietic progenitor compartment other than the number of such cells present prior to treatment. It is even possible that

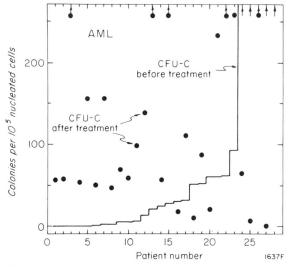

Figure 1
Results of assays for colony-forming granulopoietic progenitor cells (CFU-C) carried out on 28 marrow samples obtained from patients with AML prior to treatment (Curtis et al. 1975) and ranked in order of increasing frequencies of CFU-C per 10^5 nucleated cells (———). Results for samples obtained from the same patients 8 weeks after initiation of treatment are also shown (●). The cultures contained added colony-stimulating activity (Iscove et al. 1971).

similar factors may underlie this type of variation and the variation seen among spleen colonies.

To pursue this possibility further, we measured numbers of CFU-C and of early (BFU-E) and late (CFU-E) erythropoietic progenitors in the same marrow samples, which were taken from patients with AML at the time of diagnosis. Considerable patient-to-patient variation again was observed, and the numbers of CFU-E were significantly correlated with the numbers of BFU-E and, in addition, with the numbers of CFU-C (Table 1) (S. Lan et al. 1978). However, the mean frequencies of colony formation were rather low in comparison with those observed for marrow samples for a separate group of patients studied after treatment had been initiated (Table 2). The results obtained after initiation of treatment are shown in Figure 2. It is apparent from the figure that marrows containing high levels of CFU-E also tended to show high levels of BFU-E; the two classes of erythropoietic progenitors showed a significant positive correlation $R_s = 0.88$; Table 1). CFU-C data for the same marrows were also positively correlated with the BFU-E and CFU-E values ($R_s = 0.60$ and 0.52, respectively; Table 1). An increase in erythropoiesis in culture was associated with a reduction in percentage of blast cells (myeloblasts plus promyelocytes) as patients proceeded toward remission (Fig. 2).

To test whether or not the observed correlations between progenitor cell frequencies were confined to leukemic hemopoiesis, we carried out similar assays on marrow samples from patients without detectable hematological malignancy. Again, significant positive correlations between frequencies

Table 1
Correlation Coefficients

Progenitor	Patients[a]	CFU-E	CFU-C
BFU-E	U	0.61[b]	0.23
	T	0.88[b]	0.60[b]
	N	0.60[b]	0.54[b]
CFU-C	U	0.52[b]	–
	T	0.52[b]	–
	N	0.52[b]	–
Blasts	U	0.02	0.18
	T	−0.48[b]	−0.15

Table gives values of Spearman's rank correlation coefficients, corrected for tied values (Siegel 1956). Correlations were based on frequencies of BFU-E, CFU-E, and CFU-C per 10^5 nucleated marrow cells and on percentages of marrow myeloblasts plus promyelocytes.

[a] U: Untreated patients, 24 marrow samples; T: treated patients, 35 marrows; N: nonleukemic patients, 29 marrows.

[b] Significant at the 1% level.

of BFU-E, CFU-E, and CFU-C were observed (S. Lan et al. 1978). In addition, the correlation coefficients for BFU-E and CFU-E were consistently higher than those for BFU-E (or CFU-E) and CFU-C, either for patients with AML at the time of diagnosis, for treated patients with AML, or for patients with nonhematological malignancy (Table 1). This is the result expected for a closely related parent-progeny pair such as BFU-E and CFU-E, if it is assumed that the statistically significant correlation coefficients are a reflection of cell-lineage relationships. The correlation between the erythropoietic progenitors and CFU-C is weaker, as expected if its origin is the shared relationship of these classes of progenitors with their common ancestors, the pluripotent stem cells.

Attempts to apply this approach to marrow blast cells from patients with AML showed no evidence of positive correlations between the blast cells and colony-forming progenitors (BFU-E, CFU-E, or CFU-C) (Table 1) (S. Lan et al. 1978). Thus, no evidence for a cell-lineage relationship between these progenitors and the marrow blasts could be demonstrated.

Table 2
Levels of Colony-Forming Committed Progenitor Cells in Marrow

Progenitor	Untreated – 24 marrows		Treated – 35 marrows		Nonleukemic – 29 marrows	
	mean	95% C.L.	mean	95% C.L.	mean	95% C.L.
BFU-E	0.3	0.02–0.7	1.9	1.0– 3.1	2.4	1.2– 4.4
CFU-E	1.1	0.3 –2.4	6.7	3.4–12.6	14.1	7.9–24.8
CFU-C	3.0	1.1 –6.8	11.6	6.1–21.2	23.8	14.5–38.6

Table values represent colonies formed per 10^5 nucleated marrow cells cultured. The geometric means and 95% confidence limits are given; these were obtained using a $\log_e (x + 1)$ transformation to allow for zero values.

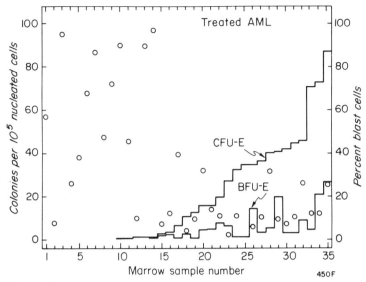

Figure 2
Results of assays for colony-forming erythropoietic progenitor cells (CFU-E and BFU-E) in 35 marrow samples from 22 patients under treatment for AML. The marrow samples are ranked in the order of increasing frequencies of CFU-E per 10^5 nucleated cells. Data for CFU-E, BFU-E, and marrow blasts (○; myeloblasts plus promyelocytes) are shown for the same marrow samples (S. Lan et al. 1978).

Table 2 gives data for the mean levels of the three classes of committed myelopoietic progenitors. Although the heterogeneity in the data was great, the mean levels for these three classes of committed progenitors was reduced in marrows from patients with AML prior to treatment. It may be concluded that, since the numbers of colony-forming progenitors were reduced in such patients at the time of diagnosis, the levels of pluripotent stem cells in marrow also were subnormal.

In summary, these results support the view that correlation analyses of the kind used to study cell-lineage relationships of cell classes found in murine spleen colonies (Gregory and Henkelman 1977) may also be useful for investigation of cell-lineage relationships in leukemic and non-leukemic human hematopoiesis. The results obtained provide no evidence for gross distortions of the lineage relationships between colony-forming progenitors in the marrow of patients with AML, although the levels of these progenitors were reduced at the time of diagnosis.

PROGNOSTIC FACTORS IN AML

Blast cells are usually considered the hallmark of leukemia and are used routinely to monitor the state of the disease. In contrast, differentiated elements are considered an index of normal hematopoietic function, even though there is good evidence that leukemic clones may retain some capacity for cytodifferentiation (Moore and Metcalf 1973; Aye et al. 1974; Blackstock and Garson 1974). It has been reported that patterns of proliferation in cell culture in assays designed to detect granulopoietic progenitors (CFU-C) may be useful in predicting response to chemotherapy (Moore et al. 1974; Spitzer et al. 1976; Knudtzon 1977; Vincent et al. 1977). A lack of evidence for a significant association between response and assays for marrow CFU-C has also been reported (Curtis et al. 1975). Because cells belonging to leukemic clones can form colonies under these conditions (Duttera et al. 1972; Moore and Metcalf 1973; Aye et al. 1974), it is not clear to what extent these assays yield indices of potential for cytodifferentiation and to what extent they provide indices of the disease burden. Other measures of disease burden (such as percentage of marrow blasts) and of potential for cytodifferentiation (such as assays for colony-forming progenitors of erythropoiesis) need to be included in the analysis. For this reason, we have assessed several such indices (attributes) for their usefulness as prognostic factors. The outcome variable used in the analysis was the proportion of patients achieving remission after treatment with one particular protocol of combination chemotherapy (J. E. Curtis et al., in prep.).

When several attributes show evidence of prognostic significance, methods of multivariate analysis are helpful for ranking these attributes in the order of their usefulness as predictors of outcome. One particular technique, stepwise regression analysis, has been used to investigate prognostic factors in acute leukemia (Gehan et al. 1976; Spitzer et al. 1976). This approach permits attributes to be selected serially; each succeeding attribute is chosen for a maximum association with outcome, after the stronger associations already identified with the preceding attributes are taken into account. A weakness of stepwise regression techniques is that they do not necessarily select the "best" set of prognostic factors, although they usually yield an acceptable one (Draper and Smith 1966). This weakness can become apparent when different attributes for a group of patients are highly correlated in the manner described above for the results of assays for colony-forming progenitor cells (Table 1). Under these conditions, important attributes may be overlooked. If a strong correlation exists between two attributes, then the more powerful prognostic factor is selected first. The other attribute, although it may also be associated significantly with outcome, may have a greatly diminished chance of being selected subsequently, since the strength of prediction outside of that related to the correlation between the attributes may be small. Thus, selection of the more powerful attributes can suppress important, but correlated, attributes.

An approach to the problems posed by correlated attributes is to utilize a criterion of goodness of fit, such as the C_p statistic (Daniel and Wood 1971) to compare regression equations involving different subsets of the set of available attributes. Instead of selecting important attributes sequentially, one at a time, as is done in stepwise procedures, subsets of several attributes are assessed simultaneously and compared with other subsets. When this approach is used, important *combinations* of attributes can be identified, even though, when assessed sequentially, these particular attributes might not have been selected first. When, using the C_p statistic, one has identified important subsets of attributes, one can then apply stepwise regression procedures to compare the relative importance of each of the members of the subset (J. E. Till et al., in prep.). The usefulness of the C_p statistic for multivariate analysis involving correlated attributes has been largely overlooked by medical statisticians, even though correlated attributes are frequently encountered in data from patients.

These approaches have been applied to data for 37 patients, using a variety of attributes assessed at the time of presentation with the diagnosis of AML. For 24 of these patients, pretreatment results of assays for CFU-E and CFU-C were available. After treatment with combination chemotherapy utilizing adriamycin and arabinosyl cytosine (J. E. Curtis et al., in prep.), the patients were assessed to learn whether or not successful induction of remission had been achieved. The objective of the statistical analysis was to identify any pretreatment factors which might be useful predictors of posttreatment remission status. After an initial review of a set of approximately 30 pretreatment attributes, an analysis involving the C_p statistic was carried out to identify a series of subsets of useful attributes. The first five subsets selected as useful prognostic factors were, in rank order: (1) normoblasts, (2) normoblasts + CFU-E, (3) normoblasts + age, (4) normoblasts + CFU-C, and (5) normoblasts + peripheral blood blasts. Of the attributes selected, the percentage of marrow normoblasts is of particular interest because of its recurrent presence in all five of these subsets.

The pretreatment attributes (percentage of early, intermediate, and late marrow normoblasts, frequency of CFU-E and CFU-C in marrow, frequency of blast cells in peripheral blood, and the age of the patient) identified in this manner were investigated further; stepwise logistic regression analysis (Gehan et al. 1976) was employed to test their association with the remission status of the patients after the patients had been treated with adriamycin-arabinosyl cytosine. The results are summarized in Table 3; only one attribute, normoblasts, showed a statistically significant association with remission status in this subgroup of 24 patients.

The larger group of 37 patients was then analyzed in a similar manner. Stepwise regression analysis indicated that each of the pairs of pretreatment

Table 3

Stepwise Regression Analysis of Response of 24 AML Patients Treated with Adriamycin-Arabinosyl Cytosine

Attribute entering regression[a]	Cumulative percentage variation explained	Chi-square test: P
Normoblasts	54	8×10^{-5} [c]
Age[b]	59	0.16
CFU-E	61	0.33

[a] Five pretreatment attributes were used (see text).
[b] Inversely related to remission status.
[c] Statistically significant contribution to regression.

attributes, normoblasts-age or marrow blasts-platelets, showed a significant association with remission status. These results illustrate the fact that stepwise regression analysis may not be able to select a single "best" subset of prognostic factors. On the basis of assessment of the results for this larger group by means of the C_p statistic, the "best" subset appeared to be normoblasts-age.

These results indicate that, although a colony assay for erythropoietic progenitors (CFU-E) may have had some predictive value for this group of patients with AML who were treated by one particular protocol of combination chemotherapy, it was a less powerful predictor than the percentage of normoblasts. The greater predictive value of an index of late-stage erythropoietic differentiation (normoblasts) rather than earlier-stage differentiation (CFU-E) is of some interest, and may reflect a decisive need for late-stage differentiated elements during the critical period of myelopoietic depletion that results from cytoreductive combination chemotherapy.

It is difficult to compare the results described here with those reported for other groups of patients when somewhat different cell-culture techniques, different sets of patient attributes, and different treatment protocols have been used (Moore et al. 1974; Spitzer et al. 1976; Vincent et al. 1977; Knudtzon 1977). It seems likely that the predictive usefulness of culture data will be found to depend markedly not only on the culture technique utilized, but also on the characteristics of the patient population being studied.

In summary, these results indicate that, for the particular group of AML patients studied, a pair of pretreatment attributes (percentage of marrow normoblasts and age) is a significant prognostic factor, and that an attribute related to an earlier stage of differentiation than normoblasts (frequency of marrow CFU-E) has less predictive value. However, another pair of pretreatment attributes (percentage of marrow blasts and platelets in the peripheral blood) is also a significant prognostic factor. This latter finding, taken together with the fact that the percentage of marrow blast cells is used routinely to monitor the state of the disease, points to the need for more information about leukemic blast cells in patients with AML.

BIOLOGY OF HUMAN LEUKEMIC BLAST CELLS

As mentioned previously, myelopoietic differentiation may persist in leukemic clones in AML. For example, cytogenetic markers in cell populations from patients with AML have been used to demonstrate erythropoiesis of leukemic origin (Blackstock and Garson 1974), and evidence for a leukemic origin for colony-forming progenitor cells committed to granulopoiesis has been obtained (Moore and Metcalf 1973; Aye et al. 1974). However, myelopoietic differentiation is often obscured by the blast-cell population. The blast cells frequently become the predominant cell type, characteristic of the disease. The relationship of these cells to known stages in the hierarchy of myelopoietic cell differentiation is often difficult to establish solely on the basis of morphological criteria, and currently available methods of cell culture have not been used exhaustively to explore the functional properties of these cells. The results outlined above indicate that measurements of myelopoiesis, by themselves, may provide an incomplete assessment of the central cellular defects in myeloblastic leukemia, and that functional measurements related to the blast-cell population are badly needed.

Previous work has indicated that blast-cell populations, defined by conventional morphological criteria, are heterogeneous, and that cell-marker systems (such as membrane markers) may reveal this heterogeneity (Janossy et al. 1976). Studies of cell proliferative capacity have also revealed heterogeneity in blast-cell populations (Buick et al. 1977; McCulloch and Till 1977). It appears that a minority subpopulation is able to proliferate in culture under appropriate conditions, whereas the majority of blast cells either are inert or serve as sources for stimulators of the proliferation of the minority subpopulation.

A colony assay for the proliferative subpopulation has been described recently (Buick et al. 1977; McCulloch et al. 1977). The technique involves culture of cells from the peripheral blood of patients with leukemia in methylcellulose, in growth medium containing fetal calf serum and 5–10% conditioned medium harvested from peripheral leukocytes cultured in the presence of phytohemagglutinin. Compact colonies of more than 20 cells per colony are visible after 5 to 9 days. The cells in the colonies are predominantly peroxidase-negative, and are negative for markers of lymphopoietic populations (surface immunoglobulin or capacity to form E rosettes with sheep erythrocytes). Colonies with properties similar to those of T-lympho-

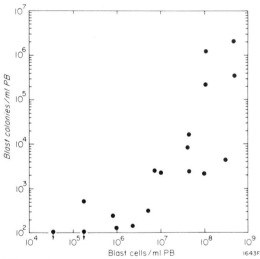

Figure 3
Results of assays for colony formation by peripheral leukocytes from 18 patients with AML prior to treatment. The cultures contained conditioned medium prepared in the presence of phytohemagglutinin (Buick et al. 1977). Results are expressed per milliliter of peripheral blood (PB) and are shown as a function of the number of blast cells per milliliter (Spearman's rank correlation coefficient $R_s = 0.90$, $P \ll 0.01$).

cyte colonies (Rozenszajn et al. 1975; Fibach et al. 1976) are often seen in cultures of leukemic blood cells in the presence of higher concentrations (40%) of conditioned medium from phytohemagglutinin-stimulated peripheral leukocytes.

The frequency of colony formation varies markedly from patient to patient, as shown in Figure 3. As is apparent from the figure, the frequency of colony formation is strongly correlated with the concentration of blast cells in the peripheral blood ($R_s = 0.90$, results for 18 patients). The efficiency of colony formation varies from 10^{-2} to 10^{-4}; only a small minority of cells in the blast-cell population had sufficient proliferative capacity in culture to be detected.

At this early stage, it is not known how much functional heterogeneity is present within this subpopulation of colony-forming cells, nor has their relationship, if any, to the classes of myelopoietic and lymphopoietic cells found in normal hematopoiesis been adequately defined. However, the positive correlation that has been observed between morphologically identified blast cells and the progenitor cells detected by colony formation, suggests that the significance of this minority subpopulation of colony-forming cells merits further investigation.

CONCLUSIONS

Cell-culture assays for colony-forming progenitors have provided a powerful approach to the investigation of cell-lineage relationships at early stages in the differentiation of hematopoietic cells. These new techniques are now being applied to studies of normal and abnormal human hematopoiesis. The examples described above have provided evidence that cell-lineage relationships in myelopoiesis are not grossly disrupted in AML, even though quantitative changes in the populations of colony-forming progenitors are apparent. Analyses of the predictive value of colony assays for these progenitors have indicated that they are less powerful than some other pretreatment attributes as predictors of successful induction of remission in a group of patients undergoing treatment for AML by combination chemotherapy. The results of these analyses emphasized the need for more detailed characterization of the main recognizable disease burden in acute leukemia, the population of blast cells. A colony assay for a proliferative subpopulation present in the peripheral blood of patients with acute leukemia has been developed. The number of colony-forming progenitors is closely correlated with the total number of morphologically recognizable blast cells; this supports the view that this colony assay may provide useful information about functional heterogeneity present within cell populations from the peripheral blood of patients with leukemia.

Successful identification of prognostic factors makes it possible to define subgroups of patients with similar characteristics. Predictions for individual patients are then possible on the basis of the experience of other patients with comparable characteristics (Shapiro 1977). Predictions of this sort are inherently probabilistic; however, reliable estimates of the probabilities could have a major impact on patient management, inasmuch as they could provide information helpful in the choice of a method of treatment best suited to the characteristics and needs of the individual patient.

Acknowledgments

This work was supported by grants from the Ontario Cancer Treatment and Research Foundation (236), the National Cancer Institute of Canada, and the Medical Research Council (MA-1420). S. L. held a fellowship from the Medical Research Council. R.N.B. is a special fellow of the Leukemia Society of America Inc. The stepwise regression analysis was based on a computer program kindly provided by Dr. E. A. Gehan.

REFERENCES

Axelrad, A. A., D. L. McLeod, M. M. Shreeve, and D. S. Heath. 1974. Properties of cells that produce erythrocytic colonies *in vitro*. In *Hemopoiesis in culture* (ed. W. A. Robinson), p. 226. U.S. Government Printing Office, Washington, D.C.

Aye, M. T., J. E. Till, and E. A. McCulloch. 1974. Cytological studies of colonies in culture derived from the peripheral blood cells of two patients with acute leukemia. *Exp. Hematol.* (Copenh.) **2:** 362.

Blackstock, A. M., and O. M. Garson. 1974. Direct evidence for involvement of erythroid cells in acute myeloblastic leukemia. *Lancet* **2:** 1178.

Bradley, T. R., and D. Metcalf. 1966. The growth of mouse bone marrow cells *in vitro*. *Aust. J. Exp. Biol. Med. Sci.* **44:** 287.

Buick, R. N., J. E. Till, and E. A. McCulloch. 1977. Colony assay for proliferative blast cells circulating in myeloblastic leukemia. *Lancet* **1:** 862.

Chervenick, P. A., P. Ernst, L. G. Lajtha, D. Metcalf, M. A. S. Moore, J. E. Till, and J. J. Trentin. 1975. Hematopoiesis [working paper]. In *Advances in the biosciences, Dahlem workshop on myelofibrosis-osteosclerosis syndrome Berlin 1974* (eds. R. Burkhardt, C. L. Conley, K. Lennert, S. S. Adler, T. Pincus, and J. E. Till), vol. 16, p. 255. Pergamon-Vieweg, Braunschweig.

Curry, J. L., and J. J. Trentin. 1967. Hemopoietic spleen colony studies. 1. Growth and differentiation. *Dev. Biol.* **15:** 395.

Curtis, J. E., D. H. Cowan, D. E. Bergsagel, R. Hasselback, and E. A. McCulloch. 1975. Acute leukemia in adults: Assessment of remission induction with combination chemotherapy by clinical and cell-culture criteria. *Can. Med. Assoc. J.* **113:** 289.

Daniel, C., and F. S. Wood. 1971. Selection of independent variables. In *Fitting equations to data*, p. 83. John Wiley & Sons, Interscience, New York.

Draper, N. R., and H. Smith. 1966. Selecting the "best" regression equation. *Applied regression analysis*, p. 163. John Wiley & Sons, New York.

Duttera, M. J., J. M. C. Bull, J. Whang-Peng, and P. P. Carbone. 1972. Cytogenetically abnormal cells *in vitro* in acute leukemia. *Lancet* **1:** 715.

Fibach, E., E. Gerassi, and L. Sachs. 1976. Induction of colony formation *in vitro* by human lymphocytes. *Nature* **259:** 127.

Gehan, E. A., T. L. Smith, E. J. Freireich, G. Bodey, V. Rodriguez, J. Speer, and K. McCredie. 1976. Prognostic factors in acute leukemia. *Semin. Oncol.* **3:** 271.

Gregory, C. J., and R. M. Henkelman. 1977. Relationships between early hemopoietic progenitor cells determined by correlation analysis of their numbers in individual spleen colonies. In *Experimental* hematology today (eds. S. J. Baum and G. D. Ledney), p. 93. Springer-Verlag, New York.

Iscove, N. N., J. S. Senn, J. E. Till, and E. A. McCulloch. 1971. Colony formation by normal and leukemic human marrow cells in culture: Effect of conditioned medium from human leukocytes. *Blood* **37**: 1.

Janossy, G., M. Roberts, and M. F. Greaves. 1976. Target cell in chronic myeloid leukaemia and its relationship to acute lymphoid leukaemia. *Lancet* **2**: 1058.

Knudtzon, S. 1977. *In vitro* culture of leukaemic cells from 81 patients with acute leukaemia. *Scand. J. Haematol.* **18**: 377.

Lan, S., E. A. McCulloch, and J. E. Till. 1978. Cytodifferentiation in the acute myeloblastic leukemias of man. *J. Natl. Cancer Inst.* (In press.)

McCulloch, E. A., and J. E. Till. 1977. Interacting cell populations in cultures of leukocytes from normal or leukemic peripheral blood. *Blood* **49**: 269.

McCulloch, E. A., R. N. Buick, and J. E. Till. 1977. Cellular differentiation in the myeloblastic leukemias of man. *Proceedings of the 30th annual symposium on fundamental cancer research, Houston, Texas.* (In press.)

Moore, M. A. S., and D. Metcalf. 1973. Cytogenetic analysis of human acute and chronic myeloid leukemic cells cloned in agar culture. *Int. J. Cancer* **11**: 143.

Moore, M. A. S., G. Spitzer, N. Williams, and T. Buckley. 1974. Agar culture studies in 127 cases of untreated acute leukemia: The prognostic value of reclassification of leukemia according to *in vitro* growth characteristics. *Blood* **44**: 1.

Pluznik, D. H., and L. Sachs. 1965. The cloning of normal "mast" cells in tissue culture. *J. Cell. Comp. Physiol.* **66**: 319.

Rozenszajn, L. A., D. Shoham, and I. Kalechman. 1975. Clonal proliferation of PHA-stimulated human lymphocytes in soft agar culture. *Immunology* **29**: 1041.

Shapiro, A. R. 1977. The evaluation of clinical predictions: A method and initial application. *N. Engl. J. Med.* **296**: 1509.

Siegel, S. 1956. *Nonparametric statistics for the behavioral sciences*, p. 202. McGraw-Hill, New York.

Spitzer, G., K. A. Dicke, E. A. Gehan, T. Smith, K. B. McCredie, B. Barlogie, and E. J. Freireich. 1976. A simplified *in vitro* classification for prognosis in acute adult leukemia: The application of *in vitro* results in remission-predictive models. *Blood* **48**: 795.

Stephenson, J. R., A. A. Axelrad, D. L. McLeod, and M. M. Shreeve. 1971. Induction of colonies of hemoglobin-synthesizing cells by erythropoietin *in vitro*. *Proc. Natl. Acad. Sci. U.S.A.* **68**: 1542.

Till, J. E., and E. A. McCulloch. 1961. A direct measurement of the radiation sensitivity of normal mouse bone marrow cells. *Radiat. Res.* **14**: 213.

Till, J. E., E. A. McCulloch, and L. Siminovitch. 1964. A stochastic model of stem cell proliferation, based on the growth of spleen colony-forming cells. *Proc. Natl. Acad. Sci. U.S.A.* **51**: 29.

Vincent, P. C., R. Sutherland, M. Bradley, D. Lind, and F. W. Gunz. 1977. Marrow culture studies in adult acute leukemia at presentation and during relapse. *Blood* **49**: 903.

The Dynamic Two-State Model of the Kinetic Behavior of Cell Populations

S. I. Rubinow*

Department of Applied Mathematics
The Weizmann Institute of Science, Rehovot, Israel

The recognition that the cell cycle could be conveniently subdivided into four phases (Howard and Pelc 1953): the DNA synthesis period S, the mitotic period M which culminates in cell division, the period G_1 between cell birth and S, and the period G_2 between S and M, was subsequently enlarged by the addition of a resting phase (Lajtha 1963; Quastler 1963), to describe the behavior of eukaryotic cells. At the present time, cells in G_0 phase are indistinguishable on a biochemical basis from cells in the G_1 phase, and the general acceptance of the G_0 phase concept has been brought about largely by two considerations. One is that the greatest source of variability in the duration of the cell cycle in many mammalian cell types is the G_1 phase, and the other is that many cells of a given population that are nonproliferating appear to be arrested in the G_1 phase. The qualitative features of cell population growth that support the G_0 state hypothesis have been summarized and reviewed in Epifanova and Terskikh (1969).

The G_0 state was first incorporated into a mathematical model of cellular proliferation by Lajtha et al. (1962), who utilized the model to represent the recovery of the erythrocytic population in mice following total body irradiation. Subsequently, it was adopted and elaborated by Lebowitz and Rubinow (1969) who used the age-time formalism (M'Kendrick 1926; Scherbaum and Rasch 1957; von Foerster 1959) to explain labeling experiments utilizing tritiated thymidine, and who deduced some simple steady-state properties of the proliferative system. In this model of cellular proliferation, there are two states or compartments in which cells are considered to exist: an active, or cycling, state A with a more or less fixed traversal time (consisting of fixed G_1, S, G_2, and M intervals), from which state, following mitosis, cells can enter a resting, or G_0, state. The latter is characterized as a random compartment, from which cells leave at random either to reenter the active state

*Permanent address: Graduate School of Medical Sciences, Cornell University, New York, New York 10021

or to leave the system. (See Fig. 1; the notation for the parameters has been altered to conform to our more recent usage.)

This same model was "rediscovered" by Burns and Tannock (1970), and supported by the representation of labeling experiments in hamster cheek-pouch epithelial cells and in rat dorsal epidermis. Further support of a similar nature was derived from the modeling of labeling experiments describing leukemic cell growth in acute myeloblastic leukemia (Rubinow et al. 1971) and in normal neutrophil growth (Rubinow and Lebowitz 1975). However, the most direct support for the G_0 state came from the remarkable observation of Smith and Martin (1973) that the generation distribution function, when viewed in the proper manner, i.e., in the form presented by their α curve, provided kinetic evidence for the existence and behavior of the G_0 state. The asymptote of this curve gave strong presumptive support to the idea that at birth all cells, even those in eukaryotic cell populations that grow rapidly, enter the G_0 state and that they leave this state at random to enter the active state that commits them to the mitotic process.

In the section entitled "Formulation," I enlarge the context of the model previously presented so that variability of transit times through the active state is taken into account. In "Steady Exponential Growth," I apply the theory to the case of steady exponential growth and then derive the equations for the fractional growth rate and for the relative numbers of cells in G_0 and A. The integral form of the model is presented in "Integral Form of the Model Equations," and its relationship to previously utilized forms of the

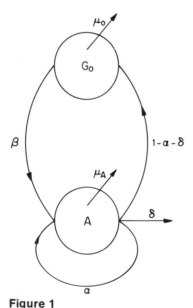

Figure 1
Model of the two-state proliferation system, from Lebowitz and Rubinow (1969) (with notation altered). Here A represents the active state, G_0 denotes the resting state, and the directed arrows with their associated labels indicate cell-flux directions and associated fractional loss rates. μ_0 and μ_A denote death rates occurring within the states.

model is shown. In "Theory of the α Curve," I apply the theory to Smith's and Martin's α curve and show how the complete curve can be parameterized. In doing so, I provide a quantitative kinetic description of cells traversing the entire proliferative state.

FORMULATION

The fundamental properties of the two-state model of cellular proliferation (Lebowitz & Rubinow 1969) are reviewed here and are considered in the wider context of an active state in which cellular transit times are variable. The cell populations in the active state and G_0 state are represented by age-density functions $n(a,t)$ and $p(a,t)$ respectively, where a represents cell age and t is the time. These functions each satisfy M'Kendrick's equation (M'Kendrick 1926)

$$\frac{\delta n}{\delta t} + \frac{\delta n}{\delta a} = -(\lambda + \mu_A)n, \tag{1}$$

$$\frac{\delta p}{\delta t} + \frac{\delta p}{\delta a} = -(\beta + \mu_0)p, \tag{2}$$

for $a,t > 0$. Here λ is the loss rate per unit time from A due to cell division, and it will be considered here to be a function of age. (It was omitted previously because of the simplifying assumption that the state A was of fixed duration T_A, so that $0 \leq a \leq T_A$.) β is the loss rate per unit time due to transfer of cells from G_0 to A, and μ_A and μ_0 are loss rates per unit time representing cell death or cell loss to the exterior, in the active and G_0 states, respectively. In what follows, the parameters β, μ_A, and μ_0 will be assumed to be constants, independent of age or time.

The total number of cells in each state at time t, designated by $N_A(t)$ and $N_0(t)$, is obtained by summation,

$$N_A(t) \equiv \int_0^\infty n(a,t)da, \tag{3}$$

$$N_0(t) \equiv \int_0^\infty p(a,t)da. \tag{4}$$

The age distributions are assumed to be given initially, e.g.,

$$n(a,0) = f(a), \; p(a,0) = q(a), \tag{5}$$

where $f(a)$ and $q(a)$ are given functions of age. The boundary conditions satisfied by n and p are as follows:

$$n(0,t) = 2\alpha \int_0^\infty n(a,t)\lambda(a)\,da + \beta N_0(t), \tag{6}$$

$$p(0,t) = 2(1 - \alpha - \delta) \int_0^\infty n(a,t)\lambda(a)\,da. \tag{7}$$

The function $n(0,t)$ is, in the absence of a G_0 state, commonly called the birthrate. Here, we see that it represents rather the rate of entry of cells into the active state, or the A-state birthrate. Similarly, $p(0,t)$ represents the G_0-state birthrate. The actual birthrate is twice the integral appearing on the right-hand side of equation (6) or (7). This integral represents the cell loss rate per unit time due to cells dividing in the A state; α is the fraction of these cells whose daughters reenter the active state, $0 \leq \alpha \leq 1$; the factor of 2 represents the number of daughters per dividing cell, and δ is the fraction of dividing cells whose daughters leave the proliferating state to go elsewhere. The remaining fraction of dividing cells whose daughters are assumed to enter the G_0 state is $1 - \alpha - \delta$. The quantity δ is introduced to allow for the possibility that in certain stem-cell systems, some cells may leave the active state to enter a differentiation pathway.

The implications of β and μ_0 being constant is that G_0 is a random compartment and that cells in G_0 are indistinguishable from one another on a biochemical or cytological basis. Hence, the variable a in equation (2) merely represents a residence time, whereas in equation (1) the age is also associated with physiological changes in the cell. In fact, in my original formulation of the model, I utilized a maturity-time description (Rubinow 1968) in place of equation (1). However, since I never actually exploited the maturity-time description, I have proceeded *ab initio* here utilizing the age-time formalism. For most practical applications of the theory that I can imagine, the function $q(a)$ will not be given, but rather the initial number of cells in G_0. This should occasion no mathematical difficulty, as these cells will usually have or can be assigned an initial residence age of zero.

The equations (1) and (2) together with the initial and boundary conditions (5-7) constitute the fundamental equations of the two-state proliferative system. Some additional clarification regarding the quantity λ is necessary. Under normal growth conditions (that is to say, excluding the presence of external chemical agents that can disturb the growth process), λ is assumed to be a function that depends on age alone. It is obtained operationally from the generation distribution function $g(a)$ (Rubinow 1968, 1978) as

$$\lambda(a) = \frac{g(a)}{\int_a^\infty g(\xi)\, d\xi}, \qquad (8)$$

where $g(a)$ is normalized to unity,

$$\int_0^\infty g(a)\, da = 1. \qquad (9)$$

A useful consequence of equation (8) is that

$$\int_a^\infty g(\xi)\, d\xi = \exp\left[-\int_0^a \lambda(\xi)\, d\xi\right]. \qquad (10)$$

We emphasize that $g(a)$ represents transit through the active state only, that is to say, it is the generation distribution function that could be observed if there is no G_0 state, or if $\beta = 0$, so that there is no contribution of cells from G_0 to the birthrate. The relationship (8) is derived on this basis. The effect of G_0 on observed generation distribution functions will be examined more closely in the last section.

The solution to equations (1) and (2) can be expressed in terms of the generation expansion (Rubinow 1968), as explained fully in Lebowitz and Rubinow (1969). Here we shall present an alternative solution to equations (1), (2), and (5), expressed in terms of the A-state and G_0-state birthrates at time $t - a$ in the following manner (Trucco 1965),

$$n(a,t) = \begin{cases} f(a-t) \exp\left[-\mu_A t - \int_{a-t}^{a} \lambda(\xi) \, d\xi\right], & t < a, \\ n(0, t-a) \exp\left[-\mu_A a - \int_0^{a} \lambda(\xi) \, d\xi\right], & t > a, \end{cases} \quad (11)$$

$$p(a,t) = \begin{cases} q(a-t) \exp\left[-(\beta + \mu_0) t\right], & t < a, \\ p(0, t-a) \exp\left[-(\beta + \mu_0) a\right], & t > a. \end{cases} \quad (12)$$

Furthermore, if the A-state birthrate is given by equation (6), then substitution of equation (11) into it shows that, with the aid of equations (4), (7), and (12), it satisfies the following integral equation,

$$n(0,t) = 2\alpha \left[\int_0^t n(0, t-a) e^{-\mu_A a} g(a) \, da + I(t)\right] + \beta N_0(t), \quad (13)$$

where

$$I(t) = e^{-\mu_A t} \int_t^{\infty} \frac{f(a-t) \, g(a)}{\int_{a-t}^{\infty} g(\xi) \, d\xi} \, da, \quad (14)$$

$$N_0(t) = e^{-(\beta + \mu_0)t} \left\{ N + 2(1 - \alpha - \delta) \int_0^t e^{(\beta + \mu_0)a} \left[\int_0^a n(0, a-\xi) e^{-\mu_A \xi} g(\xi) \, d\xi \right.\right.$$
$$\left.\left. + e^{-\mu_A a} \int_a^{\infty} \frac{f(\xi - a) g(\xi)}{\int_{\xi-a}^{\infty} g(\eta) \, d\eta} \, d\xi \right] da \right\}, \quad (15)$$

with $N_0(0) = N = \int_0^{\infty} q(a) \, da$. In equation (13), the first term represents the contribution to the birthrate of dividing cells from a time a previously, $a < t$, the term in $I(t)$ represents the contribution of dividing cells from the initial cohort of cells, described by $f(a)$, and the last term represents the contribution of cells recruited from G_0. In (15), the first term represents the contribution to $N_0(t)$ of the cells in the original cohort of cells in G_0 that have remained in G_0; the remaining terms represent cells that came from A and have spent a residence time ξ in G_0, $0 \leq \xi \leq t$. The latter are of two types, those that were in the original cohort of cells in A, and those that entered A at any time up to $t - \xi$ previously. When $\beta = 0$, equation (13) reduces to the integral equation for the birth rate given by Trucco (1965).

There are a number of important simplifications of the fundamental equation system, the principal one being that the time cells spend in the A state is of fixed duration T_A. This simplification means that the variability in cell transit times about the mean time T_A is neglected, and is readily achieved by setting

$$g(a) = \delta(a - T_A), \quad (16)$$

where $\delta(x)$ is the Dirac delta function.

Another simplification is achieved if we assume that $\beta = 0$, which effectively uncouples the G_0 state from the A state, as previously mentioned. This set of circumstances depicts pure active state growth. An alternative simplification is achieved if it is assumed that $\alpha = 0$, and, together with $\delta = 0$ and (16), is the form of the model advocated by Burns and Tannock (1970) and Smith and Martin (1973, 1974). We shall refer to it as pure G_0-state growth.

STEADY EXPONENTIAL GROWTH

One simple but important solution of the equation system that is easily investigated is the state of steady exponential growth, achieved at large times. Thus, consider the asymptotic solutions

$$n(a,t) \sim \bar{n} \exp[\gamma(t-a) - \mu_A a - \int_0^a \lambda(\xi)d\xi], \tag{17}$$

$$p(a,t) \sim \bar{p} \exp[\gamma(t-a) - (\beta + \mu_0)a], \tag{18}$$

where $\bar{n}$, $\bar{p}$, and γ are constants. Substitution of (17) and (18) into the boundary conditions yields the following condition for a nontrivial solution of these equations for the quantities $\bar{n}$ and $\bar{p}$,

$$\gamma + \beta + \mu_0 = 2[\beta(1-\delta) + \alpha(\gamma + \mu_0)]\int_0^\infty \exp[-(\gamma + \mu_A)a]g(a)da, \tag{19}$$

which is the eigenvalue equation that determines the fractional growth rate γ. Furthermore,

$$\frac{\bar{p}}{\bar{n}} = \frac{(\gamma + \beta + \mu_0)(1 - \alpha - \delta)}{\beta(1-\delta) + \alpha(\gamma + \mu_0)}. \tag{20}$$

Of greater observational significance is the ratio of cells in the G_0 and A states, in steady exponential growth. Thus, from (17) and (18) we calculate directly with equations (3), (4), and (10) that

$$N_A \sim \bar{N}_A e^{\gamma t}, \; N_0 \sim \bar{N}_0 e^{\gamma t}, \tag{21}$$

where

$$\bar{N}_A = \bar{n}\int_0^\infty \exp[-(\gamma + \mu_A)a]\int_a^\infty g(\xi)d\xi \, da, \tag{22}$$

$$\bar{N}_0 = \frac{\bar{p}}{\gamma + \beta + \mu_0}. \tag{23}$$

From (20), (22), and (23), we infer that the G_0 to A state cell ratio is

$$\frac{\bar{N}_0}{\bar{N}_A} = \frac{1 - \alpha - \delta}{[\beta(1-\delta) + \alpha(\gamma + \mu_0)]\int_0^\infty \exp[-(\gamma + \mu_A)a]\int_a^\infty g(\xi)d\xi \, da}. \tag{24}$$

INTEGRAL FORM OF THE MODEL EQUATIONS

For many purposes, it is more convenient to deal with the model equations after integration over the age variable. Thus, from equations (1) and (2), we obtain with the aid of equations (3–12) the result

$$\frac{dN_A}{dt} = (2\alpha - 1)[\int_0^t n(0,t-a)\,e^{-\mu_A a}\,g(a)da + I(t)] - \mu_A N_A + \beta N_0, \qquad (25)$$

$$\frac{dN_0}{dt} = 2(1 - \alpha - \delta)[\int_0^t n(0,t-a)e^{-\mu_A a}\,g(a)da + I(t)] - (\beta + \mu_0)N_0. \qquad (26)$$

To solve these equations, the A-state birthrate appearing on the right must first be found from its integral equation (13).

When the A state is of fixed duration so that equation (16) is applicable, equations (25) and (26) become

$$\frac{dN_A}{dt} = (2\alpha - 1)\begin{Bmatrix} f(T_A - t)e^{-\mu_A t} \\ n(0,t-T_A)e^{-\mu_A T_A} \end{Bmatrix} - \mu_A N_A + \beta N_0, \qquad \begin{matrix} t < T_A, \\ t > T_A, \end{matrix} \qquad (27)$$

$$\frac{dN_0}{dt} = 2(1 - \alpha - \delta)\begin{Bmatrix} f(T_A - t)e^{-\mu_A t} \\ n(0,t-T_A)e^{-\mu_A T_A} \end{Bmatrix} - (\beta + \mu_0)N_0, \qquad \begin{matrix} t < T_A, \\ t > T_A, \end{matrix} \qquad (28)$$

where, from equations (13–15),

$$n(0,t) = 2\alpha \begin{Bmatrix} f(T_A - t)e^{-\mu_A t} \\ n(0,t-T_A)e^{-\mu_A T_A} \end{Bmatrix} + \beta N_0(t), \qquad \begin{matrix} t < T_A, \\ t > T_A, \end{matrix} \qquad (29)$$

$$N_0(t) = e^{-(\beta+\mu_0)t}\bigg(N + 2(1 - \alpha - \delta)$$

$$\times \begin{Bmatrix} \int_0^t f(T_A - a)\exp[-\mu_A a + (\beta + \mu_0)a]da \\ \int_0^{T_A} f(T_A - a)\exp[-\mu_A a + (\beta + \mu_0)a]da + \int_{T_A}^t n(0,a - T_A)\exp[-\mu_A T_A \end{Bmatrix}$$

$$+ (\beta + \mu_0)a]da\bigg\}\bigg), \begin{matrix} t < T_A, \\ t > T_A. \end{matrix} \qquad (30)$$

Of course, equation (30) is the solution to equation (28).

When (16) is applicable, the eigenvalue equation (19) for γ also simplifies, and becomes (Lebowitz and Rubinow, 1969)

$$(\gamma + \beta + \mu_0) = 2[\beta(1 - \delta) + \alpha(\gamma + \mu_0)]e^{-(\gamma+\mu_A)T_A}. \qquad (31)$$

This equation was also given by Burns and Tannock (1970) (see also De Maertelaer and Galand 1977) for the case in which $\mu_A = \mu_0$ and $\alpha = 0$, when it reduces to a transcendental relationship between the two nondimensional parameters $(\gamma + \mu_0)T_A$ and βT_A. Equation (20) remains unchanged, but (24) becomes, with the aid of (31) above,

$$\frac{\bar{N}_0}{\bar{N}_A} = \frac{2(1 - \alpha - \delta)\,(\gamma + \mu_A)}{\beta + (2\alpha - 1)\,(\gamma + \mu_0)} \qquad (32)$$

For nonnegative exponential growth to occur, i.e., $\gamma \geq 0$ in equation (31), it is necessary that

$$\frac{\beta + \mu_0}{2[\beta(1 - \delta) + \alpha\mu_0]} \leq e^{-\mu_A T_A}, \qquad (33)$$

or,

$$\mu_0(1 - 2\alpha\,e^{-\mu_A T_A}) \leq \beta[2(1 - \delta)e^{-\mu_A T_A} - 1].$$

Hence, at the very least, $\mu_A T_A$ must be $\leq \log[2(1-\delta)]$ and δ must be $< 1/2$. When the equality in (33) holds, $\gamma = 0$. If $\beta = 0$, the solution to (31) is

$$\gamma = \frac{1}{T_A} \log(2\alpha) - \mu_A, \tag{34}$$

which shows that positive growth requires $\mu_A T_A < \log(2\alpha)$ and $\alpha > 1/2$.

A considerable simplification to the equation system (27–30) is achieved if $\alpha = 0$. This implies that all daughter cells that do not enter a differentiation pathway enter the G_0 state, and the simplification yields the pure G_0-state model. Then $n(0, t - T_A)$ can be eliminated by means of (29) from equations (27) and (28) which become

$$\frac{dN_A}{dt} = -\begin{cases} f(T_A - t)e^{-\mu_A t} \\ \beta N_0(t - T_A)e^{-\mu_A T_A} \end{cases} - \mu_A N_A + \beta N_0, \qquad \begin{matrix} t < T_A, \\ t > T_A, \end{matrix} \tag{35}$$

$$\frac{dN_0}{dt} = 2(1-\delta)\begin{cases} f(T_A - t)e^{-\mu_A t} \\ \beta N_0(t - T_A)e^{-\mu_A T_A} \end{cases} - (\beta + \mu_0)N_0 \qquad \begin{matrix} t < T_A, \\ t > T_A. \end{matrix} \tag{36}$$

These equations are essentially the form of the model utilized to represent erythrocytic recovery in mice following irradiation (Lajtha et al. 1964), to represent labeling experiments (Burns and Tannock 1970; Rubinow et al. 1971), and to describe neutrophil production in health and in acute myeloblastic leukemia (Rubinow and Lebowitz 1975, 1976a). External time-dependent causes of death have also been included in them to simulate the effects of chemotherapeutic treatment regimens (Rubinow and Lebowitz, 1976b).

THEORY OF THE α CURVE

Smith and Martin (1973) considered a cohort of initially synchronized cells that were newborn. They measured the fraction $\alpha^*(t)$ of cells remaining in interphase as a function of time, where cell loss is due exclusively to entry of cells into mitosis and division. (We have retained their use of the symbol α, but have added an asterisk, to distinguish it from the parameter α of the model.) The remarkable discovery they made is that for a variety of eukaryotic cells growing either slowly or rapidly, $\alpha^*(t)$ is, asymptotically for large times, a straight line when shown on semilog axes. This was interpreted to mean that newborn cells had initially entered the G_0 state, and that they had left the G_0 state in a random manner at a constant fractional rate to enter the active state which commits them to the division process. The transit time through the active state is, with small variability, of a fixed duration, and, hence, what is being observed at large times is the exponential decay in $\alpha^*(t)$ due to the transfer process from G_0 to A. Operationally, $\alpha^*(t)$ was obtained in the following way. Let $g^*(t)$ be the observed generation distribution function, obtained by following a cohort of newborn cells and observing the fractional number of cells $g^*(t)dt$ that divide in the time interval t to $t + dt$, assuming no cell deaths (see Fig. 2). Then

$$\alpha^*(t) = \int_t^\infty g^*(t')dt', \tag{37}$$

where, by definition, $\alpha^*(0) = 1$ (see Fig. 3).

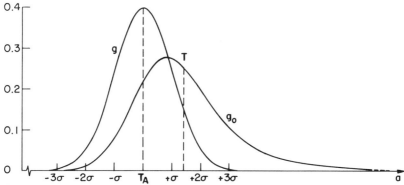

Figure 2
Theoretical generation functions $g(a)$ and $g_0(a)$, based on equations (49) and (57) with $erf(T_A/\sqrt{2}\sigma)$ and $erf(T_A + \beta\sigma^2]/\sqrt{2}\sigma)$ taken to be unity, and $\beta = 1/\sqrt{2}\sigma$. According to equation (55), the mean transit time is given essentially as $T = T_A + \sqrt{2}\sigma$. The ordinate unit is $1/\sigma$.

We have purposely assigned an asterisk to this generation distribution function to delineate it from the one described in equations (8–10). Our purpose is to distinguish between the contributions to the form of the generation distribution function of the variability in transit times in the A state and in the G_0 state. Thus, the relationship (8) is based on the assumption that cells exist in the active state only, and it is useful to retain the function $g(a)$ to represent the variability in the transit time of cells through the active state alone. If all cells at birth do indeed enter the G_0 state, and, thence, go to the active state before dividing, we shall denote the resulting generation distribution function as $g_0(a)$, which represents variability of traversal times through both G_0 and A. Thus we shall now investigate some of the consequences of pure G_0-state growth.

To find $g_0(a)$, we suppose a cell population in the active state obeys equation (1) with $\mu_A = 0$, and initial and boundary conditions

$$n(a,0) = 0,$$
$$n(0,t) = B(t), \tag{38}$$

where $B(t)$ is the prescribed A-state birth function. Then the solution to (1) is

$$n(a,t) = \begin{cases} 0, & t < a, \\ B(t-a)\exp[-\int_0^a \lambda(\xi)d\xi], & t > a. \end{cases} \tag{39}$$

The number of cells entering mitosis per unit time is

$$\int_0^\infty n(a,t)\lambda(a)da = \int_0^t B(t-a)\lambda(a)\exp[-\int_0^a \lambda(\xi)d\xi]da. \tag{40}$$

The generation function $g(t)$ is defined by the above expression with the choice $B(t) = \delta(t-a)$, which leads directly to equation (8). Suppose instead that N newborn cells are in the G_0 state at $t = 0$, from whence they leave at a fractional rate β to enter the active state. Assuming no cell death in the

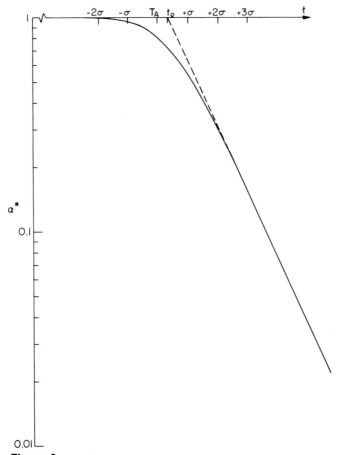

Figure 3
Theoretical curve $\alpha^*(t)$ with logarithmic ordinate scale, based on equation (58) and corresponding to the generation function $g_0(a)$ shown in Figure 2. The dashed line represents the asymptote, with slope $1/\beta = \sqrt{2}\sigma$ and intercept $t_0 = T_A + \sqrt{2}\sigma/4$, in this case.

G_0 state either ($\mu_0 = 0$), it follows directly that $N_0(t) = N\exp(-\beta t)$, and from equation (6)

$$B(t) = \beta N e^{-\beta t}. \tag{41}$$

Then, from equations (39) and (10),

$$n(a,t) = \begin{cases} 0, & t < a, \\ \beta N e^{-\beta(t-a)} \int_a^\infty g(\xi)d\xi, & t > a. \end{cases} \tag{42}$$

The associated generation distribution function $g_0(t)$ is $1/N$ times the number of cells entering mitosis, or, from equation (40),

$$g_0(t) = \beta e^{-\beta t} \int_0^t e^{\beta \xi} g(\xi) d\xi. \qquad (43)$$

This equation shows how $g_0(t)$ is obtained from $g(t)$. Conversely,

$$g(t) = \beta^{-1} e^{-\beta t} \frac{d}{dt} [e^{\beta t} g_0(t)]. \qquad (44)$$

To find $\alpha^*(t)$, we first find by integration of equation (42) with respect to age that

$$N_A(t) = \beta N \int_0^t e^{-\beta(t-a)} \int_a^\infty g(\xi) d\xi \, da, \qquad (45)$$

or, after integration by parts,

$$N_A(t) = N \left\{ \int_t^\infty g(a) da - e^{-\beta t} + e^{-\beta t} \int_0^t e^{\beta a} g(a) da \right\}. \qquad (46)$$

Hence,

$$\alpha^*(t) \equiv \frac{N_A(t) + N_0(t)}{N} = \int_t^\infty g(a) da + e^{-\beta t} \int_0^t e^{\beta a} g(a) da. \qquad (47)$$

By substituting equation (44) into the first integral on the right and by integrating by parts again, we find that

$$\alpha^*(t) = \int_t^\infty g_0(a) \, da, \qquad (48)$$

which, in fact, according to (37), is precisely the manner $\alpha^*(t)$ is determined from observation.

For definiteness, it is reasonable to suppose that $g(a)$ is represented by or is similar to a Gaussian function, as illustrated in Figure 2, e.g.,

$$g(a) = \frac{C}{\sqrt{2\pi}\sigma} \exp[-(a - T_A)^2/2\sigma^2], \qquad (49)$$

$$C = 2[1 + erf(T_A/\sqrt{2}\sigma)]^{-1},$$

$$erf \, x = \frac{2}{\sqrt{\pi}} \int_0^x e^{-\xi^2} d\xi.$$

where C has been chosen so that $g(a)$ is normalized to unity, and $erf \, x$ is the error function of argument x. Here T_A and σ represent, approximately, the mean transit time in A and its associated standard deviation, respectively. (This interpretation is only approximate because the range of a is restricted to positive values.) If we substitute equation (49) into (47) and make use of the asymptotic property of the error function (Abramowitz and Stegun 1964: 298) that $erf \, x \sim 1 - \exp(-x^2)/\sqrt{\pi} \, x, \, x \gg 1$, then the first term on the right in (47) vanishes much more rapidly than the second, as $t \to \infty$; clearly, this will hold under less restrictive conditions on the form of $g(a)$ than (49). Hence, we infer that

$$\alpha^*(t) \sim e^{-\beta t} \int_0^\infty e^{\beta a} g(a) da, \ t \gg 1, \tag{50}$$

which is the mathematical expression of Smith's and Martin's discovery. Moreover, it shows that the intercept with the t-axis of the asymptote, as well as its shape, is of quantitative significance (see Fig. 3). If, in particular, $g(a)$ is given explicitly by equation (49), then substitution into equation (50) yields

$$\alpha^*(t) \sim \exp[-\beta t + \beta T_A + \beta^2\sigma^2/2] \ [1 + erf\{(T_A + \beta\sigma^2)/\sqrt{2}\sigma\}]$$
$$\times [1 + erf(T_A/\sqrt{2}\sigma)]^{-1}. \tag{51}$$

Hence, so long as $T_A \gg \beta\sigma^2$,

$$\alpha^*(t) \sim \exp[-\beta(t - T_A - \beta\sigma^2/2)]. \tag{52}$$

This shows that the asymptote at large times of a semilog plot of $\alpha(t)$ versus t intercepts the t-axis at

$$t_0 = T_A + \beta\sigma^2/2 \tag{53}$$

If $g_0(a)$ and $\alpha^*(t)$ are known, and $g(a)$ is given by equation (49), then T_A and σ can be obtained from the knowledge of β (the negative slope of the asymptote), equation (53), and by another expression relating T_A, σ, and β to known quantities. For example, we can determine the mean intermitotic time T through the two-state system, defined as

$$T \equiv \int_0^\infty a g_0(a) da = \int_0^\infty a g(a) da + \frac{1}{\beta}, \tag{54}$$

where the last term on the right represents the mean transit time through the G_0 state, and the preceding term is the mean transit time through the active state. Using (49), we obtain the expression

$$T = \frac{1}{\beta} + T_A + \sqrt{\frac{2}{\pi}} \sigma \frac{e^{-T_A^2/2\sigma^2}}{1 + erf(T_A/\sqrt{2}\sigma)}. \tag{55}$$

If the coefficient of variation is sufficiently small, i.e., $T_A \gg \sigma$, then the last term on the right in equation (55) can be neglected.

For the explicit form of $g(a)$ given by equation (49), we find readily from equations (42), (43), (48), and (19) that

$$n(a,t) = \frac{C}{2} \beta N e^{-\beta(t-a)} erfc\left(\frac{a - T_A}{\sqrt{2}\sigma}\right), \ t > a \tag{56}$$

$$g_0(a) = [1 + erf(T_A/\sqrt{2}\sigma)]^{-1} \beta \exp[-\beta(a - T_A - \beta\sigma^2/2)]$$
$$\left\{erf\left(\frac{a - T_A - \beta\sigma^2}{\sqrt{2}\sigma}\right) + erf\left(\frac{T_A + \beta\sigma^2}{\sqrt{2}\sigma}\right)\right\}, \tag{57}$$

$$\alpha^*(t) = \frac{C}{2}\left\{erfc\left(\frac{t - T_A}{\sqrt{2}\sigma}\right) + \exp[-\beta(t - T_A - \beta\sigma^2/2)]\right.$$
$$\left.\left[erf\left(\frac{t - T_A - \beta\sigma^2}{\sqrt{2}\sigma}\right) + erf\left(\frac{T_A + \beta\sigma^2}{\sqrt{2}\sigma}\right)\right]\right\}, \tag{58}$$

$$\gamma + \beta = 2\beta \exp(-\gamma T_A + \sigma^2\gamma^2/2) \left[1 + erf \frac{T_A}{\sqrt{2}\sigma}\right]^{-1}$$
$$\times \left[1 + erf\left(\frac{T_A - \sigma^2\gamma}{\sqrt{2}\sigma}\right)\right], \quad (59)$$

where $erfc(x)$ is the error function complement of argument x, equal to $1 - erf(x)$. Equations (49), (57), and (58) are illustrated in Figures 2 and 3, assuming $T_A \gg \sigma$.

Finally, we examine the theory of the generation distribution function and the alpha curve in the general case for which $\alpha \neq 0$. Thus, with $\delta = \mu_A = \mu_0 = 0$ for the sake of simplicity, if we are given N newborn cells at time $t = 0$, then

$$n(a,0) = \alpha N \delta(a), \; n(0,t) = N_0(t), \; p(a,0) = (1-\alpha)N\delta(a), \quad (60)$$

and it follows from equations (11) and (12) that

$$n(a,t) = N \begin{cases} \alpha\delta(a-t) \dfrac{\int_a^\infty g(\xi)d\xi}{\int_{a-t}^\infty g(\xi)d\xi}, & t < a, \\ (1-\alpha)e^{-\beta(t-a)} \int_a^\infty g(\xi)d\xi, & t > a, \end{cases} \quad (61)$$

$$N_0(t) = N(1-\alpha)e^{-\beta t} \quad (62)$$

By repeating the analysis by which equations (43) et seq. were obtained, we find in an analogous fashion that the generation distribution function, which we denote by $G(t)$, is a linear combination of the generation functions for pure A-state and pure G_0-state growth, e.g.,

$$G(t) = \alpha g(t) + (1-\alpha)g_0(t) \quad (63)$$

In addition,

$$N_A(t) = N \left\{ \int_t^\infty g(a)da + (1-\alpha)e^{-\beta t}\left[-1 + \int_0^t e^{\beta a} g(a)da\right]\right\}, \quad (64)$$

$$\alpha^*(t) = \int_t^\infty g(a)da + (1-\alpha)e^{-\beta t} \int_0^t e^{\beta a} g(a)da. \quad (65)$$

Consequently, if $g(a)$ is of the form of equation (49), it follows as for equation (50) that

$$\alpha^*(t) \sim (1-\alpha)e^{-\beta t} \int_0^t e^{\beta a} g(a)da. \quad (66)$$

This result demonstrates that the asymptotically exponential decay with time of $\alpha^*(t)$ does not require that *all* cells enter the G_0 state at birth. A precise analysis of the generation distribution function and related kinetic data—for example, in the manner described above—are needed to determine whether $\alpha = 0$ for a cell population, before it can be inferred that the population obeys the pure G_0-state growth model.

Similarly, if $\alpha \neq 0$ and equation (49) is applicable, then the equations following (49) must be amended in the following manner. The quantity $(1 - \alpha)$ multiplies the right-hand side of equations (50), (51), and (52). To the right-hand side of equation (53), there must be added the term $\beta^{-1} \log (1 - \alpha)$. The quantity $(1 - \alpha)$ multiples β^{-1} in equations (54) and (55) and the exponential term in (58). Instead of equation (56) we have

$$n(a,t) = N \begin{cases} \alpha\delta(a - t) \dfrac{erfc\left(\dfrac{a - T_A}{\sqrt{2}\sigma}\right)}{erfc\left(\dfrac{a - t - T_A}{\sqrt{2}\sigma}\right)}, & t < a, \\ (1 - \alpha) \dfrac{C}{2} \beta e^{-\beta(t-a)} erfc\left(\dfrac{a - T_A}{\sqrt{2}\sigma}\right), & t > a. \end{cases} \quad (67)$$

The factor β on the right-hand side of equation (59) must be replaced by $\beta + \alpha\gamma$.

We conclude that the curve $\alpha^*(t)$ contains information pertaining to the variability of the transit time of cells through the active state, as well as through the G_0 state, and we have indicated how this information can be quantified. This qualitative feature of $\alpha^*(t)$ was, of course, already recognized by Smith and Martin (1973). Hence, the observed generation distribution function of eukaryotic cell populations, when properly interpreted, e.g., as representing either $g_0(t)$, or most generally $G(t)$, contain more quantitative information about the cell cycle than has heretofore been realized.

Acknowledgments

This work was supported in part by National Cancer Institute Grant NIH-1 RO1 CA 20610-01.

REFERENCES

Abramowitz, M., and I. A. Stegun. 1964. *Handbook of mathematical functions.* Applied Mathematics Series 55. National Bureau of Standards, Washington, D.C.

Burns, F. J., and I. F. Tannock. 1970. On the existence of a G_0-phase in the cell cycle. *Cell Tissue Kinet.* **3:** 321.

De Maertelaer, V., and P. Galand. 1975. Some properties of a 'G_0' model of the cell cycle. I. Investigation on the possible existence of natural constraints on the theoretical model in steady-state conditions. *Cell Tissue Kinet.* **8:** 11.

———. 1977. Some properties of a 'G_0' model of the cell cycle. II. Natural constraints on the theoretical model in exponential growth conditions. *Cell Tissue Kinet.* **10:** 35.

Epifanova, O. I., and V. V. Terskikh. 1969. On the resting periods in the cell life cycle. *Cell Tissue Kinet.* **2:** 75.

Howard, A., and S. R. Pelc. 1953. Synthesis of desoxyribonucleic acid in normal and irradiated cells and its relation to chromosome breakage. *Heredity,* suppl. **6:** 261.

Lajtha, L. G. 1963. On the concept of the cell cycle. *J. Cell. Comp. Physiol.* **62,** suppl. 1: 143.

Lajtha, L. G., R. Oliver, and C. W. Gurney. 1962. Kinetic model of a bone-marrow stem-cell population. *Brit. J. Haematol.* **8:** 442.

Lajtha, L. G., C. W. Gilbert, D. D. Porteous, and R. Alexanian. 1964. Kinetics of a bone-marrow stem-cell population. *Ann. N.Y. Acad. Sci.* **113:** 742.

Lebowitz, J., and S. I. Rubinow. 1969. Grain count distributions in labeled cell populations. *J. Theor. Biol.* **23:** 99.

M'Kendrick, A. G. 1926. Applications of mathematics to medical problems. *Proc. Edinburgh Math. Soc.* **44:** 98.

Quastler, H. 1963. The analysis of cell population kinetics. In *Cell proliferation* (eds. L. F. Lamerton and R. J. M. Fry), p. 18. Blackwell Scientific Publications, Oxford.

Rubinow, S. I. 1968. A maturity-time representation for cell populations. *Biophys. J.* **8:** 1055.

———. 1978. Age-structured equations in the theory of cell population. In *A study in mathematical biology*. (ed. S. Levin). Mathematical Assoc. of America, Washington, D.C.

Rubinow, S. I. and J. L. Lebowitz. 1975. A mathematical model of neutrophil production and control in normal man. *J. Math. Biol.* **1:** 187.

———. 1976a. A mathematical model of the acute myeloblastic leukemic state in man. *Biophys. J.* **16:** 897.

———. 1976b. A mathematical model of the chemotherapeutic treatment of acute myeloblastic leukemia. *Biophys. J.* **16:** 1257.

Rubinow, S. I., J. L. Lebowitz, and A.-M. Sapse. 1971. Parameterization of in vivo leukemia cell populations. *Biophys. J.* **11:** 175.

Scherbaum, O., and G. Rasch. 1957. Cell size distribution and single cell growth in *Tetrahymena pyriformis* GL. *Acta Pathol. Microbiol. Scand.* **41:** 161.

Smith, J. A., and L. Martin. 1973. Do cells cycle? *Proc. Nat. Acad. Sci. U.S.A.* **70:** 1263.

———. 1974. Regulation of cell proliferation. In *Cell cycle controls*, (eds. G. M. Padilla et al.) p. 43. Academic Press, New York.

Trucco, E. 1965. Mathematical models for cellular systems: The von Foerster equation. *Bull. Math. Biophys.* **27:** 285, 449.

von Foerster, H. 1959. Some remarks on changing populations. In *The kinetics of cellular proliferation* (ed. F. Stohlman, Jr.), p. 382. Grune & Stratton, New York.

Stem-Cell Heterogeneity: Pluripotent and Committed Stem Cells of the Myeloid and Lymphoid Systems

R. A. Phillips

Ontario Cancer Institute, and the Department of Medical Biophysics
University of Toronto, Toronto, Ontario M4X 1K9, Canada

Although several studies have shown a close relationship between the differentiation of lymphocytes and myeloid cells (erythrocytes, granulocytes, macrophages, and megakaryocytes), research in these two areas has concentrated on very different aspects of differentiation. Largely as a result of the availability of the spleen colony assay for pluripotent stem cells of the myeloid system (Till and McCulloch 1961), studies in this area have concentrated on early events and have led to the development of several colony assays for cells closely related to stem cells (e.g., Gregory and Eaves, this volume; Johnson and Metcalf, this volume). In contrast, studies on the differentiation of lymphoid cells have emphasized late stages of differentiation, especially those cellular events involved in the expression of humoral and cell-mediated immune responses by B and T lymphocytes, respectively. Relatively little is known about the early events in the differentiation of lymphocytes, primarily because of a lack of suitable quantitative assays analogous to the colony assays for immature myeloid cells.

Despite the lack of suitable, direct assays, there are many indications that a close relationship exists between the stem cells that maintain the myeloid and lymphoid systems. The evidence for a common stem cell comes from many sources, and, although none of the evidence is direct, numerous indirect approaches have led to the conclusion that such pluripotent stem cells must exist. For example, several embryological studies indicate a common stem cell. In the mouse embryo, lymphocytes do not appear in the fetal liver until after the migration of spleen colony-forming cells (CFU-S) into that tissue (Moore and Metcalf 1970). In tetraparental mice, there is always a strong correlation between the frequency of myeloid cells derived from a particular parent and the frequency of lymphocytes derived from the same parent (Mintz and Palm 1969; Gornish et al. 1972). No mice had their lymphoid system derived from one parent and their myeloid system from the other.

The most direct evidence for a close relationship between the two systems comes from analysis of the stem cells in the bone marrow of adult mice. Early experiments by Micklem et al. (1966) showed that the bone marrow of adult mice contains all of the stem cells necessary to repopulate irradiated mice with functional myeloid and lymphoid cells. In later experiments, Trentin et al. (1967) and Yung et al. (1973) showed that a limited number of spleen colonies could repopulate irradiated recipients with myeloid and lymphoid cells. Although these investigators did not demonstrate directly that the lymphoid cells in the repopulated recipients were progeny from CFU-S, these data provide strong evidence that at least some CFU-S have the potential for lymphoid differentiation.

The most direct evidence for the production of both myeloid and lymphoid progeny from a single bone-marrow stem cell comes from studies utilizing radiation-induced chromosome markers. Barnes et al. (1968) found that a proportion of the stem cells that survive a high dose of ionizing radiation have chromosome translocations which do not appear to influence the function of the stem cells. Since these translocations are induced randomly by radiation, each one is unique and, therefore, provides a specific cytogenetic marker that identifies the progeny from a single stem cell regardless of their location in the mouse. Several investigators have employed this technique to investigate stem cells. Becker et al. (1963) first used this technique to prove that the cells in a spleen colony belong to the same clone. Later, Wu et al. (1968) showed that the same radiation-induced markers could be found in high frequency in both spleen colonies and in the thymus, thereby providing direct evidence for a common stem cell for myeloid cells and at least some T lymphocytes. A similar conclusion was made by Nowell et al. (1970) who observed the same chromosome translocations in spleen colonies and in mixed lymphocyte cultures where most of the dividing cells presumably are T lymphocytes. Finally, Edwards et al. (1970) demonstrated radiation-induced translocations in spleen colonies and in antigen-binding cells in the spleen following immunization, indicating that B lymphocytes and CFU-S belong to the same clone.

However, all three of these studies with radiation-induced chromosome markers suffered from a lack of characterization of the lymphoid cell population. Because the majority of cells in the thymus do not have detectable immunological function and because there is some evidence for two independent pathways of differentiation in the thymus (Shortman et al. 1975), Wu's experiments cannot be taken as conclusive evidence for the derivation of all T lymphocytes from myeloid stem cells. In the experiments of Nowell et al., only a low proportion of dividing cells in the mixed lymphocyte culture had the marker identifying them as progeny of spleen colony-forming cells. Because myeloid cells can be stimulated to proliferate in the presence of factors released during a mixed lymphocyte culture (Parker and Metcalf 1974), the low frequency of marked cells may have indicated myeloid cells rather than lymphoid. Similarly, because only 10% of the antigen-binding cells in the experiments described by Edwards et al. (1970) secreted significant antibody, many of these cells may have been myeloid in origin and may have been antigen-binding because of the cytophilic antibody on their surface.

The lack of conclusive, direct evidence linking the differentiation of these two systems stimulated us to begin a series of experiments to examine the precise relationships between the differentiation of lymphocytes and myeloid cells. In addition, we designed our experiments to detect committed as well as pluripotent stem cells. In the context of differentiation within the myeloid and lymphoid systems, committed stem cells would have the following properties: they would have extensive self-renewal capacity, but they would have the ability to differentiate along only one or two pathways. For example, there may exist stem cells which give rise only to myeloid cells and others which differentiate only into lymphocytes. On the basis of clinical studies, one might predict the existence of such restricted stem cells. Children with severe, combined immune deficiency disease appear to have defective lymphoid stem cells. Injection of HLA-compatible bone-marrow cells often reconstitutes lymphoid function, and, in such cases, the lymphocytes are derived from the bone-marrow donor while the myeloid system continues to be of host origin. In experimental systems, there is little evidence for restricted stem cells. Although both the in vitro colony-forming assays for the precursors of granulocytes and erythrocytes exhibit heterogeneity within a colony (Johnson and Metcalf, this volume), these colony-forming cells may not have sufficient capacity for self-renewal to be called stem cells. More recently, Kadish and Basch (1976) demonstrated the existence in bone marrow of a prethymic stem cell. However, they did not measure this cell's capacity for self-renewal in adult bone marrow. The experiments described below provide direct evidence for a pluripotent stem cell and for two types of restricted stem cells, one for myeloid cells and one for T lymphocytes.

EXPERIMENTAL DESIGN

The three-stage experimental design used in these experiments is outlined in Figure 1; the details of this experiment have been described by Abramson et al. (1977). In the first stage of the experiment, lightly irradiated (250–400 rads) W/W^v recipients were injected with bone-marrow cells from co-isogenic normal donors. The normal donors were exposed to 700 rads of ionizing radiation immediately before being used as donors. This dose of radiation destroyed most of the stem cells, but, among the surviving CFU-S, 30% had a detectable translocation. Bone-marrow cells were injected in limiting numbers in an attempt to obtain repopulation of the W/W^v recipient from a limited number of stem cells. The genetic defect in these recipients gave a selective advantage to the grafted normal stem cells, and the irradiation caused a marked lymphoid depression and enhanced differentiation along this pathway. Between 8 and 11 months were allowed for the grafted stem cells to repopulate the recipients. There were two reasons for choosing such long times. First, we wished to ensure the production of sufficient differentiated progeny to make them easily detectable. Second, we wanted to select stem cells with significant capacity for self-renewal.

In the second part of the experiment, three functional assays were used to determine the distribution of chromosomally marked cells in the myeloid and the lymphoid systems. For analysis of the myeloid system, bone-marrow

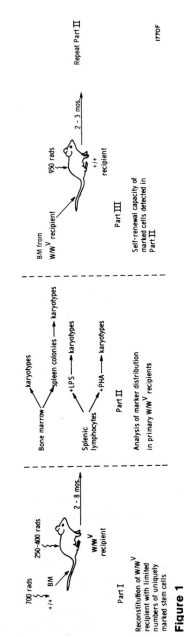

Figure 1
Three-stage experimental design used in experiments.

cells were injected into lethally irradiated recipients. Ten to 12 days later, individual colonies were dissected from the spleen and analyzed for the presence of abnormal chromosomes. The presence of a spleen colony with chromosomally marked cells was evidence that the marker was present in myeloid cells (Wu et al. 1967). To search for markers in the lymphoid system, we isolated splenic lymphocytes and placed them in culture either with lipopolysaccharide (LPS) or phytohemagglutinin (PHA) to stimulate proliferative responses in B and T lymphocytes, respectively. Observation of the same chromosomal abnormality in all three functional assays would have indicated that a translocation had been induced in a pluripotent stem cell capable of differentiating along both the myeloid and the lymphoid pathways. If, however, a marker had been detected in only one assay, we would have concluded that the initial radiation had induced a translocation in a stem cell that was already committed to differentiate along that pathway.

The third part of the experiment involved transplanting bone marrow from a reconstituted W/W^v recipient into a lethally irradiated recipient. This part of the experiment was designed to study the capacity of marked stem cells for self-renewal and to show that the restricted phenotype breeds true; that is, if marked cells were observed only in B lymphocytes in the first recipient, then bone marrow from that recipient should repopulate second recipients in which only in B lymphocytes again were marked.

RESULTS OF KARYOTYPIC ANALYSES

A total of 31 W/W^v recipients were analyzed according to the protocol described above. A high frequency of marked cells was detected in 11 mice; one mouse had two unique markers, thus giving 12 scorable markers. Ten of these markers could be placed into one of three different categories. These data are summarized in Table 1. Two markers were not included in the table. In one case, the chromosome marker was detected in bone marrow but not in spleen colonies or in the mitogen-stimulated cultures. In the other case, the marker was present in high frequency in the thymus, but was absent in spleen colonies, LPS-stimulated cultures, and PHA-stimulated cultures. This marker may have arisen in (or perhaps was the cause of) a thymic leukemia, but this possibility was not examined directly.

In four cases, the same chromosome abnormality was detected in all three assays. In these mice, the translocation must have occurred in a pluripotent stem cell. In four cases, an abnormal chromosome was detected only in the bone marrow and in bone-marrow-derived spleen colonies. No abnormal karyotypes were found in PHA or LPS blasts. In a second recipient given bone marrow from the original W/W^v recipient, the same restricted distribution of markers was observed. These data are consistent with the chromosome abnormality having been induced in a stem cell restricted for differentiation in the myeloid system only. In two mice, a high frequency of markers was observed only in PHA blasts. Insignificant numbers of marked cells were observed in LPS blasts. (Preliminary experiments indicated that up to 5% of the dividing cells in the mitogen-stimulated cultures could be contaminants. Therefore, in our analysis of the data, we ignored marker

Table 1
Distribution of Abnormal Chromosomes in Individual Mice

Cellular distribution	Mouse no.	Type of marker	Percent marked mitoses			Spleen colonies (no. marked/total)
			marrow	PHA blasts	LPS blasts	
Pluripotent	3	39 chromosomes[a]	80	90	59	7/7
	5	minute	36	61	63	12/14
	32	minute	40	13	13	7/12
	31 – primary	minute	28	9	27	10/13
	31 – secondary[b]	minute	52	39	48	10/10
Myeloid only	5	long + minute	45	0	0	2/14
	23	metacentric	53	0	0	9/12
	34	minute	73	0	0	not done
	11 – primary	2 metacentric + minute	100	0	0	4/4
	11 – secondary[b]	2 metacentric + minute	94	0	0	8/9
	11 – secondary[b]	2 metacentric + minute	92	0	0	13/13
T cell only	4	39 chromosomes	5	14	0	0/9
	16 – primary	2 minutes	0	42	5	0/10
	16 – secondary[b]	2 minutes	0	9	1	0/4

PHA, phytohemagglutinin; LPS, lipopolysaccharide.
[a] The normal karyotype has 40 telocentric chromosomes.
[b] Secondary recipients of bone marrow from primary W/W^v recipient. See Figure 1.

frequencies less than 5%.) This marker, like the myeloid marker, showed the restricted distribution in a secondary recipient of bone marrow from the original recipient.

In summary, these data provide direct evidence for the existence in adult bone marrow of a pluripotent stem cell (S_p) and two restricted stem cells, one committed to myeloid differentiation (S_m) and the other to T-cell differentiation (S_t). We assume that S_m and S_t are the progeny of S_p. It is important to emphasize, however, that we do not know if any of these stem cells form spleen colonies. Our experimental design did not allow us to study the colony-forming potential of the bone-marrow stem cells. Presumably the S_m stem cells form spleen colonies. It is difficult to predict the colony-forming potential of the S_p stem cells. If S_m are the progeny of S_p and if this transition occurs in the spleen, then one would expect S_p to form spleen colonies. In this case, one would also predict heterogeneity among spleen colonies, some producing only myeloid progeny and others producing progeny along both the myeloid and lymphoid pathways. At present, there is no evidence either for or against this sort of heterogeneity among spleen colonies.

Some of the data of Wu et al. (1968) also support the existence of restricted stem cells. Although these investigators stressed the presence in eight cases of the same chromosome marker in both bone marrow and thymus, they also found 16 other chromosome translocations that were restricted to the bone marrow. Some of these bone-marrow-restricted markers probably resulted from aberrations induced in S_m. However, it is surprising, in view of our data, that Wu and his colleagues did not find any markers restricted to the T-cell lineage. In every case, where they observed a chromosome abnormality in the thymus, they also observed the same abnormality in some spleen colonies. One difference between their experimental design and ours may explain the apparent lack of S_t stem cells in their marked populations. They usually tested their primary recipients from 2 to 3 months after grafting, whereas we analyzed the recipients from 8 to 11 months. If the S_p had a shorter average life span than did the restricted stem cells, analysis at early times could favor detection of S_p and at late times, the detection of restricted stem cells.

More recently, Fialkow and Denman (this volume) obtained evidence in humans for a close relationship between B lymphocytes and myeloid cells. T lymphocytes, according to his analysis, are less closely related. His findings suggest the presence of stem cells producing myeloid cells and B lymphocytes and of another related but distinct stem cell producing T lymphocytes.

DIFFICULTIES IN ANALYZING RESTRICTED STEM CELLS

Although radiation-induced chromosome abnormalities are good cytogenetic markers for following the differentiation of single stem cells, their use is associated with several pitfalls that can lead to misinterpretation of the data. These pitfalls deal primarily with the interpretation of restricted distributions of chromosome abnormalities. A restricted distribution creates two problems for the investigator. First, it is possible that experimental manipulations cause the restriction of differentiative potential and that the normal unperturbed cells do not express such restriction. Second, it is always difficult to prove that failure to observe markers in a particular pathway has resulted from an intrinsic inability of stem cells to produce such progeny. A stochastic variation of the differentiation process could result in only small numbers of progeny being produced in that pathway at the time of observation. In the following paragraphs, some of these possible problems and their implications are discussed more fully.

The most difficult problem to overcome is the possibility that the radiation used to induce the chromosome aberrations also created abnormal stem cells able to make progeny along only one pathway. At present this possibility is difficult to eliminate, inasmuch as the use of chromosome abnormalities of the type described here constitutes the only technique suitable for the analysis of early events both in the myeloid and lymphoid systems. The observation that leukemic cells in patients with acute myelogenous leukemia have specific chromosome abnormalities (Rowley, this volume) is also consistent with the idea that translocations can lead to clones with restricted potential for differentiation. As mentioned above, a model based on the results with radiation-induced markers predicted that spleen colonies would be heterogeneous. When we can study individual spleen colonies for lymphoid precursors, we will then be able to test directly the possibility that some colony-forming cells are restricted to myeloid differentiation whereas others are pluripotent. Such a study would confirm the existence of restricted stem cells in a normal population of unirradiated stem cells.

W/W^v mice are known to have a defective myeloid system. Inasmuch as irradiated W/W^v mice were used as primary recipients in these experiments, it was possible that some restricted stem cells resulted from translocations induced in stem cells from the W/W^v recipient. This possibility seems unlikely. Most of the restricted markers were in the myeloid system, which was defective in the W/W^v recipient. In addition, we found that doses of radiation below 400 rads produced a very low frequency of chromosome abnormalities, making it unlikely that translocations were being produced at a detectable frequency in irradiated W/W^v recipients. Finally, if this problem had been significant, we should have observed a lymphoid-restricted stem cell, which would have been a translocation induced in the defective pluripotent stem cell from the W/W^v host. Such a stem cell, because of the genetic defect, would not have formed spleen colonies, so markers would not have been detected in the myeloid system. The fact that we did not observe a lymphoid-restricted stem cell is further evidence against the contribution of marked stem cells by the W/W^v recipients.

Wallis et al. (1975) demonstrated that only a few stem cells repopulate the thymus with functional lymphocytes. Thus, a mouse containing both marked and unmarked pluripotent stem cells could have a thymus entirely repopulated by chance by stem cells from one or the other source. Our observation that the marker distributions breed true in a secondary recipient tends to rule out this possibility. Under the conditions found in the second recipient, there should have been much opportunity for the stem cells to repopulate other tissues if they had had the potential. That the myeloid-restricted stem cells did not repopulate the thymus in secondary recipients suggests that they did not have the capacity to do so.

Other more trivial explanations for restricted stem cells include such factors as the induction of leukemia. Since our mice survived for long periods of time, both in primary and secondary hosts, we feel that induction of leukemia was not a problem. Variations in the ease with which different classes of cells give analyzable chromosome spreads can also lead to incorrect conclusions regarding restriction of stem-cell potential. For example, if B-cell blasts only rarely gave scorable mitotic spreads, an unrepresentatively small proportion of B cells would be included in the analysis of LPS-stimulated cultures. This factor would make difficult the detection of chromosome markers restricted to B cells, unless the markers occurred in very high frequency. However, since we regularly obtained scorable spreads in all of the mitogen-induced cultures, it would appear that both B and T lymphocytes were readily detectable in these assays. Similarly, we had no difficulty preparing spreads either from spleen colonies or bone marrow. We found no evidence that variability in mitotic spreads could account for the class of restricted stem cells.

One of the puzzling results from our experiments was our failure to detect a lymphoid stem cell or a restricted stem cell for the B-lymphocyte pathway. The most important point to emphasize in this regard is that our experimental design did not allow us to make conclusions from negative data. There are several reasons why we may not have observed restricted stem cells even if they had existed. For example, such restricted stem cells may occur in another tissue, have a higher sensitivity to radiation, or occur at a lower frequency than do other stem cells. Examinations of more W/W^v recipients would detect the infrequent stem-cell classes. Finally, the long times used for regeneration in our experiments would select against stem cells with only a limited self-renewal potential. We are currently analyzing some recipients at shorter intervals in a search for short-lived stem cells.

REGULATION OF DIFFERENTIATION AT THE LEVEL OF STEM CELLS

There have been several reports that the thymus influences differentiation of the myeloid system. First, there is a slight decrease in the number of spleen colony-forming cells in thymectomized mice (Resnitzky et al. 1971). Genetically athymic (nude) mice also have defective hematopoiesis (Bamberger et al. 1977). Second, injection of many antigens causes CFU-S to enter a proliferative state (Frindel et al. 1976). If the antigen used to induce differentiation is a thymus-dependent antigen, then the proliferative stimulation of CFU-S is also thymus-dependent; that is, it is not possible to dissociate the immune response to the antigen from its effect on inducing proliferation on CFU-S. Third, thymus cells have been shown to reverse at least partially the repression of stem cell proliferation which is observed in some situations (Salinas and Goodman 1972). For example, parental stem cells will not proliferate or differentiate effectively in some F_1 hybrids. The grafting of large numbers of thymus cells along with CFU-S eliminates the repression. Fourth, the thymus may play a role in the reconstitution of genetically anemic mice of genotype W/W^v which lack normal CFU-S. Many investigators have shown that the anemia can be cured by grafting stem cells from a coisogenic normal donor. However, Wiktor-Jedrzejczak et al. (1977) have recently demonstrated that the removal of T lymphocytes from the donor marrow prevents the ability of marrow to reconstitute the anemia without affecting the number of detectable CFU-S. The implication of these results is that the long-term proliferation of stem cells somehow depends on the presence of a normally functioning thymus. In addition, inasmuch as defective stem-cell function can be reversed by injection of thymocytes, it is unlikely that this regulatory influence of the thymus is mediated by the thymic epithelium.

Our results are difficult to interpret in the context of thymic regulation of stem proliferation. If the results of Wiktor-Jedrzejczak et al. are generally applicable, it is difficult to understand how our experiments worked. The radiation doses used to induce translocations should have inactivated the few T cells present in the donor marrow. Nevertheless, that all of the mice were cured of their anemia was shown by increased erythrocytes and absence of the macrocytic-type red cells. Either the function of the necessary T cells in the donor bone marrow was radiation-resistant or irradiation of the W/W^v recipient changed the T-cell requirement for repopulation. In any case, we found no indication of a major role for thymus-derived cells in stem-cell proliferation and differentiation.

However, our data do suggest other interesting regulatory aspects for the differentiation of stem cells. Of greatest interest was the differentiation of cells in recipients bearing a myeloid-restricted marker. Consider, for example, mouse no. 11 (Table 1). In both the primary and secondary recipients, over 90% of the bone-marrow cells contained the unique marker; in addition, all but one of the spleen colonies examined had the same marker. Thus, the majority of the myeloid cells in these recipients was produced by one clone whose potential for differentiation was restricted to myeloid differentiation. Because these recipients had normal responses to PHA and LPS and because the recipient had one unmarked CFU-S, we can assume that the lymphocytes for these responses were derived from a pluripotent stem cell. If these recipients had contained both myeloid-restricted and pluripotent stem cells, the myeloid-restricted stem-cell pool eventually should have been diluted by the more primitive, pluripotent stem cell which one would expect to have a greater capacity for proliferation and self-renewal. The implication of this finding is that S_p do not have extensive capacity for self-renewal or that there is feedback at early stages in the differentiation of stem cells. If the entire myeloid requirement was being met by myeloid-restricted stem cells, feedback signals may have prevented differentiation by pluripotent stem cells of other myeloid-restricted stem cells. It is obviously important to confirm the existence of such feedback signals and to identify the nature of these regulatory events. Hopefully, the in vitro systems for maintenance of hematopoietic stem cells (Dexter et al., this volume) will allow a careful investigation into the factors that regulate the proliferation and differentiation of stem cells.

REFERENCES

Abramson, S., R. G. Miller, and R. A. Phillips. 1977. Identification of pluripotent and restricted stem cells of the myeloid and lymphoid systems. *J. Exp. Med.* **145**: 1567.

Bamberger, E. G., E. A. Machado, and B. B. Lozzio. 1977. Hematopoiesis in hereditary athymic mice. *Lab. Anim. Sci.* **27**: 43.

Barnes, D. W. H., E. P. Evans, C. E. Ford, and B. J. West. 1968. Spleen colonies in mice: Karyotypic evidence of multiple colonies from single cells. *Nature* **219**: 518.

Becker, A. J., E. A. McCulloch, and J. E. Till. 1963. Cytological demonstration of the clonal nature of spleen colonies derived from transplanted mouse marrow cells. *Nature* **197**: 452.

Edwards, G. E., R. G. Miller, and R. A. Phillips. 1970. Differentiation of rosette-forming cells from myeloid stem cells. *J. Immunol.* **105**: 719.

Frindel, E., E. Leuchars, and A. J. S. Davies. 1976. Thymus dependency of bone marrow stem cell proliferation in response to certain antigens. *Exp. Hematol.* (Copenh.) **4**: 275.

Gornish, M., M. P. Webster, and T. G. Wegmann. 1972. Chimerism in the immune system of tetraparental mice. *Nat. New Biol.* **237**: 249.

Kadish, J. L., and R. S. Basch. 1976. Hematopoietic thymocyte precursors. I. Assay and kinetics of the appearance of progeny. *J. Exp. Med.* **143**: 1082.

Micklem, H. S., C. E. Ford, E. P. Evans, and J. Gray. 1966. Interrelationships of myeloid and lymphoid cells: Studies with chromosome-marked cells transfused into lethally irradiated mice. *Proc. R. Soc. Lond., B, Biol. Sci.,* **165**: 78.

Mintz, B., and J. Palm. 1969. Gene control of hematopoiesis. I. Erythrocyte mosaicism and permanent immunological tolerance in allophenic mice. *J. Exp. Med.* **129**: 1013.

Moore, M. A. S., and D. Metcalf. 1970. Ontogeny of the hemopoietic system: Yolk sac origin of *in vivo* and *in vitro* colony forming cells in the developing mouse embryo. *Br. J. Haematol.* **18**: 279.

Nowell, P. C., B. E. Hirsch, D. H. Fox, and D. B. Wilson. 1970. Evidence for the existence of multipotential lympho-hematopoietic stem cells in the adult rat. *J. Cell. Physiol.* **75**: 151.

Parker, J. W., and D. Metcalf. 1974. Production of colony-stimulating factor in mixed leucocyte cultures. *Immunology* **26**: 1039.

Resnitzky, P., D. Zipori, and N. Trainin. 1971. Effect of neonatal thymectomy on hemopoietic tissue in mice. *Blood* **37**: 634.

Salinas, F. A., and J. W. Goodman. 1972. Relative effect of thymocytes from irradiated donors on hemopoiesis in P $\rightarrow$ F$_1$ chimeras. *Proc. Soc. Exp. Biol. Med.* **140**: 439.

Shortman, K., H. von Boehmer, J. Lipp, and K. Hopper. 1975. Subpopulations of T-lymphocytes: Physical separation, functional specialization and differentiation pathways of subsets of thymocytes and thymus-dependent peripheral lymphocytes. *Transplant. Rev.* **25**: 163.

Till, J. E., and E. A. McCulloch. 1961. A direct measurement of the radiation sensitivity of normal mouse bone marrow. *Radiat. Res.* **14**: 213.

Trentin, J., N. Wolf, V. Cheng, W. Fahlberg, D. Weiss, and R. Bonhag. 1967. Antibody production by mice repopulated with limited numbers of clones of lymphoid cell precursors. *J. Immunol.* **98**: 1326.

Wallis, V. J., E. Leuchars, S. Chwalinski, and A. J. S. Davies. 1975. On the sparse seeding of bone marrow of thymus in radiation chimeras. *Transplantation* **19**: 2.

Wiktor-Jedrzejczak, W., S. Sharkis, A. Ahmed, K. W. Sell, and G. W. Santos. 1977. Theta-sensitive cell and erythropoiesis: Identification of a defect in W/W^v anemic mice. *Science* **196**: 313.

Wu, A. M., J. E. Till, L. Siminovitch, and E. A. McCulloch. 1967. A cytological study of the capacity for differentiation of normal hemopoietic colony-forming cells. *J. Cell. Physiol.* **69**: 177.

———. 1968. Cytological evidence for a relationship between normal hemopoietic colony-forming cells and cells of the lymphoid system. *J. Exp. Med.* **127**: 455.

Yung, L. L. L., T. C. Wyn-Evans, and E. Diener. 1973. Ontogeny of the murine immune system: Development of antigen recognition and immune responsiveness. *Eur. J. Immunol.* **3**: 224.

Histamine H_2-Receptor and the Hematopoietic Stem Cell (CFU-S)

J. W. Byron

Department of Pharmacology and Experimental Therapeutics
University of Maryland School of Medicine Baltimore, Maryland 21201

Regulation of the growth and of the differentiation of the bone-marrow stem cell still remains a very challenging subject for investigation by experimental biologists. Approaches from several disciplines have served to unravel some of the stem cell's mysteries. It is clear, however, that the information yet to be uncovered greatly exceeds the questions answered thus far. Accordingly, it can only be with tongue in cheek and with the deepest humility that we attempt, in any manner, to discuss or to define the properties of the pluripotent stem cell.

The availability of analogues of histamine that distinguish histamine H_1-receptors from histamine H_2-receptors and the availability of chemicals that antagonize these receptors (Black et al. 1972) allow the analysis of tissues for their content of these sites. Hence, 2-methylhistamine stimulates histamine H_1-receptors and 4-methylhistamine stimulates histamine H_2-receptors. A large number of histamine H_1-receptor antagonists have been synthesized. In contrast, relatively few histamine H_2-receptor antagonists have been developed and most reports have been limited to the use of burimamide, metiamide, or cimetidine.

This report studies cell-cycle effects in the colony-forming unit—spleen (CFU-S) resulting from the stimulation of the two types of histamine receptors. The influence of imidazole-induced phosphodiesterase activation (Butcher and Sutherland 1962) on the cell-cycle effect is also examined. Any cell-cycle effect of endogenous histamine was investigated through the inhibition of histamine metabolism in bone marrow and through the use of the well-known histamine-releasing agents compound 48/80 and the divalent ionophore from *Streptomyces chartreusis* (A-23187). The effect of blocking histamine H_2-receptors on the cell-cycle actions of 4-methylhistamine, compound 48/80, or A-23187 is also reported. Finally, a comparison is made between the concentration-response curves to 4-methylhistamine of CFU-S derived from preleukemic AKR/J or from BD2F$_1$/J bone marrow.

MATERIALS AND METHOD

The technique (Byron 1971) in its more recent modification (Byron 1972a; Byron and Punchard 1974) that was used to study the action of drugs on the cell cycle of the bone-marrow stem cell has been described in detail elsewhere.

Preparation of Bone-Marrow Cell Cultures

B6D2F_1/J or AKR/J female mice, 9–11 weeks old, served either as donors or recipients of bone-marrow cells. A system was used to collect bone-marrow cells, so that a single cell suspension was achieved without the use of enzymes or severe mechanical forces. Siliconized glassware was used at all stages during which cells might come into contact with glass during the preparation of bone-marrow cell suspension.

A siliconized glass petri dish of 63-mm diameter was covered by a piece of stainless steel, no. 100 mesh gauze. A hole was cut in the gauze near the side of the dish. Mice were killed immediately prior to the removal of the bone marrow by breaking of their necks; no anesthesia was used. Bone-marrow cells from one intact femur per mouse were washed with 2 ml cold Fischer's medium onto the stainless steel gauze. The cells were allowed to filter through the gauze; through the hole in the gauze, filtrate was pipetted out of the petri dish with a Pasteur pipette, dropped gently onto the gauze, and again allowed to filter through. This process was repeated three more times. Mice were killed until enough cells had been collected. The procedure described above was repeated for each additional femur. The cells were then washed more than 20 times with the filtrate.

Nucleated cell count was performed and 600 to 700 cells were counted using both sides of an improved Neubauer counting chamber. Cell suspension was pipetted into sterile 12-by-75-mm polystyrene culture tubes (Falcon). The caps were loosened to permit aerobic culture. Cultures containing 1.55 ml of bone-marrow cells (4×10^6 nucleated cells/ml) suspended in Fischer's medium (Gibco) at pH 7.4 were incubated at 37°C in a shaking water bath; L-glutamine was added to the medium from stock solutions previously stored at −20°C. For buffering, the medium was supplemented with 10 mM hydroxyethylpiperazine ethane sulfonic acid (HEPES, Sigma) and 2.6 mM sodium bicarbonate; penicillin (500 units/ml) and streptomycin (50 μg/ml) were also added.

Addition of Drugs to Cultures

Bone-marrow cells were exposed in vitro to drugs for 1.0 hour. The dihydrochloride salts of both 2- and 4-methylhistamine and the ionophore A-23187 were studied alone over a range of concentrations. Compound 48/80 (Sigma) was studied at a concentration of 0.5 μg/ml. 4-methylhistamine, compound A-23187, and compound 48/80 were also studied in the presence either of 10^{-5} or 10^{-6} M metiamide. In addition, 4-methylhistamine was studied in the presence of 4 mM imidazole (Sigma). Metiamide or imidazole was added to cultures 5 minutes prior to the addition either of 4-methylhistamine, compound A-23187, or compound 48/80. Whether or not a buildup of endogenous histamine in bone-marrow cultures might influence the cell cycle of the stem cell was tested by treating cultures with a combination of 10^{-6} M aminoguanidine (Sigma) and 10^{-6} M chloroquine (Sigma). In some experiments, such cultures were pretreated with 10^{-7}, 10^{-6}, or 10^{-5} M metiamide.

Treatment with Hydroxyurea

Following drug treatments, cultures were treated for an additional hour with 10^{-3} M hydroxyurea (Byron, 1972b). The hydroxyurea (HU) (Sigma) was dissolved in 0.45 ml Fischer's medium, bringing the final volume of cultures to 2.0 ml. Control cultures received a comparable volume of Fischer's medium.

Assay of Cell Suspension for Bone-Marrow Stem Cells (CFU-S)

After exposure to HU, bone-marrow cultures were assayed for stem cells (CFU−S) (see Till and McCulloch 1961 for technique used). The cultures were diluted, and 6×10^4 bone-marrow cells contained in 0.25-ml medium were injected intravenously into total-body-irradiated recipients. The radiation dose to recipient mice was 820 rad. Spleen colonies were counted 9 days later. A comparison was then made between numbers of colonies per spleen for control or for HU-treated cultures, and the percentage of loss of CFU-S due to the addition of HU was calculated.

RESULTS

Figure 1 shows the response of CFU-S either to 2- or 4-methylhistamine. At concentrations above 10^{-9} M, the sensitivity of CFU-S to the cytocidal action of HU was markedly increased by 4-methylhistamine. No response to 2-methylhistamine was observed at concentrations as high as 10^{-6} M.

Figure 2 compares the effect on the concentration-response curve for 4-methylhistamine either of 10^{-6} M or 10^{-5} M metiamide. Metiamide antagonized the effect of low concentrations of 4-methylhistamine. An increase in the concentration of 4-methylhistamine reversed the antagonistic effects of the histamine H_2-receptor antagonist. The shift to the right of the response curve to 4-methylhistamine was more pronounced in the presence of 10^{-5} M metiamide.

The effects of imidazole pretreatment on the concentration-response curve to 4-methylhistamine are shown in Figure 3. The curve is shifted to the right such that a 100-fold increase in the concentration of 4-methylhistamine was required to obtain maximum increase in sensitivity again to the cytocidal effects of HU.

Table 1 shows the effect of treating bone-marrow cells with a combination of 10^{-6} M aminoguanidine and 10^{-6} M chloroquine. It also reports the effects of increasing concentrations of metiamide on any cell-cycle effects of this treatment. The sensitivity of CFU−S to the killing action of HU was increased by the treatment of cultures with aminoguanidine and chloroquine. Although no antagonism was observed following pretreatment of such

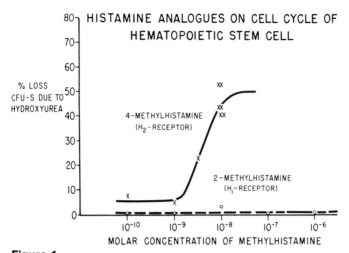

Figure 1

Response of the bone marrow stem cell (CFU−S) either to 2-methylhistamine (o———o) or to 4-methylhistamine (×———×). Bone marrow cells were taken from BD2F$_1$/J mice.

Abscissa: Molar concentrations of the methylhistamines are on a logarithmic scale. Ordinate: Percentage of loss of CFU−S due to hydroxyurea is on a linear scale. Each point is made up from a minimum of 24–32 spleens, i.e., 12–16 from cultures treated with hydroxyurea and 12–16 from control cultures.

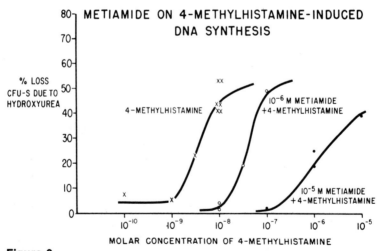

Figure 2

Antagonism by metiamide 10^{-6} M (o———o) or 10^{-5} M (●———●) of the response of the bone-marrow stem cell (CFU−S) to 4-methylhistamine (×———×). Bone marrow cells were taken from BD2F$_1$/J mice.

Abscissa: Molar concentration of 4-methylhistamine is on a logarithmic scale. Ordinate: Percentage of loss of CFU-S due to hydroxyurea is on a linear scale. Each point is made up from a minimum of 24–32 spleens, i.e., 12–16 from cultures treated with hydroxyurea and 12–16 from control cultures. Metiamide was added to cultures 5 minutes before 4-methylhistamine.

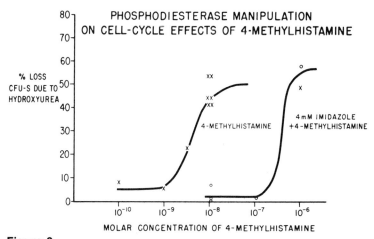

Figure 3

Suppression by 4 mM imidazole (o——o) of the response of the bone-marrow stem cell (CFU-S) to 4-methylhistamine (×——×). Bone-marrow cells were taken from $BD2F_1/J$ mice.

Abscissa: Molar concentration of 4-methylhistamine is on a logarithmic scale. Ordinate: Percentage of loss of CFU-S due to hydroxyurea is on a linear cale. Each point is made up from a minimum of 24–32 spleens, i.e., 12–16 from cultures treated with hydroxyurea and 12–16 from control cultures. Imidazole was added to cultures 5 minutes before 4-methylhistamine.

cultures with 10^{-7} M metiamide, concentrations of 10^{-6} M and 10^{-5} M metiamide antagonized the cell-cycle effects of aminoguanidine and chloroquine treatment.

The response of CFU–S to the divalent ionophore A-23187 alone or in the presence of 10^{-5} M metiamide is shown in Figure 4. At concentrations above 10^{-8} M, A-23187 increased the sensitivity of CFU-S to HU. Metiamide prevented CFU-S from responding to concentrations up to 10^{-6} M of A-23187.

Table 2 shows the response of CFU-S to 0.5 μg/ml compound 48/80 alone or in the presence of either 10^{-6} or 10^{-5} M metiamide. Compound 48/80 increased the sensitivity of CFU-S to HU, and this effect was prevented by both concentrations of metiamide.

Figure 5 compares the response to 4-methylhistamine of CFU-S derived either from AKR/J or $BD2F_1/J$ bone marrow. The concentration response curve for AKR/J-derived CFU-S is shifted to the left of that for $BD2F_1/J$-derived CFU-S.

DISCUSSION

The increased sensitivity of the hematopoietic stem cell to HU following its exposure to 4-methylhistamine and the antagonism of this response by the histamine H_2-receptor blocking agent metiamide associates a histamine H_2-receptor with CFU-S and suggests that, through histamine H_2-

Table 1
Aminoguanidine/Chloroquine-Induced DNA Synthesis in the Bone-Marrow Stem Cell – Antagonism by Metiamide

Treatment in vitro	CFU – S per spleen ± SEM		Loss CFU – S due to hydroxyurea (%)
	medium	hydroxyurea	
Aminoguanidine/chloroquine			
Experiment 1	15.9 ± 1 (14)	8.7 ± 0.7 (14)	45
Experiment 2	20.7 ± 1 (16)	14.45 ± 0.6 (13)	30
Experiment 3	15.5 ± 1 (16)	9.45 ± 0.8 (11)	29
Experiment 4	20.3 ± 1 (16)	15.3 ± 0.8 (14)	25
Mean	18.1 (62)	11.97 (52)	32.3
Aminoguanidine/chloroquine			
+ 10^{-7} M metiamide	21.9 ± 1 (9)	15.15 ± 0.8 (14)	30
+ 10^{-6} M metiamide	15.95 ± 0.9 (13)	13.7 ± 0.6 (12)	14
+ 10^{-5} M metiamide	18.85 ± 1.2 (14)	16.75 ± 0.9 (10)	11

Numerals in parentheses indicate numbers of spleens counted.

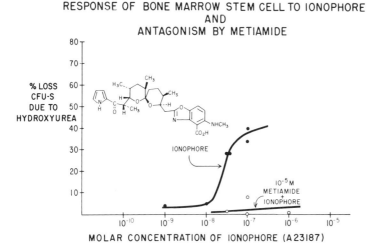

Figure 4

Response of the bone-marrow stem cell (CFU-S) to the ionophore A-23187 alone (●——●) or in the presence of 10^{-5} M metiamide (o——o). Bone marrow cells were taken from BD2F$_1$/J mice.

Abscissa: Molar concentration of A-23187 is on a logarithmic scale. Ordinate: Percentage of loss of CFU-S due to hydroxyurea is on a linear scale. Each point is made up from a minimum of 24–32 spleens, i.e., 12–16 from cultures treated with hydroxyurea and 12–16 from control cultures.

receptors, the stem cell may be triggered from the G_0 state into the S phase of the cell cycle. Recent studies have demonstrated that the histamine H_2-receptor antagonist cimetidine is a more effective antagonist of 4-methylhistamine effects on CFU-S than is metiamide (Byron, unpubl.). In the presence of imidazole, the concentration-response curve of the stem cell to 4-methylhistamine shifted significantly to the right. This suggests that the cell-cycle action of the histamine H_2-receptor agonist involves a cyclic nucleotide; hence, cyclase systems such as the adenylate cyclase system may mediate in the cell-cycle action of 4-methylhistamine. Previous studies (Byron 1971) have demonstrated that the dibutyryl derivative of adenosine $3':5'$-cyclic monophosphate (cyclic AMP, cAMP) triggers CFU-S from the G_0 into the S phase. In addition, phosphodiesterase inhibition alone Byron 1974) triggers bone-marrow stem cells into the S phase. In a manner similar to that observed for 4-methylhistamine, phosphodiesterase activation with imidazole suppresses the cell-cycle action of dibutyryl-cAMP and also that of isoproterenol, an activator of adenylate cyclase (Byron 1972a).

Exposure of bone-marrow cell suspensions to inhibitors of histamine catabolism or to known histamine-releasing agents initiated DNA synthesis in the bone-marrow stem cell. This initiation of DNA synthesis was antagonized by metiamide. These results would suggest that endogenous histamine has cell-cycle effects if made available to the stem cell. Many hematopoietic cell types, such as myeloid leucocytes, contain histamine. Whether endogenous histamine is derived from the bone-marrow stem cell

Table 2
Metiamide- and 48/80-Induced DNA Synthesis

Concentration of compound 48/80	Concentration of metiamide	CFU−S per spleen ± SEM		Loss CFU−S due to hydroxyurea (%)
		medium	hydroxyurea	
0.5 μg/ml	—	14.5 ± 0.97 (14)	9.6 ± 0.77 (14)	34
0.5 μg/ml	10^{-6} M	13.4 ± 1.0 (12)	13.1 ± 0.65 (14)	2
0.5 μg/ml	10^{-5} M	13.2 ± 0.87 (13)	12.6 ± 0.89 (12)	5

Numerals in parentheses indicate numbers of spleens counted.

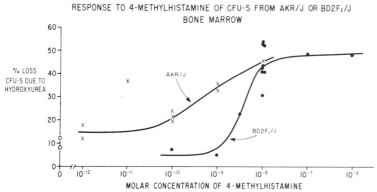

Figure 5
Response to 4-methylhistamine of the bone-marrow stem cell (CFU-S) derived from AKR/J (×——×) or B6D2F$_1$/J (●——●) bone marrow. Bone marrow donors were 9–11 weeks old.

Abscissa: Molar concentration of 4-methylhistamine is on a logarithmic scale. Ordinate: Percentage of loss of CFU-S due to hydroxyurea is on a linear scale. Each point is made up from a minimum of 24–32 spleens, i.e., 12–16 from cultures treated with hydroxyurea and 12–16 from control cultures.

or from other hematopoietic cell types is a question for future study. The data, however, support the hypothesis that histamine may play a role in regulating cell proliferation in certain tissues (Kahlson and Rosengren 1968). The presence of histamine in bone-marrow cells and the evidence reported herein that a histamine H_2-receptor is associated with the stem cell raise the strong possibility that some of the agents now known to influence the cell cycle may do so because of their ability either to release histamine from granules or to inhibit the metabolic inactivation of diffusible histamine molecules.

The ionophore A-23187 is capable of releasing histamine from mast cells (Foreman et al. 1973). The data in Figure 4 confirm an earlier report that demonstrated that A-23187 could trigger CFU−S from G_0 into the S phase (Gallien-Lartigue 1976). This result suggests strongly that the cell-cycle effect of A-23187 in bone marrow is secondary to the compound's ability to release histamine from bone-marrow cells, e.g., basophils. Hence, cell-cycle changes in CFU-S may not be the direct result of altering calcium conductance in the stem cell. This caution must be extended to other cell systems such as the peripheral blood lymphocyte system (Luckasen et al. 1974; Hovi et al. 1976), in which A-23187 has been studied for cell-cycle effects.

The concentration-response curve to 4-methylhistamine shifts to the left for CFU−S derived from high leukemia incidence AKR/J bone marrow; this is in contrast to the curve derived from B6D2F$_1$/J mice (low incidence of leukemia). This suggests either that subtriggering levels of endogenous histamine available to the histamine H_2-receptor are elevated in this strain or that the affinity of the histamine H_2-receptor for 4-methylhistamine has been increased significantly. The extension of this type of study to other strains or to F_1 hybrids with low or high incidence of neoplastic disease is required to determine whether the change in response to the histamine H_2-receptor observed in AKR/J mice is a general characteristic of early neoplastic processes.

Acknowledgments

The author thanks Wendy Resneck, Audrey Ford, and Gloria Parham for technical assistance during these experiments. Thanks are also expressed to the Radiobiology Department of the University of Maryland School of Medicine for X-irradiations. The dihydrochloride salts of 2- or 4-methylhistamine and metiamide that were added to the cultures of blood-marrow cells were a gift from Smith, Kline and French, United Kingdom. Compound A-23187 was donated by the Eli Lilly Company.

This work was supported by United States Public Health Service Grant CA-16147 from the National Cancer Institute.

REFERENCES

Black, J. W., W. A. M. Duncan, C. J. Durant, C. R. Ganellin, and E. M. Parsons. 1972. Definition and antagonism of histamine H_2-receptors. *Nature* **236**: 385.

Butcher, R. W., and E. W. Sutherland. 1962. Adenosine 3',5'-phosphate in biological materials. I. Purification and properties of cyclic 3',5'-nucleotide phosphodiesterase and use of this enzyme to characterize adenosine 3',5'-phosphate in human urine. *J. Biol. Chem.* **237**: 1244.

Byron, J. W. 1971. Effect of steroids and dibutyryl cyclic AMP on the sensitivity of haemopoietic stem cells to ^{3}H-thymidine *in vitro*. *Nature* **234**: 39.

———. 1972a. Evidence for a β-adrenergic receptor initiating DNA synthesis in haemopoietic stem cells. *Exp. Cell Res.* **71**: 228.

———. 1972b. Comparison of the action of ^{3}H-thymidine and hydroxyurea on testosterone-treated hemopoietic stem cells. *Blood* **40**: 198.

———. 1974. Molecular basis for triggering of hemopoietic stem cells into DNA synthesis. In *Hemopoiesis in culture*, DHEW Publication No. (NIH) 74–205 (ed. W. A. Robinson), p. 91.

Byron, J. W., and J. K. Punchard. 1974. *In vitro* method for studying agents that trigger CFU_s from G_0 into S. In *Hemopoiesis in culture*, DHEW Publication No. (NIH) 74–205 (ed. W. A. Robinson), p. 436.

Foreman, J. C., J. L. Mongar, and B. D. Gomperts. 1973. Calcium ionophores and movement of calcium ions following the physiological stimulus to a secretory process. *Nature* **245**: 249.

Gallien-Lartigue, O. 1976. Calcium and ionophore A-23187 as initiators of DNA replication in the pluripotent haemopoietic stem cell. *Cell Tissue Kinet.* **9**: 533.

Hovi, T., A. C. Allison, and S. C. Williams. 1976. Proliferation of human peripheral blood lymphocytes induced by A23187, a streptomyces antibiotic. *Exp. Cell Res.* **97**: 92.

Kahlson, G., and E. Rosengren. 1968. New approaches to the physiology of histamine. *Physiol. Rev.* **48**: 155.

Luckasen, J. R., J. G. White, and J. H. Kersey. 1974. Mitogenic properties of a calcium ionophore. A23187. *Proc. Natl. Acad. Sci. U.S.A.* **71**: 5088.

Till, J. E., and E. A. McCulloch. 1961. A direct measurement of the radiation sensitivity of normal mouse bone marrow cells. *Radiat. Res.* **14**: 213.

Human Myeloproliferative Disorders: Clonal Origin in Pluripotent Stem Cells

P. J. Fialkow
Medical Genetics Section, Medical Service, Veterans Administration Hospital
Seattle, Washington 98108, and Departments of Medicine and Genetics
University of Washington, Seattle, Washington 98195

A. M. Denman
Division of Immunological Medicine, Clinical Research Centre
Harrow, Middlesex HA1 3UJ, United Kingdom

J. Singer
Oncology Section, Medical Service, Veterans Administration Hospital
Seattle, Washington 98108, and Department of Medicine, University of Washington
Seattle, Washington 98195

R. J. Jacobson and M. N. Lowenthal
Baragwanath Hospital and Department of Medicine, University of the Witwatersrand
Johannesburg, South Africa

An indirect approach to the elucidation of the pathogenesis of a tumor involves the determination of the number of cells from which that tumor has arisen. For example, if a neoplasm was initiated by a rare, more-or-less random event such as "spontaneous" somatic mutation, single cell (clonal) origin would be expected. In contrast, a malignancy caused by continuous cell-to-cell spread of a virus would have had a multicellular origin.

This question can be investigated in subjects with two (or more) genetically distinct types of cells, i.e., in subjects with cellular mosaicism. For example, if a subject has cells of two types, A and B, present in equal proportions, both types will be found in normal tissues, but neoplasms with clonal origin will contain A *or* B cells. The same rationale allows delineation of stem-cell relationships. If a tumor has arisen in a pluripotent stem cell of type A, all descendants of that progenitor will type as A.

A very useful genetic probe applicable to the study of human myeloproliferative disorders is provided by the mosaicism in females that is due to inactivation of one X chromosome in each somatic cell. The X-linked glucose-6-phosphate dehydrogenase (G-6-PD) locus is an especially useful marker because it is polymorphic in many ethnic groups. Because this locus undergoes inactivation, only one of the two G-6-PD genes is active in somatic cells of females (Beutler et al. 1962; Davidson et al. 1963; De Mars and Nance 1964). Thus, women heterozygous for the usual B gene (Gd^B) and a variant allele such as Gd^A or Gd^{A-}, have two populations of cells, one producing type-B isoenzyme and the other, type-A. A tumor with clonal origin in a Gd^B/Gd^A heterozygote will display B *or* A isoenzyme (a single-enzyme phenotype), whereas one arising from multiple cells may have both B *and* A types (a double-enzyme phenotype). In this communication, studies conducted in the authors' laboratories on the myeloproliferative disorders—chronic myelocytic leukemia, polycythemia vera, and agnogenic myeloid metaplasia with myelofibrosis—are discussed. More extensive reviews appear elsewhere (Fialkow 1974, 1976).

CHRONIC MYELOCYTIC LEUKEMIA

Clonal Origin

Since the myeloproliferative disorders are hematopoietic in origin, skin, rather than blood, is used to determine if female patients are G-6-PD heterozygotes. Twelve women with Philadelphia-chromosome-positive chronic myelocytic leukemia (CML) were found to have both B and A enzymes in their skin cells (i.e., they were heterozygous for G-6-PD), but only one type of G-6-PD was observed in their granulocytes (eight patients typed as B and four as A) (Fialkow et al. 1967; Barr and Fialkow 1973; Fialkow et al. 1977; P. J. Fialkow et al., unpubl.). Conversely, in Gd^B/Gd^A and Gd^B/Gd^{A-} heterozygotes without hematopoietic neoplasms, the granulocytes almost invariably show both B and A enzyme types (Fialkow 1973). The fact that single-enzyme phenotypes occur in CML granulocytes but not in the granulocytes of normal subjects strongly favors a clonal origin of the leukemia. This postulate is supported by studies with other isoenzyme (Fialkow et al. 1969; Fialkow et al. 1972) and chromosomal markers (Fitzgerald et al. 1971; Gahrton et al. 1974; Hayata et al. 1974; Moore et al. 1974; Hossfeld 1975). Obviously, the statement that CML has a clonal origin applies only to the stage at which the disease can be recognized. Conceivably, many cells could have been affected at an earlier phase, but, by the time leukemia is clinically manifest, only one clone remains.

Stem Cell Origin

In G-6-PD heterozygotes with CML, single-enzyme phenotypes are found not only in granulocytes but also in erythrocytes, platelets, and cultured blood monocytes/macrophages (Fialkow et al. 1977). Thus, this disease involves multipotent myeloid stem cells, a suggestion supported by studies with other markers (reviewed in Fialkow 1976).

Cause and Pathogenesis

The isoenzyme and chromosome marker studies indicate that CML is a stem-cell disorder. However, microscopic examination of the marrow in this disease usually shows an overabundance of myelocytes and more mature granulocytic cells. The genetic and morphologic observations are reconciled by assuming that an abnormality has occurred in the myelogenous stem cells that has conferred upon them the ability to populate the marrow at the expense of normal stem cells. As differentiation proceeds along the granulocyte pathway, cells bearing this abnormality become overabundant and present the clinical and hematological manifestations we recognize as CML. Erythrocytes and, less frequently, platelets may also become overabundant.

Among the causative factors postulated for CML are "spontaneous" or radiation-induced genetic accidents, viruses, and physiologic defects in marrow homeostasis. The "spontaneous" mutation hypothesis predicts that the tumor will begin in one cell, and the genetic marker data are in accord with this suggestion. The aberrant homeostasis model implies that the basic abnormality is not intrinsic to the marrow cells themselves but is found in the mechanisms that regulate marrow proliferation and maturation. Since this hypothesis predicts multicellular origin, the genetic-marker data make it unlikely to be correct. If caused by a virus or by radiation, the malignancy could originate in one or many cells. For example, if the virus were only one of several factors necessary for the origination of a tumor or if the oncogenic change induced by the virus were rare, a clonal origin would be found (e.g., chromosome abnormalities in many cells might be induced, but only the rare cell that happened to have the Philadelphia chromosome (Ph[1]) would evolve into CML). Alternatively, the putative oncogenic virus might have specific affinity for the DNA in the involved regions of chromosomes 22 and 9, in which case Ph[1] could be induced in multiple cells. The probable clonal origin of CML makes this latter possibility less likely and also virtually excludes any hypothesis of pathogenesis based on continuous cell recruitment.

POLYCYTHEMIA VERA

Stem Cell and Clonal Origin

Thus far we have studied two G-6-PD heterozygotes with polycythemia vera (Adamson et al. 1976). One patient was treated with chlorambucil and the other, only with phlebotomy. Each patient had both B and A enzymes in normal tissues but only a single-enzyme phenotype (type A) in the red cells, suggesting clonal origin of the polycythemia vera. The same single-enzyme phenotype was found in the patients' granulocytes and platelets, indicating that the disease involves a multipotent marrow stem cell.

Pathogenesis

Two general theories have been advanced to explain the development of polycythemia vera: one, that the disease is a myeloproliferative neoplasia similar to CML (Gurney 1965); and, two, that it is a hyperplasia either with normal stem cells reacting to an abnormal myeloproliferative factor (Ward et al. 1974) or with abnormal cells having, for example, increased sensitivity to erythrocyte-stimulating factors (Zanjani 1976). The occurrence of single-enzyme phenotypes in G-6-PD heterozygotes with polycythemia vera virtually excludes the normal stem-cell hyperplasia hypothesis, makes the increased sensitivity of an abnormal stem cell very unlikely, and is most compatible with the concept of the disease being a neoplasm.

AGNOGENIC MYELOID METAPLASIA WITH MYELOFIBROSIS

The fundamental nature of this disorder is unknown, but it is often classified with CML and polycythemia vera as a myeloproliferative neoplasm. Neither the identity of the putative neoplastic cell nor the factors underlying the marrow fibrosis have been defined. Some investigators feel that this disorder is not a neoplasia but a generalized hyperplasia.

We studied a patient heterozygous at the G-6-PD locus who displayed both B and A isoenzymes in nonhematopoietic cells but who had only type A in granulocytes, red cells, and platelets (Jacobson et al. 1977). This observation demonstrates that the disorder affects a multipotent stem cell and provides strong evidence that it has a clonal origin. Similar conclusions were suggested by Kahn et al. (1975) in a study of a patient heterozygous for G-6-PD deficiency. The probable clonal origin of agnogenic myeloid metaplasia is more compatible with neoplasia than with hyperplasia theories of pathogenesis.

ACUTE MYELOBLASTIC LEUKEMIA

This disease is not generally classified as a myeloproliferative syndrome; however, patients with CML, polycythemia vera, and agnogenic myeloid metaplasia not infrequently undergo transition to acute leukemia.

Earlier chromosome studies indicated that AML at least sometimes involves a stem cell common to erythrocyte and myelocyte precursors. About 40% of patients with acute myeloblastic leukemia have chromosomal abnormalities in marrow cells. The suggestion that chromosomal aberrations are present in red-cell precursors as well as in myeloblasts (Jensen and Killmann 1971) was confirmed directly by Blackstock and Garson (1974), who showed that abnormal chromosomes characteristic of the patient's leukemic cells are present in radioactive-iron-incorporating red-cell precursors.

Thus far we have studied only one patient with AML (G. Wiggans et al., in prep.). The patient was a 55-year-old woman who underwent hemicolectomy in 1967 for adenocarcinoma. A pulmonary recurrence 2 years later was resected but in 1973 she suffered another recurrence. Over the next few years she was treated with irradiation and chemotherapy. Approximately 2.5 years after the first treatment, acute myeloblastic leukemia was diagnosed.

Normal skin cells had approximately equal amounts of B and A G-6-PD isoenzymes indicating heterozygosity for G-6-PD. However, only type-B enzyme was detected in the patient's leukemic white blood cells and erythrocytes. The suggestion based on the G-6-PD data that this patient's AML had clonal origin is of special interest, inasmuch as the leukemia developed after radiation and chemotherapy for disseminated colon cancer.

The data given above indicate that the myeloproliferative disorders arise in multipotent stem cells and that they have clonal origin at the time of diagnosis. Three further questions are considered below: (1) Do marrow fibroblasts arise from the multipotent stem cell involved in the myeloproliferative syndromes? (2) Do lymphocytes arise from this stem cell? (3) Are there any residual normal stem cells?

MARROW FIBROBLASTS IN MYELOPROLIFERATIVE DISORDERS
Agnogenic Myeloid Metaplasia with Myelofibrosis

Marrow fibrosis is most prominent in agnogenic myeloid metaplasia with myelofibrosis. In fact, it is this feature which dominates the clinical picture. The cause of the myelofibrosis is unknown but most theories consider it to be part of the same process as that which affects the blood cells. G-6-PD and chromosome marker studies suggest that this interpretation may not be correct.

Technical considerations and the relatively limited amount of material available for investigation preclude direct study of marrow fibroblasts. Consequently, we assayed G-6-PD in "fibroblasts" grown in culture from the patients' marrows. Since these cells grow as monolayers, have the appearance of fibroblasts, and are repeatedly passaged, they are more likely to be connective tissue elements than blood cells. It is most probable, but not definitely proven, that they are fibroblasts.

In the G-6-PD heterozygote with myeloid metaplasia and myelofibrosis who displayed both B and A enzymes in normal cells but had only type A in blood cells, the cultured marrow "fibroblasts" had both B and A isoenzymes in proportions identical to those observed in skin fibroblasts (Jacobson et al. 1977). This patient also had a distinctive chromosome abnormality (47,XX,+8) in blood cells which was not detected in cultured marrow fibroblasts. These isoenzyme and cytogenetic studies strongly suggest that the marrow fibrosis was a secondary abnormality. An alternative possibility is that marrow fibroblasts in vivo were clonally derived, but outgrowth of these cells in culture was depressed in favor of residual normal cells. However, not even one chromosomally abnormal cultured marrow fibroblast was detected, and the proportion of B:A G-6-PD after the first transfer was the same as it was in skin. Thus, any in vitro selection against abnormal marrow fibroblasts would have had to be virtually complete, even though the same cells predominated in vivo. A more likely explanation is that the cultured connective tissue cells were derived predominantly from the cells responsible for the myelofibrosis in vivo, and, therefore, that the latter cells did not arise from the abnormal hematopoietic multipotent stem cell. Thus, the myelofibrosis in this patient with agnogenic myeloid metaplasia apparently was a secondary abnormality. Similar conclusions were reached previously from study of a patient with acute myelofibrosis (Van Slyck et al. 1970).

Chronic Myelocytic Leukemia

There is one report of the Philadelphia chromosome in cultures of marrow "fibroblasts" (Hentel and Hirschhorn 1971), suggesting that such cells may also arise from the CML multipotent stem cell. However, as in several other studies (Maniatis et al. 1969; De La Chapelle et al. 1973; H. Van den

Berghe and G. David, per. comm.), we did not find Ph¹ in cultured marrow fibroblasts from a patient with CML (Fialkow et al. 1977). Similarly, in contrast to the blood cells, the marrow fibroblasts displayed both B and A enzymes. It appears, therefore, that at least in this case of CML with myelofibrosis, cultured fibroblasts did not arise from the CML stem cell and that the fibrosis might have represented a reaction secondary to the leukemic process.

BLOOD LYMPHOCYTES

Polycythemia Vera

Both enzyme types were found in blood lymphocytes from the two Gd^B/Gd^A heterozygotes with polycythemia vera who had only type-A enzyme in their red cells, granulocytes, and platelets (Adamson et al. 1976). The lymphocytes were prepared by Ficoll/Hypaque sedimentation or filtration through nylon fiber columns. The proportion of A:B G-6-PD in the homogeneous lymphocyte preparations was about 1:1, the same as that observed in normal skin cells. Thus, a major portion of blood lymphocytes do not arise from the hematopoietic stem cell involved in polycythemia vera. Studies are in progress to determine if these are T lymphocytes only or if they also include B lymphocytes.

Chronic Myelocytic Leukemia

More data are available for lymphocytes from patients with CML than from those with polycythemia vera. Such cells when stimulated to divide in vitro with phytohemagglutinin (PHA) reportedly are Ph¹-negative (Tough et al. 1963; Whang et al. 1963), but this finding does not necessarily exclude involvement of all lymphocytes. This question has been investigated in three G-6-PD heterozygotes with CML. The CML myelocytic cells in each patient showed a single type: B.

Major technical problems included isolating homogeneous lymphocyte populations, documenting their identity as lymphocytes and identifying them as to T or B type. Simple methods of preparation such as continuous and discontinuous density gradient centrifugation all failed to separate lymphocytes from granulocyte precursors and immature forms. Thus, we adopted complex multistaged protocols of preparation. These methods and the results which have been described in detail elsewhere (P. J. Fialkow et al., in prep.) are summarized here.

Lymphocyte Separation

The essential preliminary steps in the preparation of the lymphocytes were density gradient centrifugation on a Ficoll/Hypaque gradient followed by velocity sedimentation. The small cells were then rosetted. T-cell fractions were obtained by allowing rosettes to form spontaneously from neuraminidase-treated sheep red cells, i.e., E rosettes. In some experiments, selection of T lymphocytes was assisted by filtration of the cells through nylon wool columns. T-cell preparations were stimulated with PHA. In some experiments B-cell preparations consisted of non-E-rosetted small cells stimulated with pokeweed mitogen. In other experiments, B-cell preparations were obtained by allowing complement rosettes to form from the small cells. For this technique, sheep red cells sensitized with rabbit IgM antibodies were used with fresh human serum as a source of complement. The rosettes were separated on a Ficoll/Hypaque gradient and the final preparation was stimulated with pokeweed mitogen.

Lymphocyte Identification

Lymphocytes were identified as small round cells sedimenting from 3.5 to 4.7 mm/hour which were transformed by mitogens in culture. Operationally we defined a homogeneous population as one in which at least 85% of cells fulfilled the morphologic criteria conventionally accepted for small lymphocytes or for mitogen-transformed lymphoblasts. Any important admixture with other cell types was excluded. For example, in initial experiments, the lymphocyte-enriched fractions were tested for monocyte admixture by latex particle phagocytosis and by an esterase method; the percent of monocytes was always less than one. Mature granulocytes and small myeloblasts were excluded morphologically.

Criteria for Defining T and B Cells

Formation of E rosettes was used in the preparation of T lymphocytes and also as diagnostic markers of these cells. T lymphocytes also lacked such B-cell markers as complement receptors, readily demonstrable surface Ig, and intracytoplasmic Ig. Similarly, T cells did not secrete Ig in vitro but they did "help" normal B cells to synthesize Ig. B lymphocytes from patients with CML were defined as cells that did not form E rosettes and that had complement receptors and easily demonstrable Ig on their surfaces. They also contained intracytoplasmic Ig and secreted Ig in culture. They did not provide helper function to normal B cells and were transformed by pokeweed mitogen.

Not all methods for characterizing the isolated populations could be carried out in every experiment because the numbers of lymphocytes isolated were very limited. Furthermore, unusually large proportions of lymphocytes appeared as null cells (i.e., cells with neither T- nor B-lymphocyte markers); this we attributed to the deleterious effect of repeated separation maneuvers. Similar manipulations of human spleen, lymph node, and tonsil cells lead to a comparable increase in the proportion of cells lacking T- and B-cell markers (A. M. Denman and B. K. Pelton, unpubl.).

The results provide evidence for three distinct populations; three experiments are illustrated for each population.

B Lymphocytes

As shown in Table 1, cells with B-cell characteristics but lacking those of T cells displayed only one enzyme type (B), the same type as that found in the CML granulocytes. This B-cell single enzyme phenotype was maintained during remission as well as in the active phase of CML; no population of B lymphocytes with both isoenzymes was found. These data indicate that B lymphocytes arise from the CML stem cell.

Table 1
B Lymphocytes from G-6-PD Heterozygotes with Chronic Myelocytic Leukemia

Experi-ment	Morphology	E rosettes		C receptors	Ig		Ig syn-thesis	G-6-PD
		Prep	Dx	Prep	Surface	Cytoplasm		
V.M. 2	100	"0"	<1	ND	25	ND	ND	B
V.M. 8	95	"0"	ND	"100"	ND	18	yes	B
M.K. 5	100	"0"	ND	ND	ND	10	yes	B

Note: Prep, E rosettes or C rosettes used in preparation of T- and B-cell pools, respectively; Dx, E rosettes used as diagnostic marker of T cells after all preparative steps had been completed; ND, not done.

T Lymphocytes that Do Not Arise from Chronic Myelocytic Leukemia Stem Cells

The other two populations we detected had T-cell characteristics. Of these, one could not have arisen from the CML clone (Table 2), because it contained both B and A enzymes in proportions similar to those of normal tissues. This population was detected only when the patient was in clinical remission.

It is possible that these non-CML T lymphocytes were long-lived and antedated the development of leukemia in the CML stem line. An argument against this possibility was provided by experiments that were performed more than 2 years after the onset of CML. T cells isolated at that time had a proportion of A:B enzyme similar to that found in skin. If a significant proportion of T cells had arisen from the CML clone during those 2 years, there should have been a preponderance of the B type G-6-PD (i.e., the type found in the patients' CML granulocytes). We therefore

Table 2
T Lymphocytes from G-6-PD Heterozygotes with Chronic Myelocytic Leukemia in Clinical Remission

Experiment	Morphology	E rosettes		Ig	Helper function	G-6-PD
		Prep	Dx			
V.M. 5	99	"100"	27	<1	ND	B/A (2:1)
V.M. 8	100	"100"	ND	0	yes	B/A (6:4)
M.K. 2	100	"100"	74	0	ND	B/A (4:6)

Note: Prep, E rosettes or C rosettes used in preparation of T- and B-cell pools, respectively; Dx, E rosettes used as diagnostic marker of T cells after all preparative steps had been completed; ND, not done; Ig, Cell-surface Ig in experiments V.M. 5 and M.K. 3 and intracytoplasmic Ig in experiment V.M. 8.

suggest that some of these T lymphocytes arose after the development of CML and did not originate from the CML stem cell.

T Lymphocytes that May Arise from Chronic Myelocytic Leukemia Stem Cells

Some of our results suggest the existence of a third population of lymphocytes, one that has T-cell characteristics and that arises from the CML clone (Table 3). In contrast to the non-CML T cells whose presence was most easily demonstrated in clinical remission, this clonal population was detected only when the disease was under poor control or in the accelerated phase. Although this population of "T-cells" with a single-enzyme phenotype must have arisen from the CML clone, we do not know whether it represented an independently arising T-cell population or a subset of B cells that had acquired T-cell markers secondary to neoplastic alteration. The fact that some lymphocyte populations arise from the CML stem cell may explain why in many cases of blast transformation in vivo the cells have lymphoblast characteristics.

Figure 1 depicts schematically one possible interpretation of the findings in CML. An initial abnormality leads to the evolution of a clone of pluripotent marrow stem cells giving rise to erythrocytes, platelets, polymorphonuclear cells, monocytes, and B lymphocytes. The B-lymphocyte population is demonstrable throughout all phases of the disease and apparently is not accompanied by B cells of nonclonal origin. In addition, there is a population of cells with T-lymphocyte characteristics that arises from this clone. These cells could be B lymphocytes altered by the disease to give T-cell characteristics. Alternatively, they may represent independent development of clonal T lymphocytes. During remission, there is a population of T lymphocytes that does not arise from the CML stem cell but probably arises from progenitors not affected by the disease. Presumably, these progenitors are present throughout the course of the disease, but their descendants are demonstrable only in clinical remission. When the disease is under poor control, expansion of the CML clone, chemotherapy, or both decrease proliferation of this pool so that it is not detected.

Table 3

Cells with T Lymphocyte Characteristics from G-6-PD Heterozygotes with Chronic Myelocytic Leukemia When Their Disease is Poorly Controlled

| | | Lymphocyte markers (% of cells) | | | |
| | | E rosette | | | |
Experiment	Morphology	Prep	Dx	Ig	G-6-PD
V.M. 2	100	"100"	22	<1	B
V.M. 4	98	"100"	25	0	B
M.K. 5	88	"100"	ND	<1	B

Note: Prep, E rosettes or C rosettes used in preparation of T- and B-cell pools, respectively; Dx, E rosettes used as diagnostic marker of T cells after all preparative steps had been completed; ND, not done; Ig, Cell-surface Ig in experiments V.M. 2 and 4 and intracytoplasmic Ig in experiment M.K. 5.

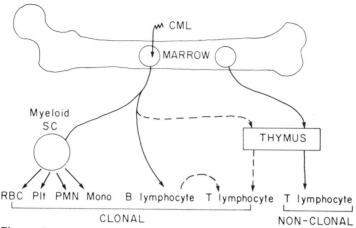

Figure 1
Stem cell origin of chronic myelocytic leukemia (CML).

RESIDUAL NORMAL STEM CELLS

Polycythemia Vera

In the two Gd^B/Bg^A heterozygotes with polycythemia vera, only type-A enzyme was found in peripheral blood granulocytes, red cells, and platelets, thereby providing no evidence for residual normal stem cells. However, it was possible that a minor G-6-PD isoenzyme component had been missed if it had contributed less than 5 percent of the total G-6-PD activity. To study this problem at a more sensitive level, we analyzed erythroid colonies growing in semisolid medium. First, it was shown that colonies obtained from normal G-6-PD heterozygotes have single-enzyme phenotypes (Prchal et al. 1976). These data and observations, made with time-lapse photography (Cormack 1976), indicate that erythroid colonies from normal subjects have clonal origin. Subsequently, colonies grown from the marrows of patients with polycythemia vera were studied (Prchal et al. 1978). All independent (i.e., spontaneously growing) erythroid colonies derived from the two G-6-PD heterozygotes with polycythemia vera manifested the same single-enzyme phenotype as that observed in the blood cells. Therefore, these colonies had the characteristics of the abnormal clone. However, when erythropoietin was added, increasing numbers of colonies were found that had B-enzyme types and, therefore, did not belong to the polycythemia vera clone. These colonies presumably arose from normal stem cells. These observations suggest that there are normal stem cells in patients with polycythemia vera but that their expression is suppressed in vivo.

Detailed statistical analyses indicate that erythropoietin stimulates not only putative normal stem cells but also erythroid progenitors which have arisen from the polycythemia vera clone (Prchal et al. 1978). These findings provide another example of a neoplastic-like cell which retains its sensitivity to hormonal regulators.

Granulocytic colonies were harvested from the same culture dishes in which the erythroid colonies had been grown. No colony-stimulating activity (CSA) was added, but granulocyte colony growth presumably was facilitated by the CSA produced by cells in the cultures. Although the majority of the colonies had the same A phenotype as was found in the polycythemia blood cell clone, some type-B colonies were detected; these were most notable in the patient who had been treated with chlorambucil (Prchal et al. 1978).

Chronic Myelocytic Leukemia

Granulocyte colonies grown from patients with CML are also being studied. The method used is a modification (Singer et al. 1977) of the one described by Pike and Robinson (1970) in which colonies are grown and harvested from a methylcellulose overlayer on an agar underlayer containing feeder cells derived from buffy coat cells obtained with a cell separator. Thus far, almost 1000 colonies from four G-6-PD heterozygotes with CML have been studied. The progenitor cells were obtained from patients when their disease was active as well as when it was in complete clinical remission. All four patients had type-B enzyme in the CML clone, but only one type-A colony was found. This virtual lack of A colonies was observed despite the fact that colonies were studied after their growth had been stimulated with exogenous CSA, graft-versus-host serum (Singer et al. 1977), and etiocholanalone. Thus, these data provide no evidence for a large population of residual normal granulocyte colony-forming cells in patients with CML, a situation which contrasts to that in polycythemia vera.

In some patients receiving cycle-active therapy and in rare patients treated with conventional regimens or recovering from marrow failure due to unusual sensitivity to busulfan, large populations of Ph^1-negative marrow cells appear (Finney et al. 1972; Galton 1972; Clarkson et al. 1974; Golde et al. 1976; Smalley et al. 1977). Similarly, in one report some Ph^1-negative granulocytic colonies grown in culture were observed (Chervenick et al. 1971). How are these chromosome observations suggesting persistence of normal stem cells in CML reconciled with the failure using G-6-PD as a marker to detect granulocyte colonies arising from non-CML progenitors? One possibility is that the Ph^1-negative cells do in fact arise from the CML clone. As discussed above, this could occur if at least two steps were involved in the development of CML, one causing clonal proliferation of pluripotent stem cells and the other inducing Ph^1 in descendants of this progenitor clone. Although the Ph^1-positive and negative cells arise from the progenitor clone, cells with Ph^1 must have a selective advantage since they predominate. Another possibility is that the Ph^1-negative cells are normal, but usually are not detected because their proliferation is suppressed by the CML clone. This question is being studied by analyses of both G-6-PD and Ph^1 in colonies grown in vitro.

SUMMARY

Application of the G-6-PD marker system to three myeloproliferative disorders: chronic myelocytic leukemia, polycythemia vera, and idiopathic myeloid metaplasia has shown that in each instance the disease involved pluripotent stem cells. Furthermore, at the time of diagnosis, the disorders almost certainly had clonal origin. According to some theories of pathogenesis, polycythemia vera and idiopathic myeloid metaplasia with myelofibrosis result from the proliferation of normal stem cells. The data presented here do not support these hypotheses and are much more compatible with neoplastic origin.

Although each disorder involves a pluripotent stem cell, the clinical manifestations are quite different. It is not known whether these manifestations differ because there are different kinds of genetic alterations governing the response of proliferating cells to regulatory factors or because different levels of stem cells are involved in each disease. These two mechanisms are not mutually exclusive, so both could be operating.

There are data to suggest that marrow fibrosis, the predominant feature in idiopathic myeloid metaplasia and a not-infrequent occurrence in chronic myelocytic leukemia and polycythemia vera, is a secondary event, not an integral part of the abnormal clonal proliferation.

Lymphocytes with B-cell characteristics do arise from the abnormal CML stem cell. Similarly, one population of T cells may arise from this abnormal stem cell, but another population of T cells definitely does not.

Finally, in polycythemia vera the presence of normal "uncommitted" stem cells has been documented but, as yet, few such cells have been detected in chronic myelocytic leukemia.

Acknowledgments

We are grateful to the many colleagues without whose help the studies described here could not have been accomplished. These individuals include Drs. John Adamson, Scott Murphy, Jaroslav Prchal, and Thalia Papayannopoulou, Mesdames Peta Basford, Jean Buiter, Armi Salo, and Laura Steinmann and Mr. Gabe Herner. We thank Dr. Eloise R. Giblett for critical review of the manuscript. Studies done in P.J.F.'s laboratory were supported by Grants CA 16448 and GM 15253 from the National Cancer Institute and the Institute of General Medical Sciences, National Institutes of Health.

REFERENCES

Adamson, J. W., P. J. Fialkow, S. Murphy, J. F. Prchal, and L. Steinmann. 1976. Polycythemia vera: Stem-cell and probable clonal origin of the disease. *New Engl. J. Med.* **295:** 913.

Barr, R. D., and P. J. Fialkow. 1973. Clonal origin of chronic myelocytic leukemia. *New Engl. J. Med.* **289:** 307.

Beutler, E., M. Yeh, and V. F. Fairbanks. 1962. The normal human female as a mosaic of X-chromosome activity: Studies using the gene for G-6-PD–deficiency as a marker. *Proc. Natl. Acad. Sci. U.S.A.* **48:** 9.

Blackstock, A. M., and O. M. Garson. 1974. Direct evidence for involvement of erythroid cells in acute myeloblastic leukaemia. *Lancet* **2:** 1178.

Chervenick, P. A., L. D. Ellis, S. F. Pan, and A. L. Lawson. 1971. Human leukemic cells: In vitro growth of colonies containing the Philadelphia (Ph¹) chromosome. *Science* **174:** 1134.

Clarkson, B. D., M. D. Dowling, T. S. Gee, I. Cunningham, S. Hopfan, W. H. Knapper, T. Vaartaja, and M. Haghbin. 1974. Radical therapy for chronic granulocytic leukemia (CGL). In *Abstracts of XV Congress of the International Society of Hematology,* Jerusalem, p. 136.

Cormack, D. 1976. Time-lapse characterization of erythrocytic colony-forming cells in plasma cultures. *Exp. Hematol.* (Copenh.) **4:** 319.

Davidson, R. G., H. M. Nitowsky, and B. Childs. 1963. Demonstration of two populations of cells in the human female heterozygous for glucose-6-phosphate dehydrogenase variants. *Proc. Natl. Acad. Sci. U.S.A.* **50:** 481.

de la Chapelle, A., P. Vuopio, and G. H. Borgström. 1973. The origin of bone marrow fibroblasts. *Blood* **41:** 783.

De Mars, R., and W. E. Nance. 1964. Electrophoretic variants of glucose-6-phosphate dehydrogenase and the single-active-X in cultivated human cells. In *Retention of functional differentiation in cultured cells* (ed. V. Defendi), p. 35. Wistar Institute Press, Philadelphia.

Fialkow, P. J. 1973. Primordial cell pool size and lineage relationships of five human cell types. *Ann. Hum. Genet.* **37:** 39.

———. 1974. The origin and development of human tumors studied with cell markers. *New Engl. J. Med.* **291:** 26.

———. 1976. Clonal origin of human tumors. *Biochim. Biophys. Acta* **458:** 283.

Fialkow, P. J., S. M. Gartler, and A. Yoshida. 1967. Clonal origin of chronic myelocytic leukemia in man. *Proc. Natl. Acad. Sci. U.S.A.* **58:** 1468.

Fialkow, P. J., R. J. Jacobson, and T. Papayannopoulou. 1977. Chronic myelocytic leukemia: Clonal origin in a stem cell common to the granulocyte, erythrocyte, platelet and monocyte/macrophage. *Am. J. Med.* **63:** 125.

Fialkow, P. J., R. Lisker, J. Detter, E. R. Giblett, and C. Zavala. 1969. 6-phosphogluconate dehydrogenase: Hemizygous manifestation in a patient with leukemia. *Science* **163:** 194.

Fialkow, P. J., R. Lisker, E. R. Giblett, C. Zavala, A. Cobo, and J. Detter. 1972. Genetic markers in chronic myelocytic leukemia: Evidence opposing autosomal inactivation and favoring 6-PGD-Rh linkage. *Ann. Hum. Genet.* **35:** 321.

Finney, R., G. A. McDonald, A. G. Baikie, and A. S. Douglas. 1972. Chronic granulocytic leukaemia with Ph¹ negative cells in bone marrow and a ten year remission after busulphan hypoplasia. *Br. J. Haematol.* **23:** 283.

Fitzgerald, P. H., A. F. Pickering, and J. R. Eiby. 1971. Clonal origin of the Philadelphia chromosome and chronic myeloid leukemia. *Br. J. Haematol.* **21:** 473.

Gahrton, G., J. Lindsten, and L. Zech. 1974. Clonal origin of the Philadelphia chromosome from either the paternal or the maternal chromosome number 22. *Blood* **43:** 837.

Galton, D. A. G. 1972. *Proceedings of the 2nd Padua Seminar on Clinical Oncology,* Piccin Medical Books, Padua, p. 95.

Golde, D. W., N. L. Bersch, and R. S. Sparkes. 1976. Chromosomal mosaicism associated with prolonged remission in chronic myelogenous leukemia. *Cancer* **37:** 1849.

Gurney, C. W. 1965. Polycythemia vera and some possible pathogenetic mechanisms. *Annu. Rev. Med.* **16:** 169.

Hayata, I., S. Kakati, and A. A. Sandberg. 1974. On the monoclonal origin of chronic myelocytic leukemia. *Proc. Jpn. Acad.* **50:** 381.

Hentel, J., and K. Hirschhorn. 1971. The origin of some bone marrow fibroblasts. *Blood* **38:** 81.

Hossfeld, D. K. 1975. Additional chromosomal indication for the unicellular origin of chronic myelocytic leukemia. *Z. Krebsforsch.* **83:** 269.

Jacobson, R. J., A. Salo, and P. J. Fialkow. 1978. Agnogenic myeloid metaplasia: A clonal proliferation of hematopoietic stem cells with secondary myelofibrosis. *Blood.* (In press.)

Jensen, M. K., and S. A. Killmann. 1971. Additional evidence for chromosome abnormalities in the erythroid precursors in acute leukaemia. *Acta Med. Scand.* **189:** 97.

Kahn, A., J.-F. Bernard, D. Cottreau, J. Marie, and P. Boivin. 1975. Gd (−) Abrami: A deficient G-6PD variant with hemizygous expression in blood cells of a woman with primary myelofibrosis. *Humangenetik* **30:** 41.

Maniatis, A. K., S. Amsel, W. J. Mitus, and N. Coleman. 1969. Chromosome pattern of bone marrow fibroblasts in patients with chronic granulocytic leukaemia. *Nature (Lond.)* **222:** 1278.

Moore, M. A. S., H. Ekert, M. G. Fitzgerald, and A. Carmichael. 1974. Evidence for the clonal origin of chronic myeloid leukemia from a sex chromosome moasic: Clinical, cytogenetic, and marrow culture studies. *Blood* **43:** 15.

Pike, B., and W. Robinson. 1970. Human bone marrow growth in agar gel. *J. Cell. Physiol.* **76:** 77.

Prchal, J. F., J. W. Adamson, L. Steinmann, and P. J. Fialkow. 1976. Human erythroid colonies: Evidence for clonal origin. *J. Cell. Physiol.* **89:** 489.

Prchal, J. F., J. W. Adamson, S. Murphy, L. Steinmann, and P. J. Fialkow. 1978. Polycythemia vera: Demonstration of normal and abnormal stem cells and characterization of the *in vitro* response to erythropoietin. *J. Clin. Invest.* (In press.)

Singer, J. W., M. C. James, and E. D. Thomas. 1977. Serum colony stimulating factor. A marker for graft-vs.-host disease in man. In *Experimental hematology today* (eds. S. J. Baum and G. D. Ledney), p. 221. Springer-Verlag, New York. (In press.)

Smalley, R. V., J. Vogel, C. M. Huguley, Jr., and D. Miller. 1977. Chronic granulocytic leukemia: Cytogenetic conversion of the bone marrow with cycle-specific chemotherapy. *Blood* **50:** 107.

Tough, I. M., P. A. Jacobs, W. M. Court Brown, A. G. Baikie, and E. R. D. Williamson. 1963. Cytogenetic studies on bone-marrow in chronic myeloid leukaemia. *Lancet* **1:** 844.

Van Slyck, E. J., L. Weiss, and M. Dully. 1970. Chromosomal evidence for the secondary role of fibroblastic proliferation in acute myelofibrosis. *Blood* **36:** 729.

Ward, H. P., R. Vautrin, and J. Kurnick. 1974. Presence of a myeloproliferative factor in patients with polycythemia vera and agnogenic myeloid metaplasia. I. Expansion of the erythropoietin-responsive stem cell compartment. *Proc. Soc. Exp. Biol. Med.* **147:** 305.

Whang, J. E., E. Frei III, J. H. Tjio, P. P. Carbone, and G. Brecher. 1963. The distribution of the Philadelphia chromosome in patients with chronic myelogenous leukemia. *Blood* **22:** 664.

Zanjani, E. D. 1976. Hematopoietic factors in polycythemia vera. *Semin. Hematol.* **13:** 1.

Section 2

ERYTHROCYTE DIFFERENTIATION
AND REGULATION

Introduction

P. A. Marks

Cancer Center and Departments of Medicine and Human Genetics and Development
Columbia University, New York, New York 10032

Regulation of the rate of erythropoiesis, that is, the rate of red-blood-cell formation, may be achieved by control mechanisms that exert their effects at a number of critical steps. These include (1) proliferation of the pluripotent hematopoietic stem cells, (2) commitment of hematopoietic stem cells to erythropoiesis, (3) proliferation of committed erythroid precursor cells, which include several recognized sequential stages responsive to erythropoietin, and (4) commitment of the erythroid precursor to express the program of biosynthetic and morphogenetic activities characteristic of terminal differentiation of this specialized cell lineage. These regulatory mechanisms have been reviewed extensively elsewhere (McCulloch 1970; Rifkind and Marks 1975; Harrison 1976).

Erythropoiesis constitutes one of the terminal differentiation pathways of a hematopoietic cell renewal system which is responsible for the production of at least three of the formed elements of the blood: granulocytes, megakaryocytes, and erythrocytes. Evidence is presented in this volume to suggest that certain lymphocyte populations may arise from this same hematopoietic precursor cell (Fialkow and Denman, this volume). The multipotency of a common hematopoietic stem cell has been amply demonstrated (McCulloch 1970). The influence of the hematopoietic microenvironment on the development potential of the stem cell is supported by considerable data (Trentin 1970). Both the intrinsic developmental potential of the hematopoietic stem cell and the influence of the hematopoietic microenvironment appear to be under genetic control (Russell 1970). Commitment of the multipotent hematopoietic stem cell to erythroid differentiation involves initiation of a series of developmental stages in the erythroid precursor cell compartment. These different stages of the erythroid precursor, as characterized in vitro, are distinguishable by proliferative capacity and responsiveness to the hormone erythropoietin (Stephenson et al. 1971; Gregory and Eaves 1977). The most immature precursor, termed the burst-forming unit—erythroid (BFU-E), has the greatest proliferative capabilities and the lowest responsiveness to erythropoietin among the erythropoietin-responsive precursors. The most mature precursor element, the colony-forming unit—erythroid (CFU-E), supports only a limited number of cell divisions, perhaps not more than three to five, but is sensitive to low concentrations of erythropoietin. The intermediate developmental stage (the day-3 BFU-E) appears to be intermediate in these properties. None of these precursor cells accumulate detectable globin mRNA or synthesize globin (Ramirez et al. 1975).

From a biochemical point of view, cultures of erythroid cells respond to erythropoietin by increasing the rate of hemoglobin synthesis (Marks and Rifkind 1972). This increase in hemoglobin synthesis is due to an increase in numbers of differentiating erythroid cells, not to augmentation of the rate of hemoglobin synthesis by individual erythroblasts. It appears likely that the principle biological effect of erythropoietin is upon the proliferation of target erythroid precursor cells and that the differentiation of progeny of these precursors is an inherent property of the cell lineage, following a probabalistic (stochastic) model (Till et al. 1964; Korn et al. 1973). Stimulation of RNA synthesis is the earliest change that has been detected in macromolecular synthesis associated with differentiation of erythroid precursor cells (Rifkind and Marks 1975). This early RNA synthesis includes ribosomal RNA and 4S and 5S RNA but not globin mRNA. Synthesis of globin mRNA is detectable only after the erythroid cells have been 6 to 10 hours in culture with erythropoietin (Ramirez et al. 1975). DNA synthesis is not required for expression of the early RNA synthesis but may be required for the later erythropoiesis-specific biosynthetic events (Marks et al. 1977). Taken together, these observations suggest that target-cell proliferation is the principle mechanism whereby erythropoiesis is regulated physiologically by erythropoietin, and that DNA synthesis is required for the expression of the differentiated function of erythroid precursor cells, namely, the transition to globin mRNA production and globin synthesis.

Experiments in which normal fetal or adult erythropoietic precurser cells in culture are utilized to elucidate regulatory mechanisms that initiate erythropoietic differentiation have been limited severely by several problems. First, culturing techniques have not been developed that permit sustained renewal of precursor cells and prolonged erythropoiesis in vitro. Secondly, it has not been possible, to date, to dissociate the effects of erythropoietin on proliferation of its target cell and initiation of erythroid cell differentiation. Furthermore, it appears difficult to synchronize normal erythroid precursors with respect to critical events in differentiation. Lastly, the availability of developmental mutants is severely limited. For these reasons, numerous laboratories have, in recent years, been exploring the potential of the virus-transformed murine erythroleukemia cell (MELC) system as a useful model for investigations into control mechanisms that regulate the expression of erythroid cell differentiation (Friend 1957; Harrison 1976; Marks et al. 1977).

MELC, derived from susceptible mouse spleens infected with the Friend virus complex, have been established in continuous culture (Friend 1957; Friend et al. 1966; Ikawa and Sugano 1966; Patuleia and Friend 1967; Singer et al. 1974; Pragnell et al. 1976). The transforming Friend virus

Table 1
Agents Active as Inducers of Murine Erythroleukemia Cell Differentiation

Strength of inducer	Polar-planar compounds	Antibiotics and antitumor agents	Purine and purine derivatives	Diamines	Fatty acids	Other
Generally strong inducers	dimethylsulfoxide 1-methyl-2-piperidone N,N-dimethylacetamide N-methyl pyrrolidinone N-methyl acetamide N,N-dimethyl formamide N-methyl formamide acetamide triethylene glycol polymethylene bisacetamides (no. = 2–8) hexamethylene bispropionamide tetramethyl urea	bleomycin N-dimethyl-rifampicin actinomycin-D actinomycin-C	hypoxanthine 1-methylhypoxanthine 2,6-diaminopurine 6-mercaptopurine 6-thioguanine 6-amino-2-mercaptopurine 2-acetylamino-6-mercaptopurine		butyrylcholine	ouabain
Generally weak inducers	2-pyrrolidinone propionamide pyridine-N-oxide piperidone pyridazine dimethylurea	vincristine 5-fluorouracil methramycin cycloheximide X-irradiation adriamycin cytosine arabinoside mitomycin-C hydroxyurea UV irradiation		cadaverine	acetate propionate butyrate isobutyrate	hemin methyl isobutyl xanthine

The relative activity as inducers assigned to the agents listed in this table is based on the best available data from our laboratory and as reported in the literature. The inducing activity of a given agent may vary with different MELC lines, culture conditions, and other as-yet-unrecognized factors. An agent is arbitrarily listed as a "generally strong inducer" if it has been reported to induce 50 percent or more of a population of MELC at optimal concentration of the inducer. "Generally weak inducers" include agents that induce 5 to 50 percent of a population of MELC at optimal concentration of the inducer. For references for each of the listed agents, see Marks and Rifkind (1977).

complex includes at least two viruses: a defective spleen focus-forming virus (SFFV) and a murine leukemia helper virus (MuLV) (Lilly and Pincus 1973). The host range of susceptibility to Friend virus complex is governed by genetic factors that control susceptibility both to SFFV and the MuLV. The target cell for virus infection appears to be an erythroid precursor cell, perhaps the erythropoietin-responsive cell (Odaka 1969; McGarry and Mirand 1973; Clarke et al. 1975; Tambourin and Wendling 1975).

MELC strains have been developed that show a low level ($<1\%$) of spontaneous erythroid differentiation in culture (Marks et al. 1977). On culture with dimethylsulfoxide, or a variety of other agents (Table 1), MELC are induced to erythroid differentiation at a much higher level. This program of differentiation has many morphological and biochemical aspects similar to those observed in normal, erythropoietin-regulated erythropoiesis (Marks and Rifkind 1972), including chromatin condensation and other morphological changes (Friend et al. 1971), terminal cell division (Gusella and Housman 1976; Fibach et al. 1977), accumulation of globin mRNA (Ross et al. 1972; Ostertag et al. 1973; Conkie et al. 1974; Ross et al. 1974; Aviv et al. 1976; Bastos and Aviv 1977; Curtis et al. 1977; Nudel et al. 1977a), and globin chain synthesis (Boyer et al. 1972; Ostertag et al. 1972), increase in heme synthesis (Friend et al. 1971; Ebert and Ikawa 1974), synthesis of characteristic erythrocyte enzymes, such as carbonic anhydrase (Kabat et al. 1975), and the appearance of erythrocyte-specific membrane antigen (Furusawa et al. 1971) and other proteins such as spectrin (Arndt-Jovin et al. 1976; Eisen et al. 1977). Although MELC differentiation shows many characteristics of normal erythropoiesis, it must be recognized that MELC are virus-infected cells that exhibit abnormal aspects in their proliferation and differentiation. MELC have the capacity to proliferate without erythropoietin (Marks et al. 1977), rarely proceed to the nonnucleated stage of differentiation (Friend et al. 1971), and may exhibit patterns of erythroid cell gene expression that differ from those seen in normal erythropoiesis (Nudel et al. 1977b).

A number of agents, including polar-planar chemicals such as hexamethylene bisacetamide, purine and purine derivatives, hemin, short-chain fatty acids, inhibitors of DNA or RNA synthesis, UV-irradiation, and X-irradiation, are inducers of MELC differentiation (Table 1). Although the mechanism of action of these agents in inducing differentiation is being investigated, it is still unknown. The variety of inducing agents, the isolation of variant MELC lines that are resistant to induction by some agents but sensitive to others, and evidence indicating a primary site of action at the cell membrane for some chemicals and on chromatin for others suggest that the initial cellular effect may differ for various agents. Presumably, all active agents affect certain common steps required for expression of the erythroid program of differentiation.

REFERENCES

Arndt-Jovin, D. T., W. Ostertag, H. Eisen, F. Klink, and T. M. Jovin. 1976. Studies of cellular differentiation by automated cell separation. Two model systems:

Friend-virus transformed cells and *Hydra attenuata*. *J. Histochem. Cytochem.* **24**: 332.

Aviv, H., Z. Volch, R. Bastos, and S. Levy. 1976. Biosynthesis and stability of globin mRNA and cultured erythroleukemia Friend cells. *Cell* **8**: 495.

Bastos, R. N., and H. Aviv. 1977. Globin RNA precursor molecules: Biosynthesis and processing in erythroid cells. *Cell* **11**: 641.

Boyer, S. H., K. D. Wuu, A. N. Noyes, R. Young, W. Scher, C. Friend, H. D. Preisler, and A. Bank. 1972. Hemoglobin biosynthesis in murine virus-induced leukemic cells in vitro: Structure and amounts of globin chains produced. *Blood* **40**: 823.

Clarke, B. J., A. A. Axelrad, M. M. Shreeve, and D. L. McLeod. 1975. Erythroid colony induction without erythropoietin by Friend leukemia virus *in vitro*. *Proc. Natl. Acad. Sci. U.S.A.* **72**: 3556.

Conkie, D., N. Affara, P. R. Harrison, J. Paul, and K. Jones. 1974. In situ localization of globin messenger RNA formation. II. After treatment of Friend virus-transformed mouse cells with dimethylsulfoxide. *J. Cell Biol.* **63**: 414.

Curtis, P. J., N. Mantei, J. van den Berg, and C. Weissmann. 1977. Presence of a putative 15S precursor to β-globin mRNA but not to α-globin mRNA in Friend cells. *Proc. Natl. Acad. Sci. U.S.A.* **74**: 3184.

Ebert, P. S., and Y. Ikawa. 1974. Induction of δ-aminolevulinic acid synthetase during erythroid differentiation of cultured leukemia cells. *Proc. Soc. Exp. Biol. Med.* **146**: 601.

Eisen, H., S. Nasi, C. P. Georgopoulos, D. Arndt-Jovin, and W. Ostertag. 1977. Surface changes in differentiating Friend erythroleukemic cells in culture. *Cell* **10**: 689.

Fibach, E., R. C. Reuben, R. A. Rifkind, and P. A. Marks. 1977. Effect of hexamethylene bisacetamide on the commitment of differentiation of murine erythroleukemia cells. *Cancer Res.* **37**: 440.

Friend, C. 1957. Cell-free transmission in adult Swiss mice of a disease having the character of a leukemia. *J. Exp. Med.* **105**: 307.

Friend, C., M. C. Patuleia, and E. deHarven. 1966. Erythrocytic maturation *in vitro* of murine (Friend) virus-induced leukemia cells. *Natl. Cancer Inst. Monogr.* **22**: 505.

Friend, C., W. Scher, J. G. Holland, and T. Sato. 1971. Hemoglobin synthesis in murine virus-induced leukemic cells *in vitro:* Stimulation of erythroid differentiation by dimethyl sulfoxide. *Proc. Natl. Acad. Sci. U.S.A.* **68**: 378.

Furusawa, M., Y. Ikawa, and H. Sugano. 1971. Development of erythrocyte membrane-specific antigen(s) in clonal cultured cells of Friend virus-induced tumor. *Proc. Jpn. Acad.* **47**: 220.

Gregory, C. J., and A. C. Eaves. 1977. Human marrow cells capable of erthropoietic differentiation *in vitro:* Definition of three erythroid colony responses. *Blood* **49**: 855.

Gusella, J. F., and D. Housman. 1976. Induction of erythroid differentiation in vitro by purines and purine analogues. *Cell* **8**: 263.

Harrison, P. R. 1976. Analysis of erythropoiesis at the molecular level. Review article. *Nature* **262**: 353.

Ikawa, Y., and H. Sugano. 1966. An ascites tumor derived from early splenic lesion of Friend's disease. *Gann* **57**: 641.

Kabat, D., C. C. Sherton, L. H. Evans, R. Bigley, and R. D. Koler. 1975. Synthesis of erythrocyte-specific proteins in cultured Friend leukemia cells. *Cell* **5**: 331.

Korn, A. P., R. M. Henkelman, F. P. Ottensmeyer, and J. E. Till. 1973. Investigations of a stochastic model of haemopoiesis. *Exp. Hematol.* (Copenh.) **1**: 362.

Lilly, F., and T. Pincus. 1973. Genetic control of murine viral leukemogenesis. *Adv. Cancer Res.* **17**: 231.

Marks, P. A., and R. A. Rifkind. 1972. Protein synthesis: Its control in erythropoiesis. *Science* **175**: 955.

Marks, P. A., and R. A. Rifkind. 1977. Erythroleukemic differentiation. *Annu. Rev. Biochem.* (In press.)

Marks, P. A., R. A. Rifkind, A. Bank, M. Terada, G. Maniatis, R. C. Reuben, and E. Fibach. 1977. Erythroid differentiation and the cell cycle. In *Growth kinetics and biochemical regulation of normal and malignant cells* (eds. B. Drewinko and R. M. Humphrey), p. 329. Williams & Wilkins Co., Baltimore.

McCulloch, E. A. 1970. Control of hematopoiesis at the cellular level. In *Regulation of hematopoiesis* (ed. A. S. Gordon). vol. 1. p. 132. Appleton-Century-Crofts. New York.

McGarry, M. P., and E. A. Mirand. 1973. Incidence of Friend virus-induced polycythemia in splenectomized mice. *Proc. Soc. Exp. Biol. Med.* **142**: 538.

Nudel, U., J. Salmon, E. Fibach, M. Terada, R. Rifkind, P. A. Marks, and A. Bank. 1977a. Accumulation of α- and β-globin messenger RNAs in mouse erythroleukemia cells. *Cell* **12**: 463.

Nudel, U., J. Salmon, M. Terada, A. Bank, R. A. Rifkind, and P. A. Marks. 1977b. Differential effects of chemical inducers on expression of β globin genes in murine erythroleukemia cells. *Proc. Natl. Acad. Sci. U.S.A.* **74**: 1100.

Odaka, T. 1969. Effect of bone marrow graft on the susceptibility of mice to Friend leukemia virus. *Jpn. J. Exp. Med.* **39**: 99.

Ostertag, W., T. Crozier, N. Kluge, H. Melderis, and S. Dube. 1973. Action of 5-bromodeoxyuridine on the induction of haemoglobin synthesis in mouse leukemia cells resistant to 5 BUdR. *Nature New Biol.* **243**: 203.

Ostertag, W., H. Melderis, G. Steinheider, N. Kluge, and S. Dube. 1972. Synthesis of mouse haemoglobin and globin mRNA in leukaemic cell cultures. *Nature New Biol.* **239**: 231.

Patuleia, M. C., and C. Friend. 1967. Tissue culture studies on murine virus-induced leukemia cells: Isolation of single cells in agar-liquid medium. *Cancer Res.* **27**: 726.

Pragnell, I. B., W. Ostertag, P. R. Harrison, R. Williamson, and J. Paul. 1976. Regulation of erythroid differentiation in normal and leukemia cells. In *Progress in differentiation research* (eds. N. Muller-Berat et al.), p. 501. North-Holland Publishing Company, Amsterdam.

Ramirez, F., R. Gambino, G. M. Maniatis, R. A. Rifkind, P. A. Marks, and A. Bank. 1975. Changes in globin messenger RNA content during erythroid cell differentiation. *J. Biol. Chem.* **250**: 6054.

Rifkind, R. A., and P. A. Marks. 1975. The regulation of erythropoiesis. *Blood Cells* **1**: 417.

Ross, J., J. Gielen, S. Packman, Y. Ikawa, and P. Leder. 1974. Globin gene expression in cultured erythroleukemia cells. *J. Mol. Biol.* **87**: 697.

Ross, J., Y. Ikawa, and P. Leder. 1972. Globin messenger-RNA induction during erythroid differentiation of cultured leukemia cells. *Proc. Natl. Acad. Sci. U.S.A.* **69**: 3620.

Russell, E. S. 1970. Abnormalities of erythropoiesis associated with mutant genes in mice. In *Regulation of hematopoiesis* (ed. A. S. Gordon), vol. 1, p. 649. Appleton-Century-Crofts, New York.

Singer, D., M. Cooper, G. M. Maniatis, P. A. Marks, and R. A. Rifkind. 1974. Erythropoietic differentiation in colonies of cells transformed by Friend virus. *Proc. Natl. Acad. Sci. U.S.A.* **71**: 2668.

Stephenson, J. R., A. A. Axelrad, D. L. McLeod, and M. M. Shreeve. 1971. Induction of colonies of hemoglobin-synthesizing cells by erythropoietin *in vitro.* *Proc. Natl. Acad. Sci. U.S.A.* **68:** 1542.

Tambourin, P. E., and F. Wendling. 1975. Target cell for oncogenic action of polycythaemia-inducing Friend virus. *Nature* **256:** 320.

Till, J. E., E. A. McCulloch, and L. Siminovitch. 1964. A stochastic model of stem cell proliferation based on the growth of spleen colony-forming cells. *Proc. Natl. Acad. Sci. U.S.A.* **51:** 29.

Trentin, J. J. 1970. Influence of hematopoietic organ stroma (hematopoietic inductive microenvironments) on stem cell differentiation. In *Regulation of hematopoiesis* (ed. A. S. Gordon), vol. 1. p. 159. Appleton-Century-Crofts, New York.

Regulation of the Population Size of Erythropoietic Progenitor Cells

A. A. Axelrad, D. L. McLeod, S. Suzuki, and M. M. Shreeve

Division of Histology, Department of Anatomy
University of Toronto, Toronto, Ontario M5S 1A8, Canada

Two subcompartments of erythropoietic differentiation in vivo are represented in culture by erythrocytic colonies (Stephenson et al. 1971) and erythropoietic bursts (Axelrad et al. 1974). These develop under different culture conditions and on different time scales. Their characterization is basic to an understanding of the mechanisms that regulate the rates of production of red blood cells and the population sizes of their progenitors.

The present article will provide evidence of clonality of erythrocytic colonies and erythropoietic bursts. It will then examine the magnitude and nature of their requirements for erythropoietin and the proliferative states of their respective progenitors under normal conditions and in response to altered erythropoietic demand. This will lead to a hypothesis regarding the site and mode of transition from one of these subcompartments to the other.

MATERIALS AND METHODS

The assay methods used in the present work have been described elsewhere: for colony-forming units—erythroid (CFU-E), the method of McLeod et al. (1974); for burst-forming units—erythroid (BFU-E), the method of Heath et al. (1976) modified as described by McLeod et al. (1976) and of Strome et al. (1978).

Pokeweed-mitogen-stimulated spleen cell conditioned medium (PWCM) was prepared as follows. Lyophilized pokeweed mitogen (Gibco) diluted in water was added at 15 μg/ml final concentration to spleen cells (3×10^6/ml) plated in medium NCTC 109 with 10% heat-inactivated fetal calf serum in 20-ml Falcon plastic bottles, harvested at 4 days by centrifugation, and stored at $-20°C$. PWCM was used at 10% or 20% final concentration, replacing culture medium.

Clonal Expansion along the Erythrocytic Line of Differentiation

When cells from murine hematopoietic tissues are plated in erythropoietin-containing plasma (McLeod et al. 1974) or methylcellulose cultures (Iscove et al. 1974) and the cultures are scored at 2 days for colonies composed of hemoglobin-containing cells, a proportional relationship can be found between the number of colonies produced and the number of cells plated; at high dilutions, the numbers of colonies per culture conform to a Poisson distribution (McLeod et al. 1974). These findings, which form the basis of a reliable quantitative assay method, are consistent with the concept that each erythrocytic colony in culture is derived from a single unit: the CFU-E, which responds to erythropoietin (Stephenson et al. 1971).

That this unit is a single cell was formally demonstrated by Cormack (1976) in our laboratory. Fetal liver cells from 12–13-day-old C3Hf/Bi mice were used because of the known high concentration of CFU-E ($25/10^3$ nucleated cells) in this tissue. The development of erythrocytic colonies was monitored by time-lapse microcinematography in violet light (Soret band, 416 nm, at which light absorption by heme is at a maximum) together with phase-contrast optics. Projection of the film in reverse, beginning with a frame in which an erythrocytic colony could be clearly recognized because it was composed of cells as dark as mature red cells in the same field, permitted Cormack to follow the development of the colony back to its origin from a single cell. The erythrocytic colony was thereby shown to be a clone. It was also found by this means that the CFU-E is a large (10.3–13.5 μm), motile cell that does not contain hemoglobin in an amount detectable by this light-absorption technique. The first sign of obvious absorption at 416 nm occurred, on the average, after two divisions. The modal generation time measured directly on the cells in the developing erythrocytic colony was 10 hours, and synchrony of division was evident between the two daughter cells of a given division but not among different branches of the clonal family tree.

CFU-E appear to constitute a relatively heterogeneous population of erythrocytic progenitors, as judged by their proliferative behavior (Cormack 1976), the wide range of erythropoietin concentrations over which they respond by erythrocytic colony formation (Iscove et al. 1974; McLeod et al. 1974; Gregory 1976), and their great range of sedimentation velocities (Heath et al. 1976). CFU-E are believed to be late erythrocytic progenitors because they appear to be capable only of five to six cell divisions; because their number within an individual spleen colony does not correlate with the number of colony-forming units-spleen (CFU-S) in the same colony, in contrast to BFU-E and CFU-C whose numbers do correlate highly with CFU-S number in the same colony (Gregory 1976; Gregory and Henkelman 1977), indicating that a relatively large number of randomizing events occurs between the initial pluripotent stem cell and the CFU-E; and because their population size is dependent upon erythropoietin, indicating the existence of an earlier erythropoietin-responsive progenitor (Gregory et al. 1973; Axelrad et al. 1974; Heath et al. 1976).

Erythropoietic bursts (Axelrad et al. 1974) are composed of cells numbering from 50 to several thousand, some or most of which contain hemoglobin.

They are produced in plasma or methylcellulose cultures (Iscove and Sieber 1975) that have been seeded with cells from the hematopoietic tissues if high concentrations of erythropoietin (2-4 U/ml) have been included in the medium and if the cultures have been incubated from 6 to 10 days. Each burst appears either as an isolated collection of more-or-less dispersed colonies or as a single, large, compact colony (Heath et al. 1976). The number of bursts produced has been found to be directly proportional to the number of nucleated cells plated, and at low cell numbers the distribution of bursts per culture is Poissonian. Thus, a single unit is responsible for the production of each erythropoietic burst.

To determine whether each erythropoietic burst is a clone, Strome et al. (1978) plated mixtures of male and female C3Hf/Bi bone-marrow cells in plasma cultures containing a high concentration of erythropoietin and carried out the G-banding procedure in situ in these cultures. They were able to score 2 to 15 metaphases in each burst for the presence either of 40 centromeric heterochromatin bands (female) or 39 centromeric heterochromatin bands and one uniformly staining Y chromosome darker than the others (male). In 26 of 26 bursts, all the metaphases were either of male or female origin but not of both.

These findings strongly indicate that each erythropoietic burst is a clone; they are incompatible with the model of a burst arising from a group of unrelated CFU-E being induced into erythrocytic colony formation at a particular site in the culture. Unit gravity sedimentation data indicating a relatively small and uniform size for the BFU-E are also against the possibility of clumping of unrelated CFU-E before culture. Whether or not cell interactions play a role in the initiation of an erythropoietic burst in vitro or in vivo was not decided by these experiments.

BFU-E are believed to be early erythropoietic progenitors because of their high proliferative capacity (6-13 divisions); because they give rise to bursts composed of colonies similar to those derived from CFU-E; because their number within an individual spleen colony is highly correlated with the number of CFU-S in the same colony, indicating a low number of randomizing events between the two (Gregory 1976; Gregory and Henkelman 1977); and because their population size does not appear to be controlled by erythropoietin (Axelrad et al. 1974; Heath et al. 1976; Hara and Ogawa 1977a; Iscove 1977) as are the late erythropoietic elements.

The High Erythropoietin Requirement for Burst Formation in Vitro

The original description of experiments pointing to the existence of an erythropoietic progenitor distinct from CFU-E noted that erythropoietic burst formation occurred in cultures repeatedly fed with medium containing a high concentration of erythropoietin (Axelrad et al. 1974). Subsequently, Iscove and Sieber (1975) showed that a single large dose of erythropoietin given at the initiation of culture sufficed for the induction of erythropoietic bursts. This facilitated quantitative investigation, and it appeared that erythropoietic burst formation by BFU-E required concentrations of erythropoietin 5 to 10 times higher than those required for erythrocytic colony formation by CFU-E. Gregory (1976) carried out a series of experiments which indicated that the requirement for erythropoietin could be used as a marker for different stages of differentiation along the erythrocytic line: the earlier the stage of differentiation, the greater the concentration of erythropoietin needed for expression of that state. She reported that early burst formation requires about 100 times more erythropoietin than that needed to stimulate CFU-E.

The next development came from another direction. Metcalf et al. (1975) reported the production of megakaryocyte colonies in agar cultures containing spleen-cell conditioned media. Because staining in situ could not be carried out satisfactorily in these cultures, Nakeff and Daniels-McQueen (1976) used plasma culture to show that acetylcholinesterase-positive megakaryocyte colony formation could be stimulated with conditioned media obtained from spleen cells exposed to either phytohemagglutinin or pokeweed mitogen.

In the meantime, McLeod et al. (1976) described the production of megakaryocyte colonies with platelet formation in plasma cultures containing erythropoietin in concentrations sufficient to induce erythropoietic burst formation. Since Metcalf et al. (1975) had seen megakaryocyte colonies in the agar cultures where granulocyte-macrophage colonies were developing in the presence of spleen-cell conditioned medium, it became of interest to investigate the effect of spleen-cell conditioned medium and erythropoietin on the production of erythropoietic bursts and megakaryocyte colonies in plasma and agar cultures. The conditioned medium used (PWCM) was prepared from spleen cells exposed for 4 days to pokeweed mitogen.

The most striking effect of PWCM on erythropoietic burst formation in bone-marrow-cell plasma cultures was seen on the survival of responsiveness to erythropoietin. Burst formation did not occur in the absence of erythropoietin, and the ability of the cells to respond to erythropoietin with burst formation was rapidly lost if erythropoietin was not provided at the time of initiation of the cultures. However, it was found that the administration of erythropoietin could be delayed for as long as 3 days and burst formation would still occur with undiminished efficiency if the cultures had been initiated in the presence of PWCM. Under these conditions, the time required for bursts to develop after exposure to erythropoietin could be reduced from 7 to 4 days. Bursts also formed, although in reduced numbers, in cultures initiated with PWCM even when erythropoietin administration had been delayed for 6 days, with only 3 days remaining for the development of the bursts. Lysis in the 9-day cultures prevented any further observation. The time required for burst formation thus seems to be divisible into an early period during which PWCM is capable of rescuing the cells responsive to erythropoietin or their progenitors, which are otherwise lost, and into a late period in which exposure to erythropoietin is mandatory for burst development.

The concentration of erythropoietin required for burst formation was also affected by early exposure of the cells to PWCM. When erythropoietin alone was given at the initiation of culture, the maximum number of bursts was reached at erythropoietin concentrations of around 3 U/ml. When both PWCM and erythropoietin were administered at the beginning of culture, the maximum number of bursts was reached at an erythropoietin concentration of 1 U/ml. Some erythropoietic burst formation occurred at concentrations of erythropoietin as low as those required for erythrocytic colony formation (0.25 U/ml), especially when PWCM was given at the start of culture and erythropoietin administration was delayed for 3 days. The maximum number of bursts at high erythropoietin concentrations was not increased. Thus, PWCM given at the initiation of culture appeared to reduce the amount of erythropoietin required for the full development of erythropoietic bursts. Pokeweed mitogen itself added to the bone-marrow cell cultures had no effect on erythropoietic burst formation.

These results suggest that BFU-E are able to survive and proliferate in the presence of a factor in pokeweed mitogen-stimulated spleen-cell conditioned medium. Differentiation of their progeny to CFU-E and the response of these CFU-E to erythropoietin would result in the production of colonies of hemoglobinized cells, thus rendering the whole clonal association of colonies recognizable as an erythropoietic burst.

One prediction from this model, that the administration of PWCM and erythropoietin should produce larger erythropoietic bursts than erythropoietin alone, was in fact realized; the effect was most marked when erythropoietin administration was delayed. The numbers both of hemoglobinized and nonhemoglobinized cells in the bursts appeared to be increased by PWCM.

This model postulates two distinctly different effects of erythropoietin preparations: (1) maintenance of survival of BFU-E in a proliferating and differentiating state, which requires a high concentration of the erythropoietin preparation and appears to be replaceable by PWCM; and (2) differentiation of CFU-E into hemoglobin-synthesizing cells, which requires a low concentration of erythropoietin and is not replaceable by PWCM. Whether both are functions of the same molecule or whether the first is due to a contaminant of erythropoietin remains to be investigated.

The Proliferative States of BFU-E

Several authors have examined the behavior of BFU-E under conditions of altered erythropoietic demand and have concluded that erythropoietic stimulation does not induce proliferation of BFU-E.

Hara and Ogawa (1977a) investigated the serial changes in number of erythropoietic progenitor cells that occur in mice after bleeding, erythropoietin injection, or hypertransfusion. They also examined the proliferative states of BFU-E and CFU-E with [^{3}H]TdR suicide in vitro. This was done in normal, phenylhydrazine-treated, and hypertransfused mice. Approximately 36% of BFU-E and 74% of CFU-E in the marrow and spleen of normal mice were found to be in the DNA-synthesis phase of the cell cycle, and neither anemia nor polycythemia induced significant changes in those percentages. Hara and Ogawa concluded that erythropoietic stimulation induces migration of BFU-E from marrow to spleen but has no effect on proliferation of BFU-E. This was unexpected in light of the behavior of BFU-E in vitro, where exposure to erythropoietin preparations results in the production of bursts containing up to 10^4 cells.

Iscove (1977) also examined the state of cycle of BFU-E and CFU-E in murine marrow after bleeding or hypertransfusion by [³H]TdR suicide in vitro. Thirty percent of the BFU-E from normal animals were killed by [³H]TdR, but this number was neither increased by bleeding nor decreased by hypertransfusion. He concluded that BFU-E do not respond to physiological levels of erythropoietin and that regulation by erythropoietin commences in cells at a stage of maturation intermediate between BFU-E and CFU-E.

We have recently reexamined the effects of bleeding on the population size and state of cycle of erythropoietic progenitor cells in the mouse. The most striking effect was seen in the spleen, where a 15- to 100-fold increase in the number of CFU-E occurred within 2 days. It was remarkable that, in contrast, only minor effects of bleeding were noted on the total number of BFU-E in spleen or marrow or on the number of marrow CFU-E. The origin of this almost explosive increase in the number of splenic CFU-E was obscure.

Because BFU-E were considered to give rise to CFU-E, it was reasonable to expect to find kinetic curves of the precursor-product type for these respective erythropoietic progenitors. However, the kinetics actually found by us, as by Hara and Ogawa (1977a), were at variance with this model, as were the published data on [³H]TdR suicide (Hara and Ogawa 1977a; Iscove 1977).

In our first [³H]TdR-suicide experiments, suspensions of bone-marrow cells were exposed to high-specific-activity [³H]TdR in vitro for 20 minutes, with appropriate controls, and the suspensions were then assayed for surviving BFU-E and CFU-E. The results showed that a high proportion of CFU-E were killed by this procedure, but the proportion of BFU-E killed by [³H]TdR varied greatly, depending on such conditions as pH in the incubation medium. Feeling that reliable conclusions could not be drawn from our [³H]TdR data obtained in vitro, we turned to experiments in vivo to try to determine the states of cycle of BFU-E and CFU-E in the hematopoietic tissues under normal conditions and under conditions of increased erythropoietic demand. Mice of the C57BL/6 strain were either bled or left as controls. At various times thereafter, the mice were injected intravenously with [³H]TdR or nonradioactive TdR as control and killed 1 hour later. Cell suspensions of bone marrow and spleen were assayed for their content of surviving BFU-E and CFU-E.

The results showed that 50–70% of the CFU-E of normal bone marrow were killed by exposure to high-specific-activity [³H]TdR in vivo but practically none of the BFU-E were killed under the same conditions. These data indicate that a high proportion of CFU-E in bone marrow are normally in cycle, whereas the vast majority of marrow BFU-E are in a noncycling (G_0) state or in a state of prolonged cell cycle indistinguishable from it. The fact that under the same conditions in which no kill of BFU-E was observed, a high proportion of CFU-E in the same tissues regularly were killed by [³H]TdR makes it unlikely that [³H]TdR could not reach the sites where BFU-E were situated. No significant BFU-E kill occurred at [³H]TdR doses of 1, 2, or 4 mCi per mouse.

Blood loss induced a remarkable increase in the number of BFU-E that were killed by exposure to [³H]TdR in vivo.

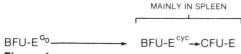

Figure 1
Scheme of transitions of early and late erythropoietic progenitor cell populations.

In the spleen, we found a striking parallelism between the total number of CFU-E and the number of BFU-E killed by [^{3}H]TdR at various times after bleeding. Both showed a transient peak at 4 hours and a second, greater rise reaching a peak at around 48 hours and then falling off gradually over the subsequent days. The CFU-E curve was about 100-fold higher than the curve of BFU-E killed by [^{3}H]TdR, and there was no obvious time lag between the curves. No such parallelism was observed between the curves of CFU-E number and the number of BFU-E not killed by [^{3}H]TdR.

The similarity between the splenic time curves of total number of CFU-E and of number of BFU-E killed by [^{3}H]TdR (BFU-E in DNA synthesis) after bleeding indicates a very close relationship between these two cell populations. Although several explanations for this phenomenon are possible, the one we now favor is that CFU-E are derived by differentiation directly from cycling BFU-E (BFU-E^{cyc}) without significant amplification (Fig. 1); therefore, the transition could occur very rapidly in large numbers of cells. The major site of this transition would be the spleen, although it could also occur in marrow. The transition would take place presumably under the influence of erythropoietin.

CFU-E do not circulate, but BFU-E do (Hara and Ogawa 1976, 1977b), and they do so in larger numbers after bleeding. Preliminary studies of ficoll-hypague separated cells from peripheral blood of C57BL/6 mice given [^{3}H]TdR indicate that a high proportion of circulating BFU-E are in cycle under these conditions. The major increase in number of splenic BFU-E killed by [^{3}H]TdR after bleeding could therefore have been derived mainly from influx of BFU-E^{cyc} from the circulation. In this model the spleen would serve as a trap for BFU-E^{cyc} in the circulation, and on entering the spleen, the vast majority of these BFU-E^{cyc} would differentiate very rapidly into CFU-E. The model accounts for the rise in number of splenic CFU-E, which occurred too rapidly to be explained entirely by local proliferation of either CFU-E or BFU-E in the spleen.

DISCUSSION

The observations described here confirm the importance of the spleen as an erythropoietic organ in the mouse, especially active under conditions of sudden demand (Boggs et al. 1969; Bozzini et al. 1974), and they provide a cellular basis for the rapid and intense response.

The population size of BFU-E does not appear to be under the direct control of erythropoietin, since bleeding does not have a pronounced or early effect on the total number of BFU-E in bone marrow or spleen (Hara and

Ogawa 1977a; Iscove 1977; present observations) nor does hypertransfusion have a major or immediate influence (Axelrad et al. 1974; Heath et al. 1976; Hara and Ogawa 1977a; Iscove 1977).

The population size of CFU-E in the spleen, where the major response to increased erythropoietic demand occurs, seems to be largely determined by the influx and transition of cycling BFU-E to CFU-E, which is presumably balanced by the conversion of CFU-E to the hemoglobin-synthesizing cells of erythrocytic colonies. The latter process is known to be exquisitely sensitive to erythropoietin. Since hypertransfusion is known to have a profound effect in reducing the total number of CFU-E in the spleen (Gregory et al. 1973), it is possible that the transition from BFU-E^{cyc} to CFU-E is also sensitive to erythropoietin stimulation.

However, the ability, at short notice, to alter the relative proportions of BFU-E in G_0 and in cycle and the capacity to maintain their progeny in a viable, cycling state seem also to constitute important mechanisms for regulating the population sizes of both early and late erythropoietic progenitors. The role of erythropoietin in these mechanisms remains to be clarified.

Acknowledgments

This work was supported by Grant MA 3969 from the Medical Research Council of Canada and by a grant from the National Cancer Institute of Canada.

REFERENCES

Axelrad, A. A., D. L. McLeod, M. M. Shreeve, and D. S. Heath. 1974. Properties of cells that produce erythrocytic colonies in plasma culture. In *Proceedings of the second international workshop on hemopoiesis in culture,* Airlie, Virginia (ed. W. A. Robinson), p. 226. Grune & Stratton, New York.

Boggs, D. R., A. Geist, and P. A. Chervenick. 1969. Contributions of the mouse spleen to post-hemorrhagic erythropoiesis. *Life Sci.* **8:** 587.

Bozzini, C. E., M. A. Martinez, C. A. Alvarez Ugarte, V. Montangero, and G. Soriano. 1974. The importance of the spleen on proliferation of erythropoietin-responsive cells induced by erythropoietin. *Exp. Hematol.* **2:** 93.

Cormack, D. 1976. Time-lapse characterization of erythrocytic colony-forming cells in plasma cultures. *Exp. Hematol.* **4:** 319.

Gregory, C. J. 1976. Erythropoietin sensitivity as a differentiation marker in the hemopoietic system: Studies of the three erythropoietic colony responses in culture. *J. Cell. Physiol.* **89:** 289.

Gregory, C. J., and R. M. Henkelman. 1977. Relationships between early hemopoietic progenitor cells determined by correlation analysis of their numbers in individual spleen colonies. In *Experimental hematology today* (eds. S. J. Baum, and G. D. Ledney), p. 93. Springer-Verlag, New York. (In press.)

Gregory, C. J., E. A. McCulloch, and J. E. Till. 1973. Erythropoietic progenitors capable of colony formation in culture: State of differentiation. *J. Cell. Physiol.* **81:** 411.

Hara, H., and M. Ogawa. 1976. Erythropoietic precursors in mice with phenylhydrazine-induced anemia. *Am. J. Hematol.* **1:** 453.

———. 1977a. Erythropoietic precursors in mice under erythropoietic stimulation and suppression. *Exp. Hematol.* **5:** 141.

———. 1977b. Erythropoietic precursors in murine blood. *Exp. Hematol.* **5:** 161.

Heath, D. S., A. A. Axelrad, D. L. McLeod, and M. M. Shreeve. 1976. Separation of the erythropoietin-responsive progenitors BFU-E and CFU-E in mouse bone marrow by unit gravity sedimentation. *Blood* **47:** 777.

Iscove, N. N. 1977. The role of erythropoietin in regulation of population size and cell cycling of early and late erythroid precursors in mouse bone marrow. *Cell Tissue Kinet.* **10:** 323.

Iscove, N. N., and F. Sieber. 1975. Erythroid progenitors in mouse bone marrow detected by macroscopic colony formation in culture. *Exp. Hematol.* **3:** 32.

Iscove, N. N., F. Sieber, and K. H. Winterhalter. 1974. Erythroid colony formation in cultures of mouse and human bone marrow: Analysis of the requirement for erythropoietin by gel filtration and affinity chromatography on agarose-concanavalin A. *J. Cell. Physiol.* **83:** 309.

McLeod, D. L., M. Shreeve, and A. A. Axelrad. 1974. Improved plasma culture system for production of erythrocytic colonies in vitro: Quantitative assay method for CFU-E. *Blood* **44:** 517.

———. 1976. Induction of megakaryocyte colonies with platelet formation in vitro. *Nature* **261:** 492.

Metcalf, D., H. R. MacDonald, N. Odartchenko, and B. Sordat. 1975. Growth of mouse megakaryocyte colonies in vitro. *Proc. Natl. Acad. Sci. U.S.A.* **72:** 1744.

Nakeff, A., and S. Daniels-McQueen. 1976. In vitro colony assay for a new class of megakaryocyte precursor: Colony-forming unit megakaryocyte (CFU-M). *Proc. Soc. Exp. Biol. Med.* **151:** 587.

Stephenson, J. R., A. A. Axelrad, D. L. McLeod, and M. M. Shreeve. 1971. Induction of colonies of hemoglobin-synthesizing cells by erythropoietin in vitro. *Proc. Natl. Acad. Sci. U.S.A.* **68:** 1542.

Strome, J. E., D. L. McLeod, and M. M. Shreeve. 1978. Evidence for the clonal nature of erythropoietic bursts: Application of an in situ method for demonstrating centromeric heterochromatin in plasma cultures. *Exp. Hematol.* (In press.)

Competitive Effects of Erythropoietin and Colony-Stimulating Factor

G. Van Zant and E. Goldwasser

Department of Biochemistry and The Franklin McLean Memorial Research Institute
University of Chicago, Chicago, Illinois 60637

One widely accepted model for hematopoietic stem-cell differentiation (Lajtha and Schofield 1974) postulates the existence of a pluripotent stem cell (PSC) that can develop, by some as-yet-unknown mechanism, into one of three unipotent, committed stem cells. These unipotent cells are considered to be target cells for the inducers of specific lines of blood-cell differentiation; for example, a granulocyte-committed cell is acted upon by granulopoietin (gpo) to initiate that line of differentiation. Similarly, a thrombocyte-committed cell is induced by thrombopoietin (tpo), and an erythrocyte-committed cell is induced by erythropoietin (epo). Evidence for the action of the specific inducers as primary stimuli for initiation of the differentiative pathways is scanty, except in the case of erythropoietin.

For some time, we have been studying the properties of rodent bone-marrow cells in primary culture, especially the mode of action of epo on epo-responsive cells (ERC), with the ultimate aim of finding an in vitro system with which to investigate the molecular mechanisms by which stem-cell proliferation and differentiation are regulated. This system has yielded a fair amount of information regarding erythroid differentiation, which has been corroborated and extended in other systems; however, there is still a paucity of information regarding the molecular nature of stem-cell regulation.

In the course of studying cell-cycle parameters of ERC in culture, we found that, after all of the cycling ERC had been killed by thymidine suicide and the label removed, the remaining cells gave rise to a new cohort of ERC in about 2.5 days (Gross and Goldwasser 1972). Since it was known that hematopoietic stem cells (defined as cells that yield spleen colonies in irradiated mice; colony-forming units – spleen [CFU-S]), for the most part were not in cycle (Lajtha et al. 1969) and, hence, would not be affected by tritiated thymidine in vitro, we concluded that the cultures had contained CFU-S throughout the culture period. These CFU-S had then generated new ERC, replacing those killed by the thymidine. It was reported, however, that CFU-S quickly disappeared from primary marrow cultures (Krantz and Fried 1968). When we repeated these experiments and extended the time of incubation, we found that, after an eclipse phase of about 3 days, there was an increase in CFU-S (to 44-89% of the original number, depending on the species) at 7 days, followed by a secondary decline to a low but significant number of CFU-S that persisted for 4 weeks (Meints and Goldwasser 1973).

On the basis of these observations, it seemed feasible to use the primary marrow-cell culture system for study both of erythroid differentiation and of the properties of stem cells. Use of this system also afforded us the opportunity to put to experimental test the hypothesis concerning unipotent stem cells. The unipotent-stem-cell concept predicts that an inducer of one pathway of differentiation should have no short-term effect either on another pathway or on PSC. Although our findings are not in agreement with that concept, they do agree with an alternative model of stem-cell differentiation (Goldwasser 1975).

METHODS

Assay of CFU-S

Recipient mice (BDF_1 females, 12–16 weeks old) were irradiated at 64 R/minute, as measured with a Victoreen meter, to a total of 810 R with X-rays from a 250 kV Maxitron, with filtration of 0.25 mm Cu plus 1 mm Al and a half-value layer of 1.05 mm Cu, at a distance of 77 cm. The mice, in plastic containers, were rotated under the X-ray source during irradiation. Recipients were injected 1 to 5 hours after irradiation as follows. Groups of 10 to 15 mice received 10^5 nucleated cells in 0.5 ml of 70% NCTC 109, 30% fetal calf serum. Irradiated mice were housed four to five per cage and were given food and Terramycin-supplemented drinking water *ad libium*. Colonies were counted by someone who did not know the key to the experimental design.

Plethoric Animals

When plethoric mice or rats were used either as hosts for CFU-S assay or as sources of marrow for experiments in vitro, they were subjected to 0.5 atm from 4 to 6 weeks. During that time, they were removed from the hypobaric chamber for about 3 hours per day, 5 days per week, to permit cage cleaning and replenishing of food and water. When the mice were used as irradiated hosts, they were removed from the chamber 1 day before irradiation and injection of cells. When they were used as sources of marrow cells, they were kept at normal pressure for 1 week to permit maturation of all of the differentiated erythroid cells.

Cultures

Suspension cultures for evaluation of short-term hemoglobin synthesis were done by the method already published from this laboratory (Goldwasser et al. 1975). Colony-forming units—erythroid (CFU-E) and burst-forming units—erythroid (BFU-E) were measured by the method of Iscove and Sieber (1975). Mouse bone-marrow cells (2×10^5 nucleated cells/ml) were cultured in 35×10 mm dishes (Lux Scientific Corp., model 5221-R) in 1 ml of alpha medium (Flow Laboratories) containing 30% fetal calf serum (Flow Laboratories), 0.8% methylcellulose (Dow Chemical Co., 4000 cps), 10^{-4} M β-mercaptoethanol, and 1% bovine serum albumin. Cultures were incubated at 37°C and 100% relative humidity in 5% CO_2, 95% air either for 2 days (CFU-E) or 8 days (BFU-E). Colonies and bursts were scored at 25 to $40 \times$ with a dissecting microscope after staining with benzidine according to the method of Ogawa et al. (1976).

Quantitative measurement of hemoglobin synthesis related to CFU-E and BFU-E involved the following changes in method (J. F. Eliason and E. Goldwasser, unpubl.): (1) Marrow cells were cultured at 7.5×10^5 cells/ml; (2) the culture volume was 0.3 ml in the 16-mm wells of culture plates (Costar model 3524); (3) the gas used was 3% CO_2, 97% air; (4) $^{59}Fe^{+++}$ bound to rat serum (0.4 μCi per well) was added to the cultures in a volume of 20 μl on day 1 for measurement of 2-day hemoglobin synthesis or on day 7 for measurement of 8-day hemoglobin synthesis. The labeling of serum and extraction of hematin were carried out as previously described (Goldwasser et al. 1975), except that alpha medium was substituted for NCTC 109.

The epo used was a preparation derived from human urine with a potency of about 70,000 U/mg and was homogeneous by electrophoretic analysis (Miyake et al. 1977). The following colony-stimulating factor (CSF) samples were used: a preparation from a medium conditioned with human embryonic kidney cells, 5300 U/mg of protein; a human urinary fraction, 70,000 U/mg of protein; and a preparation from L-cell conditioned medium, 8×10^7 U/mg of protein.

RESULTS

Using the combined in vitro–in vivo method of determining CFU-S in primary cultures, we studied the effect of added epo on stem cells in the cultures (Van Zant and Goldwasser 1977). To distinguish between erythroid and nonerythroid spleen colonies, we performed each CFU-S assay in two sets of host mice—normal and plethoric. In the former, we detected total CFU-S and in the latter, nonerythroid colonies (non-E-CFU-S); the difference is the number of erythroid colony formers (E-CFU-S). As shown in Figure 1, addition of epo caused an increase in total CFU-S at the 7-day peak, the increase being due to E-CFU-S. At the same time, epo caused a decrease in non-E-CFU-S without having any effect on the total cell number in the culture (data not shown). These findings indicate not only that epo caused an increase in erythroid colonies, but also that its action caused a decrease in the number of cells available for nonerythroid differentiation.

The possibility of competitive demands on hematopoietic stem cells by perturbations of whole animals had already been suggested (Harris et al.

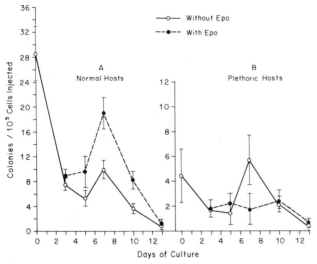

Figure 1
The effect of epo on spleen colony-forming cells. (o———o) Without epo.
(●———●) With 0.1 U of epo per ml.

1966; Bradley et al. 1967; Hellman and Grate 1967, 1968; Kubanek et al. 1973; Steinberg et al. 1976; Langdon and McDonald 1977); we used the in vitro system to study that possibility by a more direct method. The availability of a CSF that was a good candidate for being a granulopoietin permitted us to look for evidence of competition between epo and CSF. The data in Table 1 show that CSF added to control marrow cells had no effect on hemoglobin synthesis; whereas, when it was added to epo-stimulated cells, there was a significant suppression (55%) of epo-induced hemoglobin synthesis. Inactivation of CSF by heating at 100°C for 30 minutes (Stanley and Metcalf 1969) abolished its suppressive action. In a reciprocal experiment, epo added to cultures in a semisolid medium caused a 48% suppression of CSF-induced granulocyte-macrophage (GM) colonies (Table 2).

When marrow cells from normal animals are used, as in the two experiments just described, a problem arises due to the fact that epo acts on at least two classes of cells: those that are recognizably erythroid and those that have no morphological markers characteristic of erythroid cells. The

Table 1
Effect of CSF on Hemoglobin Synthesis by Normal Rat Bone Marrow in Vitro

Addition	*Hemoglobin* cpm ± SD
None	22 ± 6
epo (0.048 U)	164 ± 8
CSF (1800 U)	24 ± 6
Both	88 ± 6

Table 2
Effect of epo on CSF-Mediated GM Colony Formation by Normal Rat Bone Marrow

Addition	GM colonies ± SD
None	0
epo (0.1 U)	0
CSF (200 U)	85 ± 17
Both	44 ± 13

latter are the cells that we assumed to be the primitive ERC. Although marrow from animals with a plethora of red cells is essentially devoid of later erythroid cells, it does contain primitive ERC. The suppressive effect of CSF on epo-induced hemoglobin synthesis by marrow from plethoric rats is considerably greater than its effect on normal cells. When 0.048 U of epo was used, the suppression of its effect on normal cells by 1800 U/ml of CSF was 55% (Table 1). With cells from plethoric rats 400 U of CSF completely suppressed the effect of the same amount of epo. This finding suggests that the locus of competition between the two inducers is at the more primitive cell level.

The suppressive effect of CSF was detected with several preparations from different sources (human urine, L-cell conditioned medium, human embryonic kidney-cell conditioned medium, and serum from endotoxin-treated mice) and with several degrees of potency (ranging from 5300 U/mg to 8×10^7 U/mg). Pure CSF has been reported to have a potency of 1.6×10^8 (Stanley and Heard 1977) or 7×10^7 U/mg (Burgess et al. 1977), suggesting that it is the CSF activity that causes the suppression. Similarly, the effect of epo on GM colony formation induced by CSF is manifest when pure human epo is used. The magnitudes of both suppressive effects depend on the concentration of the competitive inducer used.

Pretreatment of normal rat bone-marrow cells with CSF caused an increase in its suppressive effect (Fig. 2). The suppression of epo-induced hemoglobin synthesis increased as the time between addition of CSF and of epo was lengthened. In a similar experiment with marrow cells from plethoric rats (Fig. 2), we found the expected increased sensitivity to suppression. When both inducers were added simultaneously, there was 90% suppression; when 910 U/ml of CSF were added 1 or more hours before 0.05 U/ml of epo, the suppressive effect of CSF was complete. The sensitivity of marrow cells from plethoric mice was even more striking; 210 U/ml of CSF added at the same time as 0.05 U/ml of epo caused complete suppression of the epo effect (data not shown).

When the order of addition was reversed, we found that CSF suppression of epo-induced hemoglobin synthesis by cells from plethoric rats was not diminished at the shorter periods (Fig. 3). A decreased suppressive effect ($P < 0.01$) was found only when epo had had about 12 hours to act before CSF was added. When CSF was added 23 hours after epo, there was, surprisingly, a significant ($P = 0.02$) augmentation of the epo effect.

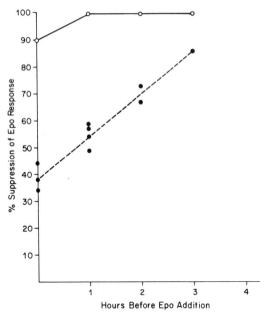

Figure 2

The effect of preincubation of marrow cells with CSF on epo-induced hemoglobin synthesis. (●———●) Marrow from normal rats. (o———o) Marrow from plethoric rats. Epo was added at 0.050 U per well (0.2 ml); CSF, at 910 U per well.

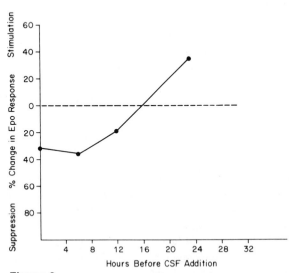

Figure 3

The effect of preincubation of marrow cells with epo on CSF suppression. Marrow cells were obtained from plethoric rats; CSF was added at 450 U per well; epo, at 0.024 U per well.

In an attempt to determine whether a brief exposure of marrow cells to CSF would be sufficient to cause suppression of the epo effect, we incubated cells from plethoric rats for 1 hour with 3600 U/ml of CSF; we washed and resuspended them in fresh medium to which 0.032 U/ml of epo had been added. Control cells were handled in the same manner without added CSF. The results (Table 3) show that the 1-hour exposure to CSF was enough to cause 68% suppression of the response to epo.

In addition to estimating the effect of CSF on epo-induced hemoglobin synthesis, we studied its effect on epo-dependent erythroid colonies grown in semisolid media. In preliminary experiments, using the method of Iscove and Sieber (1975), we found that the usual time for scoring of erythroid bursts was, in our hands, too long. We found that, by 9 to 10 days, most of the bursts no longer contained benzidine-positive cells, although the colony morphology did permit us to distinguish between bursts and GM colonies. On the basis of these observations, we chose the 8th day as the optimal time for counting of erythroid bursts and GM colonies, so that benzidine staining could be used for the identification of bursts. In the counting of colonies and distinguishing between benzidine-positive bursts and GM colonies, it is important to note that CSF causes the appearance of a small number of benzidine-positive GM colonies; some nonerythroid cells, high in peroxidase, stain with benzidine.

We examined the effect of an increasing epo dose, at a constant level of CSF (170 U/ml), on the number of erythroid bursts and GM colonies at 8 days. The results (Fig. 4), showing bursts expressed as a percentage of the total erythroid bursts plus GM colonies, indicate that increasing epo in the presence of CSF caused a small but definite increase in the precentage of erythroid bursts and, therefore, a decrease in the percentage of GM colonies. At 3.0 U/ml of epo, the deficit in the number of GM colonies was 39 and the increase in the number of erythroid bursts was 37. The effect of epo was essentially to redirect the cells toward erythroid bursts at the expense of GM colonies, while the total number of colonies was not changed significantly (CSF alone, 204; CSF + epo, 198). None of the cultures examined showed additive effects of the factors; the total number of colonies did not increase as the epo concentration increased.

In a comparable experiment, we scored erythroid bursts and GM colonies,

Table 3

Effect of 1-Hour Exposure of Cells to CSF on epo-Induced Hemoglobin Synthesis

	Hemoglobin cpm ± SD		Δ *cpm*	*Percent of epo response*
	$-epo$	$+epo$		
Control cells	154 ± 56	448 ± 80	294	100
Cells incubated with CSF for 1 hour	204 ± 26	298 ± 80	94	32

Marrow cells from plethoric rats were incubated at a density of 20×10^6 per ml. CSF was used at a concentration of 3600 U/ml for 1 hour at 37°C. After washing, cells were cultured at 15×10^6 per ml with and without 0.032 U of epo per well (0.2 ml).

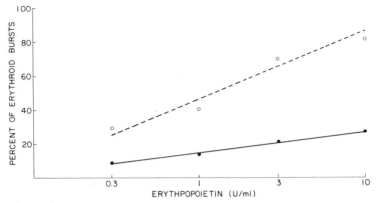

Figure 4
Simultaneous effects of epo and CSF on colonies and bursts at 8 days. (●——●) Epo + CSF. (o——o) Epo (170 U); marrow cells were from normal mice. Data are expressed as percentage of erythroid bursts.

in the same dishes, as CSF was increased from zero to 1680 U/ml and epo was kept constant at 10 U/ml, a level that causes maximal burst formation. The results (Table 4) show that (1) at the two lower amounts of CSF, GM colony formation was significantly reduced in the presence of 10 U/ml of epo; (2) at the highest level of CSF, the number of erythroid bursts was significantly reduced; (3) the lower levels of CSF caused a significant increase in the number of epo-dependent bursts but had no effect in the absence of epo; and (4) the larger amounts of CSF caused a significant decrease in the number of benzidine-positive cells within the bursts, so that, at the highest level, the greater fraction of the bursts contained less than 50% of cells that were benzidine-positive.

Because of the ambiguity introduced by the nonerythroid benzidine-positive colonies and the need for quantitative determination of hemoglobin synthesis in these cultures, we adapted the method of Iscove and Sieber (1975) to the measurement of ^{59}Fe incorporation into hemoglobin in 0.3-ml cultures. By increasing the cell density to 7.5×10^5 per ml, incubating the cultures with ^{59}Fe-labeled serum, and extracting hematin after the cultures were stopped, we were able to show that epo-induced incorporation of label into hematin at 2 days was a measure of CFU-E and that labeling at 8 days was a measure of BFU-E. Cells in this type of culture synthesize no detectable hemoglobin by the 3rd day of incubation in the absence of epo. By use of this system, we were able to determine, with the same pool of marrow cells, the effect of CSF on epo-induced 2-day hemoglobin synthesis, on epo-induced 8-day hemoglobin synthesis, and on GM colonies (Fig. 5). The results show that hemoglobin synthesis at 2 days (due to descendants of CFU-E) was not affected by CSF until its concentration was considerably above the linear region of the log dose–log response curve. Even at these high concentrations, the suppression was not greater than 40%. On the other hand, hemoglobin synthesis at 8 days, by descendants of BFU-E, was markedly suppressed at lower concentrations of CSF and, at high concentrations, was reduced to zero.

Table 4
Competition between epo and CSF in the Same Culture of Mouse Marrow

Addition	Erythroid bursts				GM colonies		Total bursts + colonies
	all B+[a]	>50% B+	<50% B+	total	all B−[b]	25–35% B+	
None	0	0	0	0	4 ± 1	0	4
epo (10 U)	13 ± 3	15 ± 4	3 ± 3	31 ± 5	4 ± 2	0	35
CSF (110 U)	0	0	0	0	80 ± 16	0	80
epo + CSF (110 U)	23 ± 5	18 ± 1	2 ± 0.5	43 ± 5	54 ± 2[c]	0	97
CSF (340 U)	0	0	0	0	186 ± 22	0	186
epo + CSF (340 U)	16 ± 5	26 ± 2	3 ± 2	45 ± 4	132 ± 23[c]	0	177
CSF (560 U)	0	0	0	0	271 ± 26	29 ± 6	300
epo + CSF (560 U)	6 ± 5[d]	18 ± 5	14 ± 6[d]	38 ± 6	311 ± 52	13 ± 2	362
CSF (1680 U)	0	0	0	0	300 ± 45	25 ± 16	325
epo + CSF (1680 U)	0	2 ± 2[d]	20 ± 7[d]	22 ± 5[d]	364 ± 19	17 ± 3	393

Colonies were counted at 8 days. Results are the means of five replicates ± SD. All concentrations are U/ml.
[a] B+, benzidine-positive.
[b] B−, benzidine-negative.
[c] Significantly different ($P < 0.05$) from CSF alone.
[d] Significantly different ($P < 0.05$) from epo alone.

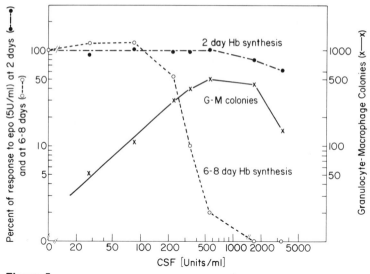

Figure 5
The effect of CSF on epo-induced hemoglobin synthesis at 2 and 8 days. (●——●) Hemoglobin synthesis at 2 days. (o——o) Hemoglobin synthesis at 8 days. (×——×) GM colonies at 8 days. Epo concentration, 5 U/ml.

To determine whether the suppressive effect of CSF could be overcome by increasing concentrations of epo, we tested the effect of a range of concentrations of both factors on 8-day hemoglobin synthesis (Table 5). At a low concentration of CSF (11 U/ml), there was significant augmentation of the epo response at the lower concentrations of epo. This small degree of potentiation of epo-induced hemoglobin synthesis by low levels of CSF was a consistent finding, for which we have no explanation at present. As the concentration of CSF was increased, the degree of suppression of hemoglobin synthesis due to low doses of epo increased, so that, at 110 U/ml of CSF and above, there was no effect of epo at 0.1 U/ml. At 560 U/ml of CSF, the only epo effects seen were at 5 and 10 U/ml, and these were very slight. Conversely, as the concentration of epo was increased, at constant CSF dose, there was increased hemoglobin synthesis, suggesting that the relationship between the two inducers was competitive.

Table 5
Effects of epo and CSF on Hemoglobin Synthesis at 8 Days

CSF concentration	Hemoglobin cpm in the presence of units of epo				
	0.1	0.5	1.5	5.0	10.0
0 U	110	380	1020	2630	3910
11 U	220	550	1130	2820	3920
45 U	40	270	710	2320	3980
110 U	0	100	420	1080	1720
220 U	0	0	140	610	940
560 U	0	0	0	46	94

Interpretation of these results might be complicated by loss of epo during this extended period of incubation. Therefore, we measured epo concentrations in culture media after incubation of mouse bone-marrow cells for 1, 3, and 7 days and found that the mean recovery of 0.048 U added at zero time was 97%. These results confirm the finding of Iscove and Sieber (1975), who found that the factor (most probably epo) responsible for erythroid colony formation in semisolid culture medium was stable for 9 days in the presence of mouse bone-marrow cells.

DISCUSSION

One important assumption inherent in the interpretation of these experiments is that CSF either is a specific granulopoietin or that it acts in a nonspecific manner to induce granulopoiesis in vitro. If GM colony formation due to CSF in semisolid media is a phenomenon unrelated to granulopoiesis, the conclusions we draw from our data will require substantial modification. Those data that are now available regarding this question, however, make the basic assumption a reasonable one.

Our results show that the responses to epo or CSF of mouse and rat bone-marrow cells, especially from plethoric animals, are modulated by the presence of the other factor. The degree of suppression of erythroid function by CSF and of GM colony appearance by epo is dependent on the relative amounts of these two substances. In the case of low concentrations of CSF in the presence of epo, there is a consistent potentiation of the action of the latter. Although we have observed this frequently, we do not as yet understand the mechanism of the potentiation. The predominant cellular locus of the suppressive effects is some population of primitive cells without any erythroid characteristics. Our data are consistent with this cell type being the BFU-E; it is fairly evident from our findings that erythroid cells (CFU-E) formed later in the pathway of differentiation and having quite limited proliferative potential, while responsive to epo, are not affected by CSF. Our data indicate that, when epo has stimulated erythroid burst formation in the presence of CSF, erythroid differentiation occurs at the expense of GM colony formation. This suggests that epo and CSF compete for a common target cell capable eventually of forming either erythroid bursts or granuloid colonies. This interpretation was strengthened by our finding that, in 8-day cultures, the suppressive effect of one inducer can be overcome by increased amounts of the other. The effect of epo on CFU-S is also consistent with this view.

Although pure CSF was not available to us for these experiments, the conclusion that the effects seen were due to CSF rather than to a contaminant are valid. As sources of CSF activity, we used a variety of preparations: unfractionated serum from endotoxin-treated mice, a relatively crude fraction of human embryonic kidney-cell conditioned medium, a fraction of L-cell conditioned medium, and a preparation of human urinary CSF estimated to be about 55% pure. Competition with epo was found for all of these, suggesting that the CSF activity, rather than some contaminant, was responsible.

If there were indeed two populations of unipotent, committed stem cells, one for epo and the other for CSF, the inductive effect of one inducer should have no short-term effect on the other. Our data show clearly that epo does affect CSF-responsive cells and that CSF does affect epo-responsive cells. It is possible that the committed stem cell has a single type of receptor, but that the "wrong" inducer can crossreact with this receptor, blocking it from interacting with the "right" inducer. We would expect in this case that, at a saturating level of the "right" inducer, the "wrong" one should have no effect. According to the results given in Table 5, at a level of epo as high as 10 U/ml, 110 U of CSF still caused a marked suppression. On the other hand, as shown in Table 4, at very high levels of CSF, epo had essentially no suppressive effect on GM colony formation. At these concentrations (10 U/ml of epo and 1680 U/ml of CSF), the molar concentration of epo is still about 15 times that of CSF. It is still possible that cross-reacting nonproductive interactions occur, but more evidence on this point is needed.

Taken together, the experiments reported do not disagree with the alternative hypothesis: that inducers can act on the same cell, which, by definition, must be a pluripotent stem cell. The surface of this cell must then have receptors for both inducers; we have suggested that the different receptors are not present simultaneously but appear during restricted parts of the cell cycle (Goldwasser 1975). There is no strong evidence in support of this suggestion, but there is an indication that epo acts on its target cell only during the post-S phase (Bedard and Goldwasser 1976).

Acknowledgments

We are grateful to Dr. Richard Stanley, Albert Einstein College of Medicine, for gifts of human urinary and L-cell CSF preparations, and to Mr. Otto Walasek, Abbott Laboratories, for human embryonic kidney cell CSF. We are also grateful to Nancy Pech for superb technical assistance. This work was supported in part by Grants CA 18375 and CA 19265 from the National Cancer Institute. The Franklin McLean Memorial Research Institute is operated by The University of Chicago for the United States Department of Energy under Contract No. EY-76-C-02-0069.

REFERENCES

Bedard, D. L., and E. Goldwasser. 1976. On the mechanism of erythropoietin-induced differentiation. XV. Induced transcription restricted by cytosine arabinoside. *Exp. Cell Res.* **102:** 376.

Bradley, T. R., W. Robinson, and D. Metcalf. 1967. Colony production *in vitro* by normal, polycythemic and anaemic bone marrow. *Nature* **214:** 511.

Burgess, A. W., J. Camakaris, and D. Metcalf. 1977. Purification and properties of colony-stimulating factor from mouse lung conditioned medium. *J. Biol. Chem.* **252:** 1998.

Goldwasser, E. 1975. Erythropoietin and the differentiation of red blood cells. *Fed. Proc.* **34:** 2285.

Goldwasser, E., J. F. Eliason, and D. Sikkema. 1975. An assay for erythropoietin *in vitro* at the milliunit level. *Endocrinology* **97**: 315.

Gross, M., and E. Goldwasser. 1972. On the mechanism of erythropoietin-induced differentiation. X. The effect of thymidine suicide on erythropoietin-responsive cells. *Cell Differ.* **1**: 287.

Harris, P. F., R. S. Harris, and J. H. Kugler. 1966. Studies of the leucocyte compartment of guinea pig bone marrow after acute haemorrhage and severe hypoxia. Evidence for a common stem cell. *Br. J. Haematol.* **12**: 419.

Hellman, S., and H. E. Grate. 1967. Haemopoietic stem cells: Evidence for competing proliferative demands. *Nature* **216**: 65.

―――. 1968. Enhanced erythropoiesis with concomitant diminished granulopoiesis in pre-irradiated recipient mice. Evidence for a common stem cell. *J. Exp. Med.* **127**: 605.

Iscove, N. N., and F. Sieber. 1975. Erythroid progenitors in mouse bone marrow detected by macroscopic colony formation in culture. *Exp. Hematol.* **3**: 32.

Krantz, S. B., and W. Fried. 1968. *In vitro* behavior of stem cells. *J. Lab. Clin. Med.* **72**: 157.

Kubanek, B., O. Bock, W. Heit, E. Bock, and E. B. Harriss. 1973. Size and proliferation of stem cell compartments in mice after depression of erythropoiesis. In *Haemopoietic stem cells, Ciba Foundation Symposium* 13. North-Holland Publishing Co., Amsterdam.

Lajtha, L. G., and R. Schofield. 1974. On the problem of differentiation in haemopoiesis. *Differentiation* **2**: 313.

Lajtha, L. G., L. V. Pozzi, R. Schofield, and M. Fox. 1969. Kinetic properties of hemopoietic stem cells. *Cell Tissue Kinet.* **2**: 39.

Langdon, J. R., and T. P. McDonald. 1977. Effects of chronic hypoxia on platelet production in mice. *Exp. Hematol.* **5**: 191.

Meints, R., and E. Goldwasser. 1973. The persistence of hemopoietic stem cells *in vitro*. *J. Cell Biol.* **56**: 429.

Miyake, T., C. K. H. Kung, and E. Goldwasser. 1977. Purification of human erythropoietin. *J. Biol. Chem.* **252**: 5558.

Ogawa, M., R. T. Parmley, H. C. Bank, and S. S. Spicer. 1976. Human marrow erythropoiesis in culture. I. Characterization of methylcellulose colony assay. *Blood* **48**: 407.

Stanley, E. R., and P. M. Heard. 1977. Factors regulating macrophage production and growth: Purification and some properties of the colony stimulating factor from medium conditioned by mouse L cells. *J. Biol. Chem.* **252**: 4305.

Stanley, E. R., and D. Metcalf. 1969. Partial purification and some properties of the factor in normal and leukaemic human urine stimulating mouse bone marrow colony growth *in vitro*. *Aust. J. Exp. Biol. Med. Sci.* **47**: 467.

Steinberg, H. N., E. S. Handler, and E. E. Handler. 1976. Assessment of erythrocytic and granulocytic colony formation in an *in vivo* plasma clot diffusion chamber culture system. *Blood* **47**: 1041.

Van Zant, G., and E. Goldwasser. 1977. The effect of erythropoietin *in vitro* on spleen colony-forming cells. *J. Cell. Physiol.* **90**: 241.

In Vitro Studies of Erythropoietic Progenitor Cell Differentiation

C. J. Gregory
Department of Biophysics, British Columbia Cancer Foundation
Vancouver, British Columbia V5Z 3J3, Canada

A. C. Eaves
Department of Medical Oncology, Cancer Control Agency of British Columbia
Vancouver, British Columbia V5Z 3J3, Canada

Abnormalities in the regulation of hematopoietic cell differentiation processes may underly many hematologic diseases, in particular the myeloproliferative disorders (McCulloch et al. 1973). Thus, definition of the hierarchy of cell types within the hematopoietic system and characterization of factors that regulate the flow of cells from one stage to the next continues to stimulate the interest and contributions of both basic science and clinical medicine.

It is now generally accepted that cells of the erythroid, granulopoietic, and platelet series all derive from a common hematopoietic progenitor—the pluripotent stem cell. Evidence for this concept was first obtained from studies of spleen colony formation in irradiated mice that showed that the spleen colony-forming unit (CFU-S) possessed both pluripotentiality (Becker et al. 1963) and the capacity for extensive self-renewal (Siminovitch et al. 1963). Subsequently, chromosomal findings in patients with chronic myelocytic leukemia (Tough et al. 1963; Whang et al. 1963; Rastrick et al. 1968) and polycythemia rubra vera (Kay et al. 1966) provided indirect evidence for the existence of a pluripotent stem cell population in man.

During the last 10 years, progress in culture technology has led to the development of a number of in vitro colony assays applicable both to mouse and human hematopoietic cells. According to the particular stimulatory factors included in the culture medium, a variety of different colony types may be obtained (Pluznik and Sachs 1965; Bradley and Metcalf 1966; Stephenson et al. 1971; Axelrad et al. 1974; Metcalf et al. 1975a,b; Rozenszajn et al. 1975). In contrast to the mixed composition of most spleen colonies (Becker et al. 1963), individual colonies that develop in vitro from marrow progenitors have not been found regularly to contain cells belonging to more than a single differentiation line, even though several different types of colony may be produced simultaneously in a given culture (Stephenson et al. 1971; Metcalf et al. 1975a). A notable exception to this is the recent report by McLeod et al. (1976) of mixed megakaryocyte-erythroid bursts.

Thus, the cell types detected by in vitro colony assays generally were thought to represent various classes of "committed" hematopoietic progenitor cells, more restricted in their differentiation potentialities than pluripotent stem cells, although closely related in some instances (Wu et al. 1968; Gregory 1976). Considerable evidence in support of such a model has now been obtained (e.g., see review by McCulloch and Till 1972). At the same time, it also has been widely recognized that the earliest differentiation events that precede the irreversible commitment of hematopoietic progenitors to specific pathways remain ill-defined. Thus, attention has come to focus on the need for an in vitro system where stem-cell proliferation and differentiation can be investigated.

Recently, Dexter and coworkers (Dexter et al. 1973, 1977) described a series of experiments in which pluripotent stem cells were maintained in culture for several weeks after an initial decline. Proliferation in the pluripotent stem-cell population was inferred from the finding that stem-cell numbers remained constant although the number of nonadherent cells in the culture flasks was halved at weekly intervals. Concomitant maintenance of granulopoietic precursor (colony-forming units – culture; CFU-C) numbers and their mature progeny in the same cultures (Dexter et al. 1973, 1977) suggested that stem-cell differentiation, i.e., production of CFU-C from CFU-S, could also occur in this in vitro system. We reasoned that if this were so we might be able to demonstrate the production of other types of myeloid progenitor cells. Previous studies had established differential scoring criteria applicable both to mouse (Gregory 1976) and human (Gregory and Eaves 1977) red-cell colonies that allow multiple stages of erythropoietic progenitor cell development (burst-forming units – erythroid [BFU-E] → colony-forming units – erythroid [CFU-E]) to be distinguished and followed separately. The red-cell pathway, therefore, seemed a particularly promising one to investigate, and preliminary studies did indeed show that the most primitive erythropoietic cell types could be detected for up to several weeks after initiation of cultures either of mouse or human marrow cells (Gregory and Eaves 1978b). In these first experiments, we deliberately chose to investigate precursor kinetics in cultures without preestablished feeders so that results obtained with mouse cells might serve directly as a model for human studies in which the same procedures might be used. We have continued with this experimental approach and now have examined the importance of a number of other culture variables with a view to optimizing conditions for the recovery of erythropoietic progenitors. This paper presents the results obtained, together with a preliminary characterization of the differentiation patterns seen in flask cultures of mouse bone-marrow cells initiated without feeders.

MATERIALS AND METHODS

Mice

The mice used in these experiments were F_1 hybrids of a cross between C3H/HeB and C57BL/6B parents. They were purchased from the barrier stock of Biobreeding Laboratories (Ottawa, Ontario) and used when 2 to 3 months old to provide marrow cells and up to 6 months as recipients in CFU-S assays.

Assays

CFU-S determinations were made using the spleen-colony assay (Till and McCulloch 1961). Recipients were given 900 rads cobalt-60 γ-irradiation (Gregory 1976) prior to intravenous injection with test cells and were sacrificed 9 days later for macroscopic spleen colony counts.

CFU-C determinations were performed using an assay medium consisting of 0.8% methylcellulose (Dow Chemical Co.), 20% horse serum (HS, lot. no. 412024 Flow Laboratories), alpha medium (Connaught Laboratories) and medium conditioned by EMT6 (Rockwell et al. 1972) tumor-cell monolayers as a source of colony-stimulating activity (CSA). This conditioned medium, prepared in the same way as L-cell conditioned medium (Worton et al. 1969) has been found to be about 10 times more active. It can be diluted to a final concentration of <1% before any decrease in colony number is detected and, for the routine CFU-C assays in the experiments reported here, was present as 2% of the final culture mixture. Assays were set up in replicate 1-ml volumes and incubated at 37°C in a humidified atmosphere of 5% CO_2 in air for 1 week prior to scoring. All colonies containing >20 cells were counted.

CFU-E and BFU-E measurements were performed also using medium made viscous with methylcellulose (0.8%). Other components of the assay medium were fetal calf serum (FCS, 30%, lot. no. 4055762 Flow Laboratories), 2-mercaptoethanol (2-ME, 10^{-4} M), erythropoietin (epo, Step III, Connaught Laboratories), alpha medium and, for BFU-E only, deionized bovine serum albumin (BSA, 1%, Sigma Chemical Co.) (Gregory 1976). For both types of assays, 1-ml aliquots were plated in duplicate and incubated as for CFU-C. Data for CFU-E are from counts of single or paired clusters of erythroblasts seen after 2 days in assay cultures containing 0.05 U of epo per ml. Data for BFU-E are from counts of bursts, i.e. colonies containing four or more clusters, seen in assay cultures containing 2.5 U of epo per ml. The actual times of burst scoring are indicated in the text of the results. Under these conditions, the plating efficiency of CFU-E and BFU-E has been shown to be constant for the range of cell concentrations used.

Flask Cultures

Flask cultures were initiated by adding to each flask 1 ml containing 3×10^7 marrow cells in 2% FCS and 14 ml of alpha medium supplemented with 20% HS, 10^{-4} M 2-ME, and 1% BSA unless specified otherwise. Tissue-culture grade, 75 cm² flasks (Lux Scientific Corp.) were used throughout. These were capped loosely and incubated either at 33° or 37°C in a humidified atmosphere of 5% CO_2 in air. At each weekly passage, flasks were shaken briefly but vigorously to detach all "nonadherent" cells. One-half of the medium (containing one-half of the nonadherent cells) was then removed and fresh medium of the same composition added back to restore the volume to 15 ml. For this procedure, every flask was handled individually. For assays, cells removed from replicate flasks were pooled. These then were centrifuged and resuspended in 2% FCS for plating, injections, and cell counts. Flasks were usually passed until CFU-S could no longer be detected, but some experiments were discontinued after an arbitary, predetermined period (usually 3 or 4 weeks).

Thus, approximately 10,000 CFU-S, 60,000 CFU-C, and 7,000 BFU-E (day 8-10) were present in the original inoculum added to each flask with some variation in these numbers between experiments (e.g., see Fig. 1). To compare results from different runs, we have expressed measured progenitor cell values in each instance as a percentage of the number originally seeded. In addition, the recovery values shown have been corrected for the serial twofold dilution of nonadherent cells performed at each weekly passage. Thus, numbers from assays at 1 week have not been changed but at 2 weeks these have been multiplied by 2; at 3 weeks by 2^2, at 4 weeks by 2^3, at 5 weeks by 2^4, etc. It should be noted, however, that this correction was based on the assumption that significant numbers of hematopoietic progenitors were not retained in the adherent layer at any time so that an accurate estimate of all early progenitors contributing cells to the nonadherent fraction was made by a simple monitoring of the nonadherent fraction itself. The validity of this assumption in the present studies was tested and confirmed by experiments in which progenitor cell recovery was shown to be unaffected by previous irradiation of the adherent cell layer. These studies are described in detail below.

RESULTS

Effect of Added BSA and 2-ME

Because of the well-established enhancing effect of BSA and thiols on the plating efficiency of erythropoietic progenitor cells in semisolid assay systems (Iscove and Sieber 1975), the effect of these additives to the medium used for flask culture was also investigated. Groups of flasks were set up with the appropriate media, incubated at 37°C, split and passed at weekly intervals, and evaluated after 3 weeks to allow any differences to be magnified. This experiment was repeated on two separate occasions and the results obtained are shown in Table 1. Addition of either BSA or 2-ME appeared to improve the recovery not only of BFU-E but also of CFU-S and CFU-C, but the best results were obtained when BSA and 2-ME were present together. Although the effects observed were not dramatic, because of their consistency BSA and 2-ME were routinely included in the medium used for marrow culture in flasks.

Comparison of HS and FCS

It has been reported that FCS is generally inferior to HS to maintain CFU-S (Dexter et al. 1977); however, preliminary studies indicated that HS is highly toxic for the most primitive class of erythropoietic progenitor cells

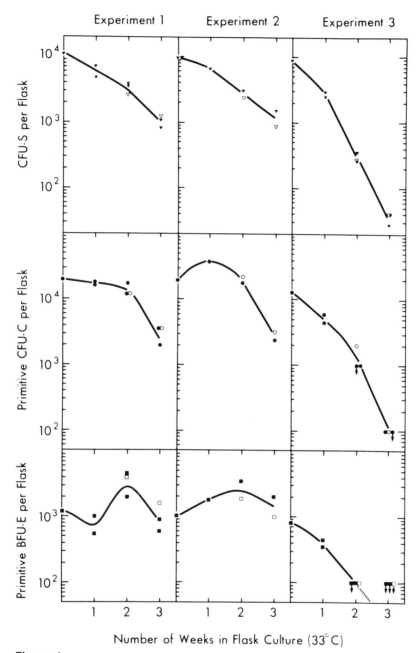

Figure 1
Progenitor cell recovery values for CFU-S and for primitive BFU-E and CFU-C in flask cultures maintained at 33°C. Three individual experiments are shown. For details, see text. Open symbols, groups of flasks in which the adherent cells (only) had been given 1000 rads 1 week previously; solid symbols, corresponding groups of flasks given no irradiation.

Table 1
Effect of BSA[a] and 2-ME in the Flask Culture Medium

Additions to alpha medium containing 20% HS	Progenitor cell recovery after 3 weeks at 37°C[b]					
	CFU-S		BFU-E[c]		CFU-C	
	I	II	I	II	I	II
BSA (1%) + 2-ME (10^{-4} M)	8.9	2.4	17.2	1.1	383	11.5
None	2.1	1.1	<1	0.6	125	5.5
BSA (1%) only	2.5	1.9	2.9	0.6	230	13.5
2-ME (10^{-4} M) only	3.1	1.2	1.0	0.6	133	18.4

[a] From the same batch as that selected and prepared for use in BFU-E assay cultures.
[b] Percentage of starting value, corrected for the serial twofold dilution performed at each weekly passage as described in materials and methods. Results for two separate experiments (I and II) are shown.
[c] Burst counts performed 9 to 10 days after plating.

in the mouse and that this effect becomes even more pronounced for later cell types (Table 2). Although only one batch of HS was tested in these experiments, the failure of HS to support erythroid colony growth by murine CFU-E has been observed previously. This phenomenon has also been seen with CFU-E and BFU-E of human origin, albeit to a lesser extent (data not shown). It was of interest, therefore, to compare the number of CFU-E and BFU-E recovered from flask cultures where FCS was used instead of HS in the flask culture medium. Accordingly, flask cultures were set up with either HS or FCS (the same batch as used in CFU-E and BFU-E assays), halved and fed at weekly intervals, and assayed 3 weeks later for erythropoietic progenitors using the standard methylcellulose-FCS assay. Table 3 shows the results obtained in three different experiments. In contrast to the differential effect of these two sera on CFU-E and BFU-E maturation in methylcellulose assays, the number of BFU-E recovered after 3 weeks in flask culture was not found to be consistently higher with either one serum or the other. However, since the best BFU-E recoveries were obtained with HS and since this was true also for CFU-S, HS was used routinely in sub-

Table 2
Toxicity of HS in BFU-E and CFU-E Assays

Serum used in the assay medium		Plating efficiency (% of control)					
		day-8 BFU-E		day-3 BFU-E		CFU-E	
FCS	HS	I	II	I	II	I	II
30%	+ 0%	100	100	100	100	100	100
29%	+ 1%	100	72.0	105.2	70.4	64.5	84.7
25%	+ 5%	112.6	79.9	75.3	72.8	30.3	36.6
0%	+ 30%	12.6	11.6	1.3	5.6	4.4	0.6

[a] Two experiments (I and II) were conducted in each of which mouse bone-marrow cells were assayed in standard 1-ml methylcellulose assay cultures (2×10^5 cells per dish) as described in materials and methods. Control (30% FCS + 0% HS) counts per dish were as follows: *experiment I* – day-8 BFU-E, 82; day-3 BFU-E, 62.5; CFU-E, 1986; *experiment II* – day-8 BFU-E, 55.5; day-3 BFU-E, 38.5; CFU-E, 726.

Table 3
Comparison of HS versus FCS in the Flask Medium

	Progenitor cell recovery after 3 weeks at 37° C[a]								
	CFU-S			BFU-E			CFU-C		
Serum supplement (20%)	I	II	III	I	II	III	I	II	III
HS	2.4	8.9	15.5	1.1	17.2	19.5	11.5	383	63.2
FCS	1.8	8.2	11.4	5.5	5.7	7.4	24.9	468	109

[a] Percentage of starting value, corrected for the serial twofold dilution performed at each weekly passage as described in materials and methods. Results from three separate experiments (I, II, and III) are shown.

sequent experiments. CFU-E have never been seen in flask cultures set up with HS and also were not detected in the present experiments with FCS.

Comparison of 33° versus 37°C Incubation

Incubation of cultures at 33°C has been found to improve and prolong CFU-S and CFU-C recovery (Dexter et al. 1977). Therefore, the effect of the lower incubation temperature on BFU-E recovery was also investigated. Parallel groups of flasks were maintained at 33°C and at 37°C in seven different experiments. The superiority of the lower temperature for all cell types monitored including BFU-E was demonstrated by the data obtained in each individual experiment as well as by the averaged results (Fig. 2). It is interesting to note that, in spite of these numerical differences, the same kinetics of CFU-S, BFU-E, CFU-C, and total cell recovery (described below) were evident at both temperatures.

Analysis of Precursor Recovery Kinetics

Particularly noteworthy in the curves shown in Figure 2 are the following features: (1) CFU-S numbers continuously *declined* to reach levels below detectability in all experiments by 7 weeks; (2) BFU-E numbers (based on burst counts on day 12) showed a consistent decrease at 1 week followed by a *rise* and then a gradual decline; (3) CFU-C numbers *increased* immediately but also declined after 2 weeks; (4) the total numbers of cells decreased and remained low for at least 2 weeks, then increased to the original value. In spite of the lack of evidence for CFU-S proliferation or even maintenance in these experiments, the BFU-E and CFU-C recovery kinetics suggested that these populations had undergone expansion. The following possible explanations were considered: (1) proliferation of BFU-E and CFU-C with the production of daughter cells that remained within the population operationally defined as BFU-E or CFU-C; (2) input of newly differentiated cells from the stem (CFU-S) compartment; and (3) input of additional undetected precursors by detachment from the adherent layer.

The role of internal proliferation in the BFU-E and CFU-C compartments was investigated by a reexamination of the recovery kinetics of subpopulations within each. A correlation between proliferative capacity and differentiated state has been well documented for progenitors on the erythro-

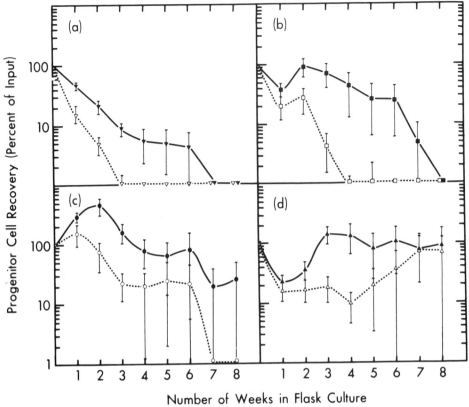

Figure 2
Effect of incubation temperature on the number of hematopoietic progenitors detected in flask cultures. (a) CFU-S; (b) BFU-E (scored 12 days after plating); (c) CFU-C; and (d) total nucleated cells. Points shown are mean values ± 1 SEM for data from seven unselected experiments in which parallel groups of flasks were incubated at the two temperatures. Flasks discontinued prior to 8 weeks because CFU-S were not detected were assumed to have zero cells of all types at later times. Solid symbols, 33°C; open symbols, 37°C.

poietic pathway (Gregory et al. 1973a; Gregory and Henkelman 1977) and similar data exist for CFU-C (Gregory et al. 1973b). In fact, there seems to be a general association between proximity to CFU-S, proliferative capacity, decreasing sensitivity to tropic stimuli, and progenitor cell size (Metcalf and MacDonald 1975; Gregory and Eaves 1977a). Therefore, in an attempt to look more selectively at the most primitive erythropoietic and granulopoietic elements, we devised scoring criteria that excluded smaller burst and colony counts in BFU-E and CFU-C assays respectively. In the case of BFU-E, this was done by scoring macroscopic bursts 2 weeks after plating. The definition used for "primitive" CFU-C was a minimum colony size of 500 cells (i.e., approximately nine divisions) at 1 week after plating. Figure 1 shows the raw data from three separate experiments in which counts for both primitive BFU-E and primitive CFU-C defined in these ways were obtained for marrow cells incubated in flask cultures at

33°C. By comparison to the patterns shown in Figure 2, primitive BFU-E after 2 weeks in culture appeared to show even greater increases in their numbers than did the total BFU-E population. For CFU-C this situation was reversed and the subpopulation of primitive CFU-C showed little, if any, absolute increase. A comparison of this differential effect of flask culture on the proportion of primitive cells in the total erythropoietic and granulopoietic progenitor populations detected as BFU-E and CFU-C respectively is shown quantitatively in Figure 3.

In the three experiments shown in Figure 1, the relative importance of the adherent cell layer as a hidden source of progenitors was also examined by exposing the adherent cells to 1000 rads cobalt-60 γ-rays at the time of the 1- and 2-week passage (open symbols). The results shown in Figure 1 serve to illustrate the degree of interexperimental variability encountered with this system in contrast to the interflask *reproducibility* within a given experiment. It can also be seen that irradiation of the adherent layer had no effect on the number of CFU-S, CFU-C, or BFU-E recovered 1 week later in the nonadherent fraction.

The relative and, in some experiments, absolute increases observed in the number of primitive BFU-E strongly suggest that an influx of cells from the stem-cell pool occurred during the first 2 weeks of culture. Apparently, under the culture conditions used, further flow of cells down this pathway was prevented. In contrast, granulopoiesis did proceed and the vast majority of cells present in the cultures were morphologically recognizable elements of this line (Dexter et al. 1973, 1977). If the kinetics of primitive CFU-C, total CFU-C (which, as shown in Figure 3, were > 90% of the more differentiated type by 1 week), and total cells (i.e., terminal granulopoietic

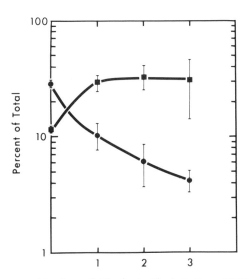

Figure 3
Effect of flask culture on the proportion of all CFU-C (●) and BFU-E (■) detected that gave rise to large colonies (>500 cells at 1 week) or bursts (macroscopic at 2 weeks) for the same three experiments as shown in Figure 1.

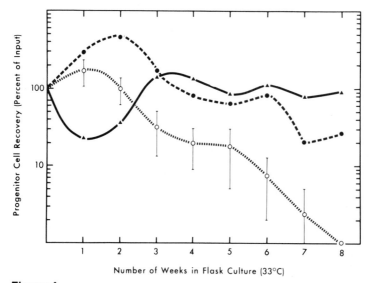

Figure 4
Sequential increases in flask cultures of granulopoietic cells at different stages of differentiation: ○, primitive CFU-C (proliferative capacity > eight divisions); ●, all CFU-C (mainly cells with a proliferative capacity of < eight but > four divisions); and ▲, total cells (mainly granulocytes, monocytes, and macrophages). Data are from the same seven experiments as shown in Figure 2 for 33° flasks.

cells) are compared (Figure 4), it can be seen that these various stages of granulopoietic cell differentiation increased and declined in successive waves. It thus appears that, under the conditions of the present experiments, there was a continuous and increasing efflux of cells down the granulopoietic pathway which exceeded the rate at which the stem cells became committed to this line.

To investigate the mechanism underlying these kinetics, we sought to test the medium from flask cultures at various times after their initiation for the presence of factors capable of stimulating granulopoietic colony formation in the standard CFU-C assay. Figure 5 shows the results for a series of medium samples taken from an experimental run in which marrow cells were maintained in flask culture at 37°C and passed in the usual way at weekly intervals. (Values for precursor recoveries in this experiment were similar to the average values shown in Fig. 2.) Initially, factors stimulating granulopoiesis were not detectable but, with time, evidence of their production became apparent. Although the rise in the numbers of colonies stimulated by sequential samples was not great, it must be remembered that this end point is related to the logarithm of the activity present. Thus, the level of granulopoietic CSA in flask cultures can be assumed to increase substantially with time, a finding consistent with the corresponding shift to more mature cell types.

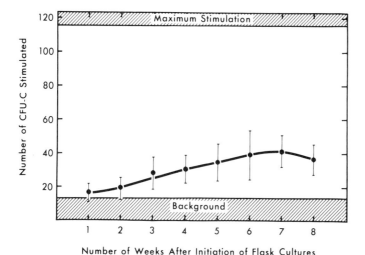

Figure 5
Granulopoietic CSA of culture media sampled at various periods after initiation of flask cultures. Values shown are the mean counts ± 1 SEM from three experiments in which 0.2 ml of three replicate flask culture medium samples were added to standard 1-ml CFU-C assay cultures each containing 5×10^4 mouse bone-marrow cells. Control assay cultures containing 2% EMT6 cell conditioned medium (maximum stimulation) and alpha medium (background) yielded counts in the ranges shown by the hatched areas.

DISCUSSION

This paper describes some recent experiments using a simple marrow-culture system in which the earliest hematopoietic cell-differentiation events can be shown to occur. Following the lead provided by the studies of Dexter et al. (1973,1977), we monitored various types of progenitor cells in cultures of mouse marrow that were incubated at 37° or 33°C, and halved (in terms of nonadherent cell number) and fed (by replacement of one-half of the medium) at weekly intervals. In our experiments however, the culture system devised by Dexter et al. (1973,1977) was modified as follows. Preestablished feeders were not used and marrow cells were placed directly into standard tissue culture flasks rather than into glass bottles. Under these conditions, BFU-E as well as CFU-S and CFU-C — but not CFU-E — could be detected for several weeks in the nonadherent cell population, although an adherent cell layer also developed. As reported previously (Dexter et al. 1977), we found that incubation at 33°C greatly increased the number of CFU-S and CFU-C detected in flask cultures. This improvement in progenitor cell recovery now has been found to apply also to early progenitor cells on the red-cell pathway. Best results for BFU-E were obtained with a medium supplemented with HS, BSA and 2-ME, although substitution of FCS and omission both of BSA and 2-ME did not prevent the detection of erythropoietic progenitors.

It has been suggested (Dexter et al. 1973) that the adherent cell layer provides the environment required for stem-cell proliferation; however, the adherent cell layer may also represent a hidden source of progenitors that are subsequently released and detected in the nonadherent cell fraction. Thus, the presence of the adherent layer can complicate the analysis of progenitor cell-recovery kinetics since only the nonadherent cell component can readily be monitored. In the flask cultures described here, a lack of a significant input of cells from the adherent layer was shown by irradiation studies (Fig. 1). Thus, sequential nonadherent cell-recovery values could be used to follow progenitor population changes as a function of time in flask culture.

We did not obtain evidence in these experiments of an increase in the CFU-S population after culture, although a rise in BFU-E numbers between the 1st and 2nd week was consistently observed. This increase could not be explained simply by the limited differentiation in vitro of more primitive BFU-E, resulting in an accumulation of later cell types still operationally classified as BFU-E, since the relative numbers of these later BFU-E cell types were found to decrease with time (Fig. 3). Moreover, an absolute increase in the number of primitive BFU-E initially placed in culture was seen in two of three experiments (Fig. 1). Even greater culture-induced increases in CFU-C numbers were noted. However, these were accompanied by a relative increase in the flux of cells into the more differentiated end of the total CFU-C population, in contrast to the pattern described above for BFU-E. Nevertheless, an absolute increase in the most primitive CFU-C elements present after 1 week of culture was frequently observed (in three of the six experiments shown in Fig. 4). These data provide evidence for the de novo production in flask culture of the most primitive erythroid and granulopoietic cells detectable. It is interesting that these increases occurred as CFU-S numbers decreased, a finding consistent with the hypothesis that they are CFU-S derived. Further support for this view was provided by comparing individual experiments in which large fluctuations in CFU-S recovery were accompanied by parallel fluctuations in the number of CFU-C and BFU-E detected (Fig. 1).

The role of regulatory molecules in relation to the progenitor cell population kinetics observed also deserves comment. Increasing levels of mouse granulopoietic CSA were found with continued incubation, a finding consistent with the observation of a concomitant acceleration in granulopoietic differentiation. In contrast, production of the most primitive species of CFU-C (and BFU-E) was maximal in the first 2 weeks when CSA had not yet reached detectable levels, suggesting a lack of effect of CSA on their differentiation from more primitive precursors. Similarly, it appears that the presence of erythropoietin also is not required for the generation of the earliest BFU-E (or CFU-C) cell types. Indeed, the point on the red-cell pathway at which erythropoietin-dependence first appears is suggested by the stage at which erythropoietic progenitor cell differentiation appears to stop in flask culture. This was evidenced by the failure to detect day-3 BFU-E and CFU-E (even when the toxic effects of HS were avoided by use of FCS) and by the correction of this differentiation arrest following the addition of erythropoietin either directly to the flask cultures (Gregory and Eaves 1978b) or in the methylcellulose assay dishes. In this respect, flask cultures appeared to simulate the conditions of the hypertransfused mouse, where studies of erythropoietic progenitors have suggested a terminal maturational role of erythropoietin in vivo also (Gregory and Eaves 1978a; Iscove 1977). Thus, molecules known to stimulate granulopoietic colony formation in vitro and erythropoietin, the accepted hormonal regulator of red-cell production in vivo, can influence the differentiation kinetics of primitive CFU-C and BFU-E generated in flask cultures. Whether the production of either of these primitive cell types from pluripotent stem cells is subject to the action of other regulatory factors remains to be established. The present system, in which such early transitions could clearly be detected, provides an approach to the future investigation of this question.

Acknowledgments

The expert technical assistance of M. Heppner, S. Thomas, and C. Smith is gratefully acknowledged. This work was supported by the National Cancer Institute of Canada and the British Columbia Cancer Foundation. C. J. G. is a Scholar of the National Cancer Institute of Canada.

REFERENCES

Axelrad, A. A., D. L. McLeod, M. M. Shreeve, and D. S. Heath. 1974. Properties of cells that produce erythrocytic colonies in vitro. In *Hemopoiesis in culture* (ed. W. A. Robinson), p. 226. U.S. Government Printing Office, Washington, D.C.

Becker, A. J., E. A. McCulloch, and J. E. Till. 1963. Cytological demonstration of the clonal nature of spleen colonies derived from transplanted mouse marrow cells. *Nature* **197:** 452.

Bradley, T. R., and D. Metcalf. 1966. The growth of mouse bone marrow cells in vitro. *Aust. J. Exp. Biol. Med. Sci.* **44:** 287.

Dexter, T. M., T. D. Allen, and L. G. Lajtha. 1977. Conditions controlling the proliferation of haemopoietic stem cells in vitro. *J. Cell. Physiol.* **91:** 335.

Dexter, T. M., T. D. Allen, L. G. Lajtha, R. Schofield, and B. I. Lord. 1973. Stimulation of differentiation and proliferation of haemopoietic cells in vitro. *J. Cell. Physiol.* **82:** 461.

Gregory, C. J. 1976. Erythropoietin sensitivity as a differentiation marker in the hemopoietic system: Studies of three erythropoietic colony responses in culture. *J. Cell. Physiol.* **89:** 289.

Gregory, C. J., and A. C. Eaves. 1977. Human marrow cells capable of erythropoietic differentiation in vitro. Definition of three erythroid colony responses. *Blood* **49:** 855.

―――. 1978a. Three stages of erythropoietic progenitor cell differentiation distinguished by a number of physical and biological properties. *Blood* **51:** 527.

―――. 1978b. Generation of early erythroid progenitors in longterm marrow cultures. (Abstr.) In Proceedings of the 16th International Congress of Hematology. (In press.)

Gregory, C. J., and R. M. Henkelman. 1977. Relationships between early hemopoietic progenitor cells determined by correlation analysis of their numbers in individual spleen colonies. In *Experimental hematology today* (eds. S. J. Baum and G. D. Ledney), p. 93. Springer-Verlag, New York.

Gregory, C. J., E. A. McCulloch, and J. E. Till. 1973a. Erythropoietic progenitors capable of colony formation in culture: State of differentiation. *J. Cell. Physiol.* **81:** 411

———. 1973b. Content of hemopoietic progenitor cells in individual spleen colonies. *Problems of hematology and blood transfusion* [in Russian] **10**: 44.

Iscove, N. N. 1977. The role of erythropoietin in regulation of population size and cell cycling in early and late erythroid precursors in mouse bone marrow. *Cell Tissue Kinet.* (Copenh.) **10**: 323.

Iscove, N. N., and F. Sieber. 1975. Erythroid progenitors in mouse bone marrow detected by macroscopic colony formation in culture. *Exp. Hematol.* **3**: 32.

Kay, H. E. M., S. D. Lawler, and R. E. Millard. 1966. The chromosomes in polycythaemia vera. *Br. J. Haematol.* **12**: 507.

McCulloch, E. A., and J. E. Till. 1972. Leukemia considered as defective differentiation: Complementary *in vivo* and culture methods applied to the clinical problem. In *The nature of leukemia* (ed. P. C. Vincent), p. 119. Proc. Int. Cancer Conf., Sydney, Blight, Sydney, Australia.

McCulloch, E. A., C. G. Gregory, and J. E. Till. 1973. Cellular communication early in hemopoietic differentiation. In *Haemopoietic stem cells,* p. 183. Symposium 13 (new series). Elsevier, Amsterdam.

McLeod, D. L., M. M. Shreeve, and A. A. Axelrad. 1976. Induction of megakaryocyte colonies with platelet formation *in vitro. Nature* **261**: 492.

Metcalf, D., and H. R. MacDonald. 1975. Heterogeneity of *in vitro* colony and cluster-forming cells in mouse marrow: Segregation by velocity sedimentation. *J. Cell. Physiol.* **85**: 643.

Metcalf, D., H. R. MacDonald, N. Odartchenko, and L. B. Sordat. 1975a. Growth of mouse megakaryocyte colonies *in vitro. Proc. Natl. Acad. Sci. U.S.A.* **72**: 1744.

Metcalf, D., N. L. Warner, G. J. V. Nossal, J. F. A. P. Miller, K. Shortman, and E. Rabellino. 1975b. Growth of B-lymphocyte colonies *in vitro* from mouse lymphoid organs. *Nature* **255**: 630.

Pluznik, D. H., and L. Sachs. 1965. The cloning of normal "mast" cells in tissue culture. *J. Cell. Comp. Physiol.* **66**: 319.

Rastrick, J. M., P. H. Fitzgerald, and F. W. Grunz. 1968. Direct evidence for presence of Ph^1 chromosome in erythroid cells. *Br. Med. J.* **1**: 96.

Rockwell, S. C., R. F. Kallman, and L. F. Fajardo. 1972. Characteristics of a serially transplanted mouse mammary tumor and its tissue-culture adapted derivative. *J. Natl. Cancer Inst.* **49**: 735.

Rozenszajn, L. A., D. Shoham, and I. Kalechman. 1975. Clonal proliferation of PHA-stimulated human lymphocytes in soft agar culture. *Immunology* **29**: 1041.

Siminovitch, L., E. A. McCulloch, and J. E. Till. 1963. The distribution of colony-forming cells among the spleen colonies. *J. Cell. Physiol.* **62**: 327.

Stephenson, J. R., A. A. Axelrad, D. L. McLeod, and M. M. Shreeve. 1971. Induction of colonies of hemoglobin-synthesizing cells by erythropoietin *in vitro. Proc. Natl. Acad. Sci. U.S.A.* **68**: 1542.

Till, J. E., and E. A. McCulloch. 1961. A direct measurement of the radiation sensitivity of normal mouse bone marrow cells. *Radiat. Res.* **14**: 213.

Tough, I. M., P. A. Jacobs, W. M. C. Brown, A. B. Baikie, and E. R. D. Williamson. 1963. Cytogenetic studies on bone marrow in chronic myeloid leukemia. *Lancet* **i**: 844.

Whang, J., E. Frei, J. H. Tjio, P. P. Carbone, and G. Brecher. 1963. The distribution of the Philadelphia chromosome in patients with chronic myelogenous leukemia. *Blood* **22**: 664.

Worton, R. G., E. A. McCulloch, and J. E. Till. 1969. Physical separation of hemopoietic stem cells from cells forming colonies in culture. *J. Cell. Physiol.* **74**: 171.

Wu, A. M., L. Siminovitch, J. E. Till, and E. A. McCulloch. 1968. Evidence for a relationship between mouse hemopoietic stem cells and cells forming colonies in culture. *Proc. Natl. Acad. Sci. U.S.A.* **59**: 1209.

Differentiation of Murine Erythroleukemia Cells: The Central Role of the Commitment Event

D. Housman, J. Gusella, R. Geller, R. Levenson, and S. Weil

Department of Biology and Center for Cancer Research
Massachusetts Institute of Technology, Cambridge, Massachusetts 02139

The in vitro differentiation of murine erythroleukemia cells (MELC) has been studied extensively at the biochemical level. The basic results of these studies can be considered in two categories: biochemical events that occur within 12 hours after addition of inducer and those that require at least 12 hours of exposure to inducer before expression of a new biochemical phenotype can be observed. Biochemical changes occurring within the first hours of the induction process include changes in membrane permeability (Dube et al. 1974; Mager and Bernstein 1978), levels of spectrin synthesis (Eisen et al. 1977), and the appearance of a nuclear protein IP25 (Keppel et al. 1977). A common element among these observations is that these biochemical parameters will return to the original level shortly after removal of the inducer. In other words, these early biochemical changes are reversible. Biochemical changes occurring after longer exposure times to inducers do not exhibit reversibility. These include heme synthesis (Preisler and Giladi 1975) and globin synthesis (Boyer et al. 1972).

A CLONAL ANALYSIS OF MELC DIFFERENTIATION

We have characterized the differentiation program at the level of the individual cell. A key element in our approach was analysis of the reversibility of differentiation-specific events. We have developed techniques for characterizing the progeny of individual MELC differentiating in vitro. A critical requirement for this analysis was a technique for cloning MELC with close to 100% plating efficiency. The data shown in Table 1 indicate that the plasma-clot method of McLeod et al. (1974) fulfilled this criterion. MELC were exposed to 1.5% dimethylsulfoxide (DMSO) for varying lengths of time from 0 to 60 hours. Two hundred cells were then cultured in the absence of inducer for an additional 84 hours. The plating efficiency of MELC is almost 100% whether or not they have been exposed to the

Table 1
Plating Efficiency and Clonal Phenotypes of Erythroleukemia Cells Treated with DMSO

Time in DMSO (hours)	Number of cells plated	Total colonies observed	Clonal phenotypes			Plating efficiency (percent)
			all benzidine-negative cells	all benzidine-positive cells	sectored	
0	200	190	173	6	11	95.0
12	200	197	185	4	8	98.5
18	200	191	136	34	21	95.5
30	200	184	48	111	23	92.0
48	200	162	13	151	0	81.0
60	200	161	1	158	2	80.5

745-PC-4 cells were seeded at a density of 5×10^4/ml in medium containing 210 mM DMSO and were maintained thereafter at a density of between 5×10^4/ml and 3×10^5/ml. Cells were plated in plasma clot in the absence of DMSO at the indicated times. The clots were incubated at 37°C for 84 hours, then fixed and stained on microscope slides. A microscope at 100× was used to score colonies.

inducer. An additional advantage of the plasma-clot system is that the progeny of individual cells can be observed and characterized with respect to heme content by benzidine staining. In general, colonies consisted either of heme-containing cells only or of non-heme-containing cells only; the proportion of colonies containing both types of cells (sectored colonies) was never more than 12% (Table 1).

A second significant observation that emerged from our analysis of clonal phenotypes was that cells that can give rise to pure colonies of differentiated cells have a limited proliferative capacity. We demonstrated previously that the proliferative capacity of such cells is limited to a maximum of five cell divisions (Gusella et al. 1976). The direct association of the ability of a cell to express a differentiated phenotype (heme synthesis) in the absence of inducer and a strict limitation on the proliferative capacity of that cell has lead us to postulate an event which fundamentally alters the control machinery of that cell. We term this event commitment. Therefore, our definition of commitment in this context is the simultaneous acquisition by a cultured MELC of (1) a limited proliferative capacity and (2) the ability to express a differentiated phenotype in the absence of inducer. The remainder of this paper will be devoted to a discussion of the characteristics of the commitment event.

SEQUENTIAL PROGRAMS IN MELC DIFFERENTIATION

Biochemical events associated with in vitro differentiation of MELC can be divided into two categories, as indicated above: (1) those which occur early in the differentiation program (approximately within the first 12 hours after addition of inducer) and appear to be reversible; and (2) those which appear later in the differentiation program and are not reversible. At the cellular level, a dichotomy between the characteristics of early and late events in the differentiation program can also be observed. Regardless of which inducer of differentiation or which cell line is chosen for study, a latent period of at least 9 hours occurs prior to the first appearance of significant numbers of committed cells in a differentiating culture of MELC. Figure 1 illustrates representative experiments in which two independently derived cell lines were challenged with thioguanine and DMSO, two potent inducers of MELC differentiation (Gusella and Housman 1976). The proportion of committed cells was measured as a function of time by the plasma-clot technique. It is clear that in all cases a latent period of at least 9 hours occurred before the appearance of significant numbers of committed cells in the culture. These observations have led us to suggest that two sequential programs are involved in the control of MELC differentiation. The first program operates during the latent period. This program presumably controls the reversible biochemical changes that precede the irreversible commitment event. At the end of the latent period, cells are capable of entry into a second program. This program involves a simultaneous limitation of proliferative capacity and an irreversible induction of biochemical parameters associated with the final differentiated state.

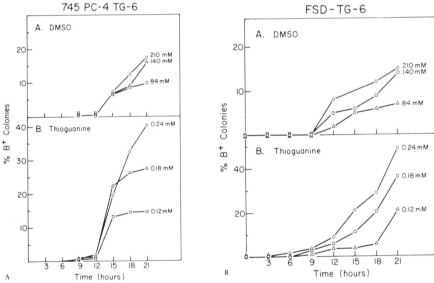

Figure 1
Time course of commitment of MELC. MELC were maintained in alpha medium + 15% fetal calf serum (FCS) containing either DMSO or thioguanine at the indicated concentration. Cell density was maintained at 1×10^5 to 4×10^5/ml. Cells were removed from liquid culture at 3-hour intervals and then grown from 84 to 96 hours in plasma culture in the absence of inducer. Plasma culture and determination of benzidine-reactive colonies were carried out as described previously (Gusella et al. 1976). (A) Cell line 745-PC-4 is a subclone of cell line 745. 745-PC-4-TG-6 is a subclone of cell line 745, selected for thioguanine resistance by culture in the presence of 0.06 mM thioguanine. (B) FSD-TG-6 is a thioguanine-resistant subclone of FSD (Gusella and Housman 1976).

A STOCHASTIC MODEL FOR COMMITMENT OF MELC

The time of commitment was not simultaneous for all cells in the culture. A significant heterogeneity could be demonstrated in the time at which individual MELC became committed to terminal erythroid differentiation (Gusella et al. 1976). We demonstrated further that for cells exposed to DMSO the kinetics of the commitment process are most consistent with a stochastic model in which a given proportion of cells becomes committed in each cell generation. To assess the potential generality of a stochastic model for MELC differentiation, we extended our analysis to induction of differentiation by a number of other inducers. Typical experimental results for three different inducers of erythroid differentiation are shown in Figure 2. It is apparent that the exposure to inducer either can reduce the apparent doubling time of the uncommitted cells (1 mM butyric acid) or cause an actual exponential decrease (210 mM DMSO and 0.18 mM thioguanine) in the numbers of uncommitted cells in the population. According to a stochastic model, whether an increase or decrease in the numbers of uncommitted cells actually occurs will depend upon whether the probability (P) with which a cell makes the transition to the committed state is greater than or less than 0.5. If $P < 0.5$, then a decreased doubling time of the uncommitted cell population should be observed when compared to an untreated culture. If $P = 0.5$, then the number of uncommitted cells in the culture should remain at a constant level indefinitely. If $P > 0.5$, then the number of uncommitted cells in the population will decline exponentially. Experimental results with a number of inducers followed this pattern and led us to develop a mathematical model that can be used to describe the in vitro differentiation of populations of cultured MELC and to predict their behavior (Fig. 3).

The model is based on the following assumptions. (1) Commitment of a cell to the differentiation program is always accompanied by a restriction in the proliferative capacity of the committed cell to four subsequent cell divisions. (2) Cells undergo a transition from the uncommitted to the committed state with probability (P) during each cell division cycle. When P is constant, the number of uncommitted cells in the population can be expressed as a function of time by the following mathematical relationship:

$$U(t) = U_0[2(1-P)]^{t/T_0},$$

where $U(t)$ is the number of uncommitted cells at time t, U_0 is the number of committed cells at $t = 0$, and T_0 is the generation time of uncommitted cells. (3) The value of P can vary between 0 and 1. In the absence of inducer and during the latent period $P = 0$. Subsequent to the latent period, P rises to an equilibrium value which is constant over many cell generations.

BIOCHEMICAL IMPLICATIONS OF A STOCHASTIC MODEL

If our model is applicable, there are important implications for biochemical studies. To illustrate, we measured cytoplasmic globin mRNA levels in a pulse-and-chase protocol of exposure to DMSO (Fig. 4). MELC exposed

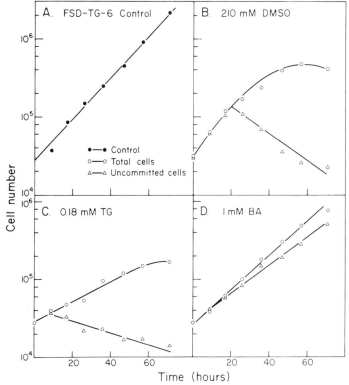

Figure 2
Rate of change of uncommitted cell number during treatment of FSD-TG-6 with DMSO, thioguanine, and butyric acid. Cultures of FSD-TG-6 were seeded at 3×10^4 cells/ml and maintained at 3×10^4 to 5×10^5 cells/ml in alpha medium + 15% FCS containing (A) no inducer, (B) 210 mM DMSO, (C) 0.18 mM thioguanine, or (D) 1 mM butyric acid. At the indicated times, total cell number was determined with a Coulter counter, model ZBI. The number of uncommitted cells was determined by plating cells in plasma clot (Gusella et al. 1976) in the absence of inducer. Clots were incubated from 84 to 120 hours, fixed, stained, and scored for large (>50 cells) benzidine-negative colonies using a microscope at 100×.

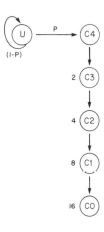

Figure 3
A stochastic model for in vitro differentiation of MELC. Uncommitted cells (U) have a probability P of becoming committed and a probability $(1 - P)$ of remaining uncommitted during any given cell generation. Newly committed cells (C4) have the potential to undergo four cell divisions. Their progeny have the potential to undergo three (C3), two (C2), one (C1), and eventually, no divisions (C0).

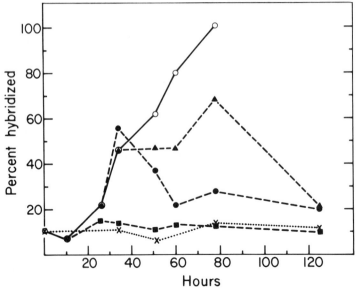

Figure 4
Globin mRNA accumulation of MELC exposed to 210 mM DMSO in a pulse-and-chase protocol. MELC were maintained at 1×10^5 to 4×10^5 cells/ml in alpha medium + 15% FCS for 0, 10, 26, and 34 hours with DMSO and then grown in the absence of inducer. At times indicated, cytoplasmic RNA was extracted as previously described (Gusella et al. 1976). Mouse globin cDNA was synthesized and hybridizations were carried out as described (Housman et al. 1974) with the following modifications: duplicate samples of 2 μg cytoplasmic RNA were incubated with constant input cDNA (3000 cpm) in a 5-μl reaction mix for 20 hours at 70°C. Percentage of hybridization was determined after S1 nuclease digestion. ×, control cells grown in alpha medium + 15% FCS only; ■, cells grown in DMSO for 10 hours, subsequent growth in the absence of DMSO; ●, cells grown in DMSO for 26 hours, subsequent growth in the absence of DMSO; ▲, cells grown in DMSO for 34 hours, subsequent growth in the absence of DMSO; o, cells grown in DMSO continuously, subsequent growth in the absence of DMSO.

to DMSO for 10 hours, a period shorter than the latent period, showed no detectable increase in cytoplasmic globin mRNA levels at any time following exposure to DMSO. This result is consistent with the view that globin mRNA accumulation rate does not increase until a cell has become committed to differentiate. In contrast, a culture of MELC exposed to DMSO for a 26-hour period showed a significant increase over the uninduced globin mRNA level. Removal of these cells from DMSO did not prevent a further increase in the cytoplasmic globin mRNA level at 34 hours. This culture, incubated in the absence of DMSO for 60 hours or more, approached the value for an uninduced culture. These results suggest that the fraction of cells committed to differentiate by a 26-hour exposure to DMSO will continue to accumulate globin mRNA in the absence of inducer. By 60 hours, however, the differentiating cells will fail to divide further and the culture will be overgrown by the descendants of uncommitted cells. Similar results were observed for cultures exposed to DMSO for longer periods of time except that as exposure to DMSO continued, a larger proportion of the population became committed and, hence, a smaller proportion of the population remained uncommitted. Two quantitative differences in the behavior of these cultures were observed. (1) The cytoplasmic globin mRNA levels reached a higher value as the length of the DMSO pulse increased, and (2) a longer period of time was required for the uncommitted cells to overgrow the committed population once the inducer had been removed. Therefore, the length of time required for the globin mRNA level to return to control values was increased.

To emphasize the critical importance of considering the probability of commitment in analyzing the biochemical behavior of the system, we calculated the theoretical proportion of heme-positive cells as a function of P value after 5 days' continuous exposure to DMSO (Fig. 5). It is clear that a wide range of probabilities between 0.3 and 0.9 will give rise to a culture containing 90% or more heme-containing cells after 6 days of continuous exposure to inducer. Nevertheless, the characteristics of a culture with P values from 0.3 to 0.9 will be very different during the critical early phases of the induction process. The cultures of high and low P values approximated each other in composition when the proportion of uncommitted cells became an insignificant portion of the population.

The techniques described in the previous section allow such questions as the following relating to the induction of differentiation of MELC to be investigated. (1) Do all compounds that cause an increased level of hemoglobin synthesis act by inducing commitment of MELC? (2) Among those compounds which do induce commitment of MELC, can a common pathway

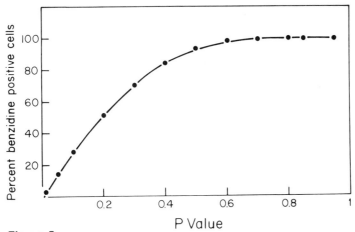

Figure 5
Theoretical proportion of benzidine-staining cells as a function of probability (P) value. The theoretical percentage of benzidine-positive cells in a culture at 6 days was calculated for different values of P, assuming an initial latent period ($P = 0$) of 16 hours, with $N = 1$. It was assumed that only cells that had undergone three or four divisions since commitment (C1 or C0 cells) had accumulated enough heme to stain positively with benzidine.

for induction be demonstrated? (3) What physiological parameters affect the commitment of MELC to the differentiation program? (4) What genetic factors influence the commitment of MELC?

THE RELATIONSHIP BETWEEN COMMITMENT AND GLOBIN mRNA SYNTHESIS

To determine whether compounds reported to increase levels of hemoglobin synthesis of MELC also cause commitment to the differentiation program, we performed a series of screening experiments. Most compounds which have been reported to induce high levels of hemoglobin synthesis in MELC, including purines (Gusella and Housman 1976), butyric acid (Leder and Leder 1975), and cryoprotective agents (Preisler and Lyman 1975), also cause extensive commitment of MEL cells (D. Housman et al., unpubl.). A clear-cut exception to this pattern, however, was the behavior of MELC exposed to hemin. We monitored levels of globin mRNA accumulation and levels of commitment by three different MELC lines treated with DMSO or with hemin. The results are shown in Table 2. As previously reported (Ross and Sautner 1976), hemin caused a five- to tenfold increase in cytoplasmic globin mRNA over basal levels. No detectable increase over the basal value of the proportion of committed cells was observed for hemin-treated cells of any cell line tested. By contrast, from 88 to 97% of the MELC treated with DMSO became committed during the 72 hours of exposure to the inducer. These observations indicate that hemin acts to increase cytoplasmic globin mRNA levels of MELC without causing commitment to a loss of indefinite proliferative capacity. In this way, hemin

Table 2

The Effects of Hemin and DMSO on Commitment and Globin mRNA Accumulation in MELC

Cell line	Treatment (72 hours)	Percent globin mRNA	Percent committed cells	Modal cell volume (μm^3)
T3C12	control	1.7×10^{-2}	2	1024
	210 mM DMSO	6.1×10^{-1}	88	800
	0.1 mM Hemin	2.0×10^{-1}	3	1024
FSD-TG-6	control	1.8×10^{-2}	<1	672
	210 mM DMSO	4.6×10^{-1}	93	464
	0.1 mM hemin	1.5×10^{-1}	1	688
745-PC-4	control	4.4×10^{-2}	<1	624
	210 mM DMSO	8.5×10^{-1}	97	352
	0.1 mM hemin	2.6×10^{-1}	<1	608

Cultures of cell lines T3C12, FSD-TG-6, and 745-PC-4 were seeded at 5×10^4 cells/ml and maintained well below saturation density for 72 hours in alpha medium + 15% FCS containing 210 mM DMSO, 0.1 mM hemin, or no inducer. Globin mRNA content, expressed as a percentage of total cytoplasmic RNA, was determined as previously described (Gusella et al. 1976). Modal cell volume was measured with a Coulter counter equipped with a channelyzer. Percent commitment was determined as described in Figure 2.

Table 3
Commitment of MELC Following a Switch of Inducers at 10.5 Hours

Experiment no.	Treatment		Percent benzidine-positive colonies (total exposure time to inducers)	
	0–10.5 hours	10.5–24 hours	19 hours	24 hours
1	0	0	1.6 ± 0.4	1.2 ± 0.1
2	0	DMSO	5.4 ± 0.6	8.7 ± 0.3
3	0	hypoxanthine	4.4 ± 0.4	7.6 ± 1.0
4	DMSO	0	8.4 ± 1.4	5.2 ± 0.6
5	hypoxanthine	0	8.4 ± 0.4	6.1 ± 0.6
6	DMSO	DMSO	20.2 ± 0.4	24.4 ± 0.1
7	DMSO	hypoxanthine	19.9 ± 0.8	24.8 ± 1.2
8	hypoxanthine	DMSO	20.9 ± 0.8	24.1 ± 0
9	hypoxanthine	hypoxanthine	21.3 ± 0.2	24.1 ± 0.7

MELC were seeded at a density of 10^5 cells/ml in alpha medium + 15% FCS containing no inducer, 210 mM DMSO, or 3.7 mM hypoxanthine. After 10.5 hours at 37°C, cells were concentrated by centrifugation at 800 rpm for 10 minutes at 25°C (IEC CRU-5000 centrifuge). The cells were resuspended in fresh medium; DMSO or hypoxanthine was then added to the culture medium as indicated; and incubation was continued at 37°C for 14 hours. After a total of 19 or 24 hours exposure to inducer, aliquots of these cultures were plated in plasma culture in the absence of inducers. These cultures were incubated at 37°C in a humidified atmosphere for 96 hours. Clots were fixed and stained as previously described (Gusella et al. 1976), and the proportion of benzidine-positive (hemoglobinized) colonies was determined.

differs fundamentally from DMSO and the other inducers mentioned above. An interesting corollary of these findings is that all inducers that caused commitment also caused a shift in the modal cell volume of treated MELC (Table 3). Hemin-treated cells, in contrast, did not exhibit any such shift in modal cell volume. These observations suggest that a reduction in modal cell volume is correlated with commitment.

A COMMON PATHWAY OF INDUCTION LEADING TO COMMITMENT

To determine whether a temporal sequence in the pathway of induction exists for those inducers known to cause commitment, we used the following experimental protocol. Two cultures of MELC were exposed to two different inducers of differentiation, and, 10.5 hours after addition of inducer, some cells from each culture were removed from the presence of one inducer and "switched" into the presence of the other. The kinetics of commitment of the switched cultures were followed and compared to control "unswitched" cultures. The results of one such experiment are shown in Figure 2. In this case, the inducers used were DMSO and hypoxanthine. The experiment was designed so that approximately equal P values would be reached by the two unswitched induced cultures. Essentially no difference in the proportion of committed cells was observed in either of the switched cultures (numbers 7 and 8) compared to the unswitched cultures (numbers 6 and 9); yet a stimulation 20 to 25 times greater than that in an untreated culture (no. 1) was observed. We interpret these results to mean that DMSO and hypoxanthine share an intracellular target and that a biochemical process initiated by one inducer during the latent period can be continued by the other inducer until commitment occurs.

RELATIONSHIP OF THE CELL CYCLE TO COMMITMENT

It has been observed (Geller et al. 1978; Terada et al. 1977) that MELC treated with DMSO exhibit an increase in the length of G1 compared to untreated cells. A question which remains is whether this lengthening of G1 is directly related to the commitment of MELC. To examine this question further, we obtained synchronized populations of MELC by unit gravity sedimentation. Details of this synchronization technique have been reported elsewhere (Geller et al. 1978). These studies suggest that MELC in G1 or G2 at the time at which DMSO was added showed a kinetic advantage with respect to commitment over MELC in early, middle, or late S phase. Data supporting this view are summarized in Table 4. MELC closer to G1 (i.e., in G1 or G2) at the time of addition of DMSO committed sooner and to a greater extent than did those MELC that had to traverse a greater proportion of the cell cycle (early, middle, or late S) before entering the G1 phase. A continuation of this analysis, to determine whether commitment can occur only at a specific point in the cell cycle, should allow us to understand further the physiology of the MELC undergoing commitment.

GENETIC ANALYSIS OF THE COMMITMENT PROCESS

Genetic analysis can be an extremely powerful tool for an understanding of the molecular events that control the commitment process. The isolation of MELC variants which exhibit an alteration in the kinetics or extent of com-

Table 4

The Commitment to Erythroid Differentiation by a Series of Synchronous Populations of MELC Fractionated by Unit Gravity Velocity Sedimentation

Cell population	Modal cell volume (μm^3)	Percent benzidine-reactive colonies		
		24 hours	27 hours	30 hours
Unfractionated	600	9	11	18.5
G1	450	11	21.5	27.5
early S	550	5	6	10.5
mid S	700	3	8	13.5
late S	900	5.5	7.5	16.5
G2	1200	14	16	21

7×10^7 MELC were separated by unit gravity velocity sedimentation into a series of synchronous cell populations. Details of this technique have been described elsewhere (Geller et al. 1978). Each cell population was adjusted to a cell concentration of 1×10^5 cells/ml in alpha medium + 15% FCS, then challenged with 210 mM DMSO. Cultures were incubated at 37°C. At the times indicated, approximately 200 cells from each culture were plated in plasma culture in the absence of DMSO. After 80 to 90 hours in plasma culture, colonies were analyzed for benzidine reactivity as previously described (Gusella et al. 1976).

mitment would be facilitated by a specific selection procedure. Because commitment leads to an irreversible loss in proliferative capacity, those MELC with a reduced level of commitment will have a selective advantage when present in a culture continuously exposed to inducer.

We exposed MELC to DMSO for a period of time sufficient to allow outgrowth of a resistant cell population. Results of a typical experiment are shown in Figure 6. Approximately 200 hours after DMSO had been added, a cell population with a doubling time similar to the untreated parent culture emerged. To ascertain whether a simultaneous selection for ability to proliferate in DMSO and failure to express differentiated functions had taken place, we sampled the culture to determine the proportion of benzidine-positive cells (Fig. 7). It was clear that the proportion of benzidine-positive cells declined sharply between 200 and 300 hours, corresponding to the time of reemergence of the DMSO-insensitive cell population.

To further characterize the cells that were selected by this procedure (Table 5), we isolated clones at various times during the selection, grew them in the absence of DMSO, and subsequently tested their ability to differentiate when rechallenged with DMSO. Clones isolated up to 137 hours after addition of DMSO gave rise to a high (>50%) proportion of differentiated cells. At 167 hours clones were first observed that exhibited reduced levels (4–30%) of benzidine positivity upon being rechallenged with DMSO. Further selection led to a population of clones that exhibited a 0.2 to 6.2% level of benzidine-positive cells when rechallenged with DMSO. It was of

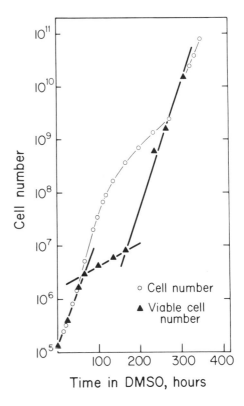

Figure 6
Growth curve of total cells and uncommitted cells in 140 mM DMSO. A culture of 745-TG-11 (Gusella and Housman 1976) was seeded in medium containing 140 mM DMSO and maintained at a density below 5×10^5 cells/ml. Total cell number was calculated from Coulter counts and the dilutions used when feeding the culture. Uncommitted cell number was determined from the plating efficiency of the culture in methylcellulose relative to that of an untreated population.

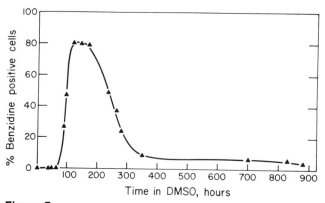

Figure 7
Presence of heme-containing cells in a culture of 745-TG-11 treated with 140 mM DMSO. Aliquots of the culture used in Figure 6 were centrifuged onto microscope slides and benzidine-stained as described in Gusella and Housman (1976). A minimum of 500 cells was counted to determine the proportion of heme-containing (orange-staining) cells in the population.

interest to note that the proportion of benzidine-positive cells in clones grown in the absence of DMSO was almost always less than 0.2%, whereas clones isolated early in the selection contained 0.8 to 6.4% benzidine-positive cells when grown in the absence of DMSO. Whether the subclonal variation of differentiation in response to DMSO represented a stable phenotype was determined by rechallenging 10 subclones with DMSO after 2 and 25 weeks of continuous growth in the absence of inducer. The results of this experiment are shown in Table 6. Both at 2 weeks and 25 weeks, each clone exhibited the same characteristic level of differentiation.

These results confirm the theory that chronic exposure to inducer does indeed lead to the selection of stable variants that exhibit a heritable alteration in the commitment response.

IS COMMITMENT A GENERAL PHENOMENON IN DIFFERENTIATING EUKARYOTIC SYSTEMS?

The studies described here were conducted with permanent cell lines of malignant erythroid cells. We have shown that these cells, when induced to differentiate, undergo a limitation of proliferative capacity to four- or five-cell generations. In this respect MELC resemble the normal erythroid precursor, the colony-forming unit−erythroid (McLeod et al. 1974). A further parallel should be noted between MELC differentiation and the differentiation of pluripotent hematopoietic stem cells in vivo (Till et al. 1964; Korn et al. 1973). In both cases, the differentiation of the system can best be modeled as a stochastic process in which commitment to differentiation occurs in a probabilistic fashion.

Analysis of this type is not confined to hematopoietic cells, however. Holliday et al. (1977) recently described an analysis of fibroblast senescence

Table 5
Response to DMSO of Clones Surviving Induction

Time removed from induced culture (hours)	Percent benzidine-positive cells		
	clone no.	control	DMSO
0	0/1	2.8	73.2
	0/2	0.8	68.4
	0/3	6.0	62.8
	0/4	2.0	89.8
	0/5	1.0	81.2
48	48/1	2.0	84.0
	48/2	2.6	76.2
	48/3	1.4	74.4
	48/4	2.4	76.0
97	97/1	6.4	86.6
	97/2	4.8	89.4
	97/3	1.2	81.4
	97/4	2.4	60.8
137	137/1	2.4	72.6
	137/2	2.6	57.2
	137/3	2.2	83.8
167	167/1	1.8	27.8
	167/2	0.2	4.2
	167/3	0.8	28.2
	167/4	1.8	80.4
236	236/1	<0.2	0.4
	236/2	<0.2	1.6
	236/3	<0.2	4.0
	236/4	<0.2	0.8
276	276/1	<0.2	1.6
	276/2	0.4	6.2
	276/3	<0.2	0.6
	276/4	<0.2	0.2
800	800/1	<0.2	3.2
	800/2	<0.2	0.4
	800/3	<0.2	0.8
	800/4	<0.2	1.2

Clones were picked from the methylcellulose plates of the experiment shown in Figure 6. They were cultured for several weeks and subsequently rechallenged for 6 days with 140 mM DMSO. Control and DMSO-treated cultures were then stained with benzidine and the percentage of benzidine-positive cells was determined (Gusella and Housman 1976).

which parallels in detail our analysis of MELC differentiation. Gunther et al. (1974) conducted an analysis of lymphocyte differentiation which similarly involved a commitment step made in a probabilistic manner. Smith and Martin (1973) applied similar criteria to analyze the transition of mammalian cells from G1 to S phase of the cell cycle. How each of these processes is

Table 6
Stability of the Low-Inducible Phenotype

Clone	Response to 140 mM DMSO (percent benzidine-positive cells) by time after cloning	
	2 weeks	25 weeks
745-DR-1	14.4	18.4
745-DR-2	4.4	5.8
745-DR-3	0.6	0.8
745-DR-4	1.4	3.2
745-DR-5	31.2	33.4
745-DR-6	13.2	18.4
745-DR-7	3.6	1.6
745-DR-8	<0.2	0.4
745-DR-9	0.2	<0.2
745-DR-10	9.0	10.8

Clones were isolated following chronic exposure of line 745 to 140 mM DMSO and were cultured in the absence of inducer. After 2 weeks and 25 weeks of culture, subclones were rechallenged with 140 mM DMSO for 6 days and the proportion of heme-containing cells was determined by benzidine-staining (Gusella and Housman 1976).

controlled at the molecular level remains unclear at this time. These widespread observations of cellular decision-making processes which involve a commitment event and which can best be modeled as stochastic processes raise the possibility that similar molecular control mechanisms underlie each of these processes.

Acknowledgments

The authors wish to thank Dr. Michael Garrick for his helpful comments and criticism during the preparation of this manuscript and Mrs. Susan Posner for patiently typing the manuscript. Cell line T3C12 was obtained from Dr. J. Ross and cell line 745, from Dr. C. Friend. The work described here was supported by Grants CA-14051 and CA 17575 from the National Institutes of Health. R. L. was supported by a postdoctoral fellowship from the American Cancer Society and S. W. by a postdoctoral fellowship from the NIH.

REFERENCES

Boyer, S. H., K. D. Woo, A. N. Noyes, R. Young, W. Scher, C. Friend, H. D. Preisler, and A. Bank. 1972. Hemoglobin biosynthesis in murine virus-induced leukemic cells *in vitro:* Structure and amounts of globin chains produced. *Blood* **40:** 823.

Dube, S. K., G. Gaedicke, N. Kluge, B. J. Weimann, H. Melderis, G. Steinheider, T. Crozier, H. Beckmann, and W. Ostertag. 1974. Hemoglobin-synthesizing mouse and human erythroleukemic cell lines as model systems for the study of differentiation and control of gene expression. In *Differentiation and control of malignancy of tumour cells* (eds. W. Nakahara et al.), p. 103. University of Tokyo Press, Tokyo.

Eisen, H., R. Bach, and R. Emery. 1977. Induction of spectrin in erythroleukemia cells transformed by Friend virus. *Proc. Natl. Acad. Sci. U.S.A.* **74**: 3898.

Geller, R., R. Levenson, and D. Housman. 1978. Significance of the cell cycle in commitment of murine erythroleukemia cells to erythroid differentiation. *J. Cell. Physiol.* **95**: 213.

Gunther, G. R., J. L. Wang, and G. M. Edelman. 1974. The kinetics of cellular commitment during stimulation of lymphocytes by lectins. *J. Cell Biol.* **62**: 366.

Gusella, J., and D. Housman. 1976. Induction of erythroid differentiation by purines and purine analogues. *Cell* **8**: 263.

Gusella, J., R. Geller, B. Clarke, V. Weeks, and D. Housman. 1976. Commitment to erythroid differentiation by Friend erythroleukemia cells: A stochastic analysis. *Cell* **9**: 221.

Holliday, R., L. I. Huschtscha, G. M. Tarrant, and T. B. L. Kirkwood. 1977. Testing the commitment theory of cellular aging. *Science* **198**: 366.

Housman, D., A. Skoultchi, B. G. Forget, and E. J. Benz, Jr. 1974. Use of globin cDNA as a hybridization probe for globin mRNA. *Ann. N.Y. Acad. Sci.* **241**: 280.

Keppel, F., B. Allet, and H. Eisen. 1977. Appearance of a chromatin protein during the erythroid differentiation of Friend virus-transformed. *Proc. Natl. Acad. Sci. U.S.A.* **74**: 653.

Korn, A. P., R. M. Henkelman, F. P. Ottensmeyer, and J. E. Till. 1973. Investigations of a stochastic model of haemopoiesis. *Exp. Hematol.* (Copenh) **1**: 362.

Leder, A., and P. Leder. 1975. Butyric acid, a potent inducer of erythroid differentiation in cultured erythroleukemia cells. *Cell* **5**: 319.

Mager, D. L., and A. Bernstein. 1978. Early transport changes during erythroid differentiation of Friend leukemic cells. *J. Cell. Physiol.* **94**: 275.

McLeod, D. L., M. Shreeve, and A. A. Axelrad. 1974. Improved plasma culture system for production of erythrocytic colonies *in vitro:* Quantitative assay method for CFU-E. *Blood* **44**: 517.

Preisler, H. D., and M. Giladi. 1975. Differentiation of erythroleukemic cells *in vitro:* Irreversible induction by dimethyl sulfoxide (DMSO). *J. Cell. Physiol.* **85**: 537.

Preisler, H. D., and G. Lyman. 1975. Differentiation of erythroleukemic cells *in vitro:* Properties of chemical inducers. *Cell Differ.* **4**: 179.

Ross, J., and D. Sautner. 1976. Induction of globin mRNA by hemin in cultured erythroleukemic cells. *Cell* **8**: 513.

Smith, J., and L. Martin. 1973. Do cells cycle? *Proc. Natl. Acad. Sci. U.S.A.* **70**: 1263.

Terada, M., J. Fried, U. Nudel, R. Rifkind, and P. Marks. 1977. Transient inhibition of S-phase associated with dimethyl sulfoxide induction of murine erythroleukemia cells to erythroid differentiation. *Proc. Natl. Acad. Sci. U.S.A.* **74**: 248.

Till, J. E., E. A. McCulloch, and L. Siminovitch. 1964. A stochastic model of stem cell proliferation, based on the growth of spleen colony-forming cells. *Proc. Natl. Acad. Sci. U.S.A.* **51**: 29.

Erythroleukemia Cells: Commitment to Differentiate and the Role of the Cell Surface

R. A. Rifkind, E. Fibach, R. C. Reuben, Y. Gazitt, H. Yamasaki, I. B. Weinstein, U. Nudel, I. Sumida, M. Terada, and P. A. Marks

Cancer Center and Departments of Medicine and Human Genetics and Development
Columbia University, College of Physicians and Surgeons, New York, New York 10032

The rate of normal erythropoiesis is governed in large part by the titer of a hormone, erythropoietin, responsible for the proliferation of its target cell, an immature precursor cell several generations removed from the multipotent hematopoietic stem cell (Stephenson et al. 1971; Marks and Rifkind 1972; Rifkind et al. 1974; Gregory 1976; Gregory and Eaves 1977). The murine erythroleukemia cell (MELC) appears to be an erythroid precursor cell (perhaps the erythropoietin-responsive target cell) transformed by the Friend virus complex (Friend et al. 1966; Mirand 1967; Steeves et al. 1969; Tambourin and Wendling 1971; Lilly 1972; Lilly and Pincus 1973; McGarry and Mirand 1973; Fredrickson et al. 1975; Steeves 1975; Tambourin and Wendling 1975; Nasrallah and McGarry 1976). One component of this virus complex, spleen focus-forming virus, is responsible for uncoupling cellular replication from erythropoietin, the physiological regulator of this function. By some mechanism, perhaps selection of cells with the greatest growth potential, cell lines become established in which the normal process of erythropoietic differentiation (and concomitant terminal cell division) becomes suppressed. In addition, a wide variety of agents have been discovered that can alter significantly the rate at which MELC express the program of erythropoietic differentiation (Friend et al. 1971; Takahashi et al. 1975). This combination of events—proliferation without erythropoietin and the availability of agents that modulate differentiation—provides an interesting model system for exploring the cellular events which are of significance in the regulation of the expression of differentiated functions in erythroid cells. Many of the properties of MELC are discussed in this volume (Marks et al., this volume) and have been reviewed in considerable detail elsewhere (Ikawa et al. 1973b; Harrison et al. 1976; Pragnell et al. 1976; Reuben et al. 1978; Marks and Rifkind 1978). This paper will focus on two aspects of the problem of regulation of MELC erythropoiesis. First, we will review the evidence that indicates that there are two components to induced differentiation: the commitment of the MELC to erythropoietic differentiation, and the program that coordinates the series of biosynthetic and morphogenetic events that constitute the actual expression of the erythropoietic developmental process. Secondly, we will present and review some evidence that indicates that alterations at the plasma membrane may be implicated in the induction of differentiation by some agents.

COMMITMENT TO DIFFERENTIATION

MELC (strain DS19 derived from strain 745 of C. Friend) display less than 0.5% of spontaneously differentiating erythroblasts in culture (Singer et al. 1974). When MELC are exposed to increasing concentrations of the potent polar-planar inducing reagent, hexamethylene bisacetamide (HMBA), an increasing proportion of cells are induced to differentiate, a process detectable by the benzidine reaction for hemoglobin (Reuben et al. 1976). The total amount of hemoglobin detectable in these cultures is directly proportional to the number of benzidine-reactive differentiating erythroblasts. Furthermore, the mean amount of hemoglobin (Fig. 1) in each differentiating erythroblast is constant; that is, it is independent both of the total number of erythroblasts triggered to differentiate and of the concentration of HMBA. These observations provided the first substantive clue in our laboratory that two distinct programs were necessary for erythropoietic differentiation: commitment, which determines the number of differentiating cells, and expression, which achieves the biochemical and morphogenetic changes characteristic of erythropoiesis. The first appears to be responsive to the concentration of inducer, whereas the second does not.

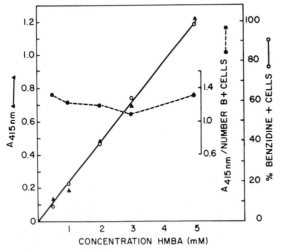

Figure 1
Effect of HMBA concentration on the induction of differentiation. Cultures were grown in HMBA (0.5–5 mM) and at 5 days the percentage of benzidine-reactive cells (B$^+$) (0) and the hemoglobin per 10^7 cells per ml (A_{415nm}) (▲) were determined. The average hemoglobin content per B$^+$ cell was calculated from these data (●). Data reprinted from Reuben et al. 1976.

To study the kinetics of recruitment of MELC to differentiation induced by HMBA, that is, the kinetics of commitment, we devised experiments designed to examine the duration of exposure to inducer required to commit MELC to express differentiation after they have been removed from the inducing agent (Fibach et al. 1977). The kinetics of induced differentiation were compared under three sets of experimental conditions (Fig. 2). When MELC were exposed continuously to an optimal inducing concentration (5 mM) of HMBA, benzidine-reactive, hemoglobin-containing cells were detectable in culture from about 48 hours, and the proportion of benzidine-reactive cells increased to over 90% by 96 hours. A second set of MELC were cultured in suspension with 5 mM HMBA for various times (5 to 100 hours), then transferred to fresh suspension medium free of HMBA. Commitment to differentiation in these "transfer-out" studies was assayed by determining the proportion of cells which had become benzidine-reactive after a total of 120 hours of culture. Under these conditions, commitment to differentiation above the background level (about 0.5%) was detected after 24 hours of exposure to HMBA. Commitment, in this assay, could be detected at least 24 hours before the cells displayed hemoglobin accumulation by the benzidine reaction. The proportion of committed cells increased linearly with time in HMBA, from 24 to 80 hours. As might be predicted, the progeny of those MELC committed to differentiate after only a short

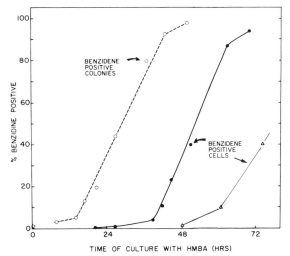

Figure 2
The kinetics of recruitment to differentiate induced by HMBA. Three independent assays of induced differentiation were compared. (△——△) Cells were grown continuously in 5 mM HMBA and the proportion of B$^+$ cells scored at the time points indicated. (●——●) Cells were cultured in 5 mM HMBA for the periods indicated, then washed and transferred to fresh HMBA-free suspension culture; the proportion of B$^+$ cells was determined after a total of 5 days in culture. (o-----o) Cells were cultured in 5 mM HMBA for the periods indicated, then washed and transferred to HMBA-free semi-solid (methylcellulose) cloning media (Fibach et al. 1977); the proportion of colonies containing B$^+$ cells was determined after a total of 5 days in culture.

period (24–48 hours) of exposure to HMBA became equally mature, by the criteria of morphology and the benzidine reaction, as did MELC exposed continuously to HMBA for 120 hours. These results suggest that there is a stage in the process of differentiation, prior to the accumulation of hemoglobin or other morphological and biochemical features of differentiation, at which MELC become irreversibly committed to continue the developmental process in the absence of inducer.

A third set of conditions was designed to measure quantitatively the commitment to differentiation at the single cell level. To determine the kinetics of commitment to differentiation and the capacity for replication of single MELC cells following culture with HMBA, we washed the cells and transferred them to inducer-free semisolid media (Fig. 2). Under these conditions, both induced and uninduced cells produced colonies which could be scored for differentiation after 5 days in culture by staining the culture in situ with benzidine. The background for spontaneously induced colonies, under these conditions, was consistently less than 3% and the cloning efficiency was over 85%, measured as the proportion of inoculated cells detectable as colonies or small clusters on day 5 of culture. Under these conditions, commitment was first detected in cells that had been exposed to 5 mM HMBA from 12 to 16 hours; the percentage of colonies containing benzidine-reactive cells increased linearly with duration of exposure to inducer, reaching virtually 100% by 50 hours. Commitment was also proportional to concentration of HMBA, from 0.5 to 5 mM; the optimal concentration for this inducer was found to be 5 mM HMBA, and only at concentrations higher than this was there toxic suppression of cloning efficiency.

Colonies derived from cells exposed in suspension to HMBA are of three types, as assayed by the benzidine reaction (Gusella et al. 1976; Fibach et al. 1977): (1) colonies containing uniformly benzidine-negative cells, (2) colonies containing uniformly benzidine-reactive cells, and (3) colonies containing a mixture of benzidine-reactive and nonreactive cells (mixed colonies). The contribution of mixed colonies to the total population of differentiated colonies was highest at suboptimal concentrations of HMBA or after short periods of exposure to the inducing agent (Fig. 3). These observations suggest that, under these conditions, a committed cell may give rise both to differentiated and undifferentiated progeny; that is, the differentiated state may be unstable, or the decision to differentiate may be expressed, in a statistical fashion, at a time subsequent to the exposure to inducing agent.

The number of cells in the colony was related to the proportion of benzidine-reactive cells in the colony. Colonies without benzidine-reactive cells, as well as mixed colonies, continued to increase in size throughout the period of culture. Uniformly benzidine-reactive colonies were smaller, generally showing no increase in size after day 4 and containing not over 16 to 32 cells. This suggests that induction to differentiation is associated with a limitation in the potential for cell division, consistent with the pattern of terminal differentiation characteristic of normal erythropoiesis (Marks and Rifkind 1972; Rifkind et al. 1974; Gusella et al. 1976; Fibach et al. 1977).

Taken together, these studies strongly implicate two phases in chemically

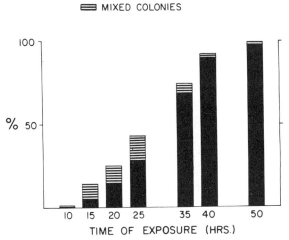

Figure 3
The effect of time of exposure to HMBA on the number and heterogeneity of colonies derived from MELC. Cells were cultured in suspension with HMBA (5 mM) for various periods of time, then washed, transferred to HMBA-free semisolid medium; the colonies were scored, by the benzidine reaction, as undifferentiated, fully differentiated, and mixed colonies containing both benzidine-reactive and undifferentiated cells.

induced differentiation: commitment and biochemical development. Considerable work in this and other laboratories is being aimed toward a definition of those cellular and biochemical events which are specifically related to the commitment phase of erythropoietic differentiation in MELC.

CELL SURFACE EFFECTS AND COMMITMENT TO DIFFERENTIATION

Following exposure to inducing agents MELC undergo a variety of alterations at the plasma membrane. Some of these are characteristic of normal erythropoiesis and, presumably, reflect expression of the program of erythropoietic development. These include changes in membrane-associated erythrocyte-specific proteins and antigens (Furusawa et al. 1971; Ikawa et al. 1973a, b) including spectrin (Arndt-Jovin et al. 1976) and receptors for transferin, the iron-binding serum protein (Hu et al. 1977). Other changes, including an early decrease in cell volume (Loritz et al. 1977) and an increase in plant lectin agglutinability (Eisen et al. 1977), may occur early enough to be implicated in the process of commitment. Evidence implicating cell-surface-mediated functions in the induction of differentiation has been derived also from the differentiation-inducing and differentiation-suppressing activities of agents with known or postulated plasma-membrane activities; these include the inducer ouabain (Bernstein et al. 1976) and the inhibitors procaine and tetracaine (Bernstein et al. 1975). These agents—the cardiac glycosides and the local anesthetics—have well-documented biological effects at the plasma membrane.

To develop further evidence implicating the plasma membrane or plasma-membrane-related functions in commitment, we examined the levels of cyclic nucleotide (cAMP) during chemically mediated commitment. Cells of MELC strain DS19 (a cell line sensitive to induction by all the inducing agents which we tested) and strain DR10 a cell line (Ohta et al. 1976) selected for resistance to induction by dimethylsulfoxide (DMSO) were placed in culture and exposed to each of the following inducing agents: HMBA (5 mM), DMSO (280 mM), or sodium butyrate (1.5 mM), and the cellular content of cAMP was determined during the first 18 hours in culture (Fig. 4). During the first 3 to 6 hours of culture of DS19 cells, a distinct rise occurred in cAMP content in cells exposed to any of the three inducing agents; a small rise occurred in cAMP content in control cells during the same period. By 12 hours, the cAMP content of these cells had returned to the initial value. Cells of strain DR10 failed to show a significant increase in cAMP content, compared to control cells, when exposed to DMSO but did show an increase in cAMP when exposed to HMBA or butyric acid, two inducing agents that can induce this DMSO-resistant cell line to differentiate.

Because cyclic nucleotide values are closely influenced by the progression of cells through the cell division cycle (Zeilig et al. 1976), we redesigned experiments on the effects of chemical inducers on cAMP levels, using MELC synchronized with respect to the cell cycle by the sequential application of the double thymidine blockade procedure and hydroxyurea (HU) (Levy et al. 1975; C. Schildkraut, per. comm.) (Fig. 5). With this

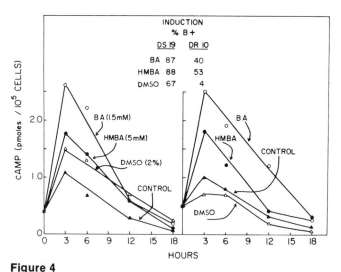

Figure 4
The effect of inducing agents on the cAMP content of an DMSO-sensitive (DS19) and DMSO-resistant (DR10) MELC line. Cells were cultured with the indicated concentrations of each inducer (BA, butyric acid; HMBA, hexamethylene bisacetamides; DMSO, dimethylsulfoxide) and samples were removed for radioimmunoassay of cAMP content (Cailla et al. 1973; Schultz et al. 1973) at the times indicated. At 5 days of culture the cultures were scored for the proportion of benzidine-reactive cells (% B^+).

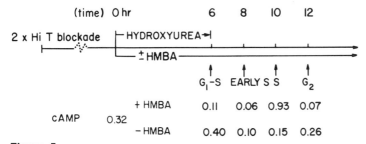

Figure 5
Effect of HMBA on cAMP content of MELC synchronized with respect to the cell cycle. Synchronization was achieved by two sequential exposures to 2 mM thymidine (Levy et al. 1975) and to 0.5 mM hydroxyurea. cAMP, measured in pmoles/10^5 cells, was assayed (as in Fig. 4) at the times indicated. Progress through the cell cycle was monitored by flow microfluorometry (Coulter model TPS-1; Coulter Electronics) after staining with propidium iodide.

procedure, control cells and cells exposed to 4 mM HMBA proceeded through the first cell cycle, following release from HU, synchronously and with the same kinetics. A brisk, almost tenfold increase in cAMP level was detected in cells exposed to HMBA, in mid-S-phase of the first cell cycle. Taken together with the data derived from studies on nonsynchronized MELC, these observations suggest that an early plasma-membrane-related effect of HMBA during commitment involves the accumulation of cAMP. This effect was not observed in the variant cell line (DR10) resistant to induction by DMSO, when exposed to that agent.

As an alternative approach, the effects of a series of such tumor-promoting plant diterpenes as 12-0-tetradecanoyl-phorbol-13-acetate (TPA) on spontaneous and chemically induced differentiation have been studied. Both spontaneous (Rovera et al. 1977; Yamasaki et al. 1977) and HMBA-mediated differentiation (Yamasaki et al. 1977) were inhibited by simultaneous exposure of MELC to TPA. Spontaneous differentiation was more sensitive to inhibition by TPA (0.5 to 1.0 ng/ml) than was HMBA-induced differentiation (10–100 ng/ml). Other diterpenes such as phorbol-12-,13-didecanoate, mezereine, and ingenol dibenzoate, which are active tumor-promoting agents, were likewise effective inhibitors both of spontaneous and HMBA-induced differentiation. Related compounds which are ineffective as tumor promotors were ineffective as differentiation inhibitors. Inhibition of differentiation was not a nonspecific toxic effect of these agents. MELC could be incubated, by repeated passages, for prolonged periods of time in the presence of HMBA and TPA, without displaying either toxicity or differentiation; when the TPA was removed, these cells (Table 1) displayed their normal responsiveness to HMBA, attaining over 80% benzidine-reactive cells.

To determine the relationship between TPA-mediated inhibition of differentiation and commitment to differentiation, TPA was added to MELC cultures at various times after HMBA had been added. Aliquots were assayed at selected times for commitment to differentiation. The addition

Table 1
Effect of Prolonged TPA + HMBA on
Inducible Differentiation

Compound	Time in HMBA + TPA		
	0 (%)	55 days (%)	75 days (%)
HMBA	94	82	86
HMBA + TPA	1	1	4

MELC were grown continuously in 4 mM HMBA and 100 ng/ml TPA for the periods noted, with passages every 4–5 days into fresh medium containing both agents. After 55 days (13 passages) and 75 days (18 passages), the cells were washed and resuspended in media containing either HMBA alone or HMBA + TPA. The proportion of benzidine-reactive cells was determined after 4 days of culture.

of TPA could be delayed for up to 24 hours after addition of HMBA and still achieve maximal inhibition of differentiation. When the addition of TPA was delayed beyond this point, TPA-mediated inhibition of differentiation progressively declined. This decline in inhibition was concomitant with the appearance of MELC committed to differentiate. The addition of TPA to cells already committed to differentiate had no effect on their ability to express differentiation, as manifested by the accumulation of benzidine-reactive cells. These observations suggest that TPA can only inhibit differentiation of cells which have not yet passed a critical step related to commitment to differentiation.

Although the cellular site of action of the tumor-promoting plant diterpenes with respect either to carcinogenesis or to differentiation inhibition has not been ascertained, several lines of evidence strongly suggest an effect of these compounds at the plasma membrane. First, MELC became adherent to the culture dish when incubated with TPA for several hours (Yamasaki et al. 1978). This TPA-mediated change in cell-surface characteristics was displayed only by TPA-susceptible MELC; two TPA-resistant MELC lines, in which differentiation was minimally influenced by exposure even to high doses of TPA, failed to show this response. Second, the tumor promotors characteristically induce plasminogen activator (Weinstein et al. 1977), and TPA induces plasminogen activator in TPA-sensitive MELC as well (Yamasaki et al. 1978). Despite these alterations we failed to detect any changes in plasma-membrane structure (by freeze-fracture electron microscopy) that could be ascribed to TPA (I. Sumida and R. A. Rifkind, unpubl.). Taken together, these observations suggest that the cellular target for TPA-mediated inhibition of differentiation may be located at the plasma membrane and that the TPA-mediated effect is directed at a step involved in the commitment phase of erythropoietic differentiation.

SUMMARY AND CONCLUSION

A wide variety of agents have been discovered which can alter significantly the probability for commitment to and expression of the program of erythro-

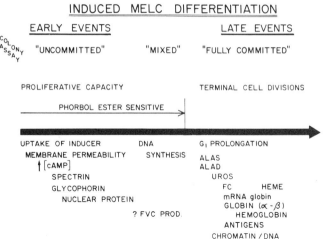

Figure 6
Hypothetical schema for the sequence of events during chemically induced commitment of MELC to differentiate. Exposure to inducer increases the probability for MELC to become committed to express differentiation. Commitment is defined, operationally, by the colony assay, as described in the legend to Figure 2. Cells which become committed lose the capacity for unlimited cell proliferation, initiate their terminal cell divisions, and accumulate hemoglobin (and, probably, red-cell-specific membrane antigens). "Early events" are those which occur, in cells cultured with inducer, prior to commitment, and "late events" are those which occur associated with or after commitment. FVC, Friend virus complex; ALAS, aminolevulinic acid synthetase; ALAD, aminolevulinic acid dehydrase; UROS, uroprotophorphyrin I synthetase; FC, ferrochelatase; Chromatin/DNA, alterations in structure or configuration of these constituents.

poietic differentiation in MELC. The diversity of these agents makes it difficult to provide a single generalizing statement as to the essential property of biologically active agents. Activities which involve effects on the cell-division cycle and on chromatin and DNA are discussed elsewhere (Marks et al. this volume). The present evidence (Fig. 6) suggests that at least some agents may have a primary effect at the level of the plasma membrane, which is manifested by alterations in cyclic nucleotide metabolism and by the inhibitory effects of a variety of agents including local anesthetics and the tumor-promoting phorbol esters. Presumably all inducing agents affect in common certain steps required for setting into action the erythropoietic program of differentiation; the unique properties of individual inducing agents may reflect various potential sites of interaction between inducing agents and the sequential steps of the common pathway for the regulation of erythropoietic differentiation.

Acknowledgments

Work by the authors of this review was supported in part by grants and contracts from the National Institutes of Health (GM-14552, CA-13696,

CA-18316, NO1-CB-4-4008, NO1-CP-6-1008) and the National Science Foundation (NSF-PCM-75-08696).

REFERENCES

Arndt-Jovin, D. J., W. Ostertag, H. Eisen, F. Klimek, and T. M. Jovin. 1976. Studies of cellular differentiation by automated cell separation. Two model systems. Friend-virus transformed cells and *Hydra attenuata*. *J. Histochem. Cytochem.* **24:** 332.

Bernstein, A., A. S. Boyd, V. Crichley, and V. Lamb. 1975. Induction and inhibition of Friend leukemic cell differentiation: The role of membrane-active compounds. In *Biogenesis and turnover of membrane macromolecules* (ed. J. S. Cook), p. 145. Raven Press, New York.

Bernstein, A., D. M. Hunt, V. Crichley, and T. W. Mak. 1976. Induction by ouabain of hemoglobin synthesis in cultured Friend erythroleukemic cells. *Cell* **9:** 375.

Cailla, H. L., M. S. Racine-Weisbuch, and M. A. Dellaage. 1973. Adenosine $3',5'$ cyclic monophosphate assay at 10^{-15} mole level. *Anal. Biochem.* **56:** 394.

Eisen, H., S. Nasi, C. P. Georgopoulos, D. Arndt-Jovin, and W. Ostertag. 1977. Surface changes in differentiating Friend erythroleukemic cells in culture. *Cell* **10:** 689.

Fibach, E., R. C. Reuben, R. A. Rifkind, and P. A. Marks. 1977. Effect of hexamethylene bisacetamide on the commitment to differentiation of murine erythroleukemic cells. *Cancer Res.* **37:** 440.

Fredrickson, T., P. Tambourin, F. Wendling, C. Jasmin, and F. Smajda. 1975. Target cell of the polycythemia-inducing Friend virus: Studies with myleran. *J. Natl. Cancer Inst.* **55:** 443.

Friend, C., M. C. Patuleia, and E. deHarven. 1966. Erythrocytic maturation *in vitro* of murine (Friend) virus-induced leukemia cells. *Natl. Cancer Inst. Monogr.* **22:** 505.

Friend, C., W. Scher, J. G. Holland, and T. Sato. 1971. Hemoglobin synthesis in murine virus induced leukemic cells *in vitro:* Stimulation of erythroid differentiation by dimethylsulfoxide. *Proc. Natl. Acad. Sci. U.S.A.* **68:** 378.

Furusawa, M., Y. Ikawa, and H. Sugano. 1971. Development of erythrocyte membrane-specific antigen(s) in clonal cultured cells of Friend virus-induced tumor. *Proc. Jpn. Acad.* **47:** 220.

Gregory, C. J. 1976. Erythropoietin sensitivity as a differentiation marker in the hemopoietic system: Studies of three erythropoietic colony responses in culture. *J. Cell. Physiol.* **89:** 289.

Gregory, C. J., and A. C. Eaves. 1977. Human marrow cells capable of erythropoietic differentiation *in vitro*. Definition of three erythroid colony responses. *Blood* **49:** 855.

Gusella, J., R. Geller, B. Clarke, V. Weeks, and D. Housman. 1976. Commitment to erythroid differentiation by Friend erythroid leukemia cells: A stochiastic analysis. *Cell* **9:** 221.

Harrison, P. R., N. Affara, D. Conkie, T. Rutherford, J. Sommerville, and J. Paul. 1976. Regulation of erythroid differentiation in Friend erythroleukaemic cells. In *Progress in differentiation research* (eds. N. Muller-Berat et al.), p. 135. North-Holland Publishing Co., Amsterdam.

Hu, H.-Y. Y., J. Gardner, P. Aisen, and A. I. Skoultchi. 1977. Inducibility of transferrin receptors on Friend erythroleukemic cells. *Science* **197:** 559.

Ikawa, Y., M. Furusawa, and H. Sugano. 1973a. Erythrocyte membrane-specific antigens in Friend virus-induced leukemia cells. *Bibl. Haematol.* **39:** 955.

Ikawa, Y., J. Ross, P. Leder, J. Gielen, S. Packman, P. Ebert, K. Hayashi, and H. Sugano. 1973b. Erythrodifferentiation of cultured Friend leukemia cells. In *Proceedings of the 4th international symposium of Princess Takomatsu cancer research fund* (eds. W. Nakahara et al.), p. 515. University of Tokyo Press, Tokyo.

Levy, J., M. Terada, R. A. Rifkind, and P. A. Marks. 1975. Induction of erythroid differentiation by dimethylsulfoxide in cells infected with Friend virus: Relationship to the cell cycle. *Proc. Natl. Acad. Sci. U.S.A.* **72:** 28.

Lilly, F. 1972. Mouse leukemia: A model of a multiple-gene disease. *J. Natl. Cancer Inst.* **49:** 927.

Lilly, F., and T. Pincus. 1973. Genetic control of murine viral leukemogenesis. *Adv. Cancer Res.* **17:** 231.

Loritz, F., A. Bernstein, and R. G. Miller. 1977. Early and late volume changes during erythroid differentiation of cultured Friend leukemic cells. *J. Cell. Physiol.* **90:** 423.

Marks, P. A., and R. A. Rifkind. 1972. Protein synthesis: Its control in erythropoiesis. *Science* **175:** 955.

———. 1978. Erythroleukemic differentiation. *Annu. Rev. Biochem.* (In press.)

McGarry, M. P., and E. A. Mirand. 1973. Incidence of Friend virus-induced polycythemia in splenectomized mice. *Proc. Soc. Exp. Biol. Med.* **142:** 538.

Mirand, E. A. 1967. Virus-induced erythropoiesis in hypertransfused-polycythemic mice. *Science* **156:** 832.

Nasrallah, A. G., and M. P. McGarry. 1976. Brief communication: In vivo distinction between a target cell for friend virus (FV^p) and murine hematopoietic stem cells. *J. Natl. Cancer Inst.* **57:** 443.

Ohta, Y., M. Tanaka, M. Terada, O. J. Miller, A. Bank, P. A. Marks, and R. A. Rifkind. 1976. Erythroid cell differentiation: Murine erythroleukemia cell variant with unique pattern of induction by polar compounds. *Proc. Natl. Acad. Sci. U.S.A.* **73:** 1232.

Pragnell, I. B., W. Ostertag, P. R. Harrison, R. Williamson, and J. Paul. 1976. Regulation of erythroid differentiation in normal and leukaemic cells. In *Progress in differentiation research* (eds. N. Muller-Berat et al.), p. 501. North-Holland Publishing Co., Amsterdam.

Reuben, R. C., R. L. Wife, R. Breslow, R. A. Rifkind, and P. A. Marks. 1976. A new group of potent inducers of differentiation in murine erythroleukemia cells. *Proc. Natl. Acad. Sci. U.S.A.* **73:** 862.

Reuben, R. C., P. A. Marks, R. A. Rifkind, M. Terada, E. Fibach, U. Nudel, Y. Gazitt, and R. Breslow. 1978. Induction of erythroid differentiation in Friend cells. In *Oji international seminar on genetic aspects of Friend virus and Friend cells* (ed. Y. Ikawa). Academic Press, New York. (In press.)

Rifkind, R. A., A. Bank, and P. A. Marks. 1974. Erythropoiesis. In *The red blood cell* (ed. D. McN. Surgenor), vol. 1, p. 51. Academic Press, New York.

Rovera, G., T. G. O'Brien, and L. Diamond. 1977. Tumor promoters inhibit spontaneous differentiation of Friend erythroleukemia cells in culture (phorbol diesters/terminal differentiation/hemoglobin subunits/mouse erythroleukemia). *Proc. Natl. Acad. Sci. U.S.A.* **74:** 2894.

Schultz, G., J. G. Hardman, K. Schultz, J. W. Davis, and E. W. Sutherland. 1973. A new enzymatic assay for guanosine $3':5'$-cyclic monophosphate and its application to the ductus deferens of the rat. *Proc. Natl. Acad. Sci. U.S.A.* **70:** 1721.

Singer, D., M. Cooper, G. M. Maniatis, P. A. Marks, and R. A. Rifkind. 1974. Erythropoietic differentiation in colonies of Friend virus transformed cells. *Proc. Natl. Acad. Sci. U.S.A.* **71:** 2668.

Steeves, R. A. 1975. Spleen focus-forming virus in Friend and Rauscher leukemia virus preparations. *J. Natl. Cancer Inst.* **54:** 289.

Steeves, R. A., E. A. Mirand, S. Thomson, and L. Avila. 1969. Enhancement of spleen focus formation and virus replication in Friend virus-infected mice. *Cancer Res.* **29:** 1111.

Stephenson, J. R., A. A. Axelrad, D. L. McLeod, and M. Shreeve. 1971. Induction of colonies of hemoglobin-synthesizing cells by erythropoietin in vitro. *Proc. Natl. Acad. Sci. U.S.A.* **68:** 1542.

Takahashi, E., M. Yamada, M. Saito, M. Kuboyama, and K. Ogasa. 1975. Differentiation of cultured Friend leukemia cells induced by short-chain fatty acids. *Gann* **66:** 577.

Tambourin, P., and F. Wendling. 1971. Malignant transformation and erythroid differentiation by polycythaemia-inducing Friend virus. *Nat., New Biol.* **234:** 230.

———. 1975. Target cell for oncogenic action of polycythemia-inducing Friend virus. *Nature* **256:** 320.

Terada, M., J. Banks, and P. A. Marks. 1971. RNA synthesized during differentiation of yolk sac erythroid cells. *J. Mol. Biol.* **62:** 347.

Weinstein, I. B., M. Wigler, and C. Pietropaolo. 1977. The action of tumor-promoting agents in cell culture. In *Origins of human cancer* (eds. H. H. Hiatt et al.), p. 751. Cold Spring Harbor Laboratory, Cold Spring Harbor, New York.

Yamasaki, H., E. Fibach, U. Nudel, I. B. Weinstein, R. A. Rifkind, and P. A. Marks. 1977a. Tumor promoters inhibit spontaneous and induced differentiation of murine erythroleukemia cells in culture. *Proc. Natl. Acad. Sci. U.S.A.* **74:** 3451.

Yamasaki, H., E. Fibach, I. B. Weinstein, U. Nudel, R. A. Rifkind, and P. A. Marks. 1977b. Inhibition of murine erythroleukemia cell differentiation by tumor promoters. In *Oji international seminar on genetics aspects of Friend virus and Friend cells* (ed. Y. Ikawa). Academic Press, New York. (In press.)

Zeilig, C. E., R. A. Johnson, E. W. Sutherland, and D. L. Friedman. 1976. Adenosine 3':5'-monophosphate content and actions in the division cycle of synchronized HELA cells. *J. Cell Biol.* **71:** 515.

Induction of Murine Erythroleukemia Cells to Differentiate: Cell-Cycle-Related Events in Expression of Erythroid Differentiation

P. A. Marks, M. Terada, E. Fibach, Y. Gazitt, U. Nudel, R. Reuben, A. Bank, and R. A. Rifkind

Cancer Center and Departments of Medicine and Human Genetics and Development
Columbia University, College of Physicians and Surgeons, New York, New York 10032

Murine erythroleukemia cells (MELC), infected with Friend virus complex (Friend 1957), are being employed in our laboratory to study the regulation of expression of differentiated characteristics as well as the block in differentiation associated with viral infection. A variety of chemicals of known structure have been shown to induce MELC to express a program characteristic of erythroid cell differentiation (Marks, this volume). Induced differentiation is characterized by morphological changes and by the appearance of newly synthesized proteins characteristic of erythroid differentiation (Friend et al. 1971; Ross et al. 1972; Harrison 1976; Marks and Rifkind 1978), including hemoglobins.

The present investigations were designed primarily to evaluate the relationship of cell-cycle-related events to induced MELC expression of the program of erythroid differentiation. A major issue with respect to induced MELC differentiation is the nature of events determining the transition from a self-renewing, nondifferentiating population of cells to one committed to expression of the characteristic erythroid program. Earlier studies from our laboratories provided evidence that MELC, synchronized with respect to the cell division cycle by exposure to high levels of thymidine (Levy et al. 1975), require the presence of dimethylsulfoxide (DMSO) during at least one round of DNA synthesis. McClintock and Papaconstantinou (1974) and Harrison (1976) have provided evidence, as well, that MELC differentiation is dependent on at least one or more rounds of DNA synthesis. On the other hand, Leder et al. (1975) reported that butyric acid induces MELC in the presence of inhibitors of DNA synthesis, hydroxyurea (HU) or cytosine arabinoside, and in the absence of cell division. They concluded that globin gene expression does not require DNA synthesis for cell division. These conclusions are open to question because, under the conditions of their experiments, neither HU nor cytosine arabinoside completely inhibits thymidine incorporation into DNA (Marks et al. 1977), and both of these agents are themselves inducers of MELC differentiation (Ebert et al. 1976).

We have further investigated the relationship between cell cycle events and transition to hemoglobin production in induced MELC by examining the pattern of cell-cycle transit of MELC cultured without and with inducers (Terada et al. 1977).

PROLONGATION OF G_1 OR TRANSIENT INHIBITION OF INITIATION OF S PHASE IN MELC CULTURED WITH INDUCERS

MELC cultured with DMSO or other inducing agents developed a prolongation of G_1 or a transient block in initiation of DNA synthesis. This was demonstrated by measurement of the rate of DNA synthesis, proportion of cells in S phase, and pattern of DNA accumulation in MELC grown in nonsynchronous cultures with and without DMSO, butyric acid, or dimethylacetamide (Terada et al. 1977). In MELC cultured without inducer, there was an initial rise in the rate of [^{3}H]-thymidine incorporation, with a maximum value achieved by about 10 hours (Fig. 1). The initial increase in rate of thymidine incorporation probably reflected entry into S phase of

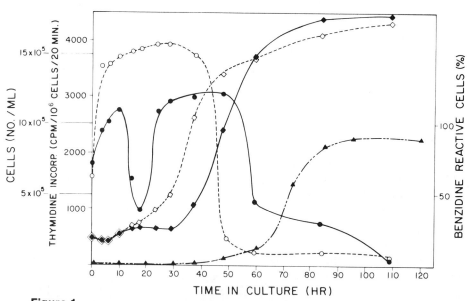

Figure 1
Cell growth and rate of synthesis of DNA in MELC cultured without and with DMSO. MELC after 60 hours of culture were transferred to medium without and with 280 mM DMSO at an initial cell concentration of 2×10^5 cells/ml. At each time indicated, an aliquot of cell suspension was removed to determine cell number (◇-----◇, control; ◆——◆, DMSO), benzidine-positive cells (▲-----▲) in cultures with DMSO, and [^{3}H]-thymidine incorporation (20-minute incubations) into trichloroacetic-acid-insoluble material in cells cultured without (○-----○) and with (●——●) DMSO. For details of methods, see Terada et al. (1977).

cells partially synchronized in the postlogarithmic growth phase of the previous cell passage. Between 10 and 40 hours, this rate remained relatively constant, decreasing thereafter to less than 10% of the peak value by 60 hours. The plateau level observed between 10 and 40 hours reflected a constant proportion of cells in S phase (loss of the partial synchronization). The fall in DNA synthesis after 60 hours coincided with the onset of stationary growth phase culture. In comparison, although there was an initial rise in the rate of [^{3}H]-thymidine incorporation in cells cultured with DMSO, a difference between cultures with and without inducers was observed as early as 4 to 6 hours after initiation of cultures (Fig. 1). In the population of cells cultured with DMSO, a decrease in the rate of thymidine incorporation was observed between 10 and 20 hours. At 20 hours, the rate of [^{3}H]-thymidine incorporation was at its lowest, about 25% of the rate in control cultures. The initial rise in rate of thymidine incorporation in cells cultured with inducer indicated that some cells were proceeding through S phase prior to the prolonged G_1. As cells passed into G_1, the proportion of cells whose DNA synthesis was blocked increased. In the induced cultures, the rate of [^{3}H]-thymidine incorporation rose between 20 and 30 hours to a peak value of about 75% of the highest rate in control cultures, where it remained until 50 to 60 hours and then decreased. This latter decrease coincided with the stationary growth phase and with terminal differentiation. These results indicate a prolongation of G_1 in induced MELC cultures and were confirmed by three other types of studies evaluat-

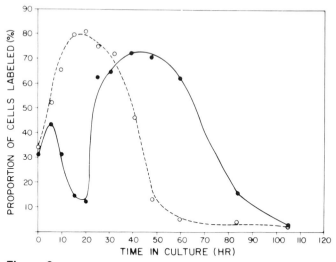

Figure 2
Thymidine-labeling index during erythroid cell differentiation of MELC induced by DMSO. MELC were transferred to medium without and with 280 mM DMSO as described in the legend for Figure 1. At each time indicated, an aliquot of cell suspension was removed for incubation with [^{3}H]-thymidine for 20 minutes to obtain the thymidine-labeling index by radioautography (o-----o, no DMSO; •——•, with DMSO) according to the methods described in Terada et al. (1977).

ing the pattern of cell cycle. Thus, determination of the proportion of cells in S phase by radioautography of cells labeled during 20 minutes' exposure to [^{3}H]-thymidine (Fig. 2) and determination of the relative DNA content per cell, with both fluorescent Feulgen assay and propidium-iodide binding measured with flow microfluorometery (Fig. 3) being used, provide evidence for a decrease in the proportion of cells in S phase or prolongation of G_1 during the early period of culture with inducing agent. The decrease in the proportion of cells in S is most marked in comparison with uninduced cultures at about 20 hours under these culture conditions.

The effects of inducers on transit of MELC through the cell division cycle was examined further. MELC were synchronized by culture through sequential periods with 2 mM thymidine for 9 hours, no inhibitor for 5 hours, again a 9-hour period with 2 mM thymidine, no inhibitor for 5 hours, and 0.5 mM HU for 6 hours (Manduca et al. 1977). During this latter 6 hours, an inducer was present. After culture with HU, cells were transferred to fresh medium with the same inducer. Cells without or with inducer proceed through S, G_2, and M in synchronous manner and with similar transit times. Thereafter, MELC cultured with inducers remained in G_1 6.5 to 8 hours, compared with 4 hours for cells cultured without inducer (Table 1; Fig. 4).

CHANGES IN DNA ASSOCIATED WITH INDUCTION OF MELC DIFFERENTIATION

In addition to prolongation of G_1, there was evidence that structural changes in chromatin occur in the course of induced MELC differentiation. DNA from MELC cultured with 280 mM DMSO showed a decrease in sedimentation rate in alkaline sucrose gradients after alkali lysis of the cells (Fig. 5) (Terada et al. 1978). These changes were detected as early as 27 hours after beginning of culture with nonsynchronous cells. No change in the coefficient of sedimentation of DNA was observed when DNA prepared from MELC cultured from 27 or 60 hours with DMSO was analyzed on neutral sucrose gradients. The decrease in rate of sedimentation of DNA on alkaline sucrose gradients was also seen with preparations of DNA from MELC cultured with 1.8 mM butyric acid or 30 mM dimethylacetimide (Fig. 6). When DNA from a DMSO-resistant MELC line (DR10) (Ohta et al. 1976) was analyzed after culture with 280 mM DMSO, no changes in the DNA sedimentation pattern on alkaline sucrose gradient were detected (Fig. 6) (Terada et al. 1978).

To summarize this aspect of our investigations—several pieces of evidence support the hypothesis that these changes in DNA are related to the initiation of erythroid differentiation of MELC. The changes can be detected as early as 27 hours after beginning of culture, at approximately the same time at which a portion of the cells in these conditions of nonsynchronous cultures begin to show commitment to differentiate (Levy et al. 1975) and to accumulate globin mRNA (Singer 1975). Further, this reduction of apparent S value on alkaline sucrose gradient can be detected after culture with other inducers but cannot be detected in DNA from the

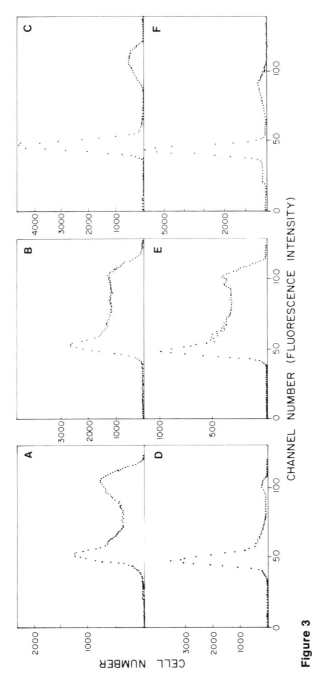

Figure 3

Distributions of DNA content per cell measured by propidium-iodide staining and flow microfluorometry (FMF). Peak corresponding to G_1 cells was adjusted to be about channel 50. After beginning of culture without and with 280 mM DMSO, an aliquot of cell suspension was removed at the times indicated below to determine the DNA distribution histogram by FMF according to the methods described in Terada et al. (1977). (*A*) cells cultured without DMSO for 19 hours; (*B*) cells cultured without DMSO for 38 hours; (*C*) cells cultured without DMSO for 88 hours; (*D*) cells cultured with DMSO for 19 hours; (*E*) cells cultured with DMSO for 38 hours; (*F*) cells cultured with DMSO for 88 hours.

Table 1
Effects of Inducers of MELC on Transit through Cell Cycle Phases

		Time (hours)		
Source of cells	Inducers	G_1	S	$G_2 + M$
Nonsynch	None	4	6.8	2
Nonsynch	DMSO	6.5	6.2	2.3
Synch	None	4	6	2
Synch	DMSO	8	5.5	2
Synch	HMBA	6.5	6	2
Synch	B.A.	6.5	6	2

The duration of different phases of the cell cycle was measured for nonsynchronous cells by determining the proportion of cells in each phase of the cycle (from the distribution of DNA content per cell as measured by propidium iodide staining and flow microfluorometry analysis), and the doubling time of the cells in log-phase growth in culture (Terada et al. 1977). The duration of different phases of the cell cycle for synchronized cells was determined by analysis of the distribution of DNA content per cell (measured by propidium iodide staining and flow microfluorometry), as illustrated in Figure 4. B.A., butyric acid.

DMSO-resistant MELC variant DR10 after culture with DMSO. One explanation for the change in alkaline sucrose gradient sedimentation pattern accompanying MELC differentiation may be that single-strand breaks accumulate in the DNA of cells cultured with inducers (Cleaver 1975; Ormerod 1976). Such single-strand breaks may permit rapid unwinding of complementary strands during exposure to alkali.

UV-IRRADIATION AND ACTINOMYCIN-D-INDUCED MELC DIFFERENTIATION

UV-irradiation of mammalian cells produced single strand breaks in DNA and reduction of the sedimentation rate of DNA in alkaline sucrose gradient centrifugation (Cleaver 1975). Actinomycin D in low concentrations also caused reduction of sedimentation rate of DNA in alkaline sucrose gradient and prolongation of G_1 (M. Terada et al., in prep.).

Brief UV-irradiation of MELC prior to inoculation of cultures induced up to 10% of the cells to differentiate (Table 2) (Terada et al. 1978). It was not possible to increase the proportion of MELC which differentiated, since exposure of cells to UV doses exceeding 300 erg/mm² resulted in cell death. An increased proportion of MELC could be induced to differentiate with UV-irradiation when suboptimal concentrations of DMSO were present in the culture media after irradiation. MELC cultured for 5 days with 28 mM DMSO, after exposure to UV-irradiation, showed up to 25% benzidine-reactive cells. Without prior UV-irradiation, cells cultured in this concentration of DMSO showed no detectable induction to differentiate. Addition of 70 mM DMSO in culture after UV-irradiation of cells resulted in up to

HMBA EFFECT ON SYNCHRONIZED MELC CELL CYCLE

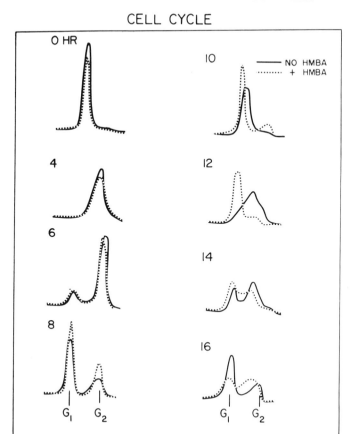

Figure 4
Distributions of DNA content per cell measured by propidium-iodide staining and by FMF (Terada et al. 1977) in cells synchronized according to the method of Manduca et al. (1977) and then cultured in medium without or with 5 mM hexamethylene bisacetamide (HMBA) as described in the text. Aliquots of the cultures were removed at the times indicated for analysis of distributions of DNA content per cell. Zero indicates the time at which cells were removed from medium with HU and HMBA and resuspended in fresh medium with HMBA. Cells cultured without HMBA (———) and with HMBA (-----) progressed through S, G_2, and M synchronously, with identical transit times. Cells in culture with HMBA remained in G_1 from 6 hours through, at least, 12 hours, whereas cells in culture without HMBA entered a second S phase by 10 hours.

70% induction; without prior UV-irradiation, only 8 to 10% benzidine-reactive cells accumulated after 5 days in culture (Table 2).

Actinomycin D, 1 to 2 ng/ml, induced over 90% of MELC to differentiate after 5 days in culture (Fig. 7) (M. Terada et al., in prep.). At concentrations below 1 ng/ml or above 3 ng/ml, actinomycin D was less effective as an in-

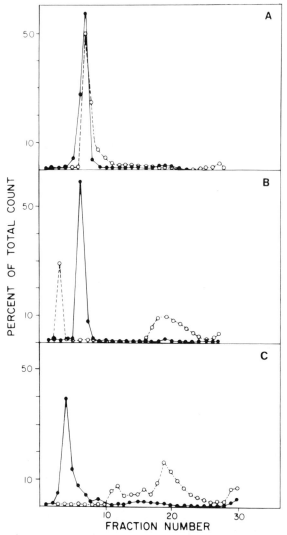

Figure 5
Alkaline sucrose density gradient sedimentation pattern of DNA from MELC after culture without and with 280 mM DMSO. MELC were cultured with 2-[¹⁴C]-thymidine, 57 µCi/mole at a final concentration of 0.2 µCi/ml, without and with DMSO, for the times indicated below. Cells were collected, washed, and then incubated in fresh culture medium for 2 to 5 hours, and again the cells were collected, washed, layered, and lysed on the alkaline sucrose gradient. Centrifugation, collection of fractions, and determination of radioactivity in each fraction were performed according to the methods described in Terada et al. (1978). To each fraction was applied 5,000 to 15,000 cpm. The percentage of counts recovered in each fraction of the gradient is presented in this figure. Direction of centrifugation is from the right to left. Duration of culture without and with DMSO was (A) 19 hours, (B) 27 hours, and (C) 64 hours; (●———●) without DMSO; (○-----○) with DMSO. Fractions 5 to 7 correspond to a coefficient of sedimentation of approximately 300S; fractions 18 to 25 correspond to a coefficient of sedimentation of approximately 120S to 150S.

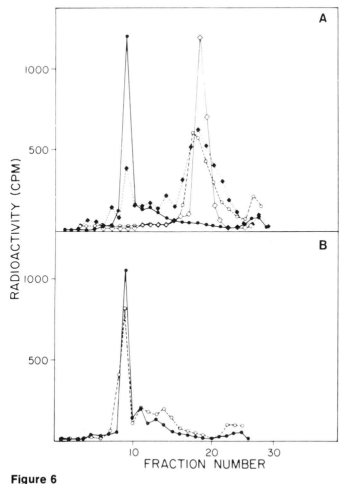

Figure 6
Alkaline sucrose gradient sedimentation pattern of DNA from MELC and a variant MELC line which is resistant to induction by DMSO (DR10). (*A*) MELC were cultured with 2[^{14}C]-thymidine and 280 mM DMSO, 30 mM dimethylacetamide or 1.8 mM butyric acid for 60 hours, and alkaline sucrose gradient analysis of DNA performed as described in legend to Figure 5. (●——●) No inducer, (o-----o) DMSO, (◇——◇) dimethylacetamide, (◆-----◆) butyric acid. (*B*) MELC (DR10) were cultured with 2-[^{14}C]-thymidine and without and with DMSO for 60 hours, and alkaline sucrose gradient analysis performed (●——●) without DMSO, and (o-----o) with DMSO.

ducer. Analysis of the time course of commitment to differentiate of MELC cultured with 1.5 ng/ml actinomycin D revealed that within 5 hours, 36% of the cells had yielded colonies with benzidine-reactive cells and by 96 hours in culture 99% of the cells had yielded differentiated colonies. Globin mRNA accumulation began within 20 hours in MELC cultured with actinomycin D (M. Terada et al., in prep.).

The above described alterations of DNA associated with induced MELC differentiation may be critical in the expression of this program of differentia-

Table 2
Effect of UV-Irradiation on Erythroid Differentiation of MELC

Concentration of DMSO	UV doses (erg/mm^2)		
	0	120	240
mM	*percentage of benzidine-reactive cells*		
0	<0.5	4	12
28	<0.5	15	25
70	10	35	67

Logarithmically growing MELC were exposed to UV-irradiation at the doses indicated in the table. The cells were continued in culture for 5 days without or with DMSO and the percentage of benzidine-reactive cells was determined. For details of procedures see Terada et al. (1978).

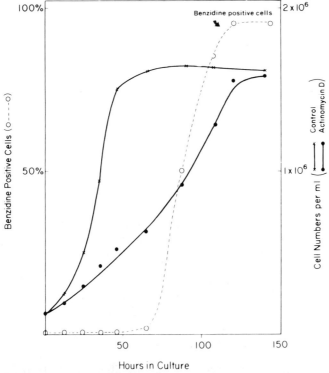

Figure 7
Cell growth and accumulation of benzidine-positive cells in MELC cultured with actinomycin D. After 60 hours of culture, cells were transferred to medium without and with 1.5 ng/ml actinomycin D at an initial cell concentration of 2 × 10^5 cells/ml. At each time indicated, an aliquot of cell suspension was removed to determine cell numbers per milliliter (x———x) without actinomycin D, (●———●) with actinomycin D, and (o-----o) benzidine-positive cells. The methods used were those described by M. Terada et al. (in prep.)

tion. Alternatively, changes in the properties of DNA could be a result of differentiation, subsequent to the events that are critical to the commitment of MELC to express the program of differentiation. The finding that UV-irradiation and actinomycin D can induce MELC to erythroid differentiation supports the former hypothesis.

SUMMARY AND HYPOTHESIS OF THE MECHANISM OF INDUCED MELC DIFFERENTIATION

MELC appear to be erythroid precursor cells transformed by the Friend virus complex. One component of this virus complex, the spleen focus-forming virus, is responsible for uncoupling cellular replication from erythropoietin, the physiological regulator of erythroid precursor cell proliferation. By some mechanism, perhaps selection of cells with the greatest growth potential, cell lines become established in which the normal process of erythropoietic differentiation is suppressed. This suppression may be accomplished by changes in the cell-cycle kinetics, including shortening of the normal G_1. A wide variety of agents have been discovered that can alter significantly the probability for commitment to an expression of the program of erythropoietic differentiation. The diversity of these agents makes it difficult to provide a single generalizing statement as to the essential property of biologically active agents. Indeed, the present evidence suggests that different agents may have different primary sites of action, for example, at the plasma membrane (e.g., polar-planar compounds and oubain) or at the level of chromatin (e.g., UV-irradiation and actinomycin D) (Bernstein et al., this volume; Rifkind et al., this volume). Presumably all inducing agents affect in common certain steps required for expression of the erythroid program of differentiation. Evidence has been reviewed which suggests that at least one round of DNA synthesis in culture with inducers occurs prior to expression of differentiated function in induced MELC. In addition, all agents studied to date with respect to their effects on the cell cycle have been shown to cause a prolongation of G_1. Associated with the induction process are alterations in chromatin structure, including possible accumulation of single-strand breaks in DNA and altered chromatin proteins (Keppel et al. 1977; Riggs et al. 1977). The precise relationships among DNA synthesis, prolongation in G_1, alteration in chromatin structure, and the commitment of MELC to differentiate and accumulate globin mRNA are not known.

We suggest that a chemical inducer is taken up by the MELC throughout the cell cycle. When a critical concentration is achieved, alterations in the cell occur, perhaps at the level of chromatin, which commit it to differentiate. The kinetics of the rate-limiting step in this process appear to be stochastic (Gusella et al. 1976; Fibach et al. 1977). The prolongation of G_1 may be important for the expression of certain of the characteristics of the erythroid phenotype, as, for example, accumulation of globin mRNA. In nonsynchronized cultures, the temporal sequence of commitment, prolongation of G_1, and accumulation of globin mRNA are consistent with this hypothesis. A more critical test of the hypothesis awaits the analysis of suitably synchronized MELC induced to differentiate.

Acknowledgments

The present investigations were supported, in part, by grants from the National Institute of General Medical Sciences (GM-14552), National Cancer Institute (CA-13696, CA-18314), and National Science Foundation (PCM 75-08696) and contract NO1-CP-6-1008. R. Reuben is a Special Fellow of the Leukemia Society, M. Terada is a Hirschl Trust Scholar, and E. Fibach is a Schultz Foundation Scholar.

REFERENCES

Cleaver, J. E. 1975. Methods for studying repair of DNA damage by physical and chemical carcinogens. In *Methods in cancer research* (ed. H. Busch), vol. 2, p. 123. Academic Press, New York.

Ebert, P. S., I. Wars, and D. N. Buell. 1976. Erythroid differentiation in cultured Friend leukemia cells treated with metabolic inhibitors. *Cancer Res.* **36:** 1809.

Fibach, E., R. C. Reuben, R. A. Rifkind, and P. A. Marks. 1977. Effects of hexamethylene bisacetamide on the commitment of differentiation of murine erythroleukemia cells. *Cancer Res.* **37:** 440.

Friend, C. 1957. Cell-free transmission in adult Swiss mice of a disease having the character of a leukemia. *J. Exp. Med.* **105:** 307.

Friend, C., W. Scher, J. G. Holland, and T. Sato. 1971. Hemoglobin synthesis in murine virus induced leukemic cells *in vitro*: Stimulation of erythroid differentiation by dimethylsulfoxide. *Proc. Natl. Acad. Sci. U.S.A.* **68:** 378.

Gusella, J., R. Geller, B. Clarke, V. Weeks, and D. Housman. 1976. Commitment to erythroid differentiation by Friend erythroid leukemia cells: A stochastic analysis. *Cell* **9:** 221.

Harrison, P. R. 1976. Analysis of erythropoiesis at the molecular level. Review article. *Nature* **262:** 353.

Keppel, F., B. Allet, and H. Eisen. 1977. Appearance of a chromatin protein during the erythroid differentiation of Friend virus-transformed cells. *Proc. Natl. Acad. Sci. U.S.A.* **74:** 653.

Leder, A., S. Orkin, and P. Leder. 1975. Differentiation of erythroleukemia cells in the presence of inhibitors of DNA synthesis. *Science* **190:** 893.

Levy, J., M. Terada, R. A. Rifkind, and P. A. Marks. 1975. Induction of erythroid differentiation by dimethylsulfoxide in cells infected with Friend virus: Relationship to the cell cycle. *Proc. Natl. Acad. Sci. U.S.A.* **72:** 28.

Manduca, P., I. Balaz, E. Brown, H. Crissman, E. Friedman, and C. Schildkraut. 1977. Synchronization of DNA synthesis in cultures of Friend erythroleukemia cells. *Cell.* (In press.)

Marks, P. A., and R. A. Rifkind. 1978. Erythroleukemic differentiation. *Annu. Rev. Biochem.* (In press.)

Marks, P. A., R. A. Rifkind, A. Bank, M. Terada, G. Maniatis, R. C. Reuben, and E. Fibach. 1977. Erythroid differentiation and the cell cycle. In *Growth kinetics and biochemical regulation of normal and malignant cells* (eds. B. Drewinko and R. M. Humphrey), p. 329. Williams & Wilkins Co., Baltimore.

McClintock, P. R., and J. Papaconstantinou. 1974. Regulation of hemoglobin synthesis in a murine erythroblastic leukemic cell: The requirement for replication to induce hemoglobin synthesis. *Proc. Natl. Acad. Sci. U.S.A.* **71:** 4551.

Ohta, Y., M. Tanaka, M. Terada, O. J. Miller, A. Bank, P. A. Marks, and R. A. Rifkind. 1976. Erythroid cell differentiation: Murine erythroleukemia cell variant with unique pattern of induction by polar compounds. *Proc. Natl. Acad. Sci. U.S.A.* **73:** 1232.

Omerod, M. 1976. Radiation induced strand breaks in the DNA of mammalian cells. In *Biology of radiation carcinogenesis* (eds. J. M. Yulras, R. W. Tennent, and J. Regan), p. 67. Raven Press, New York.

Riggs, M. G., R. G. Whittaker, J. R. Neumann, and V. M. Ingram. 1977. Appearance of a chromatin protein during the erythroid differentiation of Friend virus-transformed cell. *Nature.* (In press.)

Ross, J., Y. Ikawa, and P. Leder. 1972. Globin messenger-RNA induction during erythroid differentiation of cultured leukemia cells. *Proc. Natl Acad. Sci. U.S.A.* **69:** 3620.

Singer, D. S. 1975. Erythropoietic differentiation in murine erythroleukemic cells. Ph.D. thesis, Columbia University, New York.

Terada, M., J. Fried, U. Nudel, R. A. Rifkind, and P. A. Marks. 1977. Transient inhibition of initiation of S-phase associated with dimethylsulfoxide induction of murine erythroleukemia cells to erythroid differentiation. *Proc. Natl. Acad. Sci. U.S.A.* **74:** 248.

Terada, M., U. Nudel, E. Fibach, R. A. Rifkind, and P. A. Marks. 1978. Changes in DNA associated with induction of erythroid differentiation by dimethylsulfoxide in murine erythroleukemia cells. *Cancer Res.* **35:** 835.

Modulation of in Vitro Erythropoiesis: Hormonal Interactions and Erythroid Colony Growth

J. W. Adamson, W. J. Popovic, and J. E. Brown

Hematology Research Laboratory, Veterans Administration Hospital,
Seattle, Washington 98108, and
Division of Hematology, Department of Medicine,
University of Washington, Seattle, Washington 98195

Erythropoiesis is regulated primarily by the humoral factor erythropoietin (epo) (Krantz and Jacobson 1970). This hormone has an estimated molecular weight of 46,000 daltons and is thought not to enter cells to carry out its function. Although direct evidence is sparse (Chang et al. 1974), the concept has evolved that epo interacts with specific receptors on target cells to initiate the terminal differentiative steps in erythroid development. Because of the lack of sufficient quantities of pure epo and because of insufficient homogeneity of epo-responsive cells, little is known about the mechanism by which the hormone triggers cell activity.

In other cellular systems, however, the growth and development of individual cells and tissues may be influenced by small molecules, particularly cyclic nucleotides (Holley 1975; Tompkins 1975). First described by Sutherland and coworkers, this class of compounds has been shown to be the intracellular "second messenger," mediating the effects of a number of primary regulatory hormones that act initially at the cell surface (Robison et al. 1971). In addition, hormones of different types may interact to influence the same target cell or cells which bear receptors for those hormones. For example, exposure of established fibroblast lines such as 3T3 cells to insulin has been reported to alter the response of these cells to other hormones such as prostaglandin $F_{2\alpha}$ (de Asua et al. 1977). In addition, in some models thyroid hormones are thought to regulate the number of adrenergic receptors on selected cells, such as myocardial cells (Williams et al. 1977). Because these cells have important functions that are mediated by catecholamines, it is clear that hormonal interactions may significantly modulate tissue function.

Because of their widespread effects on cell function, a number of cyclic nucleotides and their various congeners have been studied in culture for their effect on erythropoiesis. Initial work failed to demonstrate that these compounds have any positive effect on in vitro erythropoiesis when rodent marrow cells were used in suspension culture (Graber et al. 1972; Goldwasser 1975). These findings were confirmed in our laboratory; however, the absence of response depended upon the assay system chosen and the species of animals used. Thus, when hematopoietic cells from humans, rabbits, and dogs were cultured, enhancement of hemoglobin synthesis by cyclic nucleotides was seen (Brown and Adamson 1977a). The limitations of these studies, which focused on hemoglobin synthesis, included the fact that proliferation-dependent events were not clearly identified and that the precise function of a single class of responsive cell could not be assessed. To clarify the mechanism of action of nucleotides, we used the erythroid colony-forming assay to assess the influence of these compounds. It was found that cyclic nucleotides did not initiate colony growth by themselves, but that epo was an obligate component of the culture system. Consequently, we concluded that if cyclic nucleotides were to play a physiological role in erythropoiesis, then it should come through the modulation of the interaction of the target cell and its primary regulatory hormone, epo (Brown and Adamson 1977b).

These observations prompted the following in vitro studies which were designed to examine the effects on erythroid colony growth of various compounds, including catecholamines, thought to act primarily through cell-surface receptors to activate adenyl cyclase. The results of these studies have confirmed hormonal interactions in the enhancement of colony growth and have demonstrated receptor specificity. Finally, the interaction of thyroid hormone on erythroid colony-forming units (CFU-E) has been examined, since thyroid hormones significantly affect catecholamine receptor numbers and responsiveness in other systems.

METHODS

All cultures were carried out with freshly aspirated bone-marrow cells obtained from random-bred dogs lightly anesthetized with sodium pentobarbital. The cells were aspirated in a sterile manner from the femurs or iliac crests. The cells were immediately suspended in ice-cold Hanks' balanced saline solution (BSS; Microbiological Associates) containing 10 U/ml of preservative-free heparin and 2% fetal calf serum (FCS; Reheis Chemical Co.). The cells were centrifuged, washed, spun a second time, and the buffy coat layer was removed; a hemocytometer was then used to count the trypan-blue-dye-excluding nucleated cells.

Assay for Erythroid Colony-Forming Units

A modification of the technique of Stephenson et al. (1971) was used to culture marrow cells for erythroid colony growth. Each culture dish contained a total volume of 1 ml with final concentrations of reagents as follows: 30% FCS, 10% beef embryo extract (Grand Island Biological Co.), 10% bovine serum albumin (BSA; Miles Laboratories), 40% alpha medium (Flow Laboratories), 10% bovine citrated plasma (Grand Island Biological Co.), 5 μmoles of beta mercaptoethanol, 20 U of penicillin, 20 mcg streptomycin, and appropriate concentrations of epo. The epo used in the experiments was a commercial preparation obtained from Connaught Laboratories that had a specific activity of approximately 3 International Reference Preparation (IRP) U/mgm of protein. The final concentration of epo in the stock solution was 0.1 IRP U/ml. Dog marrow cells were suspended in a concentration of 2×10^5 per ml and initially were mixed with all of the reagents with the exception of the bovine citrated plasma.

The various compounds to be tested were added to appropriate culture dishes which contained the bovine citrated plasma. The remainder of the reagents, including the cells, was then added and the contents of the plate were rapidly mixed and allowed to clot. The cultures were then incubated at 37°C in a high-humidity, 5% CO_2–95% air, tissue-culture incubator. The clots were fixed 2 days later with 5% glutaraldehyde in phosphate-buffered saline (PBS) and colonies were counted with an inverted tissue culture microscope. To confirm the identity of hemoglobin-containing cells and thereby to verify the counting results, we removed some of the clots from the dishes and stained them with benzidine. Only colonies containing eight or more obvious erythroid cells were scored.

Studies with Catecholamines

To examine their influence on erythroid colony numbers, we dissolved various β-adrenergic agonists and blocking agents in BSS and added them to the cultures in microliter quantities. The various catecholamines tested with their receptor specificities included: L-phenylephrine (α; Becton-Dickinson & Co.), L-norepinephrine ($\alpha\beta_1$; Sigma Chemical Co.), L-epinephrine ($\alpha\beta_1\beta_2$; Schwartz-Mann), L-isoproterenol ($\beta_1\beta_2$; Sigma Chemical Co.), albuterol (β_2; Schering Corp.), and metaproterenol (β_2; Boehringer-Ingelheim). The antagonists tested included: phentolamine (α; Ciba-Geigy Corp.), D-propranolol and L-propranolol ($\beta_1\beta_2$; Ayerst Laboratories), practolol (β_1; Ayerst Laboratories), and butoxamine (β_2, Burroughs-Wellcome Co.).

Other compounds thought to participate in the adenyl cyclase-cyclic AMP system were also evaluated and included: dibutyryl cyclic AMP (Sigma Chemical Co.), the phosphodiesterase inhibitor RO-20-1724 (Hoffman-LaRoche), and the adenyl cyclase stimulator cholera enterotoxin. All compounds except RO-20-1724 were dissolved in BSS; RO-20-1724 was dissolved in ethanol.

Studies with Thyroid Hormones

To evaluate their effect on erythroid colony growth, we added a number of thyroid hormones and their analogues to cultured dog marrow cells. These compounds included L-thyroxine (L-T_4), D-thyroxine (D-T_4), L-triiodothyronine (L-T_3), D-triiodothyronine (D-T_3), all obtained from Sigma Chemical Co. L-T_4 and D-T_4 were dissolved in 70% ethanol with 1N NaOH; L-T_3 and D-T_3 were dissolved in 95% ethanol with 2N HCl. All compounds were initially dissolved in a 10 mM solution and then subsequently diluted with BSS.

Studies of Catecholamine-Thyroid Hormone Interaction

To evaluate possible relationships between thyroid hormone and β-adrenergic effects, we established cultures with L-T_4 both with and without the addition of the β-adrenergic antagonist propranolol. Additional cultures were carried out with L-T_4 in the presence of selected catecholamine blocking agents including practolol and butoxamine. We cocultured L-isoproterenol and L-T_4 in various concentrations to determine whether any enhancing effects were additive or provided evidence of a synergistic effect in the action of these two hormones.

Analysis of Responsive Cells by Separation Techniques

The populations of CFU-E responsive to the various enhancing agents were analyzed physically by separation at unit gravity (Miller and Phillips 1969). The methods employed in this laboratory have been documented previously (Singer and Adamson 1976).

RESULTS

When canine marrow cells were cultured under the above conditions, colonies of hemoglobin-synthesizing cells appeared at 24 hours and reached maximum numbers from 48 to 56 hours in culture. These colonies contained from 8 to 64 well-hemoglobinized cells. No colonies formed in the absence of epo but, as epo was added to cultures, a clear concentration-dependent increase in colony formation was observed (Fig. 1). The response was log-

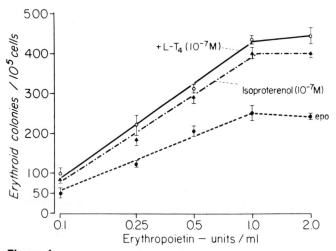

Figure 1
The effect on erythroid colony growth of increasing concentrations of epo alone and in the presence of 10^{-7} M L-T_4 and 10^{-7} M isoproterenol. Plateau colony numbers were routinely obtained with epo from 0.5 to 1.0 IRP U/ml. No erythroid colonies were formed in the absence of epo.

linear in nature with plateau colony formation occurring routinely from 0.5 to 1.0 IRP U/ml.

In other studies, it was shown that colony numbers were also linearly related to the cell concentrations plated with a regression that passed through the origin. This gives strong indication that single cells, the CFU-E, are giving rise to each colony.

Studies with Catecholamines

Although incapable of initiating colony formation alone, L-isoproterenol, a $\beta_1\beta_2$ agonist, consistently enhanced erythroid colony numbers over the full range of epo concentrations. As shown in Figure 1, this effect persisted even at optimal concentrations of epo. Reproducible enhancement by as much as 50 to 100% in colony numbers was observed.

As shown in Figure 2, the enhancing effect of isoproterenol was strictly concentration-dependent. Routinely, 10^{-7} M proved to be the most effective concentration of this compound. This concentration of isoproterenol was used in all subsequent studies.

Several experiments were carried out to determine the specificity of the effect of the catecholamines on erythroid colony growth. As shown in Figure 3, the enhancing effects of a number of compounds related to the adenyl cyclase-cAMP mechanism were tested. In addition, the effect of the β-adrenergic blocker propranolol was also evaluated. As shown in Figure 3A,

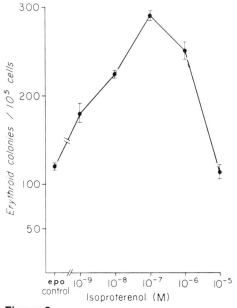

Figure 2
The influence of varying concentrations of isoproterenol on canine marrow erythroid colony growth. In this and all subsequent studies, unless specified otherwise, the concentration of epo was 0.5 IRP U/ml. Optimal enhancement by isoproterenol was routinely observed at 10^{-7} M.

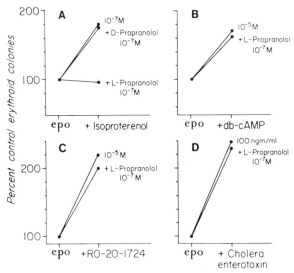

Figure 3
The influence of a number of compounds related to the adenyl-cyclase–cAMP mechanism on erythroid colony growth. In addition, the effects of the stereoisomers of the β-adrenergic blocking compound propranolol were tested. (*A*) The enhancing effect of 10^{-7} M isoproterenol was unaffected by an equimolar concentration of D-propranolol whereas the biologically active L-isomer completely inhibited β-adrenergic-enhanced colony growth. (*B*) (*C*) (*D*) Enhancement of colony growth by dibutyryl cAMP (10^{-5} M), RO-20-1724 (10^{-5} M), and cholera enterotoxin (100 ng/ml), was unaffected by propranolol.

10^{-7} M isoproterenol enhanced erythroid colony numbers almost twofold under the culture conditions employed. D-propranolol, the biologically inactive optical isomer, had no inhibiting effect on colony growth, whereas L-propranolol (10^{-7} M) completely blocked isoproterenol-enhanced colony numbers. Neither blocker at this or 100× higher concentrations reduced the number of colonies formed with epo alone. As shown in Figure 3B, C, and D, the enhancing effects of dibutyryl-cAMP, RO-20-1724, and cholera enterotoxin were unaffected by the addition of 10^{-7} M L-propranolol. These results suggest that these other compounds act intracellularly or, in the case of cholera enterotoxin, at membrane receptors which are distinct from those for β-adrenergic agonists (Fishman and Brady 1976). The absence of effect of propranolol on epo-dependent colonies also demonstrates the separateness of β-receptors and the putative receptors for epo.

Catecholamines can be classified as to the type of receptor with which they interact (Lefkowitz et al. 1976). Figure 4 demonstrates the results of experiments carried out with a variety of adrenergic agents. Consistent enhancement of erythroid colony growth was observed only with compounds having at least partial specificity for β_2 receptors. Thus, isoproterenol, epinephrine, metaproterenol, and albuterol all consistently enhanced colony growth, whereas phenylephrine and norepinephrine were inactive. These

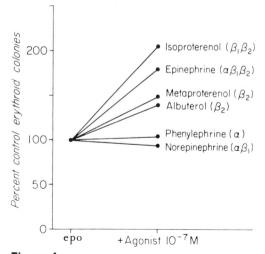

Figure 4
The influence of various catecholamines of known receptor specificity on erythroid colony growth. All compounds were tested at 10^{-7} M. Only compounds having β_2 activity were capable of enhancing colony growth.

results are consistent with the fact that in the dog CFU-E bear functional β_2 receptors.

To establish the receptor specificity, we carried out experiments in the presence of characterized adrenergic blockers. The results shown in Figure 5 confirm the findings obtained with specific agonists. Those blockers which do not possess β_2 specificity, such as phentolamine and practolol, were incapable of reducing significantly the increment of isoproterenol-induced erythroid colonies. In contrast, equimolar concentrations of butoxamine completely inhibited isoproterenol-induced colony growth. As shown in Figure 5D, appropriate blocking was obtained in cultures containing propranolol. Again, no inhibition of epo-dependent colonies was observed with blocker concentrations as high as 10^{-5} M.

Studies with Thyroid Hormones

In a manner similar to β-adrenergic agonists, all thyroid hormones tested enhanced erythroid colony growth over the full range of the epo concentration/response curve. As shown in Figure 1, this enhancement was similar in profile to that obtained with isoproterenol with increments in colony numbers persisting even at optimal concentrations of epo. Again, colony growth was not initiated by the most active concentrations of any of the thyroid hormones in the absence of epo.

As shown in Figure 6, there was clear concentration-dependence of the various thyroid hormones in their effect on erythroid colony growth. The order or potency for these various compounds paralleled their known calorigenic activities in vivo (Popovic et al. 1977). Because L-T_4 was consistently the most effective compound in its influence on colony growth, it was utilized in all subsequent experiments.

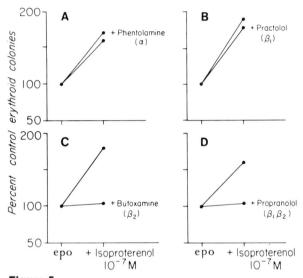

Figure 5

The influence of adrenergic blocking compounds of known specificity on isoproterenol-enhanced colony growth. Isoproterenol was tested at 10^{-7} M and all blocking agents were tested in equimolar concentrations. Blocking agents without β_2 specificity had no inhibiting effect on isoproterenol-enhanced colony growth (A, B), and inhibition was observed only with antagonists having β_2 specificity (C, D).

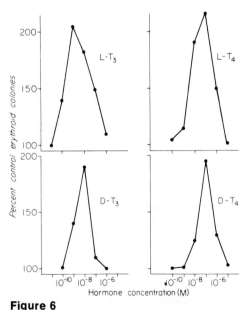

Figure 6

The concentration-dependent effects of various thyroid hormones on erythroid colony growth. The concentration optima were relatively sharp but reproducible.

Studies of Catecholamine-Thyroid Hormone Interaction

Because thyroid hormones are known to influence β-adrenergic receptor activity in other systems, the effect of L-T_4 on colony growth was tested in the presence of various adrenergic blocking agents of known specificity. As shown in Figure 7, those blockers with β_2 receptor specificity were capable of inhibiting L-T_4-enhanced colony numbers completely. In contrast, phentolamine and practolol were incapable of blocking thyroid-hormone-induced colony numbers. Of significance is the fact that the concentration of propranolol required for blockade was 10^3 times higher for L-T_4 than it was for isoproterenol-induced colonies. Consequently, although in vitro erythropoietic effects of both thyroid hormones and isoproterenol appeared to be mediated by a receptor with β-adrenergic properties, there was a reproducible difference in receptor sensitivity to the antagonist.

The specificity of the blocking effect was confirmed by studying the stereoisomers of propranolol. Only the L-isomer was capable of inhibiting L-T_4-enhanced colony growth. A 10^3 times greater concentration of butoxamine was required to block L-T_4-enhanced colony growth than was required to block isoproterenol-induced colony numbers. Additional experiments carried out with L-T_3, D-T_4, and D-T_3 confirmed the specificity of the β-like receptor for thyroid hormone activity.

To test the possibility that thyroid hormones were influencing the sensitivity of β-receptors to catecholamine stimulation, we carried out experiments using suboptimal concentrations of both classes of hormones.

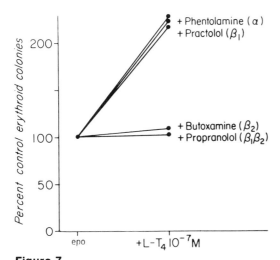

Figure 7

The influence of various adrenergic blocking agents of known receptor specificity on L-T_4 (10^{-7} M)-enhanced erythroid colony growth. As with adrenergic agonists, those blockers without β_2 specificity had no influence in concentrations up to 10^{-5} M on L-T_4-enhanced colony numbers. In contrast, butoxamine and propranolol, both with β_2 receptor specificity, were capable of blocking thyroid-hormone-induced colonies completely. The concentration requirements for blockade, however, were 10^2 to 10^3 times higher than found for isoproterenol-induced colonies (see text).

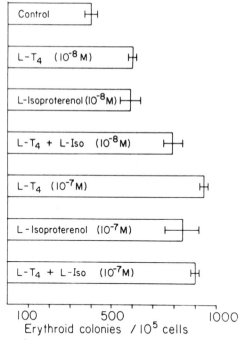

Figure 8
The effects of suboptimal concentrations of thyroid hormones and isoproterenol on erythroid colony growth. At 10^{-8} M, L-T_4 and isoproterenol enhance erythroid colony growth by approximately 50%. When added together, erythroid colony growth doubled, an almost exact additive effect. That maximal erythroid colony growth had not been reached was demonstrated in cultures containing optimally active (10^{-7} M) concentrations of L-T_4 and isoproterenol. Finally, when these concentrations were added together, no increase above plateau colony numbers was observed.

As shown in Figure 8, when 10^{-8} M L-T_4, and 10^{-8} M isoproterenol were tested individually and then together, in the presence of 0.5 U/ml of epo, only an additive increment in numbers of erythroid colonies was observed. In addition, if optimally effective concentrations of these hormones were tested individually and then together, no increase above total colony-forming ability was found. Consequently, these results indicate only additive effects when these compounds are together in culture, and the absence of synergism suggests that thyroid hormones are not affecting erythroid colony growth by enhancing the sensitivity of β-receptors on the CFU to suboptimal concentrations of catecholamine.

Analysis of Responsive Cells by Separation Techniques

Marrow cells were separated by velocity sedimentation at unit gravity and the functional capacity of the cell fractions was assessed. A broad peak of nucleated cells was obtained as shown in Figure 9. Peak numbers of CFU-E sedimented relatively rapidly with a mean sedimentation rate of 8.6 mm/

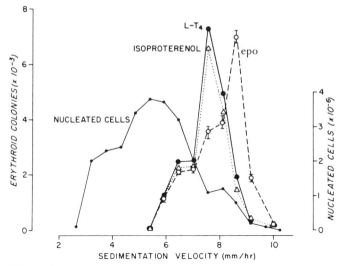

Figure 9
The velocity sedimentation profile of dog marrow cells and CFU-E. A sharp peak of epo-dependent CFU was observed at the most rapidly sedimenting portion of the gradient. Isoproterenol- and thyroid hormone-enhanced colony growth occupied a nearly identical peak which formed a subpopulation of the total CFU-E. The majority of marrow nucleated cells sedimented somewhat more slowly with a modal sedimentation rate of approximately 6 mm/hour.

hour. The peaks of CFU-E responsive to isoproterenol and L-T_4 were virtually identical and sedimented somewhat more slowly, at 7.2 mm/hour. Consequently, both classes of hormones appear to influence physically identical subpopulations of the total number of CFU-E.

DISCUSSION

These studies clearly demonstrate the influence of other hormones—in this case, β-adrenergic agonists and thyroid hormones—on erythropoiesis in culture. Previous work has demonstrated the presence of β-receptors on a variety of hematopoietic progenitors, including pluripotent stem cells and mature erythroid cells. Byron (1972) has demonstrated a functional response of mouse spleen CFU to adrenergic agonists, and the results of the present studies suggest a functional receptor on a subpopulation of cells capable of giving rise to erythroid colonies in culture. The possibility that these agents influence a second population of cells which then release an enhancing factor is made less likely by the fact that removal of adherent cells has no influence either on adrenergic or thyroid hormone effect and the time course of release of a factor such as the colony stimulating factor, which might influence erythroid colony growth, seems too short. The observations are most consistent with the canine CFU-E having a variety of surface receptors, some of which are coupled to membrane-bound adenyl cyclase. Consonant with this model, β-agonists enhance colony growth, an enhancement which is blocked by propranolol. Other compounds that influence the adenyl cyclase-cyclic AMP system at other levels such as cyclic nucleotides themselves and the phosphodiesterase inhibitor RO-20-1724 are not influenced by propranolol. Cholera enterotoxin, thought to have its own receptor system on the membrane, is also active.

In many respects, thyroid hormones and catecholamines appear to act similarly. Their effect on erythroid colony growth over the full range of epo concentrations is nearly identical, and although they have additive effects when tested at suboptimal concentrations in culture, there is neither synergism nor the ability to enhance colony growth beyond that seen with the most active concentration of either agent alone. Another point of similarity is the fact that velocity sedimentation profiles of the CFU influenced by these compounds are virtually overlapping, suggesting that they both affect a selected (and presumably the same) subpopulation of erythroid progenitor. Finally, the enhancing effect of both agents is blocked by selective β-adrenergic antagonists. Manipulation of both agonists and antagonists demonstrates that the receptors have β_2 specificity. Of interest, however, is that 10^2 to 10^3 times higher concentrations of blockers are required to inhibit thyroid-hormone-enhanced colony growth as opposed to β-adrenergic-enhanced colony numbers. Whether this represents a thyroid-hormone-mediated alteration in the binding properties of the β-receptor or whether thyroid hormones somehow interact with a β-receptor distinct from that being acted on primarily by catecholamines cannot be distinguished by the present results.

The nature of the subpopulation of CFU-E responsive to these classes of hormones is not clear. Velocity sedimentation separates cells not only on the basis of size but also on the basis of their stage in the cell cycle. Consequently, it cannot be determined whether more or less well-differentiated erythroid cells are being acted upon by these hormones or whether the hormones are influencing a particular phase of the progenitor cell cycle, altering it in such a way that CFU become responsive to epo. This latter possibility is supported by the fact that cAMP may arrest the transit of cells through G_1 of the cell cycle (Pastan et al. 1975). Cells of a given population in G_1 sediment more slowly than do cells in other stages of the cell cycle (Omine and Perry 1972), and the positioning of adrenergic and thyroid-hormone-responsive CFU in our studies is consistent with such an interpretation.

Perhaps most surprising of the results is the fact that the thyroid-hormone effect in culture may be blocked, apparently specifically, by an agent only known to act at the cell surface at adrenergic receptor sites. Although a good deal of evidence indicates that thyroid hormones primarily influence cell function by acting intracellularly, perhaps directly on the nucleus (Oppenheimer et al. 1973; Samuels and Tsai 1973), the possible interaction with the cell membrane or cytosol may occur prior to or in addition to binding to nuclear sites.

Among the known mechanisms by which thyroid hormones and catecholamines interact are augmentation of the catecholamine–adenyl-cyclase system (Nelson and Stouffer 1972) and the direct increase in the number of β-receptors (Williams et al. 1977). These studies suggest that thyroid hormonal status may influence the response of various tissues to adrenergic stimulation. However, augmentation of the response of CFU to the low levels of catechols in the culture is unlikely, and the results suggest that thyroid hormones may interact directly with membrane-bound systems linked to adenyl cyclase. Consistent with this is the marked difference in propranolol concentration required to block the L-T_4 effect. Although such a high concentration might suggest that the mechanism of propranolol blockade was independent of $β_2$ receptors, such a non-β effect has not been described previously as having stereospecificity. Thus, the action on in vitro erythropoiesis of catecholamines and thyroid hormones appears to be mediated by an adrenergic-like receptor with $β_2$ specificity.

A strong clinical parallel exists between thyroid disease states and adrenergic activity, and this appears to be reflected in the thyroid-hormone–catecholamine interactions in these culture studies. Because disorders of thyroid function in man are associated with defective erythropoiesis, one is tempted to speculate that alterations in hormonal interactions may be found which are of clinical significance and which quantitatively affect red-cell production in the intact organism. The results summarized here suggest strategies based on studies in culture that will allow assessment of these interactions both in normal and abnormal thyroid function. In addition, the characterization of hormonal interactions on the CFU-E provides a model by which the effects of other hormones and small molecules may act on common intracellular pathways to influence both differentiation and maturation in this and other cell systems.

Acknowledgments

The technical assistance of Ms. Christine Reichgott and Ms. Faith Shiota is gratefully acknowledged. The following compounds were gifts: RO-20-1724, from Roche Diagnostics Division, Hoffman-LaRoche, Inc., Nutley, New Jersey; albuterol, from Schering Corp., Bloomfield, New Jersey; metaproterenol from Boehringer-Ingelheim, Elmsford, New York; practolol and the optical isomers of propranolol from Ayerst Laboratories, Montreal, Quebec, Canada; and pure cholera enterotoxin, lot 0172, was prepared by Dr. Richard Finkelstein and made available through the National Institute of Allergy and Infectious Disease.

These studies were supported by designated research funds of the Veterans Administration and National Institutes of Health Grant AM-19410 and by a Research Career Development Award (AM-70222) to Dr. Adamson from the National Institute of Arthritis, Metabolism and Digestive Diseases.

REFERENCES

Brown, J. E., and J. W. Adamson. 1977a. Studies of the influence of cyclic nucleotides on in vitro hemoglobin synthesis. *Br. J. Haematol.* **35:** 193.

———. 1977b. Modulation of in vitro erythropoiesis: Enhancement of erythroid colony growth by cyclic nucleotides. *Cell Tissue Kinet.* **10:** 289.

Byron, J. W. 1972. Evidence for a β-adrenergic receptor initiating DNA synthesis in haemopoietic stem cells. *Exp. Cell Res.* **71:** 228.

Chang, S., D. Sikkema, and E. Goldwasser. 1974. Evidence for an erythropoietin receptor protein on rat bone marrow cells. *Biochem. Biophys. Res. Commun.* **57:** 399.

de Asua, L. J., M. O'Farrell, D. Bennett, D. Clingan, and P. Rudland. 1977. Interaction of two hormones and their effect on observed rate of initiation of DNA synthesis in 3T3 cells. *Nature* **265:** 151.

Fishman, P. H., R. O. Brady. 1976. Biosynthesis and function of gangliosides. *Science* **194:** 906.

Goldwasser, E. 1975. Erythropoietin and the differentiation of red blood cells. *Fed. Proc.* **34:** 2285.

Graber, S. E., M. Carrillo, and S. B. Krantz. 1972. The effect of cyclic AMP on heme synthesis by rat bone marrow cells in vitro. *Proc. Soc. Exp. Biol. Med.* **141:** 206.

Holley, R. W. 1975. Control of growth of mammalian cells in cell culture. *Nature* **258:** 487.

Krantz, S. B., and L. O. Jacobson. 1970. *Erythropoietin and the regulation of erythropoiesis,* p. 330. University of Chicago Press, Chicago.

Lefkowitz, R. J., L. E. Limbird, C. Mukherjee, and M. D. Caron. 1976. The β-adrenergic receptor and adenylate cyclase. *Biochim. Biophys. Acta* **457:** 1.

Miller, R. G., and R. A. Phillips. 1969. Separation of cells by velocity sedimentation. *J. Cell. Physiol.* **73:** 191.

Nelson, T. E., and J. E. Stouffer. 1972. Thyroxine modulation of epinephrine stimulated secretion of rat parotid amylase. *Biochem. Biophys. Res. Commun.* **48:** 480.

Omine, M., and S. Perry. 1972. Use of cell separation at 1 g for cytokinetic studies in spontaneous AKR leukemia. *J. Natl. Cancer Inst.* **48:** 697.

Oppenheimer, J. H., H. L. Schwartz, W. H. Dillman, and M. I. Surks. 1973. Effect of thyroid hormone analogues on the displacement of ^{125}I-L-triiodothyronine from hepatic and heart nuclei: Possible relationship to hormonal activity. *Biochem. Biophys. Res. Commun.* **55:** 544.

Pastan, I. H., G. S. Johnson, and W. B. Anderson. 1975. Role of cyclic nucleotides in growth control. *Annu. Rev. Biochem.* **44:** 491.

Popovic, W. J., J. E. Brown, and J. W. Adamson. 1977. The influence of thyroid hormones on in vitro erythropoiesis. Mediation by a receptor with beta adrenergic properties. *J. Clin. Invest.* (In press.)

Robison, G. A., R. W. Butcher, and E. W. Sutherland. 1971. *Cyclic AMP.* Academic Press, New York.

Samuels, H. H., and J. S. Tsai. 1973. Thyroid hormone action in cell culture: Demonstration of nuclear receptors in intact cells and isolated nuclei. *Proc. Natl. Acad. Sci. U.S.A.* **70:** 3488.

Singer, J. W., and J. W. Adamson. 1976. Steroids and hematopoiesis. II. The effect of steroids on in vitro erythroid colony growth: Evidence for different target cells for different classes of steroids. *J. Cell. Physiol.* **88:** 135.

Stephenson, J. R., A. A. Axelrad, D. L. McLeod, and M. M. Shreeve. 1971. Induction of colonies of hemoglobin-synthesizing cells by erythropoietin in vitro. *Proc. Natl. Acad. Sci. U.S.A.* **68:** 1542.

Tompkins, G. M. 1975. The metabolic code. *Science* **189:** 760.

Williams, L. T., R. J. Lefkowitz, A. M. Watanabe, D. R. Hathaway, and H. R. Besch, Jr. 1977. Thyroid hormone regulation of β-adrenergic receptor number. *J. Biol. Chem.* **252:** 2787.

The Cell Surface and the Control of Friend-Cell Differentiation

A. Bernstein, D. L. Mager, M. MacDonald, M. Letarte, F. Loritz, M. McCutcheon, R. G. Miller, and T. W. Mak

The Ontario Cancer Institute and the Department of Medical Biophysics
University of Toronto, Toronto, Ontario M4X 1K9, Canada

The genetic and biochemical events culminating in those changes in gene expression that are recognized at the cellular level as differentiation are largely unknown. This is particularly true of the hematopoietic system. It comprises cells in one of several possible pathways of differentiation, each containing cells at different positions along the pathway. This cellular heterogeneity precludes the kinds of genetic approaches that have been used so successfully in microbial systems and that now seem possible with mammalian cells in culture (Thompson and Baker 1973; Till et al. 1973; Siminovitch 1976). The development of permanent, clonal culture systems capable of erythroid (Friend et al. 1971) or granulocytic (Sachs 1974) differentiation raises the possibility that a molecular-genetic approach to hematopoiesis might be feasible, at least in these model systems.

The usefulness of the Friend-cell system as a model of normal erythropoiesis has been greatly strengthened by the demonstration that the induction of Friend-cell erythroid differentiation in culture is accompanied by a wide variety of cellular changes which strikingly parallel normal erythropoiesis. However, there is one major difference between Friend-cell differentiation and normal erythropoiesis: Friend cells differentiate in response to a wide variety of chemical inducers (Friend et al. 1971; Leder and Leder 1975; Preisler and Lyman 1975; Tanaka et al. 1975; Bernstein et al. 1976; Gusella and Housman 1976; Reuben et al. 1976) but do not respond to erythropoietin, the in vivo regulator of normal erythropoiesis. Although molecular basis for the inability of erythropoietin to stimulate Friend-cell differentiation is unknown, it probably is related to the observation that the disease induced in vivo by Friend leukemia virus is also characterized by erythropoietin-independent erythroid differentiation (Liao and Axelrad 1975).

Given the broad similarities between the late events in both Friend-cell differentiation and normal erythropoiesis, it seems reasonable to assume that there will also be common steps which precede the overt expression of differentiation in both these systems. Because cell-surface changes have been associated with changes in gene expression in other mammalian cell systems, we sought to determine whether membrane changes also accompany Friend-cell differentiation. In this paper, recent experiments from this laboratory are summarized which suggest that membrane changes, both early and late, are an integral part of Friend-cell differentiation.

METHODS

Tissue Culture Procedures

Methods used in this laboratory for general tissue culture procedures, including cloning in methylcellulose, isolation of mutants, and cell volume determinations, have been described previously (Bernstein et al. 1976; Loritz et al. 1977).

Radioimmune Binding Assay for Detection of Erythrocyte Membrane Antigens

Immunological procedures for the detection of erythrocyte membrane antigens have been described (MacDonald 1977; MacDonald et al. 1978). In brief, target cells were incubated with rabbit antiserum to mouse erythrocyte ghosts at 4°C for 1 hour in a 50-μl volume. After washing with 0.1% isotonic bovine serum albumin (BSA) in phosphate buffered saline (PBS), the antibody-coated cells were then incubated at 4°C with 100 μl of ^{125}I-labeled, purified, pepsin-degraded, horse F(ab')$_2$, antirabbit IgG (Jensenius and Williams 1974).

The rabbit antierythrocyte ghost serum was analyzed by absorption-inhibition. Together with varying numbers of cells in 0.5% isotonic BSA in PBS, 50 μl of serum were incubated at 4°C for 6 hours, and the unabsorbed serum was recovered by centrifugation at 18,000 g for 8 minutes. For complete thymocyte absorption of the antiserum, pelleted erythrocyte-free thymocytes were resuspended in a 1:1250 dilution of rabbit antierythrocyte ghost serum at 3×10^9 cells/ml of antiserum at 4°C for 3 hours. This procedure was performed twice.

RESULTS

Early and Late Volume Changes during Friend-Cell Differentiation

The induction of Friend-cell differentiation is accompanied by two discrete cell-volume reductions (Fig. 1; Loritz et al. 1977). The first volume shift occurs at around 10 hours after induction and consists of a 20-25% decrease in volume compared to untreated control populations. This shift persists for 2 days under our inducing conditions, and its magnitude and time of appearance correlate well with the number of differentiated cells observed 5 days later (Loritz et al. 1977). For example, Friend-cell clone AMA-2B, an alpha-amanitin-resistant mutant derived from line 745A, differentiates more slowly and to a lesser extent than do wild-type Friend

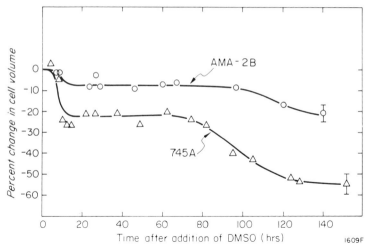

Figure 1
Changes in cell volume after induction by 1.5% DMSO of wild-type clone 745A (△——△) and AMA-2B (o——o), an alpha-amanitin resistant mutant, derived from 745A. The number of hemoglobin-positive cells, measured on day 5, was 70% for 745A and 14% for AMA-2B.

cells (according to our observations [unpubl.]). This altered response to dimethylsulfoxide (DMSO) is reflected in a two- to threefold reduction in the extent of both the early and late volume shifts and a delay in the onset of the late volume shift (Fig. 1).

The late volume shift begins at around the same time as hemoglobin-positive cells begin to appear in the cultures and reduces the modal cell volume to around 50% of that of control cultures. The magnitude of the late volume shift makes it possible to enrich for pure populations of hemoglobin-containing Friend cells by employing separation techniques based on volume differences, such as velocity sedimentation (Loritz et al. 1977) or cell sorting based on light-scattering information (Arndt-Jovin et al. 1976).

Early Transport Changes during Friend-Cell Differentiation

Considerable experimental data (reviewed in MacKnight and Leaf, 1977) exist to suggest that cell volume regulation and the permeability of the cell membrane to Na^+ and K^+ ions are interrelated. The membrane-bound enzyme Na^+/K^+ ATPase, by active inward transport of K^+ ions coupled with active extrusion of Na^+ ions from the cell (Schwartz et al. 1975; Glynn and Karlish 1975), is largely responsible for maintaining the high intracellular levels of K^+ ions. Thus, inhibition of this enzyme—by the addition of cardiac glycosides such as ouabain, for example—leads to changes in cell volume in many different cell types (reviewed in MacKnight and Leaf 1977).

Because of this interrelationship between cell volume and ATPase function, we sought to determine whether the early volume shift which

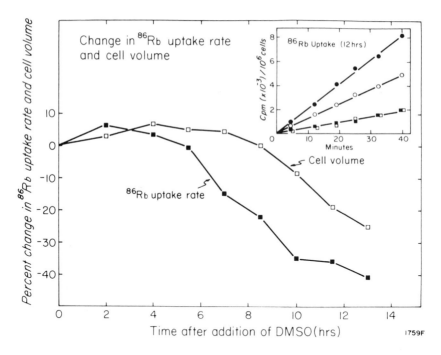

Kinetics of the change in (■——■) ^{86}Rb uptake rate and (□——□) cell volume in response to 1.5% DMSO. The inset shows the effect of preincubation (5 minutes) with ouabain on ^{86}Rb influx in control and DMSO-treated cells. (●——●) Control; (○——○) DMSO treated; (■——■) control + 1.0 mM ouabain; (□——□) DMSO-treated + 1.0 mM ouabain.

accompanies Friend-cell differentiation (Fig. 1; Loritz et al. 1977) was preceded by changes in cell-membrane transport. It can be seen from Figure 2 that early changes in the initial uptake rates of ^{86}Rb (an analogue of K$^+$) do precede the early cell-volume shift. This decrease in ^{86}Rb transport is due to a change in Na$^+$/K$^+$ ATPase function inasmuch as a brief preincubation of the cells with a concentration of ouabain sufficient to inhibit completely the Na$^+$/K$^+$ ATPase reduces the rate of uptake of ^{86}Rb to the same level in both control and DMSO-treated cells (inset to Fig. 2; Mager and Bernstein 1978). A similar decrease in the rates of uptake of α-aminoisobutyric acid, a nonmetabolized amino acid, has also been observed after induction by DMSO (Mager and Bernstein 1978).

A central question raised from these studies is whether the transport changes represent a necessary step in Friend-cell differentiation. Several experimental observations (Mager and Bernstein 1978) suggest that they do. First, such widely different inducers as DMSO, dimethylacetamide (DMA), hypoxanthine, and ouabain (see below) all reduce ^{86}Rb uptake to roughly the same levels in differentiating Friend cells. Second, xanthine, a compound which does not induce Friend-cell differentiation even though it is structurally similar to hypoxanthine (Gusella and Housman 1976), does not bring about these early transport changes. Third, as summarized below and as discussed in more detail by Bernstein et al. (1976), ouabain, a specific inhibitor of the Na^+/K^+ ATPase, is also a potent inducer of differentiation in these cells.

Early Transport Changes in Noninducible Friend-Cell Clones

Friend-cell clones can be isolated which have partially or totally lost the capacity to synthesize hemoglobin in response to one or more of the chemical inducers. To establish whether the early membrane transport changes observed after growth in the chemical inducers are an integral part of the differentiation process in these cells, we measured transport rates in two independently isolated noninducible variants and isolated a spontaneous inducible revertant from one of these clones. The results, shown in Table 1, indicate that such clones can be divided into at least two phenotypic groups: those that exhibit the early transport changes after growth in DMSO (clone TG-13) and those that show only limited changes in transport (M18D1). Significantly, the inducible revertant M18D1^R, isolated form M18D1, also demonstrated the characteristic early change in the initial rate of transport of ^{86}Rb.

Induction by Ouabain of Friend-Cell Differentiation

The cell volume and early transport changes summarized above suggest that changes in the Na^+/K^+ ATPase are an integral part of the early events in Friend-cell differentiation. This conclusion is further strengthened by the observation, described in this section, that ouabain is a good inducer of Friend-cell differentiation (Bernstein et al. 1976). Ouabain is a cardiac glycoside that binds specifically to the Na^+/K^+ ATPase (reviewed by Schwartz et al. 1975). Under normal physiological conditions, this binding results in a rapid inhibition of the Na^+/K^+ ATPase, leading to changes in the intracellular levels of Na^+ and K^+ ions (Glynn and Karlish 1975; Schwartz et al. 1975) and to cell death. Friend-cell mutants, resistant to the cytotoxic action of ouabain, can be isolated after a single-step selection procedure (Bernstein et al. 1976). These ouabain-resistant (OUA^r) Friend-

Table 1
Transport Changes in Noninducible Friend-Cell Clones

Clone	Benzidine-positive cells (%)	Change in ^{86}Rb uptake (%)
745aJG	85.5 ± 3.1 (7)	−42.6 ± 1.6 (7)
TG-13	1.2 ± 0.4 (3)	−35.3 ± 2.3 (2)
M18D1^R	52.2 ± 1.8 (2)	−31.4 ± 2.9 (2)
M18D1	11.0 ± 2.3 (3)	−12.8 ± 1.2 (3)

The initial transport rates of ^{86}Rb were measured 12 to 14 hours after the addition of 1.5% DMSO to the culture medium. Results are expressed as the percent difference between control and DMSO-treated cells measured at the same time. Benzidine staining was done on day 5. Entries are the mean and the standard deviation of the mean; numbers in parentheses indicate numbers of experiments used for each determination.

cell mutants, like OUA^r mutants of other mammalian cell lines (Baker et al. 1974), have an Na$^+$/K$^+$ ATPase with the same transport activity as wild-type OUA^s cells. However, the Na$^+$/K$^+$ ATPase activity in OUA^r cells is relatively (but not absolutely) resistant to inhibition by ouabain (Baker et al. 1974; Bernstein et al. 1976).

During the course of characterizing these mutants, we observed that the presence of ouabain resulted in the induction of hemoglobin synthesis, accompanied by an increase in the steady-state levels of globin mRNA (Bernstein et al. 1976). The ability of ouabain to induce hemoglobin synthesis is not restricted to OUA^r cells, for at tenfold lower concentrations ouabain induces approximately 25% of OUA^s cells over a 5-day period. It is of interest to note that ouabain induces hemoglobin synthesis in a concentration range where ^{86}Rb uptake is inhibited to the same extent in OUA^s and OUA^r cells (30–50%) and to the same levels observed after induction by several other Friend-cell inducers (Fig. 2; Mager and Bernstein 1978).

Given the high degree of specificity of binding of ouabain to the Na$^+$/K$^+$ ATPase (Schwartz et al. 1975), it seemed likely that ouabain induction occurred via its interaction with this membrane-bound enzyme. We obtained direct evidence for this possibility by making use of the observation that K$^+$ ions antagonize the binding of ouabain to the Na$^+$/K$^+$ ATPase (Baker and Willis 1970). If ouabain binding to the Na$^+$/K$^+$ ATPase is required for the stimulation of erythroid differentiation, then the dose-response curves for ouabain induction should be markedly sensitive to the concentration of K$^+$ ions in the growth medium. Figure 3 shows that this is the case. This effect of K$^+$ ions was specific for ouabain induction, as the dose-response curves for induction by DMSO or DMA were independent of the extra-

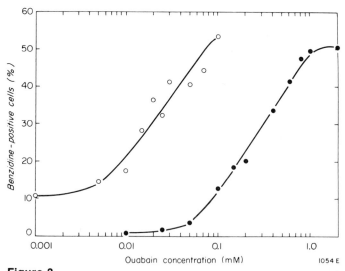

Figure 3

Dose-response for ouabain induction in (o———o) medium containing 1.8 mM K$^+$ and (●———●) 6.8 mM K$^+$. The number of hemoglobin-containing (benzidine-positive) cells was determined 5 days after the addition of ouabain. (Reprinted, by permission, from Bernstein et al. 1976.)

cellular K$^+$ ion levels over the range 1–7 mM K$^+$ ions (Bernstein et al. 1976).

These results suggest that the initial and necessary step in the induction of Friend cells by ouabain is the binding of ouabain to the Na$^+$/K$^+$ ATPase. The simplest interpretation of this observation is that ouabain inhibition of the Na$^+$/K$^+$ ATPase somehow leads to those cellular changes that are associated with Friend-cell differentiation.

Induction of Friend-Cell Differentiation by Growth in Low-Na$^+$, High-K$^+$ Medium

The addition of ouabain to mammalian cells resulted in rapid changes in the intracellular levels of Na$^+$ and K$^+$ ions (Schwartz et al. 1975). Therefore, we sought to determine whether the control of Friend-cell differentiation could also be affected by the extracellular levels of these ions. Normal tissue culture medium contains high Na$^+$ (141 mM) and low K$^+$ (5.4 mM) levels, and, under these conditions, most Friend-cell clones express a very low level of differentiation. Increasing the K$^+$ ion levels and decreasing the Na$^+$ ion levels in the growth medium, while keeping the osmotic pressure and chloride ion concentrations constant, were alone sufficient to induce hemoglobin synthesis in Friend cells (Fig. 4; D. Mager et al., in prep.). This observation provides additional experimental confirmation that

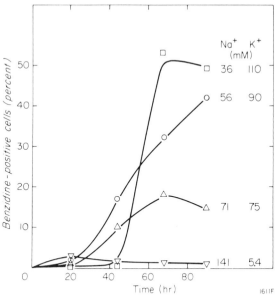

Figure 4

Induction of Friend cell differentiation in low Na$^+$, high K$^+$ medium. The numbers of benzidine-positive cells were determined at various times after Friend cells were seeded in alpha medium containing the indicated Na$^+$ and K$^+$ concentrations. (▽——▽) 141 mM Na$^+$, 5.4 K$^+$; (△——△) 71 mM Na$^+$, 75 mM K$^+$; (o——o) 56 mM Na$^+$, 90 mM K$^+$; (□——□) 36 mM Na$^+$, 110 mM K$^+$.

the levels of either or both of these monovalent cations play an important role in the regulation of gene expression in Friend cells.

Expression of Erythrocyte Membrane Antigens during Friend-Cell Differentiation

Normal erythropoiesis in vivo is accompanied by the appearance of erythrocyte antigens on the surface of nucleated erythroid cells in a manner which parallels the increase in hemoglobin synthesis (Borsook et al. 1969; Minio-Paluello et al. 1972). Similarly, using an indirect-immunofluorescence assay, others have reported that induced Friend cells have an increased reactivity to a heteroantiserum raised against fractionated erythrocyte membranes (Ikawa et al. 1973). We report here the results of a quantitative radioimmune assay for the study of erythrocyte membrane antigen expression during Friend-cell differentiation, similar to one described by Acton et al. (1974). Figure 5a demonstrates that absorption of a rabbit antierythrocyte

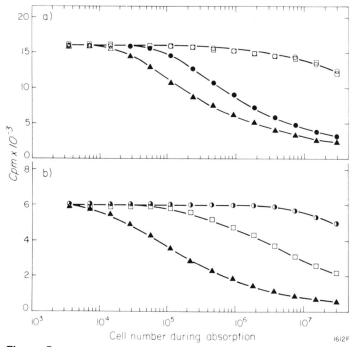

Figure 5

Expression of erythrocyte membrane antigens on differentiating Friend cells. Rabbit antimouse (DBA/2J) erythrocyte ghost serum was preabsorbed with mouse thymocytes as described in Methods. Aliquots of the serum were then further absorbed with (a) (▲——▲) erythrocytes, (●——●) DMSO-induced Friend (745A) cells, (□——□) mouse thymocytes, or (o——o) uninduced Friend cells; (b) (▲——▲) erythrocytes, (□——□) DMA-induced clone TG-13, (◖——◖) DMSO-treated TG-13, or (◐——◐) untreated TG-13. The binding of any unabsorbed serum was then determined by a cellular radioimmune assay in which mouse erythrocytes were used as targets.

membrane antiserum with induced Friend cells quantitatively inhibits the binding of this serum to erythrocytes. However, absorption with thymocytes or with uninduced Friend cells results in little inhibition even at high cell numbers (Fig. 5a), confirming that crossreactive erythrocyte membrane antigens are present on induced, but not on uninduced, Friend cells. Similar results were obtained when induced Friend cells were used as target cells for binding (MacDonald, 1977; MacDonald et al. 1978).

Figure 5b demonstrates that the expression of erythrocyte membrane antigens correlates with the induction of hemoglobin synthesis in at least one noninducible variant clone, TG-13. Although 745-TG-13 is a Friend-cell variant which is not stimulated by DMSO to produce hemoglobin, it is weakly stimulated by DMA, another Friend-cell inducer. As shown in Figure 5, absorption of the rabbit antierythrocyte ghost serum with TG-13 cultured in the presence or absence of DMSO did not result in significant inhibition of binding of this antiserum to red cells. In contrast, a DMA-induced population of 745-TG-13 (containing 30% hemoglobin-positive cells) did inhibit binding, although a higher cell number was needed to obtain the same level of inhibition produced by absorption with wild-type Friend cells containing 90% hemoglobin-positive cells.

DISCUSSION

The induction of Friend-cell differentiation can be divided into two stages. The first stage, which includes the first 36 to 48 hours after the initiation of differentiation, requires the continuous presence of inducer to allow the cells to proceed to the second stage, which is inducer-independent. This maturation process can be described by a stochastic model (Gusella et al. 1976) in which each cell has a certain probability of changing from a rapidly proliferating, undifferentiated cell to a "committed" differentiated cell which has limited or no proliferative capacity.

Both the early (inducer-dependent) and late (inducer-independent) stages of Friend-cell differentiation are accompanied by characteristic cell-surface changes, as summarized in this study. The observation that the expression of erythrocyte membrane antigens and the induction of hemoglobin synthesis appear to be correlated suggests that there may be one "master switch," turned on by the presence of inducer, which, in turn, activates such diverse events as hemoglobin and erythrocyte membrane antigen synthesis. Alternatively, Friend-cell differentiation may occur via a series of cascading events, such that the later inducer-independent events are controlled directly by cellular changes that occurred during the first inducer-dependent stage of differentiation, and, therefore, are controlled only indirectly by the presence of inducer.

Genetic and biochemical studies with a variety of noninducible Friend-cell clones should help to decide between these two alternative models. For example, although the noninducible clone TG-13 does not express the late events in Friend-cell differentiation, it does undergo the early transport changes in response to DMSO. These observations imply that the differentiation of TG-13 is blocked after the early transport changes at some step required for both hemoglobin synthesis and erythrocyte membrane antigen expression.

The results summarized in this paper suggest that early inhibition of the Na^+/K^+ ATPase is a common step in Friend-cell differentiation. This inhibition results in a decrease in the transport of a number of Na^+-dependent compounds and may affect other cellular processes which are dependent on the Na^+/K^+ ATPase. An important aspect of this model is that this early inhibition in the Na^+/K^+ ATPase activity is an integral part of the cellular changes which lead to the later, inducer-independent events in Friend-cell differentiation. The evidence for this model can be summarized as follows: (1) Ouabain, a specific inhibitor of the Na^+/K^+ ATPase, induces Friend-cell differentiation via its binding to the Na^+/K^+ ATPase. (2) Early changes in the transport of a number of compounds which require a functioning Na^+/K^+ ATPase occur during induction by a variety of different Friend-cell inducers. Furthermore, the occurrence and magnitude of these early transport changes have a strong predictive value in terms of the number of differentiated cells observed 5 days later. (3) Growth of Friend cells in medium in which the relative concentrations of Na^+ and K^+ ions have been reversed is sufficient to induce differentiation in a significant proportion of cells.

This model for Friend-cell differentiation raises several unanswered questions: First, how do the chemical Friend-cell inducers bring about these early changes in transport? Second, how do these early membrane changes ultimately lead to the transcriptional and translational changes which are recognized later as erythroid differentiation? Finally, does the differentiation of normal erythroid cells or other mammalian cells depend on similar cell-surface changes?

Acknowledgments

The authors wish to thank G. Cheong, V. Crichley, and G. Rutledge for excellent technical assistance. This work was supported by grants from the Medical Research Council and National Cancer Institute of Canada. D. Mager and M. MacDonald are recipients of Studentships from the Medical Research Council of Canada. M. Letarte is a Research Scholar of the National Cancer Institute of Canada. The ^{125}I-labeled, purified, pepsin-degraded, horse F(ab')$_2$, antirabbit IgG was a generous gift from Dr. A. F. Williams, Oxford, United Kingdom.

REFERENCES

Acton, R. T., R. J. Morris, and A. F. Williams. 1974. Estimation of the amount and tissue distribution of rat Thy-1.1 antigen. *Eur. J. Immunol.* **4:** 598.

Arndt-Jovin, D. J., W. Ostertag, H. Eisen, F. Klimek, and T. M. Jovin. 1976. Studies of cellular differentiation by automated cell separation. Two model systems: Friend-virus transformed cells and *Hydra attenuates*. *J. Histochem. Cytochem.* **24:** 332.

Baker, P. F., and J. S. Willis. 1970. Potassium ions and the binding of cardiac glycosides to mammalian cells. *Nature* **226:** 521.

Baker, R. M., D. M. Brunette, R. Mankovitz, L. Thompson, G. Whitmore, L. Siminovitch, and J. E. Till. 1974. Ouabain-resistant mutants of mouse and hamster cells in culture. *Cell* **1**: 9.

Bernstein, A., D. M. Hunt, V. Crichley, and T. W. Mak. 1976. Induction by ouabain of hemoglobin synthesis in cultured Friend erythroleukemic cells. *Cell* **9**: 375.

Borsook, H., K. Ratner, and B. Tattrie. 1969. Differential immune lysis of erythroblasts. *Nature* **221**: 1261.

Friend, C., W. Scher, J. G. Holland, and T. Sato. 1971. Hemoglobin synthesis in murine virus-induced leukemic cells *in vitro:* Stimulation of erythroid differentiation by dimethyl sulfoxide. *Proc. Natl. Acad. Sci. U.S.A.* **68**: 378.

Glynn, I. M., and S. J. D. Karlish. 1975. The sodium pump. *Annu. Rev. Physiol.* **37**: 13.

Gusella, J., and D. Housman. 1976. Induction of erythroid differentiation *in vitro* by purines and purine analogues. *Cell* **8**: 263.

Gusella, J., and R. Geller, B. Clarke, V. Weeks, and D. Housman. 1976. Commitment to erythroid differentiation by Friend erythroleukemia cells: A stochastic analysis. *Cell* **9**: 221.

Ikawa, Y., M. Furusawa, and H. Sugano. 1973. Erythrocyte membrane-specific antigens in Friend virus-induced leukemia cells. *Bibl. Haematol.* **39**: 955.

Jensenius, J. C., and A. F. Williams. 1974. The binding of anti-immunoglobulin antibodies to rat thymocytes and thoracic duct lymphocytes. *Eur. J. Immunol.* **4**: 91.

Leder, A., and P. Leder. 1975. Butyric acid, a potent inducer of erythroid differentiation in cultured erythroleukemic cells. *Cell* **5**: 319.

Liao, S. K., and A. A. Axelrad. 1975. Erythropoietin-independent erythroid colony formation *in vitro* by hemopoietic cells of mice infected with Friend virus. *Int. J. Cancer* **15**: 467.

Loritz, F., A. Bernstein, and R. G. Miller. 1977. Early and late volume changes during erythroid differentiation of cultured Friend leukemic cells. *J. Cell. Physiol.* **90**: 423.

MacDonald, M. E. 1977. Immunological analysis of the surface of Friend erythroleukemic cells. M.Sc. thesis, University of Toronto, Toronto.

MacDonald, M. E., M. Letarte, and A. Bernstein. 1978. Erythrocyte membrane antigen expression during Friend cell differentiation: Analysis of two noninducible variants. *J. Cell. Physiol.* (In press.)

MacKnight, A. D. C., and A. Leaf. 1977. Regulation of cellular volume. *Physiol. Rev.* **57**: 510.

Mager, D., and A. Bernstein. 1978. Early transport changes during erythroid differentiation of Friend leukemic cells. *J. Cell. Physiol.* **94**: 275.

Minio-Paluello, F., C. Howe, K. C. Hsu, and R. A. Rifkind. 1972. Antigen density on differentiating erythroid cells. *Nat. New Biol.* **237**: 187.

Preisler, H. D., and G. Lyman. 1975. Differentiation of erythroleukemia cells *in vitro:* Properties of chemical inducers. *Cell Differ.* **4**: 179.

Reuben, R. C., R. L. Wife, R. Breslow, R. A. Rifkind, and P. A. Marks. 1976. A new group of potent inducers of differentiation in murine erythroleukemia cells. *Proc. Natl. Acad. Sci. U.S.A.* **73**: 862.

Sachs, L. 1974. Control of growth and differentiation in normal hematopoietic and leukemic cells. In *Control of proliferation in animal cells* (eds. B. Clarkson and R. Bascrga), p. 915. Cold Spring Harbor Laboratory, Cold Spring Harbor, New York.

Schwartz, A., G. E. Lindenmayer, and J. C. Allen. 1975. The sodium-potassium adenosine triphosphatase: Pharmacological, physiological and biochemical aspects. *Pharmacol. Rev.* **27**: 3.

Siminovitch, L. 1976. On the nature of hereditable variation in cultured somatic cells. *Cell* **7:** 1.

Tanaka, M., J. Levy, M. Terada, R. Breslow, R. A. Rifkind, and P. A. Marks. 1975. Induction of erythroid differentiation in murine virus infected erythroleukemia cells by highly polar compounds. *Proc. Natl. Acad. Sci. U.S.A.* **72:** 1003.

Thompson, L. H., and R. M. Baker. 1973. Isolation of mutants of cultured mammalian cells. In *Methods in cell biology* (ed. D. M. Prescott), vol. 6, p. 209. Academic Press, New York.

Till, J. E., R. M. Baker, D. M. Brunette, V. Ling, L. Thompson, and J. A. Wright. 1973. Genetic regulation of membrane function in mammalian cells in culture. *Fed. Proc.* **32:** 29.

Chromatin Changes and DNA Synthesis in Friend Erythroleukemia and HeLa Cells during Treatment with DMSO and n-Butyrate

J. R. Neumann, M. G. Riggs, H. K. Hagopian, R. G. Whittaker,[*] and V. M. Ingram

Department of Biology, Massachusetts Institute of Technology
Cambridge, Massachusetts 02139

The mouse erythroleukemia produced by Friend virus transformation is a useful model system for the study of erythroid cell differentiation. Erythroleukemia cells are thought to be arrested in development at the proerythroblast stage, but can be induced by the addition of dimethylsulfoxide (DMSO) to differentiate in culture to the normoblast stage (Friend et al. 1971). This differentiation appears to follow a normal course. One of the early biochemical effects is the accumulation of globin mRNA (Ross et al. 1972). The appearance of characteristic red-cell-specific protein markers, including globin peptide chains (Friend et al. 1971), follows the enzymatic machinery for heme metabolism (Friend et al. 1971; Sassa et al. 1975) and red-cell-specific membrane antigens (Ikawa et al. 1973).

Numerous recent studies have suggested an important role for nonhistone chromosomal proteins (NHCP) in the regulation of transcriptional activity (Barrett et al. 1974; Tsai et al. 1976). Furthermore, NHCP can be modified extensively by phosphorylation (Langan 1968). Changes in the phosphorylation levels of both NHCP and histones have been correlated with the transcriptional state of the nucleus, the stage of the cell cycle, and with DNA-binding capacity in vitro (Adler et al. 1971; Marks et al. 1973; Stein et al. 1974; Kostraba et al. 1975; Kleinsmith et al. 1976). There is evidence that among the NHCP certain enzymes, including the RNA polymerases (Martelo 1973; Jungmann et al. 1974; Dahmus 1976), are modified by phosphorylation.

Because of the likelihood that phosphorylation-mediated controls may be operating at the nuclear level in developing red blood cells (Vidali et al. 1973), we investigated the phosphorylation of nuclear proteins in DMSO-stimulated erythroleukemia cells. The cell line 745-PC-4 (Gusella et al. 1976) was used in these studies. We chose to examine nuclear proteins

---

[*] Present address: University of New South Wales, School of Biochemistry, P.O. Box 1, Kensington, New South Wales 2033, Australia

during the first 2 days of DMSO treatment, when 90% of the cells have become committed to erythroid differentiation.

Eukaryotic chromatin consists of DNA complexed with histones and nonhistone chromosomal proteins. In terms of DNA and histones, the structure of the repeating unit of chromatin, the nucleosome, is well documented (Hewish and Burgoyne 1973; Noll 1974; Olins and Olins 1974). The nucleosome core has approximately 140 base pairs of DNA wound around a histone complex consisting of two molecules each of histones H2A, H2B, H3, and H4 (Kornberg 1974; Kornberg and Thomas 1974). Nonhistone chromosomal proteins are enriched in "active" chromatin fractions (Montagna et al. 1977). The occurrence of nonhistone chromosomal proteins in monomer nucleosomes has been reported (Lacy and Axel 1975; Liew and Chan 1976; Paul and Malcolm 1976; Mullins et al. 1977; Sanders and Hsu 1977), but their role in nucleosome structure is not understood. We have studied the changes in the distribution of nonhistone proteins in a nucleosome population generated by the action of micrococcal nuclease on nuclei from Friend erythroleukemia cells after exposure to DMSO for 48 hours.

We have also examined the effects of another inducer, n-butyrate, on histone modification in Friend erythroleukemia and in HeLa cells and on DNA synthesis in HeLa cells. At low concentrations (1-2 mM), n-butyrate causes Friend erythroleukemia cells to begin hemoglobin synthesis (Leder and Leder 1975). Prasad and Sinha (1976) have summarized the effects of n-butyrate on neuroblastoma, HeLa, and other cell types. They and others (Ginsburg et al. 1973; Wright 1973; Griffin et al. 1974; Ghosh et al. 1975; Simmons et al. 1975; Altenburg et al. 1976; Fishman et al. 1976; Henneberry and Fishman 1976) have seen morphological modifications, reversible inhibition of proliferation, decrease of DNA content, and increases in the activity of specific enzymes, such as adenylate cyclase, alkaline phosphatase, and a sialytransferase. We have observed rapid, extensive, and reversible increases in histone acetylation caused by low concentrations of n-butyrate. Some of these results have been reported elsewhere (Riggs et al. 1977). We have also studied in vivo and in cell-free systems the reversible inhibition of DNA synthesis in chick embryo fibroblasts and in HeLa cells (Hagopian et al. 1977).

METHODS

HeLa cells were maintained at 37°C in suspension culture at $0.1-0.5 \times 10^6$/ml in Joklik's modified Eagle's medium containing 7% horse serum. Friend erythroleukemia cells (clone 745-PC-4) (Gusella et al. 1976) were grown at 37°C in suspension in alpha medium (Stanners et al. 1971) lacking nucleosides and supplemented with 15% fetal calf serum. They were maintained at $0.05-0.5 \times 10^6$/ml to permit logarithmic growth.

Aliquots of a stock solution of 1 M Na n-butyrate in Ca^{++}- and Mg^{++}-free phosphate-buffered saline (final pH 7.3) were added to cultures 24 hours before histone extraction. Preparations of nuclei and histones were carried out at 0-4°C. Nuclei were prepared from washed cells by Dounce homogenization in 0.1-0.5% Triton X-100 and histones were extracted with 0.4 N

H$_2$SO$_4$ (Panyim et al. 1971). Histones were precipitated overnight at −40°C with 10 volumes of acetone, dried, and electrophoresed on acid-urea 15% acrylamide gels (Moss et al. 1973) at 20°C for 5 hours; the gels were prepared in 0.125% ammonium persulfate. Alternatively, histones prepared as before were run on 15% gels containing 6 M urea and 0.38% Triton DF-16, according to the method of Alfageme et al. (1974). Staining was with 0.1% Amido Black in 40% methanol–7% acetic acid. The gels were scanned at 615 nm.

RESULTS

Phosphorylation of Nuclear Proteins in DMSO-Induced Friend Cells

To characterize further the nuclear proteins of induced and uninduced erythroleukemia cells, we examined their patterns of phosphorylation. Figure 1A shows the phosphoprotein pattern of nuclei from Friend cells pulsed for varying periods of time with [^{32}P]-orthophosphate. This type of experiment yields data on the turnover rate of protein phosphate. Phosphoprotein bands prominent at early time points represent the most actively phosphorylated species. Those proteins which did not turn over their phosphate during the course of the pulse are, of course, not represented in the autoradiograms shown. For this reason, highly phosphorylated but stable proteins might not appear on the gels.

At least 30 phosphorylated peptides were resolved when cells were labeled with [^{32}P]-orthophosphate (Figure 1A). The pattern from nuclei treated with proteinase K demonstrates that the phosphorylated bands represent proteins (Figure 1B). The nonhistones as a whole were more highly phosphorylated than the histones. A relatively minor nuclear phosphoprotein of high mobility turned over rapidly in erythroleukemia cells, as shown by comparison of its relative intensity after 30 and 120 minutes of labeling. This band, believed to be histone H2A, was the only nuclear protein displaying significantly increased levels of phosphorylation after 24 and 48 hours of DMSO treatment (Figure 1C).

The nature of the heavily phosphorylated material that ran faster than the globin marker is unknown. We conclude that it was not protein, since this band was not seen either in Coomassie Blue staining patterns or in autoradiograms of cells pulsed for 3 hours with [^{14}C]-leucine (data not shown).

Cytoplasmic phosphoprotein patterns are also shown in Figure 1A. With the exception of the 125,000-dalton protein and the doublet at 30,000, the nuclear phosphoproteins represent a different spectrum of proteins from those in the cytoplasm. There is a group of heavy bands in the cytoplasm which appears to comigrate with the nuclear histones in the photograph shown. The identity of these bands is unknown.

We also examined histones from uninduced and DMSO-induced Friend cells on gels containing nonionic detergents and high concentrations of urea in order to identify those fractions that were phosphorylated. Figure 2 shows the scans of autoradiograms of gels loaded with ^{32}P-labeled histones isolated from mononucleosomes and also scans of the staining patterns. It can be seen that histone H1 was much more extensively phosphorylated in the in-

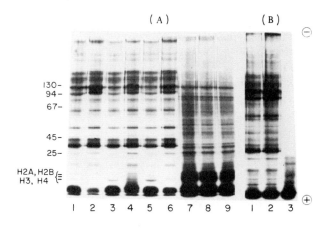

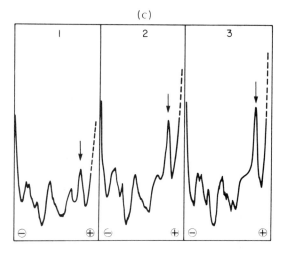

Figure 1
(A) Autoradiogram of $^{32}PO_4$-labeled nuclear proteins of Friend cells. Cells at 10^7/ml were pulsed with 300 μCi/ml $^{32}PO_4$ (carrier free, New England Nuclear). Nuclei and cytoplasms were processed as described (Neumann et al. 1978). All sample wells were loaded with equal amounts of hot PCA-insoluble radioactivity. (1, 3, 5) Nuclear phosphoproteins of uninduced, 24-hour DMSO, and 48-hour DMSO cells labeled for 30 minutes; (2, 4, 6) nuclear phosphoproteins of uninduced, 24-hour DMSO and 48-hour DMSO cells labeled for 120 minutes; (7, 8, 9) cytoplasmic phosphoproteins of uninduced, 24-hour DMSO, and 48-hour DMSO cells labeled for 120 minutes. Positions of molecular weight markers are shown at the left.

(B) Effect of proteinase K treatment on the pattern of nuclear phosphoproteins. (1) Autoradiogram of nuclear proteins labeled for 3.5 hours with $^{32}PO_4$. (2, 3) same nuclear protein preparation as in (1), except that prior to electrophoresis the sample was incubated for 3 hours at 37°C either in the absence (2) or presence (3) of 0.05 mg/ml proteinase K (sp. act. 20 U/mg, E.M. Laboratories)

(C) Scans of the autoradiograms of (1) uninduced cell nuclear phosphoproteins; (2) 24-hour DMSO; (3) 48-hour DMSO. These samples are identical to 1, 3, and 5, respectively, in (A), i.e., all represent cells that were pulsed for 30 minutes with $^{32}PO_4$. Position of histone H2A is indicated by arrows.

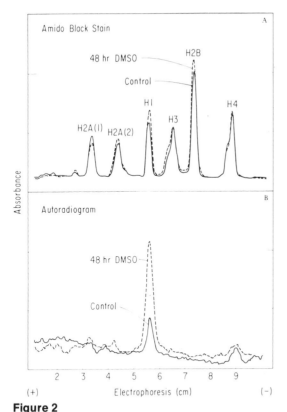

Figure 2
Phosphorylation pattern of nucleosome-associated histones from control and DMSO-stimulated cells. Nuclei of $^{32}PO_4$-labeled control and induced cells were digested for 2 minutes with micrococcal nuclease (7 and 13% acid-soluble A_{260}, respectively) and fractionated on 5–28.8% sucrose gradients. Fractions corresponding to the nucleosome monomer were pooled. The histones were extracted with 0.4 N H_2SO_4 and electrophoresed on gels containing Triton DF-16 as described in Methods. After staining with Amido Black and scanning, the gels were sliced longitudinally, dried under vacuum, and autoradiographed. (A) Amido Black staining pattern; (B) densitometric tracing of autoradiogram. The gels were loaded with equal amounts of protein.

duced cells. H1 phosphorylation was apparent also in conventional histone gels (not shown), although it was obscured in gels containing sodium dodecylsulfate (SDS) (Figure 1A). The increased phosphorylation of H2A reported in Figure 1 was obscured in the background noise of Figure 2, where histones from mononucleosomes were examined. A similar histone gel derived from oligonucleosomes, however, did show the increased phosphorylation of H2A seen in Figure 1, which represents total nuclear proteins.

Nucleosome-Associated Proteins in DMSO-Induced Friend Cells

After digestion with micrococcal nuclease for 2 minutes, the nuclei were removed from the digestion buffer by centrifugation, yielding the supernatant fraction S1. Extraction of the digested nuclei with low ionic strength buffer (Noll et al. 1975) produced a second supernatant fraction, S2, containing approximately 50% of the nuclear DNA, which was then fractionated on a sucrose gradient. The sucrose gradients of nucleosomes and the gel electropherograms of isolated DNA show the expected subunit profiles (Axel et al. 1974; Finch et al. 1975; Sollner-Webb and Felsenfeld 1975).

Fractions S1 and S2 and portions of the sucrose gradients containing nucleosome monomers, dimers, trimers, and oligomers were analyzed for protein by SDS-polyacrylamide gel electrophoresis (Fig. 3). Samples of

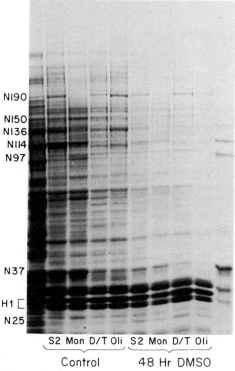

Figure 3
Nucleosome-associated proteins of control and DMSO-treated cells. Nuclei were digested for 2 minutes with micrococcal nuclease. S2 and pooled sucrose gradient fractions corresponding to monomeric, dimeric plus trimeric, and oligomeric nucleosomes were dialyzed against 0.1 mM phenyl methylsulfonyl fluoride-2 mM Na$_2$EDTA and then lyophilized. Dried samples were dissolved in 0.1% SDS-50 mM Tris-HCl, pH 8.6. Aliquots corresponding to 1 A$_{260}$ U of chromatin were made 1% 2-mercaptoethanol and 10% glycerol, heated, and electrophoresed on 5–15% acrylamide gradient slab gels (Laemmli 1970). After staining with Coomassie Blue, the slab was scanned at 550 nm. Marker proteins used for molecular weight determination were chymotrypsinogen A (25,000), ovalbumin (45,000), bovine serum albumin (67,000), phosphorylase b (94,000), β-galactosidase (130,000), and myosin heavy chain (200,000).

intact nuclei, fractions from the top and bottom of the gradients, and residual digested nuclei were also examined (data not shown). Sham-digested nuclei gave negligible release of DNA or protein into fractions S1 and S2. A third of the total extractable nonhistone protein was rapidly released into the S1 fraction with 2 minutes of digestion. With further digestion there was only a small increase in protein in this fraction, resulting in negligible differences between the 2- and 10-minute S1 profiles (data not shown). As the S1 nonhistones were capable of diffusing from the nucleus during digestion, they presumably were released as free proteins or very small nucleoprotein fragments.

The nucleosome monomers from a 2-minute digest were considerably enriched in nonhistone proteins as compared to dimers and oligomers. All nucleosome fractions contained the expected complement of core histones, but the mononucleosomes reproducibly had strikingly greater amounts of the proteins which we call N37, N97, N114, N136, and N150 (Fig. 3). Band N97 was especially notable, as it was only a minor component of the proteins of fraction S2. All of these proteins were greatly reduced or entirely absent from the oligomer fractions. Protein N190 followed the opposite pattern, being more prominent among the oligomeric nucleosomes; N190 was actually absent from the monomers. Apparently this protein requires at least a dimeric nucleosome structure for binding; it might even stabilize a subpopulation of oligomers. The chromatin-associated enzyme poly(ADP-ribose) polymerase (Mullins et al. 1977) had a distribution similar to that of N190 among nucleosomes and may have had similar structural requirements for binding. Further incubation of nuclei with micrococcal nuclease for a total of 10 minutes produced a nucleosome population with a low content of nonhistone proteins (data not shown); only N37 remained prominent in the mononucleosome fraction. The missing nonhistones were released from the mononucleosomes, since they were enriched in the top fraction of the sucrose gradient. On the contrary, protein N190 apparently was not released, since it did not appear in this top fraction, in the S1 fraction, or on the mononucleosomes.

The pattern of nucleosome-associated histones and nonhistone proteins from a 2-minute digest of DMSO-induced nuclei resembled the 10-minute digest from uninduced nuclei. Most of the nonhistone proteins, except N190, were released. However, protein N25, which might be the IP_{25} of Keppel et al. (1977), was much increased after induction (Fig. 3).

n-Butyrate-Induced Histone Acetylation

Nuclei from *n*-butyrate-treated HeLa and Friend erythroleukemia cells contained greatly increased amounts of modified forms of histone H4 (Figs. 4, 5). Total histones from nuclei of cells after 24 hours of treatment with 1 mM *n*-butyrate showed several strong, slower moving H4 bands when examined in acid-urea gels and in Triton DF16–acid-urea gels. Their electrophoretic positions corresponded to the mono-, di-, tri-, and tetraacetylated H4 histones which occur during normal histone biosynthesis (Ruiz-Carillo et al. 1975). For this reason and because of the experiments described elsewhere (Riggs et al. 1977), we believe that the modified H4 histones were

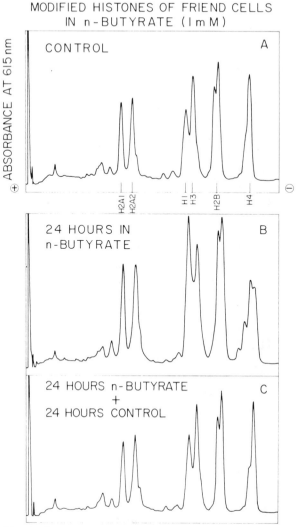

Figure 4
Reversal of *n*-butyrate effect on histones from Friend erythroleukemia cells examined on Triton-acid urea gels. The scans of the gels were normalized to peak H2B. (*A*) Untreated Friend cells; (*B*) cells exposed to 1 mM *n*-butyrate for 24 hours; (*C*) 24-hour *n*-butyrate-treated cells, washed, and resuspended in fresh control medium for 24 hours.

acetylated. In control cells only a small amount of mono-acetyl-H4 was present, whereas this is the dominant form in treated cells. In HeLa cells the extent of modification, although marked at 1 mM, increased at higher *n*-butyrate concentrations. H4 acetylation and H3 modification in HeLa cells were already noticeable after only 1 hour in 5 mM *n*-butyrate (data not shown), the earliest effect of *n*-butyrate reported. In both HeLa and Friend erythroleukemia cells the histone H4 pattern reverted completely to normal when cells treated with *n*-butyrate for 24 hours were shifted back to control medium for a further 24 hours (Fig. 4).

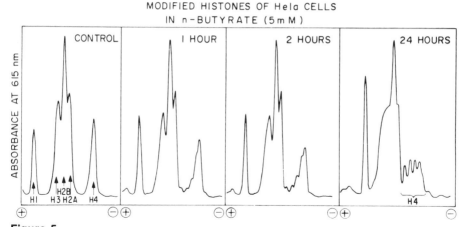

Figure 5
Time course of *n*-butyrate effect on HeLa histones, examined on acid urea gels. The scans of the gels were normalized to peak H2B.

Both cell lines also showed a clear alteration of the histone H3 band with an increase of slower moving material. This was seen as a broadening of the H3 band on 10-cm gels; on 16-cm gels, distinct slow bands behind H3 were seen. There was an increase in the relative amounts of H1 (Fig. 4), which might have been due to a lowering of the dye-binding capacity of the other (acetylated) histones. All these changes were reversible within 24 hours after removal of *n*-butyrate.

Quantitative analysis of the scans of gels shown in Figure 5 indicates that the sum of the H4 components in *n*-butyrate-treated cells was equal to the sum of the H4 components in control cells and that no H4 was lost. The proportion of acetylated H4 in HeLa cells reached approximately 80% of total H4 after 24 hours in 5 mM *n*-butyrate (Fig. 4). In addition we believe that the modification was acetylation (Riggs et al. 1977) and not phosphorylation, since incubation of the modified histones with *Escherichia coli* alkaline phosphatase (Thomas and Hempel 1976) left the modified H4 pattern intact.

Inhibition of DNA Synthesis in HeLa Cells Treated with *n*-Butyrate

The rate of incorporation of thymidine, presumably into DNA, was strongly inhibited by *n*-butyrate in chick embryo fibroblasts and HeLa cells after 25 hours of exposure (Hagopian et al. 1977). It took about 15 hours for DNA synthesis to be fully inhibited in fibroblasts.

The early time course of inhibition of DNA synthesis in HeLa cells was measured (Fig. 6). After only 1 hour, the rate of thymidine incorporation during a 15-minute pulse had fallen to 68% of control. After 24 hours the cells were at their lowest level of DNA synthesis, somewhere between 2 and 10% of control. In Figure 5 we show the extensive acetylation shown by HeLa histones during 24 hours in 5 mM *n*-butyrate. This acetylation was particularly striking for histone H4, which was found as Ac_1-H4, Ac_2-H4, Ac_3-H4, and Ac_4-H4. For the purposes of the present paper, we estimated from the time course of progressive histone acetylation (Fig. 5) the propor-

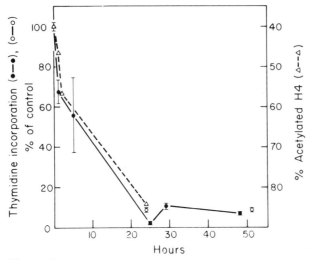

Figure 6
Time course of inhibition of thymidine incorporation and of increased histone H4 acetylation in 6.5 mM and 5 mM n-butyrate, respectively, in HeLa cells. HeLa cells were labeled in suspension culture with [³H]-thymidine for a 15-minute pulse (3.5 µCi/2 ml culture containing 0.2–0.6 × 10⁶ cells/ml). Thymidine incorporation in control cells was 91,700 cpm/10⁶ cells for a 15-minute pulse in one experiment (●——●) and 199,600 cpm/10⁶ cells in the other (o——o). The extent of H4 acetylation was calculated from the area under the curve of the acid urea gels of Figure 4.

tion of histone H4 which existed in acetylated form. These percentages are also plotted in Figure 6. The proportion of acetylated H4 found in control cells was 39.8%; the maximum reached at 24 hours was 84.0%. Clearly both effects of n-butyrate are expressed quite early and have a similar time course, suggesting a close link between the two processes.

Inhibition of Cell-Free DNA Synthesis by Prior n-Butyrate Treatment

The reconstruction experiment of Figure 7 shows that cytosol is essential for DNA synthesis in purified nuclei as reported by Fraser and Huberman (1977). Cytosol from n-butyrate-treated cells began at once to shut down DNA synthesis in control nuclei; inhibition was complete by 15 minutes. On the other hand, nuclei from n-butyrate-treated cells remained inactive in control cytosol. This result is strange in view of the reversibility of inhibition of cell proliferation. However, the reversibility of whole cells takes 24 hours and the short incubations shown in Figure 7 were probably not long enough for the reversal process to occur.

DISCUSSION

The overall pattern of nuclear phosphoproteins is similar in uninduced Friend cells and in DMSO-stimulated cells. There are, however, some qualitative and quantitative changes during induction.

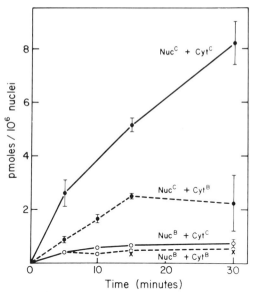

Figure 7
Reconstitution of DNA synthesizing system from purified nuclei and cytosol. The two experimental points (×-----×) were obtained from the incubation of purified nuclei from untreated cells with the buffer H of Fraser and Huberman (1977) only. Nuc^C, Cyt^C = nuclei, cytosol from control cells; Nuc^B, Cyt^B = nuclei, cytosol from cells treated with 5 mM n-butyrate for 24 hours. Purified HeLa nuclei were prepared from control cultures or from cultures treated with 5 mM n-butyrate for 24 hours by the method of Fraser and Huberman (1977). The purified nuclei ($20-40 \times 10^6$) were mixed with the appropriate clarified cytosol fraction (450–900 µl, made according to Fraser and Huberman, 1977) prior to the addition of the components of the assay mix. Membranes and membranous organelles presumably were not present in this incubation. The radioactive label used was α-[^{32}P]-dATP (5 µCi/ml).

The most striking change is the overall decrease in protein synthesis and phosphorylation after 50 hours of DMSO treatment (Neumann et al. 1978). The decrease begins about the time that cell division ceases in the first committed cells; those cells which commit to differentiation at 18 to 24 hours will have completed their final replication at 60 to 66 hours (Gusella et al. 1976). Protein synthesis and phosphorylation may be coupled to the DNA synthetic phase of the cell cycle in erythroleukemia cells.

Keppel et al. (1977), using another clone (F4N) of erythroleukemia cells, found that the only nuclear protein change paralleling DMSO induction was the appearance of a 25,000-dalton protein (IP_{25}), perhaps the same protein reported by Peterson and McConkey (1976). We observed a band of similar molecular weight, whose level in oligomeric nucleosomes was increased after DMSO induction (Figure 3). The increase was not observed when we examined total nuclear proteins.

We have shown that H1 and H2A are the major phosphorylated histone species both in uninduced and induced erythroleukemia cells. In addition, H1 and H2A appear to be the only histones that are affected by DMSO induction, showing a large increase in their rate of phosphorylation. This increase may be related to the process of induction of differentiation or to the cessation of cell division or to both these phenomena.

The mononucleosomes generated by a 2-minute digestion of nuclei with micrococcal nuclease have a much greater content of nonhistone proteins than do the dimers and oligomers in the same digest or the mononucleosomes produced by longer digestion. We speculate that nonhistone proteins in certain regions of chromatin might facilitate a conformational change which renders the "linker" DNA more susceptible to digestion. Only protein N190 is definitely absent from the nucleosome core and is distributed in a fashion that suggests that it might stabilize some particular oligonucleosomes against digestion.

The pattern of proteins associated with mono- and oligonucleosomes after only 2 minutes of digestion of the nuclei shows a striking change after DMSO induction. The pattern now resembles that of the much longer digestion of uninduced cells; most of the nonhistone chromosomal proteins have been released. This finding indicates that the DNA of DMSO-induced Friend cells is much more accessible to nuclease digestion. This is interesting, because Weintraub and Groudine (1976) found that transcribed DNA sequences are digested more rapidly by DNase I.

We are particularly interested in the effect of n-butyrate on the proliferative capacity of the malignant Friend erythroleukemia cell and of the HeLa cell, which is also derived from a malignant cell. In the Friend cell differentiation is blocked. n-Butyrate induces the continuation of differentiation and the production of specific products, as well as the striking accumulation of acetylated histones. We do not know whether there is a causal relationship between histone acetylation and these observations. Darzynkiewicz et al. (1976) and Nudel et al. (1977), using either DMSO or n-butyrate as inducing agents, reported subtle changes in the structure of Friend cell chromatin.

It is possible that the extensive histone acetylation in the presence of n-butyrate is involved in the shutdown of DNA synthesis, although we do not as yet have direct proof of such a mechanism. However, in HeLa cells the time courses for extent of histone acetylation and for decrease of DNA synthesis (to 10% or less) are roughly parallel during the first 24 hours.

Moore and Chalkley (1977), using cultured hepatoma cells, suggested an interdependence of the level of histone H4 acetylation and the rate of cell division. We did not observe acetylation changes in the histone pattern of DMSO-treated Friend cells (Neumann et al. 1978); perhaps DMSO and n-butyrate act at different points in these complex processes. Further experimental work may show a causal relationship between the accumulation of polyacetylated histones in the presence of n-butyrate on the one hand and the inhibition of DNA synthesis and the switching of malignant cells to nonmalignant differentiating cells on the other.

We conclude from our experiments that cytosol from n-butyrate-treated cells contains an inhibitor or dilutes factors needed by the nuclei for DNA synthesis. n-Butyrate itself has no effect on the cell-free system. In addition, there is a permanent or semipermanent alteration in nuclei from n-butyrate-treated cells which prevents them from carrying out DNA synthesis. For example, the complete inability of nuclei from *n*-butyrate-treated HeLa cells to function in control cytosol might be related to the presence of modified histones.

These investigations promise to clarify the control of DNA synthesis and the relationship between this process and the production of specific gene products such as the induction of globin mRNA synthesis in Friend erythroleukemia cells.

REFERENCES

Adler, A. J., B. Schaffhausen, T. A. Langan, and G. D. Fasman. 1971. Altered conformational effects of phosphorylated lysine-rich histone (f1) in f1-deoxyribonucleic acid complexes. *Biochemistry* **10**: 909.

Alfageme, C. R., A. Zweidler, A. Mahowald, and L. H. Cohen. 1974. Histones of *Drosophila* embryos. *J. Biol. Chem.* **249**: 3729.

Altenburg, B. C., D. P. Via, and S. H. Steiner. 1976. Modification of the phenotype of murine sarcoma virus-transformed cells by sodium butyrate. *Exp. Cell Res.* **102**: 223.

Axel, R., W. Melchior, Jr., B. Sollner-Webb, and G. Felsenfeld. 1974. Specific sites of interaction between histones and DNA in chromatin. *Proc. Natl. Acad. Sci. U.S.A.* **71**: 4101.

Barrett, T., D. Maryanka, P. Hamlyn, and H. Gould. 1974. Nonhistone proteins control gene expression in reconstituted chromatin. *Proc. Natl. Acad. Sci. U.S.A.* **71**: 5057.

Dahmus, M. E. 1976. Stimulation of ascites tumor RNA polymerases II by protein kinase. *Biochemistry* **15**: 1821.

Darzynkiewicz, A., F. Traganos, T. Sharpless, C. Friend, and M. R. Melamed. 1976. Nuclear chromatin changes during erythroid differentiation of Friend virus induced leukemia cells. *Exp. Cell Res.* **99**: 301.

Finch, J. T., M. Noll, and R. D. Kornberg. 1975. Electron microscopy of defined length of chromatin. *Proc. Natl. Acad. Sci. U.S.A.* **72**: 3320.

Fishman, P. H., R. M. Bradley, and R. C. Henneberry. 1976. Butyrate-induced glycolipid biosynthesis in HeLa cells: Properties of the induced sialyltransferase. *Arch. Biochem. Biophys.* **172**: 618.

Fraser, J. M. K., and J. A. Huberman. 1977. *In vitro* HeLa cell DNA synthesis: Similarity to *in vivo* replication. *J. Mol. Biol.* **117**: 249.

Friend, C., W. Scher, J. G. Holland, and T. Sato. 1971. Hemoglobin synthesis in murine virus-induced leukemic cells *in vitro*: Stimulation of erythroid differentiation by dimethyl sulfoxide. *Proc. Natl. Acad. Sci. U.S.A.* **68**: 378.

Ghosh, N. K., S. I. Deutsch, M. J. Griffin, and R. P. Cox. 1975. Regulation of growth and morphological modulation of HeLa$_{65}$ cells in monolayer culture by dibutyryl cyclic AMP, butyrate and their analogs. *J. Cell. Physiol.* **86**: 663.

Ginsburg, E., D. Salomon, T. Sreevalsan, and E. Freese. 1973. Growth inhibition and morphological changes caused by lipophilic acids in mammalian cells. *Proc. Natl. Acad. Sci. U.S.A.* **70**: 2457.

Griffin, M. H., G. H. Price, K. L. Bazzell, R. P. Cox, and N. K. Ghosh. 1974. A study of adenosine 3':5'-cyclic monophosphate, sodium butyrate and cortisol as inducers of HeLa alkaline phosphatase. *Arch. Biochem. Biophys.* **164**: 619.

Gusella, J., R. Geller, B. Clarke, V. Weeks, and D. Housman. 1976. Commitment to erythroid differentiation by Friend erythroleukemia cells: A stochastic analysis. *Cell* **9**: 221.

Hagopian, H. K., M. G. Riggs, L. A. Swartz, and V. M. Ingram. 1977. The effect of *n*-butyrate on DNA synthesis in chick fibroblasts and HeLa cells. *Cell* **12**: 855.

Henneberry, R. C., and P. H. Fishman. 1976. Morphological and biochemical differential in HeLa cells. *Exp. Cell Res.* **103**: 55.

Hewish, D. R., and L. A. Burgoyne. 1973. Chromatin sub-structure: The digestion of chromatin DNA at regularly spaced sites by a nuclear deoxyribonuclease. *Biochem. Biophys. Res. Commun.* **52**: 504.

Ikawa, Y., M. Furasawa, and H. Sugano. 1973. Erythrocyte membrane-specific antigens in Friend virus-induced leukemia cells. *Bibl. Haematol.* **39**: 995.

Jungmann, R. A., T. C. Hiestand, and J. S. Schweppe. 1974. Mechanism of action of gonadotropin. IV. Cyclic adenosine monophosphate-dependent translocation of ovarian cytoplasmic cyclic adenosine monophosphate-binding protein and protein kinase to nuclear acceptor sites. *Endocrinology* **94**: 168.

Keppel, F., B. Allet, and H. Eisen. 1977. Appearance of a chromatin protein during the erythroid differentiation of Friend virus-transformed cells. *Proc. Natl. Acad. Sci. U.S.A.* **74**: 653.

Kleinsmith, L. J., J. Stein, and G. Stein. 1976. Dephosphorylation of nonhistone proteins specifically alters the pattern of gene transcription in reconstituted chromatin. *Proc. Natl. Acad. Sci. U.S.A.* **73**: 1174.

Kornberg, R. D. 1974. Chromatin structure: A repeating unit of histones and DNA. *Science* **184**: 868.

Kornberg, R. D., and J. O. Thomas. 1974. Chromatin structure: Oligomers of the histones. *Science* **184**: 865.

Kostraba, N. C., R. A. Montagna, and T. Y. Wang. 1975. Study of the loosely bound non-histone chromatin proteins. *J. Biol. Chem.* **250**: 1548.

Lacy, E., and R. Axel. 1975. Analysis of DNA of isolated chromatin subunits. *Proc. Natl. Acad. Sci. U.S.A.* **72**: 3978.

Laemmli, U. K. 1970. Cleavage of structural proteins during the assembly of the head of bacteriophage T4. *Nature* **227**: 680.

Langan, T. A. 1968. Phosphorylation of proteins of the cell nucleus. In *Regulatory mechanisms for protein synthesis in mammalian cells* (eds. A. San Pietro, M. R. Lamborg, and F. T. Kenny), p. 101. Academic Press, New York.

Leder, A., and P. Leder. 1975. Butyric acid, a potent inducer of erythroid differentiation in cultured erythroleukemic cells. *Cell* **5**: 319.

Liew, C. C., and P. K. Chan. 1976. Identification of nonhistone chromatin proteins in chromatin subunits. *Proc. Natl. Acad. Sci. U.S.A.* **73**: 3458.

Marks, D. B., W. K. Paik, and T. W. Borun. 1973. The relationship of histone phosphorylation to deoxyribonucleic acid replication and mitosis during the HeLa S-3 cell cycle. *J. Biol. Chem.* **248**: 5660.

Martelo, O. J. 1973. Protein phosphorylation in control mechanisms. In *Miami Winter Symposium*, vol. 5, p. 199. Academic Press, New York.

Montagna, R. A., L. V. Rodriquez, and F. F. Becker. 1977. A comparative study of the nonhistone proteins of rat liver euchromatin and heterochromatin. *Arch. Biochem. Biophys.* **179**: 617.

Moore, M. R., and R. Chalkley. 1977. The effect of perturbations of the environment on histone acetylation and its turnover in hepatoma tissue culture cells. Abstract no. 2708. *Fed. Proc.* **36**: 785.

Moss, B. A., W. G. Joyce, and V. M. Ingram. 1973. Histones in chick embryonic erythropoiesis. *J. Biol. Chem.* **248**: 1025.

Mullins, D. W., Jr., D. P. Giri, and M. Smulson. 1977. Poly(adenosine diphosphate-ribose) polymerase: The distribution of a chromosome-associated enzyme within the chromatin substructure. *Biochemistry* **16**: 506.

Neumann, J. R., D. Housman, and V. M. Ingram. 1978. Nuclear protein synthesis and phosphorylation in Friend erythroleukemia cells stimulated with DMSO. *Exp. Cell Res.* **111:** 277.

Noll, M. 1974. Subunit structure of chromatin. *Nature* **251:** 249.

Noll, M., J. O. Thomas, and R. D. Kornberg. 1975. Preparation of native chromatin and damage caused by shearing. *Science* **187:** 1203.

Nudel, U., J. E. Salmon, M. Terada, A. Bank, R. A. Rifkind, and P. A. Marks. 1977. Differential effects of chemical inducers on expression of β globin genes in murine erythroleukemia cells. *Proc. Natl. Acad. Sci. U.S.A.* **74:** 1100.

Olins, A. L., and D. E. Olins. 1974. Spheroid chromatin units (ν bodies). *Science* **183:** 330.

Panyim, S., D. Bilek, and R. Chalkley. 1971. An electrophoretic comparison of vertebrate histones. *J. Biol. Chem.* **246:** 4206.

Paul, J., and S. Malcolm. 1976. A class of chromatin particles associated with non-histone proteins. *Biochemistry* **15:** 3510.

Peterson, J. L., and E. H. McConkey. 1976. Proteins of Friend leukemia cells. *J. Biol. Chem.* **251:** 555.

Prasad, K. N., and P. K. Sinha. 1976. Effect of sodium butyrate on mammalian cells in culture: A review. *In Vitro* (Rockville) **12:** 125.

Riggs, M. G., R. G. Whittaker, J. R. Neumann, and V. M. Ingram. 1977. *n*-Butyrate causes histone modification in HeLa and Friend erythroleukemia cells. *Nature* **268:** 462.

Ross, J., Y. Ikawa, and P. Leder. 1972. Globin messenger-RNA induction during erythroid differentiation of cultured leukemia cells. *Proc. Natl. Acad. Sci. U.S.A.* **69:** 3620.

Ruiz-Carrillo, A., L. J. Wangh, and V. G. Allfrey. 1975. Processing of newly synthesized histone molecules. *Science* **190:** 117.

Sanders, M. M., and J. T. Hsu. 1977. Fractionation of purified nucleosomes on the basis of aggregation properties. *Biochemistry* **16:** 1690.

Sassa, S., S. Granick, C. Chang, and A. Kappas. 1975. Sequential induction of enzymes of the heme biosynthetic pathway in Friend erythroleukemia cells in culture. *Fed. Proc.* **34:** 705.

Simmons, J. L., P. H. Fishman, E. Freese, and R. M. Brady. 1975. Morphological alterations and ganglioside sialyltransferase activity induced by small fatty acids in HeLa cells. *J. Cell Biol.* **66:** 414.

Sollner-Webb, B., and G. Felsenfeld. 1975. A comparison of the digestion of nuclei and chromatin by staphylococcal nuclease. *Biochemistry* **14:** 2915.

Stanners, G. P., G. L. Eliceiri, and H. Green. 1971. Two types of ribosome in mouse-hamster hybrid cells. *Nat. New Biol.* **230:** 52.

Stein, G. S., T. C. Spelsberg, and L. J. Kleinsmith. 1974. Nonhistone chromosomal proteins and gene regulation. *Science* **183:** 817.

Thomas, G., and K. Hempel. 1976. Correlation between histone phosphorylation and tumor aging in Ehrlich ascites tumor cells. *Exp. Cell Res.* **100:** 309.

Tsai, M-J., H. C. Towle, S. E. Harris, and B. W. O'Malley. 1976. Effect of estrogen on gene expression in the chick oviduct. *J. Biol. Chem.* **251:** 1960.

Vidali, C., L. C. Boffa, V. C. Littau, K. M. Allfrey, and V. G. Allfrey. 1973. Changes in nuclear acidic protein complement of red blood cells during embryonic development. *J. Biol. Chem.* **248:** 4065.

Weintraub, H., and M. Groudine. 1976. Globin genes are digested by deoxyribonuclease I in red blood cell nuclei but not in fibroblast nuclei. *Science* **193:** 848.

Wright, J. A. 1973. Morphology and growth rate changes in Chinese hamster cells cultured in presence of sodium butyrate. *Exp. Cell Res.* **78:** 456.

Biochemical and Genetic Analysis of Erythroid Differentiation in Friend Virus-Transformed Murine Erythroleukemia Cells

H. Eisen,* F. Keppel-Ballivet, and C. P. Georgopoulos[†]

Department of Molecular Biology, University of Geneva, Geneva CH1211, Switzerland

S. Sassa and J. Granick

The Rockefeller University, New York, New York 10021

I. Pragnell

Beatson Institute for Cancer Research, Glasgow G3 6UD, Scotland

W. Ostertag

Max Planck Institute for Experimental Medicine, Göttingen 0034, West Germany

Although mammalian erythropoiesis has been the subject of intense study for many years, little is known about the genetic regulation of this differentiation program or about the molecular mechanisms involved in its expression. Although the earliest step in erythropoiesis, involving the commitment of pluripotent stem cells to erythroid precursors, constitutes one of the most intriguing problems in modern biology, it still remains inaccessible to analysis at the molecular level and to easy manipulation. Furthermore, subsequent steps in normal erythroid differentiation have eluded molecular analysis because pure populations of precursor cells cannot be prepared or cultivated for long periods. Thus, the properties of these precursors can be discerned only indirectly by the ability of the cells to form erythroid colonies in vitro (Stephenson et al. 1971). Nevertheless the final product of erythropoiesis, the erythrocyte, is probably the simplest and best understood mammalian cell. All of its major constituents have been purified and characterized and the regulation of its metabolism has been extensively studied.

In recent years, the development of Friend virus-transformed murine erythroleukemic cells has provided a unique opportunity for the in vitro study of many of the later steps in erythroid differentiation (Friend et al. 1971). These cells, which are easily cloned and cultivated in vitro, provide homogeneous populations of transformed erythroblasts or proerythroblasts which can be induced to undergo terminal erythroid differentiation by addition of a multitude of apparently unrelated compounds. Thus, the FL cell system provides experimental material well suited to both biochemical and genetic analysis (see Harrison 1977).

Treatment of the FL cells with inducing agents results in the accumula-

---

* Present address: Department of Molecular Biology, Pasteur Institute 75015, Paris, France.
[†] Present address: Department of Microbiology, University of Utah School of Medicine, Salt Lake City, Utah 84112.

tion over a period of 3 to 6 days of such erythrocyte-specific proteins as hemoglobin (Friend et al. 1971), spectrin (Eisen et al. 1977a), acetylcholinesterase (Conscience et al. 1977), heme synthetic enzymes (Sassa 1975), and surface antigens (Furusawa et al. 1971). There is a decrease in the amount of the major histocompatibility antigen (H2) presented on the cell surface (Arndt-Jovin et al. 1976). The appearance of a new histonelike protein (IP25) in the chromatin of the differentiating FL cells has been described, although it has not yet been demonstrated that this protein is completely specific to erythroid tissue (Keppel et al. 1977).

The erythroid differentiation of the FL cells is accompanied by a reduction of globin RNA and protein synthesis and eventually by the arrest of nuclear activity and cell death (Dube et al. 1974; Kabat et al. 1975). During this time, the volume of the cells is reduced by a factor of 4 to 5. However, nuclear extrusion and erythrocyte formation are rarely observed.

Addition of inducers of erythroid differentiation to the FL cells increases the probability that the cells will commit themselves to the terminal events of erythroid differentiation (Gusella et al. 1976). This commitment, which results in the synthesis of globin mRNA and globin and in the decreased proliferative capacity of the cells, appears to be irreversible. As yet, nothing is known about the events leading to the commitment or the mechanisms by which the various inducers bring about these events.

We attempted to study the induced erythroid differentiation of the FL cells by biochemical and genetic analysis. Specifically we addressed ourselves to the following questions. (1) Is the complete sequence of events during the differentiation dependent upon a unique signal or is the expression of one set of events dependent upon the prior expression of another genetic program? In other words, are the various events expressed in parallel or in series? (2) Are there measurable events that occur before commitment to the terminal phase takes place, and, if so, is their expression dependent upon the continued presence of inducer, or, once their expression has been triggered by inducer, are they expressed even after the inducer has been removed? (3) Can the events occurring during differentiation be ordered in a temporal sequence and, eventually, in a causal sequence?

We have approached these questions by analyzing as many phenotypic characteristics of the induced cells as possible in wild-type FL cells and in variants unable to complete the differentiation. Our results demonstrate that (1) the events occurring during the induction of the FL cells take place in an ordered temporal sequence and not in parallel; (2) the induction of spectrin, IP25, the early heme biosynthetic enzymes, Friend virus complex, and certain changes in the cell membrane occur before commitment to the final stages of differentiation takes place and that the synthesis of these characters is reversible and compatible with proliferation; (3) the synthesis of globin mRNA and globin is associated with the loss of proliferative capacity. The expression of these characters depends on the presence of heme, but heme, although necessary, is not alone sufficient for their synthesis. Our results are consistent with the hypothesis that treatment of the FL cells with inducing agents will result in the induction of an early erythroid program consisting of spectrin, IP25, and heme synthetic enzymes, and that heme, plus some other component(s) of this program, are responsible for the induction of a late program consisting of globin and directly or indirectly for the arrest of cell proliferation.

MATERIALS AND METHODS
Cells and Culture Conditions

Cells and culture conditions have been described previously (Keppel et al. 1977). For measurement of δ-aminolevulinic acid synthetase (ALAS), δ-aminolevulinic acid dehydrase (ALAD), and uroporphyrinogen I synthetase (UROS), cells were grown in modified Ham's F12 medium containing 10% heat-inactivated fetal bovine serum (Sassa and Kappas 1977).

Assays

ALAS, ALAD, and UROS were assayed by methods previously described by Sassa (1975). Intracellular heme was assayed by the method of Orkin et al. (1975), and extracted heme was determined fluorometrically (Sassa 1975). IP25 was extracted as described previously (Keppel et al. 1977). Intracellular spectrin was visualized with indirect immunofluorescence (Eisen et al. 1977a). Spectrin was purified from FL cells by immune precipitation as described (Eisen et al. 1977a). Sodium dodecyl sulfate–polyacrylamide gel electrophoresis was performed as described by Laemmli (1970).

Cytoplasmic globin mRNA was determined as described previously by Young et al. 1974. Extracts for globin determination were prepared as described by Dube et al. (1974), and globin chains were purified by the method of Clegg et al. (1965).

Complement-mediated cytotoxicity was performed as follows. Cells to be tested were washed once in Earle's balanced salts solution containing 2.5% heat-inactivated fetal bovine serum and once in veronal-buffered saline (VBS) pH 7.3 containing 1 mM $MgCl_2$ and 2.5% heat-inactivated fetal bovine serum. The cells were then resuspended at 10^5 cells/ml in VBS containing 1 mM $MgCl_2$, and 25-µl samples were distributed under mineral oil in cells of hemagglutination trays. Twenty-five µl of a 0.5 dilution of freshly pooled human serum which had been previously absorbed with an equal volume of FL cells (clone F4N) were then added. Incubation was carried out for 30 minutes at 37°C, after which cell viability was assayed by the trypan-blue dye exclusion test (Cikes 1970).

RESULTS
Wild-Type FL Cells

Before attempting to analyze the phenotypes of variant clones of FL cells, we examined the characteristics of wild-type strains. Those characteristics examined were membrane changes, including sensitivity to agglutination by plant lectins; sensitivity to lytic human complement in the absence of specific antibodies; the induction and metabolism of the erythrocyte membrane-associated protein spectrin; the induction of the induced histonelike protein IP25; induction of the heme synthetic enzymes ALAS, ALAD, and UROS; and the appearance of heme, globin mRNA, and globin.

Membrane Changes

Although the membrane changes which we shall describe were not shown to be involved in normal in vivo erythroid differentiation in the mouse, they nevertheless appeared to accompany the induction of erythroid differentiation in all of the FL cells examined, independent of the inducing agent used. Furthermore, these changes did not appear to be artifacts provoked by inducing agents, inasmuch as newly isolated tumors and established lines recently passaged through animals express the induced characteristics constitutively.

As we have previously described (Eisen et al. 1977b), induction of FL cells with dimethylsulfoxide (DMSO) or butyric acid resulted in an increase of 10 to 100 times in their sensitivity to agglutination by several plant lectins including concanavalin A, phytohemagglutinin, ricin agglutinin, and the agglutinin from *Carragena arborescence*. This increased sensitivity to agglutination was detectable from 12 to 24 hours after the inducer had been added and did not correspond to an increase of lectin receptors on the surface of the cells. Although the cells remained highly sensitive to agglutination by the lectins for 1 to 2 days more, they became relatively insensitive as differentiation went to completion. Surprisingly, this decrease in sensitivity to agglutination by the lectins at the end of the induction was accompanied by an increase in the number (and density) of lectin receptors present at the cell surface. The increase in sensitivity of the inducer-treated FL cells appeared to precede the induction of globin by at least 24 hours and could be reversed at early times during the induction (up to 36 hours) by removal of the inducing agent. This shows that the increase in agglutinability of the cells was expressed prior to their commitment to the terminal phase of the differentiation and that the regulation of this phenomenon was programmed differently from that of commitment and globin synthesis. Whether the increase was necessary for commitment and globin synthesis is not known.

We also observed that within 6 to 12 hours of inducer addition FL cells became sensitive to lysis by rabbit, guinea pig, and human complement in the absence of added antibody. This appeared to be due to the activation by the cells of the alternate pathway as evidenced by (1) the requirement for Mg^{++} but not Ca^{++}, (2) the loss of lytic activity of human complement previously treated for 30 minutes at 52°C, and (3) the insensitivity of the reaction to antibodies directed to the C1 component of human complement. The kinetics of induction of complement sensitivity in DMSO-treated FL cells of clone F4N are shown in Figure 1. It can be seen that the proportion of cells lysed by complement increased during the first 48 hours of treatment but decreased after 3 days of treatment. This decrease in sensitivity could not have been due to growth of uninduced cells in the population because, after 4 days of inducer treatment, more than 90% of the cells produced hemoglobin as measured by benzidine staining. The kinetics of complement activation by the induced FL cells strongly resembled those observed for sensitivity to lectin agglutinability of the cells. Both changes appeared early

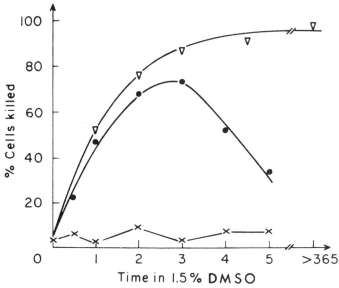

Figure 1
Sensitivity of FL cells (clone F4N) to human complement during erythroid differentiation. (▽——▽) clone F4N+2; (●——●) clone F4N; (×——×) clone F4N treated with heat-inactivated (52°C for 30 minutes) serum. Time is expressed in days.

during the induction and were reversed at the end. The mechanisms involved in these changes are completely unknown. However, both these changes occurred during the period of active virus production by the induced cells. Furthermore, it has been demonstrated that RNA tumor viruses are capable of activating the alternate pathway of complement (Welsh et al. 1975). It has also been shown that protein p15 of Rous sarcoma virus either is itself a protease or is a cofactor for a host protease (von der Helm 1977). It is tempting to attribute both the increased agglutinability of the cells and their ability to activate complement to a membrane-bound virus-associated protease. Although several nonvirus-producing subclones of F4N still demonstrate the two phenomena, it has not been shown that these cells lack the p15 component of the virus.

Spectrin

Spectrin is the major protein of the erythrocyte membrane. This peripheral protein, when isolated from mature erythrocytes and reticulocytes, contains two subunits of molecular weight 250,000 daltons (band 1) and 220,000 daltons (band 2) which are related antigenically (Bjerrum et al. 1974). In the mature erythrocyte, spectrin has been attributed a role in the organization of the cell membrane and in the regulation of cell shape (Elgsaeter and Branton 1974; Sheetz and Singer 1977). Synthesis of spectrin in vivo has been shown to precede that of globin and is complete by the reticulocyte stage of development. Thus, we felt it of interest to examine the metabolism of spectrin in cultured FL cells and to compare it with that of globin. We examined spectrin metabolism by three methods employing antibodies against purified mouse spectrin. These comprise indirect immunofluorescent examination, radioimmune assay, and pulse labeling of cells followed by extraction, immune precipitation of the spectrin, and polyacrylamide gel electrophoresis (Eisen et al. 1977a). Using these methods, we found that (1) spectrin composed of bands 1 and 2 was present at low levels in all untreated FL cells (0.01–0.03% of the cell protein) and that band 2 existed at least partially phosphorylated as it does in erythrocytes; (2) addition of inducing agents resulted in a rapid increase in the rate of synthesis of spectrin and in its intracellular accumulation, which reached a maximum after 48 to 60 hours of treatment (0.2–0.3% of the cell protein), at which time the synthesis of spectrin was arrested and the intracellular level fell to 0.1 to 0.15% of the cell protein. One rather striking feature of spectrin metabolism in the FL cells was the appearance of a new antigen band (spectrin band 3) with a molecular weight of 235,000 daltons in induced, but not in uninduced, FL cells. Fingerprint analysis showed that the new band was structurally related to band 2 of spectrin. Spectrin band 3 was detectable within 6 hours of inducer treatment, and its concentration appeared to be inversely correlated with that of band 2. Although spectrin band 3 appeared very early during the induction, its synthesis was arrested after 24 to 36 hours, and the majority of the label entering spectrin during a pulse of 1 to 2 hours at that time was found in bands 1 and 2. We did not find the third spectrin band in extracts of mouse bone marrow or of fetal liver cells. This suggests that the presence of this band in induced FL cells represented a difference between erythroid differentiation in normal tissue and that in the leukemic cells. However, the possibility that the FL cells arose from a minor population of erythroid precursors has not been ruled out. Although pulse-chase experiments with radioactive amino acids did not reveal any precursor-product relationships between the three spectrin bands, fingerprint analysis suggests that bands 2 and 3 were produced by cleavage of some as yet unidentified common precursor. It is possible that spectrin band 3 was formed by a cleavage which competed with that which formed band 2. It is again tempting to ascribe this cleavage to some virus-associated protease, although there is as yet little evidence to support this hypothesis.

IP25

The fact that nuclear activity was arrested both during normal erythroid differentiation and that of the FL cells suggests that changes occurred within the chromatin of the differentiating cells. In fact, changes in the binding of intercalating dyes to the chromatin of induced FL cells have been reported by Darzynkiewicz et al. (1977). It has also been reported that treatment of cells with butyrate results in a rapid and striking increase in the degree of acetylation of the histones (Riggs et al. 1977). We examined the protein composition of chromatin extracted from nontreated and induced FL cells (Keppel et al. 1977). Surprisingly, we observed only one major change. Within 24 hours of inducer addition, a basic protein of apparent molecular weight of 25,000 daltons appeared in the chromatin of the cells. This protein, IP25, could be eluted from the chromatin by dilute acid treatment or by salt concentration greater than 0.45 M. In this respect, IP25 resembles histone

H1. However, IP25 differs from the known histones in its amino-acid composition. IP25 becomes a major constituent of the chromatin proteins in the induced FL cells (about one molecule per 200 DNA base pairs). Although its appearance coincided with the changes in dye binding to the chromatin of the induced cells, we do not know whether IP25 was responsible for these changes.

The synthesis of IP25 preceded that of globin in the induced FL cells as did the synthesis of spectrin and the heme biosynthetic enzymes (see below). As for spectrin and two of the heme biosynthetic enzymes, IP25 synthesis was arrested after 2 to 3 days of inducer treatment. It thus appears that IP25 synthesis, like that of spectrin and the heme enzymes, is programmed differently from that of globin.

The role of IP25 in the induced FL cells is not known. Its appearance coincided with a general decrease in RNA and protein synthesis in the induced cells and a decrease in their volume. Furthermore, cells containing IP25 had a longer doubling time (16 hours) than did the uninduced FL cells (12 hours). It is possible that IP25 played a role in provoking the decrease in globin RNA and protein synthesis by stabilizing the attachment of histone complexes to the "nonerythroid" segments of the DNA.

Heme Synthesis

Several enzymes in the heme biosynthetic pathway have been examined in uninduced and induced FL cells (Ebert and Ikawa 1974; Sassa 1975). The regulation of the synthesis of these enzymes appears to be complex and somewhat noncoordinate. ALAS, ALAD, and UROS were detectable at low levels in the uninduced cells. Addition of DMSO resulted in the sequential induction of the three enzymes. ALAS appeared first and its activity per cell rose linearly until the end of the induction. ALAD appeared shortly after ALAS and its activity reached a maximum after 48 hours of DMSO treatment, after which the activity fell slightly and then remained constant. UROS appeared about 12 hours after ALAD and, like ALAD, its activity per cell reached a maximum at 48 hours, after which it fell slightly and then remained constant. On the other hand, the cellular activity of ferrochelatase, the final enzyme in the heme pathway, did not increase until the fourth day of the inducer treatment, when the synthesis both of ALAD and UROS had ceased. In this respect, ALAD and UROS behaved like spectrin.

The Role of Heme

Although heme has long been recognized to be important for globin mRNA translation in reticulocytes, only recently has it been implicated in the differentiation of the FL cells. Dabney and Beaudet (1977) have reported that 5×10^{-5} M hemin, in the presence of suboptimal doses of DMSO, produced a near-optimal induction both of globin mRNA and globin. Furthermore, they showed that addition of 3-amino-1,2,4-triazole (AT), an inhibitor of ALAD, to DMSO-treated FL cells blocked the production both of globin mRNA and globin and that this inhibition was relieved by addition of hemin. J. Granick and S. Sassa (in prep.) have demonstrated that hemin addition in the absence of DMSO resulted in increase of two to four times in ALAD and UROS activity. We examined the effects of hemin addition on the induction of spectrin and of IP25. In the case of spectrin, hemin alone at 10^{-4} M provoked no increase in cellular content nor did it induce the synthesis of spectrin band 3. Hemin was also without effect on IP25 induction. Addition of hemin in the presence of optimal doses of DMSO did not alter the amount of spectrin or of IP25. On the other hand, addition of hemin in the presence of DMSO resulted in a more pronounced appearance of erythrocyte surface antigens on the cells after 3 to 4 days of treatment. Addition of 15 mM AT in the presence of DMSO had no effect on the induction of spectrin and IP25 but did block the appearance of significant numbers of cells staining positive for benzidine. Thus, hemin appeared to have a strong effect on those events that occurred late during the induction of the FL cells although it did not alter early events. The time of expression of the various characters studied is shown schematically in Figure 2.

Variant Clones

We attempted to analyze the differentiation of the FL cells by examining variant clones that were unable to complete the differentiation. All of the clones were selected for their ability to grow indefinitely both in DMSO and hexamethylene bisacetamide (HMBA), and all were selected from clone F4N.

Physiological Variants

When cells of clone F4N were treated with optimal doses of inducing agents such as DMSO, HMBA, butyric acid, or hypoxanthene, not all the cells in the population completed the differentiation; resistant cells always grew up. Although it is impossible to estimate the frequency with which these resistant cells arise in mass culture, their frequency can be estimated when the induction is performed under cloning conditions in a viscous or semisolid medium. In this case, we estimate that the resistant clones arose with a fre-

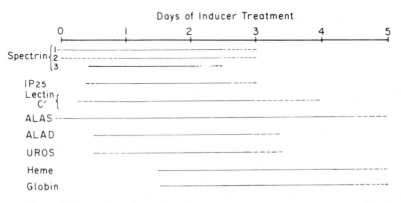

Time of Expression of Various Characteristics during Induction of FL Cells

Figure 2
Time of expression of characteristics during induced differentiation of FL cells. (———) expression at maximal levels; (---) low levels of expression.

quency of about 1-2.5% per generation. The cloning efficiency of the cells in inducer (large clones) was 5-10% of that found without inducer, and the input cells underwent an average of four divisions when cloned in inducer. This high frequency of appearance of resistant clones indicates a physiological rather than a genetic origin of the resistance. This view is supported by the following properties of the clones that were isolated in medium containing optimal concentrations of inducers. (1) The resistance appeared to depend on the addition of inducer, since the spontaneous fraction of resistant clones isolated from an untreated population of cells was < 1/5,000. (2) The frequency of reversion of the resistant clones, either in the presence or absence of inducer, was extremely high (up to 3% per generation), and the revertants appeared to be identical to the original wild-type cells of clone F4N. The properties of one resistant clone, F4N + 2, selected in 1.5% DMSO are presented in Table 1. The cells grew with a doubling time of 12 hours in inducer-free medium and 16.2 hours in medium containing 1.5% DMSO. In the presence of inducer, the cell volume was 0.5 to 0.25 that found during growth in the absence of inducer. Treatment of the resistant cell with inducing agents resulted in the normal expression of those characteristics found early in the induction of the normal wild-type F4N cells. The cells became sensitive to agglutination by lectins and to lysis by complement in the absence of added antibody (see Fig. 1). Spectrin synthesis and the appearance of the new band, spectrin band 3, were induced normally, as was the synthesis of IP25. ALAS, ALAD, and UROS were induced normally. Continued growth in the presence of inducer resulted in the continued expression of all these characteristics. Thus, the turnoff of the expression seen on induction of the wild-type cells did not occur in the resistant clones. Removal of inducer from the resistant cells resulted in their reversion over a period of 3 to 5 days to the uninduced phenotype, demonstrating the requirement for the continued presence of inducer for the expression of these early events. It is clear from these results that there was no commitment involved with the expression of these early events.

Cells of clone F4N+2 did not accumulate appreciable amounts of globin mRNA or of globin in the presence of inducer. The population of cells grown for long periods in the presence of inducer (1.5% DMSO or 2.5 mM HMBA) contained 10% of the levels both of globin mRNA and globin found in maximally induced cells of clone F4N. Furthermore, the population contained a constant proportion of cells that were benzidine positive. This suggests that the population was in a steady state with respect to production and death of hemoglobin-containing cells. Although we cannot rule out the possibility that the constant fraction of benzidine-positive cells in the population was due to an equilibrium between cells that did not produce hemoglobin and those that did, this explanation appears to be unlikely since analysis of colonies containing only benzidine-positive cells in methylcellulose was consistent with the hypothesis that cells which become benzidine positive also become committed to terminal differentiation (Gusella et al. 1976).

It appears, therefore, that the physiological variants of the FL cells represented by clone F4N+2, although expressing early functions normally, were blocked at some step just prior to their commitment to the terminal

Table 1
Properties of Clone F4N and Its DMSO-Resistant Variants

| Clone | Time in DMSO (days) | Concentration (%, v/v) | Sensitivity | | Spectrin[a] | | IP25 | ALAS[a] | ALAD[a] | UROS[a] | Heme[a] | Hemoglobin | |
			cells killed by complement	cells aggregated by lectins	band 3	accumulation						mRNA (ppm)	globin (% total protein)
F4N	0	1.25–1.5	—	—	—	—	—	±	±	±	—	10	0.5–1
	2	"	++	++	++	++	++	++	±	++	++	—	250	"
	5	"	±	±	±	+	++	++	++	++	+++	++	1250	25–30
F4N+2	0	1.5	—	—	—	—	—	±	±	+	—	7	1
	5	"	++	++	++	++	++	++	±	+++	+++	±	125	"
	>365	"	++	++	++	++	++	+						3
F4+	0	2	++	++	+	—	—	—	+	+	++	—	1	<1
	2	"	++	++	+	—	—	—	+	+	++	—	3	3
	5	"	++	++	+	—	—	—	+	+	++	—	2	"
	>365	"	++	++	+	—	—	—						"
F4D–5	0	2	—	—	—	—	—	±	—	±	±	—		<1
	5	"	—	—	±	—	+±	—	—	±	±	—		"
	>365	"	—	—	—	—	—	—						"
F4–61	0	2	—	—	—	—	—	—	—	—	—	—		<1
	2	"	—	—	±	—	—	—						"
	2	"	—	—	—	—	—	—						"
	>365	"	—	—	—	—	—	—						"

Note: Clone F4N is the wild type. Levels of various products expressed by these cells are considered as 100% and denoted by ++. Levels between 25 and 50% are denoted by +. Levels below 25% are expressed as ±. Levels higher than those expressed by cells of clone F4N are denoted by +++. Nondetectable levels are denoted by —.
[a] Data taken from H. Eisen, S. Sassa, J. Granick, I. Pragnell, and W. Ostertag, mss. in prep.

events of the differentiation. This probably was a late step in the synthesis of heme, inasmuch as the cells were normally inducible for ALAS, ALAD, and UROS but did not accumulate appreciable quantities of heme. To test this, hemin (10^{-4} M) was added to cells of clone F4N+2 in the presence and absence of DMSO or HMBA. Although the addition of hemin had little or no effect in the absence of inducer, its addition produced striking effects if inducer was present. After three to four divisions, cell proliferation ceased and synthesis of spectrin and IP25 was arrested as was that of globin DNA, RNA, and protein. During that time the percentage of benzidine-staining cells rose from 8.5 to 43%, and 80 to 90% of the cells became brightly staining with antibody directed to erythrocyte surface antigens. Often what appeared to be the beginning of nuclear extrusion was seen (see Fig. 3).

Other Variants

The fact that physiological variants such as F4N+2 are killed in the presence of inducer and of hemin provides a way of selecting other variants blocked at different stages of development. To test this, we plated cells of clone F4N in methylcellulose-containing medium to which was added nothing, 1.5% DMSO, or 1.5% DMSO + 10^{-4} M hemin, and after 10 days scored the number of large colonies. In the presence of DMSO alone, 6.3% of the input cells gave rise to resistant colonies, but only 2.7×10^{-5} of the cells plated both in DMSO and hemin formed large colonies. Similar low survival was seen when the cells were in medium containing DMSO at

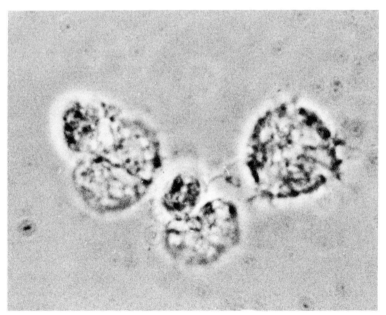

Figure 3
Cells of clone F4N+2 after 48 hours of hemin treatment (10^{-4} M) in the presence of 1.5% DMSO. The cells were cultivated for 10 days in 1.5% DMSO prior to heme addition.

higher concentration (2.5–3%) and no hemin. Mutagen treatment of the cells increased the proportion of resistant colonies by up to 40 times when selection was done in the presence of 1.5% DMSO + 10^{-4} M hemin or of 2.5% DMSO alone, but no effect of mutagen was seen when resistant colonies were selected in 1.5% DMSO alone. Therefore, it is likely that variants selected in high concentrations of DMSO or DMSO + hemin represented mutant clones rather than physiological variants like F4N+2.

The properties of three independent variants selected from clone F4N in 3% DMSO or in 1.5% DMSO + 10^{-4} M hemin are presented in Table 1. The three clones all behaved differently with respect to their erythroid characteristics.

Clone F4+

Clone F4+ was selected in 3% DMSO and was resistant to all inducers tested. The cells were tetraploid (70–85 chromosomes) and larger than the parent cells. Cells of clone F4+ were constitutive for the expression of many, but not of all, of the early characteristics seen on induction of clone F4N. They were very agglutinable by lectins and sensitive to complement, whether inducer was present or not. They also constitutively synthesized spectrin band 3; however, they seemed to make only uninduced levels of spectrin, and addition of inducer had no effect. IP25 was not present in the cells and was not inducible. Of the heme enzymes examined, UROS was made constitutively at a high level, whereas ALAS and ALAD were found at two to three times the uninduced levels and were not inducible. It is not clear whether this represented constitutive synthesis of these enzymes or an increase in the basal levels due to the polyploidy of the cells. Nevertheless, low levels of heme were detectable in F4+ cells, although intracellular heme content was, if anything, diminished in the presence of DMSO. Although globin mRNA was present in cells of clone F4+, there was less than that found in uninduced wild-type cells, and the level was not increased by inducers. Cells of clone F4+ did not appear to contain erythrocyte band 3 and the expression of this antigen was not altered by inducer treatment. Finally, cells of clone F4+ were insensitive to exogenous hemin.

The phenotype of clone F4+ cells is obviously complex. Whereas some elements of the early program of induction were expressed constitutively, other elements were not expressed or were expressed at basel level and were not inducible. Whether this indicates that the various elements were regulated independently or that F4+ contained multiple alterations or mutations is not clear.

Clone F4D-5

Cells of clone F4D-5 (isolated after selection in 3% DMSO) were completely refractible to inducer treatment. Although the untreated cells exhibited basal levels of spectrin, ALAD, and UROS, they did not induce for any of these proteins when inducer was added. However, they did synthesize low levels of IP25.

Clone F4-61

Clone F4-61 was isolated by selection in 1.5% DMSO + 10^{-4} M hemin. Populations of this clone contained 1–2% benzidine-positive cells when grown in inducer. When the cells were cultivated in the continued presence of inducer they presented no other detectable induced characteristics. When inducers were added after long periods of growth in inducer-free medium, a slight induction of spectrin band 3 was seen at 12 to 24 hours, after which this band disappeared. This could have represented a small number of revertant cells accumulating in the population. In uninduced cells of clone F4-61, the ratio of bands 1 and 2 of spectrin was different from that found in their wild-type parent. Whereas in wild-type cells the ratio of band 1 to band 2 was 0.6, in F4-61 this ratio was close to 2. This might indicate that these cells are defective in spectrin processing, although whether this defect was responsible for their failure to differentiate is not known.

DISCUSSION

The erythroid differentiation exhibited in vitro by the FL cells is clearly an extremely complex phenomenon involving multiple regulatory elements. The approach which we and others have taken of analyzing this differentiation in normal and variant clones of FL cells should ultimately be useful in unraveling the mechanisms involved.

The events that occurred during the induced differentiation of the FL cells can be grossly divided into two groups, or programs. The first program, which includes the induction of membrane changes such as sensitivity to lectins and complement, the induction of spectrin and the appearance of the new component spectrin band 3, the appearance of the chromatin protein IP25, and the induction of at least those enzymes necessary for the synthesis of uroporphyrinogen I, is expressed early during the induction of the cells, prior to their commitment to the terminal events of the differentiation. It has also been demonstrated that ALAS appears before hemoglobin in developing fetal mouse liver (Freshney and Paul 1971) and chick blastoderm (Irving et al. 1976). The regulation of this program is seen best in the physiologically variant clones which continue to grow in the presence of inducers. These clones, which are almost certainly not mutants, are fully inducible for all the events of the early program and fail to produce substantial amounts of heme and to commit to the terminal differentiation. The expression of the early program is arrested in these cells after removal of the inducer. The reversion cannot be due to the presence of a subpopulation of uninduced cells which grow faster than the induced cells, because such cells would be expected to overgrow the population even in the presence of inducer. Furthermore, repeated cloning of the cells in the presence of inducer does not affect the shutoff of the early program after removal of inducer. It seems clear, therefore, that expression of the early program is reversible and requires the continued presence of inducer. In this sense, the early program is similar to the adaptive regulatory circuits found in microorganisms.

The nature of the regulation of those characters expressed in the early program is unknown. The fact that clone F4+ is constitutive for some characters and uninducible for others indicates that many of the characters may be regulated independently. However, inasmuch as the cells are polyploid and may contain many chromosomal rearrangements, it is difficult to draw conclusions. The sequential appearance of the heme biosynthetic enzymes also indicates that each may be regulated differently. Heme is not necessary for the normal expression of the early program although exogenous hemin does induce ALAD. It is clear that many more variants must be analyzed before any conclusions can be drawn. It is also clear that variant selections other than ability to proliferate in the presence of inducer must be devised.

Another interesting aspect of the early program expressed by the FL cells is that the expression of many of the components is shut off before the end of the differentiation. It is likely that this shutoff occurs at or shortly after the time at which commitment to the final stages takes place. The physiological variants which fail to commit continue to express all of the characters of the early program. Furthermore, addition of hemin to these cells in the presence of inducers appears to result in terminal differentiation and in shutoff of the early program.

Heme itself appears later during the induction of the FL cells than do the enzymes necessary for the initial steps of its synthesis. Normally, the rate-limiting step in heme biosynthesis is the formation of ALAS (Granick and Sassa 1971); however, in the FL cells, the rate of heme formation seems to be limited by a block at a later step. It is possible that this step is the insertion of iron, inasmuch as ferrochelatase activity in the induced cells does not increase until after synthesis of ALAD and UROS has ceased. It is interesting that all of the variant clones of FL cells, although selected for their ability to grow in the presence of inducer, failed to accumulate heme. This suggests that heme plays an important role in the commitment to the terminal phase of the differentiation. This idea is also supported by the evidence that heme is necessary for the accumulation of globin mRNA in the cytoplasm of the induced FL cells (Dabney and Beaudet 1977). Furthermore, cells of the F4N+2 type which are blocked just before commitment to the terminal phase of the differentiation can be induced to enter this phase by addition of exogenous hemin. Harrison (1977) has reported similar findings with a mutant (FWT) derived from clone 745A. Although heme appears to be necessary for commitment to the terminal phase of differentiation, it alone is not sufficient, for addition to exogenous hemin in the absence of inducer has little effect on the appearance of hemoglobin-containing cells. Some induction by hemin alone has been reported (Ross and Sautner 1975; J. Granick and S. Sassa, in prep.), but this induction is slight when compared with that provoked by DMSO. It is possible that the exogenous hemin provokes the terminal phase of differentiation in those cells which become spontaneously induced for the early program.

Expression of the early program of erythroid differentiation and of a suitable level of intracellular heme result in the initiation of the terminal phase of erythroid differentiation by the FL cells. This terminal phase is characterized by a limitation of the proliferative capacity of the cells to four to five generations, the accumulation of globin mRNA and globin in the cytoplasm, the appearance of erythrocyte band 3 and acetylcholinesterase on the cell surface, and, finally, the shutoff of nuclear activity. It is not clear whether heme plays a direct role in the limitation of proliferative capacity and nuclear activity or whether some other element induced in the terminal phase of the differentiation is responsible for these effects. It is not yet possible to distinguish these possibilities, as no variants exist which uncouple globin synthesis from the arrest of cell division and nuclear activity.

On the basis of our results and those of others, we propose the following scheme for the induced erythroid differentiation of FL cells (see Fig. 4). Addition of inducers results in the expression of what we shall call the early erythroid program (EP^e). Those characters included in this program are spectrin accumulation; the appearance of spectrin band 3, IP25, ALAS, ALAD, and UROS; membrane alterations; and virus induction. Either as part of this program or as a result of its expression, heme is accumulated. Because we cannot distinguish between these possibilities, we shall call the final steps of heme biosynthesis (notably ferrochelatase) the intermediate erythroid program (EP^i). Expression of EP^e is reversible and depends on the continued presence of inducer. Although its expression results in a diminished growth rate of the cells, it does not prevent or limit cell proliferation.

Expression of EP^e and heme production would then trigger the irreversible commitment to the terminal phase of the differentiation (EP^l), resulting in the limitation of proliferative capacity, accumulation of globin mRNA and globin, the shutoff of EP^e, and, finally, the arrest of nuclear activity.

This model suggests that the induced differentiation of the FL cells involves a temporal and causal sequence of events rather than the parallel expression of the various characters. This view is supported by the fact

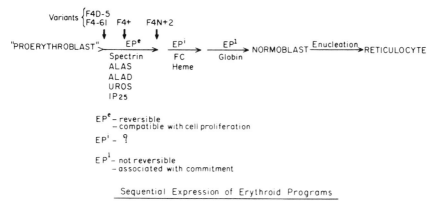

Figure 4
Scheme for final steps of erythroid differentiation.

that physiological variants such as clone F4N+2 blocked in the production of heme are unable to express the terminal erythroid program in the absence of exogenous hemin. The molecular basis for the block in heme production in these variants is unknown. The block is at a late stage of heme biosynthesis, possibly at the ferrochelatase step. As it is thought that ferrochelatase is a mitochondrial enzyme (Granick and Sassa 1971), the physiological flip-flop exhibited by these variants may be the result of some regulatory events occurring in the mitochondria. We are presently investigating this possibility.

The model which we have proposed, if applicable to normal in vivo erythropoiesis, would permit flexibility in the final amplification of erythroid precursors (proerythroblast to normoblast).

Acknowledgments

This work was supported in part by grants from the Swiss National Foundation to H. Eisen and American Cancer Society BC-180A to S. Sassa.

REFERENCES

Arndt-Jovin, D. J., W. Ostertag, H. Eisen, F. Klimek, and T. M. Jovin. 1976. Studies of cellular differentiation by automated cell separation. Two model systems: Friend virus-transformed cells and *Hydra attenuata. J. Histochem. Cytochem.* **24**: 332.

Bjerrum, O. J., S. Bhakdi, H. Knufermann, and T. C. Bog-Hanse. 1974. Immunoelectrophoretic heterogeneity and cross-reactions of individual "spectrin" components isolated by preparative solium dodecylsufate-polyacrylamide-gel electrophoresis. *Biochem. Biophys. Acta* **373**: 44.

Cikes, M. 1970. Antigenic expression of a murine lymphoma during growth *in vitro. Nature* **225**: 645.

Clegg, J. B., M. A. Naughton, and P. J. Weatherall. 1965. An improved method for the characterization of human haemoglobin mutants: Identification of $\alpha_2\beta_2^{95glu}$ haemoglobin (Baltimore). *Nature* **207**: 945.

Conscience, J. F., R. A. Miller, J. Henry, and F. H. Ruddle. 1977. Acetylcholinesterase, carbonic anhydrase and catalase activity in Friend erythroleukemic cells, non-erythroid mouse cell lines and their somatic hybrids. *Exp. Cell Res.* **105**: 401.

Dabney, B. J., and A. L. Beaudet. 1977. Increase in globin chains and globin mRNA in erythroleukemia cells in response to hemin. *Arch. Biochem. Biophys.* **179**: 106.

Darzynkiewicz, Z., F. Tranganos, T. Sharpless, C. Friend, and M. R. Melamed. 1977. Nuclear chromatin changes during erythroid differentiation of Friend virus induced leukemic cells. *Exp. Cell Res.* **99**: 301.

Dube, S. K., G. Gaedicke, N. Kluge, B. J. Weimann, H. Melderis, G. Steinheider, T. Crozier, H. Beckmann, and W. Ostertag. 1974. Hemoglobin synthesizing mouse and human erythroleukemic cell lines as model systems for the study of differentiation and control of gene expression. Proc. Int. Symp. Princess Takamatsu Cancer Res. Fund, p. 99.

Ebert, P. S., and Y. Ikawa. 1974. Induction of δ-aminolevulinic acid synthetase during erythroid differentiation of cultured leukemia cells. *Proc. Soc. Exp. Biol. Med.* **146**: 601.

Eisen, H., R. Bach, and R. Emery. 1977a. Induction of spectrin in Friend virus-transformed erythroleukemic cells. *Proc. Natl. Acad. Sci. U.S.A.* **74**: 3898.

Eisen, H., S. Nasi, C. P. Georgopoulos, D. Arndt-Jovin, and W. Ostertag. 1977b. Surface changes in differentiating Friend erythroleukemic cells in culture. *Cell* **10**: 689.

Elgsaeter, A., and D. Branton. 1974. Intramembrane particle aggregation in erythrocyte ghosts. I. The effects of protein removal. *J. Cell Biol.* **63**: 1018.

Freshney, R. I., and J. Paul. 1971. The activities of three enzymes of haem synthesis during hepatic erythropoiesis in the mouse embryo. *J. Embryol. Exp. Morphol.* **26**: 313.

Friend, C., W. Scher, J. G. Holland, and T. Sato. 1971. Hemoglobin synthesis in murine virus-induced leukemic cells *in vitro*: Stimulation of erythroid differentiation by dimethyl sulfoxide. *Proc. Natl. Acad. Sci. U.S.A.* **68**: 378.

Furusawa, M., Y. Ikawa, and H. Sugano. 1971. Development of erythrocyte membrane-specific antigen(s) in clonal cultured cells of a Friend virus-induced tumor. *Proc. Jpn. Acad.* **47**: 220.

Granick, S., and S. Sassa. 1971. δ-aminolevulinic acid synthetase and the control of heme and chlorophyll synthesis. In *Metabolic regulation-metabolic pathways* (ed. Vogel), vol. 5, p. 79. Academic Press, New York.

Gusella, J., R. Geller, B. Clarke, V. Weeks, and D. Housman. 1976. Commitment to erythroid differentiation by Friend erythroleukemia cells: A stochastic analysis. *Cell* **9**: 221.

Harrison, P. R. 1977. The biology of the Friend cell. In *International review of biochemistry* (ed. J. Paul), vol. 15, p. 227. University Park Press, Baltimore.

Irving, R. A., W. I. P. Mainwaring, and P. M. Spooner. 1976. The regulation of haemoglobin synthesis in cultured chick blastoderms by steroids related to 5β-androstane. *Biochem. J.* **154**: 81.

Kabat, D., C. Sherton, L. Evans, R. Bigley, and R. D. Koler. 1975. Synthesis of erythrocyte-specific proteins in Friend leukemia cells. *Cell* **5**: 331.

Keppel, F., B. Allet, and H. Eisen. 1977. Appearance of a chromatin protein during the erythroid differentiation of Friend virus-transformed cells. *Proc. Natl. Acad. Sci. U.S.A.* **74**: 653.

Laemmli, U. K. 1970. Cleavage of structural proteins during the assembly of the head of bacteriophage T4. *Nature* **227**: 680.

Orkin, S. H., F. I. Harosi, and P. Leder. 1975. Differentiation in erythroleukemic cells and their somatic hybrids. *Proc. Natl. Acad. Sci. U.S.A.* **72**: 98.

Riggs, M. G., R. G. Whittaker, J. R. Neumann, and V. M. Ingram. 1977. *n*-butyrate causes histone modification in HeLa and Friend erythroleukemia cells. *Nature* **268**: 462.

Ross, J., and D. Sautner. 1975. Induction of globin mRNA accumulation by hemin in cultured erythroleukemic cells. *Cell* **8**: 513.

Sassa, S. 1975. Sequential induction of heme pathway enzymes during erythroid differentiation of mouse Friend leukemia virus-infected cells. *J. Exp. Med.* **143**: 305.

Sassa, S., and A. Kappas. 1977. Induction of δ-aminolevulinate synthase and porphyrins in cultured liver cells maintained in chemically defined medium: Permissive effects of hormones on induction process. *J. Biol. Chem.* **252**: 2428.

Sheetz, M., and S. J. Singer. 1977. On the mechanism of ATP-induced shape changes in human erythrocyte membranes. I. The role of the apectrin complex. *J. Cell. Biol.* **73**: 638.

Stephenson, J. R., A. A. Axelrad, D. L. McLeod, and M. M. Shreeve. 1971. Induction of colonies of hemoglobin-synthesizing cells by erythropoietin *in vitro*. *Proc. Natl. Acad. Sci. U.S.A.* **68**: 1542.

von der Helm, K. 1977. Cleavage of Rous sarcoma viral polypeptide precursor into internal structural proteins *in vitro* involves viral protein p15. *Proc. Natl. Acad. Sci. U.S.A.* **74**: 911.

Welsh, R. M., Jr., N. R. Cooper, F. C. Jensen, and M. B. A. Oldstone. 1975. Human serum lyses RNA tumor viruses. *Nature* **257:** 612.

Young, B. D., P. R. Harrison, R. S. Gilmour, G. D. Birnie, A. Hell, S. Humphries, and J. Paul. 1974. Kinetic studies of gene frequency. II. Complexity of globin complementary DNA and its hybridization properties. *J. Mol. Biol.* **84:** 565.

Regulation of the Individual Globin Genes

A. W. Nienhuis, D. Axelrod, J. E. Barker, E. J. Benz, Jr., R. Croissant, D. Miller, and N. Young

Clinical Hematology Branch, National Heart, Lung, and Blood Institute
National Institutes of Health, Bethesda, Maryland 20014

The genomes of human and other species contain a complex of closely linked globin genes of similar sequence that are selectively expressed during the development of the organism (Nienhuis and Benz 1977). For example, in humans, this complex contains two (or more) genes for γ-globin, one for δ-globin, and one for β-globin. Expression of the γ genes in fetal life results in production of Hb F ($\alpha_2\gamma_2$), whereas expression of the β gene beginning in neonatal life results in production of Hb A ($\alpha_2\beta_2$). The δ gene is also expressed in adults but Hb A_2 ($\alpha_2\delta_2$) amounts to only 1–3% of the total hemoglobin in adult red-blood cells. The α globin genes are on a separate chromosome and are continuously expressed beginning early in gestation. γ and β genes are present in sheep and, again, differential expression of these genes results in production of Hb F or Hb A (or Hb B) in fetal or adult sheep, respectively. In addition, the genome of certain sheep (those having a β^A gene) contains an additional β-globin gene, β^C, which is expressed during states of erythropoietic stress such as anemia and hypoxia and also during the neonatal period. These animals exhibit a reversible Hb A ($\alpha_2\beta_2^A$) to Hb C ($\alpha_2\beta_2^C$) switch which appears to be mediated directly by an action of erythropoietin (epo).

Erythropoiesis is the continuous process of red-cell production whereby early erythroid progenitor cells with considerable proliferative capacity ultimately give rise to erythroblasts which express specific globin genes during terminal erythroid maturation. Early observations suggested that regulation of the β^C-globin gene in sheep occurred early during erythroid stem-cell differentiation (Gabuzda et al. 1968). Subsequent studies utilizing direct assays for erythroid stem cells (colony-forming units—erythroid [CFU-E] and burst-forming units—erythroid [BFU-E]) substantiated this impression with regard both to the Hb A to Hb C switch in sheep (Barker et al. 1975, 1976) and to the production of Hb F in cultures of adult human bone marrow (Papayannopoulou et al. 1977). Conversely, molecular analysis of hemoglobin switching has of necessity focused on study of hemoglobin synthesis and mRNA production in maturing erythroblasts. These studies have suggested that production of a specific hemoglobin (e.g., Hb C $-\alpha_2\beta_2^C$) may be related to the quantity of its specific mRNAs in a population of erythroid cells (Benz et al. 1977b,c). Thus, regulation of expression of the individual globin genes occurs by a mechanism that affects the quantity of the individual mRNA species.

Our goal has been to define the molecular mechanism of regulation of the globin genes as it occurs in erythroid stem cells. Recent studies have focused on the analysis of an Hb F ($\alpha_2\gamma_2$) to Hb C ($\alpha_2\beta_2^C$) switch in fetal sheep and on defining the temporal relationship of commitment of erythroid stem cells to expression of the β^C-globin gene and the initial accumulation of β^C-globin mRNA in maturing erythroblasts. The structure of globin genes in chromatin has been examined during erythroid maturation both in mouse erythroleukemia cells in culture and in normal erythroid cells.

EXPERIMENTAL PROCEDURES

Cell Culture

Suspensions of bone-marrow and liver erythroid cells were prepared from fetal or neonatal animals as previously described (Barker et al. 1976, 1977) and were cultured in the plasma-clot system originally described by Axelrad and coworkers (Stephenson et al. 1971). Erythropoietin (Step III, Connaught Medical Research Laboratories) was added in concentrations ranging from 0.01 to 5 U/ml and the cultures were incubated for 1 to 5 days at 37°C. Colonies were examined after staining with benzidine. Alternatively 10 μCi of [^{3}H]leucine (specific activity, 30,000–50,000 mCi/mmole) were added to each 0.1 ml clot and incubation was continued for 18 to 24 additional hours. The clots were exposed to trypsin, the cells lysed, and a globin extract was prepared for column chromatography. Mouse erythroleukemia cells (MELC) were maintained in suspension culture and induced to undergo erythroid maturation by addition of dimethylsulfoxide (DMSO) as described by Axelrod et al. (1978).

Analysis of Globin Synthesis

Labeled globin standards for carboxymethylcellulose (CMC) chromatography were prepared by incubation of reticulocytes of fetal and neonatal animals with [^{14}C]leucine (Barker et al. 1976). Appropriate amounts of hemolysates from these reticulocyte incubations were added to the extracts from plasma clot culture so that a [^{14}C]globin marker for the γ, β^A, β^B, and β^C chains would be present on the chromatogram. Heme was removed by acid-acetone extraction and the precipitated globin was dissolved in 8 M urea with 0.05 M phosphate buffer, pH 6.8, and applied to a CMC column. The several globins were resolved with a nonlinear gradient of sodium phosphate (pH 6.8) (Benz et al. 1977c).

Phlebotomy and Injection of epo in Fetal Sheep

Fetuses ranging in gestational age from 60 to 140 days (duration of gestation 135–145 days) received epo (Step III, Connaught Medical Research Laboratories) at a dose of 1 U/gm of fetal body weight on one or several occasions over 4 to 11 days (Barker et al. 1977). The in vivo biological activity of the epo preparations was verified by the posthypoxic mouse assay. Certain animals were alternatively bled (1–5 ml) on one or several occasions from 2 to 11 days. A few animals both received epo and were bled. Manipulations were performed via a plastic catheter placed in an umbilical or jugular vein. The surgical technique and method of selecting fetuses of known gestational age have been described in detail (Barker et al. 1977).

Bone-Marrow Aspirations

A 50-day-old neonatal animal was injected with epo (100 U/kg) after an initial bone-marrow sample had been obtained. Sampling of the bone marrow after induction of anesthesia was accomplished by surgical exposure of one of the humerii or femurs. Holes were drilled into the bone-marrow cavity and the cells aspirated via a soft polyethylene catheter into minimal essential Eagle's medium containing heparin (100 U/ml). The details of these techniques have been described by Barker et al. (1976). The animal was allowed to recover and maintained in a sling. Additional samples were obtained at 12, 24, 48, and 72 hours, at which point the animal was sacrificed.

Bone-Marrow Fractionation

"Early" or "late" fractions of erythroid cells, composed of pro- and basophilic erythroblasts or poly- and orthochromatophilic erythroblasts, respectively, were prepared. Five ml of bone-marrow cells, suspended at a concentration of 2×10^7/ml, were layered over 15 ml of Ficoll Paque (Pharmacia) and spun at 15°C in the HB-4 Sorvall rotor at 6000 rpm for 10 minutes. Cells remaining at the interface constituted the early fraction and included many myeloid cells in addition to early erythroblasts. Pelleted cells were employed as the "late" fraction. For certain experiments, a linear gradient was generated by mixing Ficoll Paque with Dulbecco's phosphate buffered saline (Miller et al. 1978). After centrifugation to equilibrium, the cells remaining in the gradient were utilized as the early fraction. The pelleted cells were further fractionated by 1-g sedimentation (Barker et al. 1976). The most slowly sedimenting nucleated cells were utilized as the late fraction.

Preparation of Nuclear and Cytoplasmic RNA

Nuclei were obtained from early and late fractions of cells prepared by the first procedure described above. After lysis in 1 mM mgCl$_2$ and addition of citric acid to 0.1 M, the initial supernatant obtained after pelleting of the nuclei was used for preparation of cytoplasmic RNA. The nuclei were pelleted two times through a 0.32-M sucrose cushion to reduce cytoplasmic contamination. Extraction and recovery of RNA from the nuclei and cytoplasmic fractions were performed as previously described (Nienhuis et al. 1977).

Preparation of Complementary DNA and mRNA × cDNA Hybridizations

Sheep reticulocyte mRNA was prepared and used as a substrate for RNA-directed DNA polymerase (reverse transcriptase) for synthesis of mixed ($\alpha + \beta$ or γ) complementary DNAs (cDNA) (Benz et al. 1977c). [^{3}H]dCTP (specific activity 17,000-25,000 mCi/mmole) was used to label cDNAs, giving a specific activity of 10,000-12,000 cpm/ng. The α component of each of the mixed probes was removed as follows. A heterologous mRNA (e.g., Hb B, $\alpha + \beta^B$) was annealed to Hb C ($\alpha + \beta^C$)-cDNA at 50°C for 2 hours. The reaction mixture was then incubated for 5 minutes at 68°C, above the melting temperature of the imperfect β^B-mRNA:β^C-cDNA duplexes but below the melting temperature of the more stable α:α duplexes. The single stranded β^C-cDNA was recovered by chromatography on hydroxylapatite (Benz et al. 1977c). Mouse and rabbit globin cDNAs were prepared and employed as the unresolved mixture of α and β components. For certain experiments, sheep Hb C mRNA was synthesized with [^{32}P]dCTP (sp act 250,000 mCi/mmole), giving a probe with specific activity of approximately 325,000 cpm/ng.

Hybridization reactions were performed in 50% formamide containing 0.5 M NaCl (Benz et al. 1977c). Incubation was for 60 hours at 67.7°C, a temperature only 2-4° below the melting temperature of homologous sheep β- or γ-mRNA:cDNA duplexes. This stringency was required to prevent cross-reaction of the specific probes with the closely related heterologous mRNA sequences (e.g., β^A-cDNA with β^C-mRNA). Analyses of mouse or rabbit RNA samples with the corresponding probes were made at 52°C. Duplex formation was detected by assay with S_1 nuclease (Benz et al. 1977c).

Digestion of Nuclei with Pancreatic DNAse I

These experiments were performed using the techniques described by Weintraub and Groudine (1976) with a number of modifications (Miller et al. 1978) as indicated below. Nuclei were prepared from adult and fetal mouse liver, MELC in the uninduced state and after growth in 2% DMSO, and from bone-marrow and spleen from rabbits and sheep. These animals were injected with phenylhydrazine to induce hemolytic anemia and thereby to ensure erythroid hyperplasia in the bone marrow (Benz et al. 1977c). Nuclei were recovered essentially as described by Marzluff et al. (1973). The nuclei were washed once in 10 mM NaCl, 10 mM Tris-HCl, pH 7.5, and 3 mM MgCl$_2$ before being resuspended in this buffer at a final DNA concentration of 0.5 to 1.0 mg/ml. Pancreatic DNAse I (Sigma Chemical Co.) was added to a concentration of 10 µg/ml after the nuclear suspension had been prewarmed to 37°C. Incubation periods ranging from 15 seconds to 2 minutes rendered 5–10% of the total DNA soluble in 7% perchloric acid at 4°C. EDTA was added to a concentration of 6 mM. Deproteinization was accomplished by addition of proteinase K (Beckman Instruments) (100 μg/ml). After a 2-hour digestion, NaCl was added to 0.5 M and incubation was continued for an additional 6 to 14 hours. Addition of sodium dodecylsulfate to 0.5% was followed by phenol:chloroform extraction. Subsequent steps included precipitation with cold ethanol, digestion with ribonuclease A, another phenol:chloroform extraction, and reconcentration of the DNA with ethanol essentially as described by Weintraub and Groudine (1976). Nuclei incubated without pancreatic DNAse I served as a source of control DNA. Prior to the final ethanol precipitation, this control DNA was sonicated to give a fragment size of 200 to 400 base pairs (Deisseroth et al. 1977), approximately equivalent to the average size of the DNA fragments from nuclei exposed to pancreatic DNAse I.

Certain DNA-annealing reactions were performed with [^{3}H]cDNA in approximately a fivefold excess over the globin DNA sequences in the control DNA as determined by preliminary titrations (Miller et al. 1978). Other experiments utilized the highly radioactive [^{32}P]cDNA to insure an excess of globin DNA sequence over cDNA. Incubation was at 52° or 58°C (for analysis of individual sheep globin genes) in 50% formamide containing three times standard saline citrate (SSC) (SSC 0.15 M NaCl, 0.015 M sodium citrate, pH 7.0) for variable time periods. Duplex formation was detected by analysis with single-strand-specific S_1 nuclease (Benz et al. 1977c; Deisseroth et al. 1977).

RESULTS

A Switch from Hb F to Hb C Synthesis in Vitro but Not in Vivo

The potential for β^C-globin synthesis by sheep erythroid colonies derived from midgestation liver cells (85-day-old fetus) was established by the data shown in Figure 1. Only γ-globin was made in peripheral blood reticulocytes obtained at the time of sacrifice, but a substantial amount of β^C-globin was synthesized in vitro between 96 and 120 hours in colonies generated at an epo concentration of 5 U/ml. This fetus was known to be heterozygous at the adult β locus but neither β^A- nor β^B-globin synthesis was detected in culture (Barker et al. 1977). The time course and epo-concentration-dependence of Hb C synthesis in culture of fetal liver is shown in Table 1. Minimal β^C-globin synthesis occurred in colonies that had developed at an epo concentration of 0.05 U/ml but, by 72 to 96 hours at 5.0 U/ml, 22% of the non-α-globin synthesis was β^C. This increased to 64% during the 96- to 120-hour labeling period.

Several unsuccessful attempts were made to induce the synthesis of Hb C in midgestation sheep in utero by epo injection, bleeding, or by a combination of both (Barker et al. 1977). For example, a 90-day fetal lamb was injected with epo and sacrificed 5 days later. Peripheral blood obtained at the time of injection was incubated with [^{3}H]leucine and the labeled globins chromatographed (Fig. 2a). No synthesis of β^A- or β^C-globin was detected.

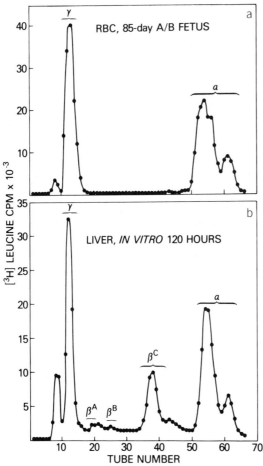

Figure 1
Globin synthesis in cells derived from an 85-day-old fetus that was heterozygous for the β^A- and β^B-globin genes. (a) Peripheral blood cells obtained at the time of sacrifice incubated with [^{3}H]leucine for 2 hours. (b) Fetal liver cells cultured in plasma clot for 120 hours. [^{3}H]leucine was added at 96 hours.

Table 1
Time Course of Induction of β^C-Globin Synthesis in Cultures of Liver from a 90-Day-Old Fetus

Hours in vitro	5 U epo/ml (% β^C)	0.05 U epo/ml (% β^C)
0– 24	0	0
24– 48	1.0	0
48– 72	6.9	0
72– 96	22.0	0
96–120	63.8	2.3

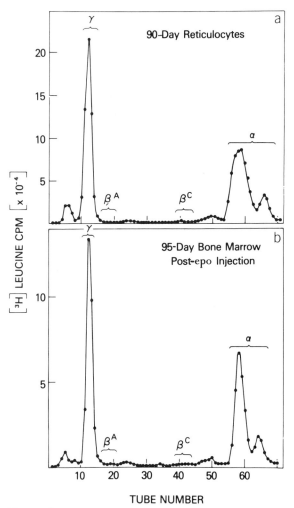

Figure 2
Globin synthesis in erythroid cells obtained from a 90-day-old fetus that had been injected with epo (Barker et al. 1977). (*a*) Peripheral blood reticulocytes incubated for 2 hours with [^{3}H]leucine. (*b*) Bone-marrow cells obtained at the time of sacrifice 5 days after the injection of epo. These cells were incubated for 12 hours with [^{3}H]leucine. The position of individual globin peaks was ascertained by including [^{14}C]leucine-labeled globin in each of the samples (Barker et al. 1976 and Methods).

Bone-marrow cells obtained at the time of sacrifice (95 days' gestation) were also incubated with [^{3}H]leucine and the globin products analyzed. Again, no β^A or β^C synthesis was detected. An equivalent dose of epo, when given to neonatal animals, regularly induces synthesis of β^C-globin (Barker et al. 1976,1977).

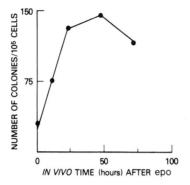

Figure 3
The concentration of CFU-E in bone marrow of a neonatal sheep following injection of epo (100 U/kg). A bone-marrow sample was obtained prior to injection and at 12, 24, 48, and 72 hours thereafter and was cultured in plasma clot at an epo concentration of 0.01 U/ml for 104 hours. Each benzidine-positive colony containing eight or more cells was counted.

Effect of *epo* Injection in a Neonatal Sheep

On CFU-E

A 50-day-old neonatal sheep was injected with epo (100 U/kg) and serial bone-marrow samples were obtained as detailed in the Methods. The number of erythroid colonies per 10^5 inoculated cells was determined after incubation for 104 hours in vitro. The concentration of CFU-E at 12 hours was 2.5 times higher than it was in the bone marrow obtained prior to injection. There was a further twofold increase by 24 hours (Fig. 3). These colonies developed at an epo concentration of 0.01 U/ml; this is sufficient to support colony growth but insufficient to induce synthesis of significant

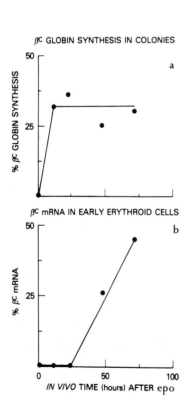

Figure 4
Comparison of the kinetics of commitment of CFU-E to generate colonies making β^C-globin and the kinetics of accumulation of β^C-globin mRNA in bone-marrow erythroblasts. (*a*) A neonatal sheep was injected with epo and serial bone-marrow samples were obtained as described in the legend to Figure 3. [^{3}H]leucine was added to plasma-clot cultures after 84 hours in vitro and incubation continued for an additional 20 hours. Analysis of globin synthesis is described in Methods. The percentage of β^C was calculated by dividing cpm in β^C-globin by cpm in non-α-globin ($\beta^A + \beta^C$). (*b*) Relative amounts of β^C-globin mRNA and β^A-globin mRNA in early erythroblasts following epo injection in a neonatal sheep. Serial bone-marrow samples were obtained as described in the legend to Figure 3. RNA was prepared and annealed to β^A- and β^C-globin mRNA (Figs. 5 and 6).

amounts of β^C-globin (Barker et al. 1976). Nonetheless, the colonies derived from the bone marrow 12 hours after the epo injection made 30% β^C-globin (of the total non-α) and a similar amount was made in colonies derived from bone marrow obtained at 24, 48, and 72 hours (Fig. 4a). The control colonies derived from the bone marrow prior to epo injection made only β^A-globin. Thus the epo injection had increased the number of CFU-E in the bone marrow and resulted in commitment of these cells to expression of the β^C-globin gene in vitro.

On β^C-Globin mRNA Accumulation

Nuclear and cytoplasmic RNA recovered from the bone marrow obtained prior to the epo injection was annealed to β^A- or β^C-cDNA (Fig. 5). Within the limits of the sensitivity of these measurements, β^C-globin mRNA was absent. Similar data were obtained for the 12- and 24-hour samples; β^C-globin mRNA was not detected until 48 hours after the epo injection and then was found in both the nuclear and cytoplasmic RNA fractions (Fig. 6). No difference in time of appearance of β^C-globin mRNA sequences was found in early or late cell fractions (data not shown).

The temporal relationship between commitment of CFU-E to generate colonies making Hb C and the induction of β^C-globin mRNA is summarized in Figure 4. Committed CFU-E were detected 12 hours after epo injection but β^C-globin mRNA appeared only at 48 hours.

Sensitivity of Globin Genes to Pancreatic DNAse I

In Mouse Liver

Nuclei were prepared from adult and 15-day fetal mouse liver. The fetal liver was composed primarily of late-stage erythroblasts. DNA recovered after exposure of these nuclei to DNAse I was annealed to mouse-globin cDNA. Solubilization of 10% of the total DNA in the fetal liver nuclei

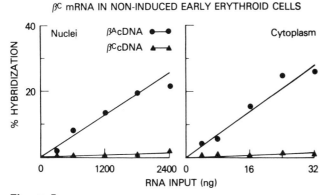

Figure 5
Analysis of nuclear and cytoplasmic RNA from bone-marrow cells of a neonatal sheep. Each 10-μl annealing reaction included 0.2 ng either of β^A- or β^C-[^{3}H]cDNA. Incubation was for 60 hours at 68°C (see Methods).

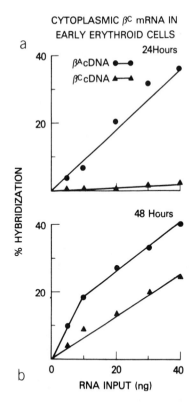

Figure 6
Analysis of cytoplasmic RNA from early erythroid cells (a) obtained 24 hours or (b) 48 hours following injection of epo into a neonatal sheep (see legend to Fig. 3). Individual 10-μl annealing reactions included 0.2 ng either of β^A- or β^C-[^{3}H]cDNA.

destroyed virtually all of the globin gene sequences whereas those genes in nuclei from adult liver were preserved (Fig. 7).

In MELC

DNA recovered from the nuclei of uninduced and induced MELC after exposure of these nuclei to DNAse I was virtually devoid of globin gene sequences (Fig. 8). Again only 8–12% of the total DNA was solubilized by the nuclease treatment. No significant difference was found between the sensitivity of globin sequences in induced and uninduced cells despite an increase of 50 times in mRNA content after exposure to DMSO (Fig. 8).

In Rabbit Erythroid Cells

Rabbits were injected with phenylhydrazine to induce bone-marrow erythroid hyperplasia. These cells were fractionated into early and late populations by sequential isopynic centrifugation and 1-g sedimentation. Nuclei were prepared from unfractionated bone marrow, early and late fractions of cells, and from spleen. These were exposed to DNAse I and the DNA was recovered. The globin gene content was reduced only 50% in the unfractionated marrow by DNAse I treatment. In the late cells, 80% of the gene sequences were lost. However, the globin genes in nuclei derived from the early erythroid cells were not DNAse-I sensitive in these experiments (Fig. 9). Nor were the globin genes in spleen nuclei sensitive to DNAse I (data not shown).

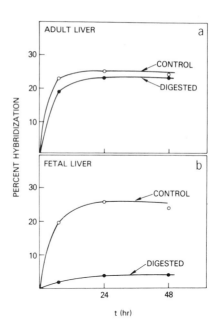

Figure 7
Sensitivity of the globin genes in nuclei (*a*) from adult liver or (*b*) from fetal liver to digestion by pancreatic DNAse. I. Incubation of each nuclear sample with DNAse I resulted in solubilization of 10 to 15% of the total DNA. The control DNA was prepared from nuclei that had not been exposed to DNAse I. Individual 20-μl annealing reactions included 120 μg of DNA and 0.5 ng of cDNA.

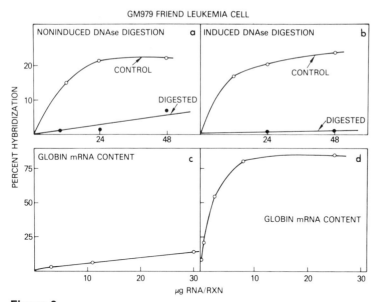

Figure 8
(*a,b*) DNAse I sensitivity of globin genes in nuclei from (*a*) uninduced MELC and (*b*) induced MELC. After induced cells had been grown for 4 days in 2% DMSO, 90% of the cells were positive for benzidine staining, whereas no cells were benzidine-positive in the uninduced population. Recovery of DNA and hybridization analysis with mouse globin cDNA were performed as described in the Methods and in the legend to Figure 7. (*c,d*) Globin mRNA concentration in (*c*) noninduced MELC and (*d*) induced MELC. Each 10-μl reaction included 0.25 ng of [^{3}H]cDNA. Incubation was for 48 hours at 52°C.

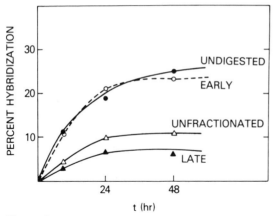

Figure 9

Relative DNAse I sensitivity of globin genes in nuclei from early and late rabbit bone-marrow erythroid cells. Bone-marrow cells, obtained from rabbits previously injected with phenylhydrazine, were fractionated by sequential isopycnic centrifugation and then 1 g sedimentation (see Methods). Nuclei were recovered and exposed to DNAse I. Subsequent extraction of DNA and annealing to rabbit [^{3}H]globin cDNA (0.5 ng/reaction mixture) was as described in the Methods. Each 20-μl reaction mix included 120 μg of DNA. The control (undigested) DNA was derived from unfractionated rabbit bone-marrow cells.

In Sheep Erythroid Cells

The DNAse I sensitivity of the β^C-globin gene in nuclei from sheep bone marrow and spleen are compared in Figure 10. The animal was injected with phenylhydrazine for 10 days (and rested for 3 days) to induce erythroid hyperplasia and to ensure expression of the β^C-globin gene. The DNA samples were annealed to a purified β^C-cDNA probe. β^C-globin gene sequences were not sensitive to DNAse I in spleen nuclei but were markedly reduced in concentration in DNA derived from bone-marrow cells in which this gene was transcriptionally active.

DISCUSSION

Although these studies provide certain insights into the mechanism of hemoglobin switching and of regulation of the individual β and γ genes, they also raise a number of potentially interesting questions. Our data unequivocally demonstrate the presence in fetal sheep of erythroid progenitor cells that are capable of giving rise in vitro to erythroblasts which express the β^C-globin gene and, therefore, make Hb C. Nonetheless, erythropoietic stress in utero, either by direct injection of epo or by bleeding, did not result in an Hb F to Hb C switch. Our interpretation of these data is that a factor(s) necessary for development of erythroid progenitor cells committed to Hb C synthesis is missing in the fetus and was supplied in the

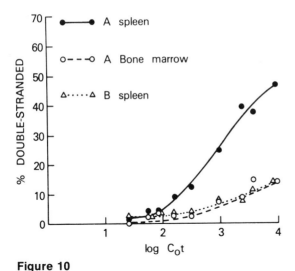

Figure 10
DNAse I sensitivity of the sheep β^C-globin gene. Nuclei were prepared from spleen and bone marrow of an anemic sheep homozygous for the β^A gene. These were exposed to DNAse I to obtain solubilization of 8 to 15% of the total DNA. Each 18-μl annealing reaction contained 250 μg of DNA and approximately 10 pg of purified β^C-globin cDNA which had been synthesized with [^{32}P]dCTP to give a probe with a specific activity of 350,000 cpm/ng. Incubation was at 58°C, only 2–4° below the melting temperature of globin DNA:cDNA duplexes. DNA obtained from spleen nuclei of an animal homozygous for the β^B-globin gene served as a control for the specificity of the annealing reaction since the β^C-globin gene is not found in these animals (Benz et al. 1977a). This DNA did not anneal to the β^C-cDNA as shown but did anneal to β^B-cDNA (not shown).

culture media used to generate colonies in vitro. Alternatively, there may be inhibiting factors in the fetus that preclude the development of these cells. Various hormones known to influence erythropoiesis (Adamson et al., this volume) may be implicated. Cell-to-cell interaction at the microenvironmental level may also influence the development of progenitor cells with varying commitment with respect to expression of the β and γ genes. Our analysis suggests that the apparent direct effect between the expression of the β^C gene and epo concentration in neonatal animals is more complex during fetal erythropoiesis.

The concept of commitment of progenitor cells to expression of the individual globin genes is now defined only operationally. In our experiments, injection of epo into a young neonatal animal resulted in "commitment" of colony-forming cells to generate erythroblasts that made Hb C as revealed by the subsequent formation of colonies that made β^C-globin at low epo concentration in vitro. A significant lag was found between this commitment event in the progenitor cells and the appearance of β^C-globin mRNA in the nuclei of early erythroblasts. Enhanced sensitivity in detecting the initial transcription of the β^C-globin gene might be provided by analysis of pulse-labeled RNA (Aviv et al. 1976; Curtis and Weissman

1976), thereby affording the opportunity to examine more directly the effect of epo on initiation of transcription of this gene or on modulating the processing of the initial transcription product. The erythroid progenitor cells of greatest interest constitute only a very small fraction of the total number of cells in the bone marrow. Various cell-fractionation schemes might afford enrichment of these cells, again presumably enhancing the ability to detect directly the relevant molecular events involved in the commitment process.

An alternative approach to examining the molecular mechanism of regulation of hemoglobin switching is to study the process of gene expression in maturing erythroblasts. Our studies have excluded regulation at the translational level since β^C-globin mRNA sequences were not detectable either in cytoplasm or in the nuclei of cells not making β^C-globin. When these sequences appeared after epo injection, as revealed by the molecular hybridization assay, they seemed to be translated immediately, inasmuch as β^C-globin synthesis was detected. The remaining mechanisms whereby selective expression of the closely linked β and γ genes might be achieved include: (1) a difference in the conformation of these genes in chromatin; (2) differential rates of transcription of the individual genes, or (3) selective processing of the initial transcription product of the individual genes. Recent advances in knowledge regarding the structure of chromatin have stimulated us to attempt to examine directly the first of these possibilities.

Genes which are transcriptionally active appear to have a conformation in chromatin that renders them exquisitely sensitive to digestion by pancreatic DNAse I. Thus, incubation of nuclei from chick erythroid cells with DNAse I results in destruction of the globin and other active genes (Weintraub and Groudine 1976), whereas incubation of chick oviduct nuclei with DNAse I leads to digestion of the gene for ovalbumin (Garel and Axel 1976). Selective sensitivity of the active genes is implicit in the observation that solubilization of only 10–15% of nuclear DNA results in complete (or nearly complete) loss of the active genes' sequences. This selective sensitivity originates in the structure of nucleosomes associated with active genes or in the more "open" conformation in the chromatin of active genes. In any event, DNAse I provides a probe by which we can examine the structure of globin genes during erythroid maturation and also approach the question of the structure of the individual globin genes in sheep erythroid cells.

The anticipated specificity of DNAse I for active genes was revealed by our study of mouse liver cells (Fig. 7). Globin genes in nuclei derived from adult liver were not rendered soluble by DNAse I, whereas incubation of nuclei from the poly- and orthochromatophilic erythroid cells present in 15-day mouse fetal liver led to digestion of the globin genes. We next compared the relative sensitivity of the mouse globin genes in MELC before and after chemical induction of erythroid maturation by DMSO. Globin gene sequences were virtually absent in DNA extracted from nuclei of uninduced and induced cells after these nuclei had been exposed to DNAse I. Upon induction, cellular mRNA sequences increased by 50 times. We infer that the accumulation of mRNA in the induced cells occurs because of accelerated transcription of the globin genes (Orkin and Swerdlow 1977) or because of altered processing or stability of RNA molecules containing globin mRNA sequences.

The experiments with fractionated rabbit bone marrow suggest that normal erythroid cells may behave somewhat differently than do MELC. Early erythroblasts, despite the fact that they already had begun to accumulate globin mRNA (Clissold et al. 1977), contained globin genes that were not sensitive to digestion by DNAse I. Globin genes in nuclei from late cells exhibited the anticipated sensitivity. The paradoxical results obtained on analysis of the early rabbit erythroid cells seem to imply that the conformation of the globin genes as sensed by DNAse I does not relate directly to the potential for transcription. Perhaps there is an equilibrium between the DNAse-I-sensitive conformation during which transcription occurs and a conformation that renders the globin genes insensitive to DNAse I. Further experiments are needed to clarify the significance of this observation.

The β^C-globin gene in bone-marrow nuclei obtained from an anemic sheep were sensitive to DNAse I although these sequences were not sensitive in nuclei from the spleen, a nonerythroid organ. The β^C-globin gene was transcriptionally active in the bone marrow of this animal because a switch from Hb A to Hb C production had been caused by anemic stress. Of more relevance to the issue of regulation of the individual globin genes will be the study of the sensitivity of the nonexpressed γ gene in adult bone-marrow cells and of the β^A- and β^C-globin genes in fetal erythroid cells. Such studies may yield insight into the role of chromatin structure in regulating the individual globin genes.

Acknowledgments

We wish to thank P. Turner, C. Geist, and B. Kefauver for their excellent technical assistance, and E. Murray, M. Motter, and L. Garza for their invaluable assistance in the preparation of this manuscript. We have enjoyed and appreciated many helpful discussions with Dr. W. F. Anderson. Thanks are also given to Drs. J. Beard and B. Beard for providing RNA-directed DNA polymerase through the Office of Program Resources and Logistics, Viral Oncology, National Cancer Institute.

REFERENCES

Aviv, H., Z. Voloch, R. Bastos, and S. Levy. 1976. Biosynthesis and stability of globin mRNA in cultured erythroleukemic Friend cells. *Cell* **8:** 495.

Axelrod, D., T. V. Gopalakrishnan, M. Willig, and W. F. Anderson. 1978. Maintenance of hemoglobin inducibility in somatic cell hybrids of tetraploid (25) mouse erythroleukemia cells with mouse or human fibroblasts. *Somatic Cell Genetics.* **4:** 157.

Barker, J. E., W. F. Anderson, and A. W. Nienhuis. 1975. Hemoglobin switching in sheep and goats: Effect of erythropoietin concentration on *in vitro* erythroid colony growth and globin synthesis. *J. Cell Biol.* **64:** 515.

Barker, J. E., J. E. Pierce, and A. W. Nienhuis. 1976. Hemoglobin switching in sheep and goats: Commitment of erythroid colony-forming cells to synthesis of beta^C globin. *J. Cell Biol.* **71:** 715.

Barker, J. E., J. E. Pierce, B. C. Kefauver, and A. W. Nienhuis. 1977. Hemoglobin switching in sheep and goats: Induction of hemoglobin C synthesis in cultures of sheep fetal erythroid cells. *Proc. Natl. Acad. Sci. U.S.A.* **74:** 5078.

Benz, E., Jr., P. Turner, J. Barker, and A. Nienhuis. 1977a. Stability of the individual globin genes during erythroid differentiation. *Science* **196:** 1213.

Benz, E. J., Jr., J. E. Barker, P. Turner, J. Pierce, and A. W. Nienhuis. 1977b. Induction of β^C globin mRNA in sheep erythroid cells by erythropoietin. *Clin. Res.* **25:** 333.

Benz, E. J., Jr., C. E. Geist, A. W. Steggles, J. E. Barker, and A. W. Nienhuis. 1977c. Hemoglobin switching in sheep and goats: Preparation and characterization of complementary DNAs specific for the alpha-, beta-, and gamma-globin messenger RNAs of sheep. *J. Biol. Chem.* **252:** 1908.

Clissold, P. M., A. R. V. Arnstein, and C. J. Chesterton. 1977. Quantitation of globin mRNA levels during erythroid development in the rabbit and discovery of a new β-related mRNA species in immature erythroblasts. *Cell* **11:** 353.

Curtis, P. J., and C. Weissmann. 1976. Purification of globin messenger RNA from dimethylsulphoxide-induced Friend cells and detection of a putative globin messenger RNA precursor. *J. Mol. Biol.* **106:** 1061.

Deisseroth, A., A. Nienhuis, P. Turner, R. Velez, W. F. Anderson, F. Ruddle, J. Lawrence, R. Creagan, and R. Kucherlapati. 1977. Localization of the human α-globin structural gene to chromosome 16 in somatic cell hybrids by molecular hybridization assay. *Cell* **12:** 205.

Gabuzda, T. D., M. A. Schumann, R. K. Silver, and H. B. Lewis. 1968. Erythrokinetics in sheep studied by means of induced changes in hemoglobin phenotype. *J. Clin. Invest.* **47:** 1895.

Garel, A., and R. Axel. 1976. Selective digestion of transcriptionally active ovalbumin genes from oviduct nuclei. *Proc. Natl. Acad. Sci. U.S.A.* **73:** 3966.

Marzluff, W. F., E. C. Murphy, and R. C. C. Huang. 1973. Transcription of ribonucleic acid in isolated mouse myeloma nuclei. *Biochemistry* **12:** 3440.

Miller, D. M., D. Axelrod, R. Croissant, and A. W. Nienhuis. 1978. Structure of the globin genes in differentiating erythroid cells. In *Proceedings of the 4th International Conference on red cell metabolism* (ed. G. Brewer), Ann Arbor, Michigan. (In press.)

Nienhuis, A. W., and E. J. Benz, Jr. 1977. Regulation of hemoglobin synthesis during development of the red cell. *N. Engl. J. Med.* **297:** 1318, 1371, 1430.

Nienhuis, A. W., P. Turner, and E. J. Benz, Jr. 1977. Relative stability of α- and β-globin messenger RNAs in homozygous β^+ thalassemia. *Proc. Natl. Acad. Sci. U.S.A.* **74:** 3960.

Orkin, S. H., and P. S. Swerdlow. 1977. Globin RNA synthesis *in vitro* by isolated erythroleukemic cell nuclei: Direct evidence for increased transcription during erythroid differentiation. *Proc. Natl. Acad. Sci. U.S.A.* **74:** 2475.

Papayannopoulou, Th., M. Brice, and G. Stamatoyannopoulou. 1977. Hemoglobin F synthesis *in vitro:* Evidence for control at the level of primitive erythroid stem cells. *Proc. Natl. Acad. Sci. U.S.A.* **74:** 2923.

Stephenson, J. R., A. A. Axelrad, D. L. McLeod, and M. N. Shreeve. 1971. Induction of colonies of hemoglobin-synthesizing cells by erythropoietin *in vitro. Proc. Natl. Acad. Sci. U.S.A.* **68:** 1542.

Weintraub, H., and M. Groudine. 1976. Chromosomal subunits in active genes have an altered conformation. *Science* **193:** 848.

The State of the Globin Gene in Friend Erythroleukemia Cells

S. K. Dube*
Department of Biological Chemistry, Harvard Medical School
Boston, Massachusetts 02115

R. B. Wallace and J. Bonner
Division of Biology, California Institute of Technology
Pasadena, California 91125

Erythropoiesis in susceptible mice infected with the Friend virus complex (FVC) (Friend 1957; Axelrad and Steeves 1964) is freed of the regulatory constraints of erythropoietin (Mirand et al. 1968; Sassa et al. 1968). This results in unregulated production of erythroid precursor cells, although erythroid maturation appears not to be affected (Liao and Axelrad 1975). Eventually, however, the erythropoietic system is inundated with autonomously growing precursor cells that are blocked at the proerythroblast stage in erythroid maturation (Tambourin and Wendling 1975). These transformed proerythroblasts are the Friend erythroleukemia cells or Friend cells, which can be induced to differentiate in tissue culture by addition of dimethylsulfoxide (DMSO) (Friend et al. 1971) or of certain other chemicals to the culture medium (Dube et al. 1973; Leder and Leder 1975; Tanaka et al. 1975; Ebert et al. 1976; Gusella and Housman 1976; Reuben et al. 1976; Marks et al. 1978). The induced cells stop dividing and synthesize hemoglobin. They contain globin mRNA and hemoglobin in amounts comparable to those present in normal reticulocytes (Dube et al. 1973), in contrast to uninduced cells which contain little or no globin mRNA and hemoglobin (Ostertag et al. 1972; Ross et al. 1972; Scher et al. 1973; Gilmour et al. 1974).

How erythroid differentiation is blocked by the FVC or how it is induced by DMSO and other chemicals is not clear. What is clear, however, is that the inability of Friend cells to synthesize hemoglobin lies in the unavailability of globin mRNA, a deficiency rectified by the inducers. This raises the possibility, among others, that the globin gene is switched off during the transformation event and is switched on by the inducers. We investigated this possibility by using methods capable of distinguishing expressed and nonexpressed portions of the genome.

The results of these studies, presented below, rule out the possibility

---
* Present address: School of Medicine, Yale University, New Haven, Connecticut 06510

that the globin gene is switched off during transformation and is switched on by the inducers. Furthermore, in cells resistant to induction, the globin-gene sequence appears not to be associated with the template-active fraction of chromatin.

METHODS

These methods are based on the digestion of chromatin with DNase I (Garel and Axel 1976; Weintraub and Groudine 1976) or DNase II (Marushige and Bonner 1971; Billing and Bonner 1972; Gottesfeld et al. 1974, 1976; Wallace et al. 1977). The underlying principle of these methods is that the expressed portion of the genome is more rapidly attacked by nucleases than is the nonexpressed portion. Although, in general, both methods lead to similar conclusions, the DNase II procedure has the added advantage of yielding an enriched fraction of transcriptionally active chromatin. We have, therefore, used this procedure more extensively.

Analysis of Chromatin from Uninduced and Induced Friend Cells

Clone FSD-3, a highly inducible clone of Friend cells (Dube et al. 1973), was grown in suspension culture as described by Ostertag et al. (1972). For induction, DMSO at a final concentration of 1.5% was added to an exponentially growing culture at a cell density of $2-5 \times 10^5$ cells/ml. After 24 hours in DMSO, cells were harvested for chromatin isolation. At this time, the globin gene was being actively transcribed (Dube et al. 1973; Singer 1975). Cells were washed in phosphate-buffered saline and suspended in TKMPE: 10 mM Tris-HCl, pH 7.5, 6 mM KCl, 5 mM magnesium acetate, 1 mM phenyl methyl sulfonyl fluoride (PMSF), and 0.1 mM ethylene glycol bis (β-amino ether)' tetra acetic acid (EGTA) containing 0.5% NP 40. Nuclei were pelleted by centrifugation at 1000 g for 5 minutes, washed once in TKMPE containing 0.5% NP 40, once in TKMPE alone, and once in 0.075 M NaCl, 0.024 M EDTA, pH 7.5, and 1 mM PMSF. The nuclear pellet was resuspended with a glass-Teflon homogenizer in TPE: 10 mM Tris-HCl, pH 8.0, 1 mM PMSF, and 0.1 mM EGTA. Chromatin from uninduced and induced cells was prepared by lysis of nuclei in low-ionic-strength buffer and was digested with DNase and fractionated as described by Gottesfeld et al. (1974). Chromatin was pelleted by centrifugation at 10,000 g for 10 minutes, washed three times in TPE, suspended in TPE at an A_{260} of 10 as measured in 0.1 N NaOH. The chromatin suspension was dialyzed overnight against 25 mM sodium acetate, pH 6.6, 0.1 mM EGTA. Digestion of chromatin with DNase II, centrifugation at 10,000 g to obtain the supernatant, precipitation of this supernatant with 2 mM magnesium acetate to obtain the soluble template-active fraction and the insuluble template-inactive fraction, and the preparation of DNA were performed as described by Gottesfeld et al. (1974).

Whole-cell DNA was prepared by established procedure (Marmur 1961) and sheared in a Virtis 60 homogenizer to 350 base pairs. ^{3}H-labeled globin cDNA was prepared as described previously (Harrison et al. 1974). Titration of the globin-gene sequence was performed as follows. DNA at 5 mg/ml in 0.41 M phosphate buffer, pH 6.8 (Britten et al. 1974), was mixed with ^{3}H-labeled globin cDNA (2×10^7 dpm/mg) in mass ratio varying from 5×10^3 to 5×10^7. ^{3}H-labeled cDNA, 250 to 6250 dpm, was used. DNA was denatured by boiling for 5 minutes and allowed to renature at 70°C for 40 hours (C_0t 10,000) with respect to driver DNA. Hybridization of the ^{3}H-labeled globin cDNA was detected by hydroxyapatite (HAP) chromatography as described by Britten et al. (1974). The fraction of ^{3}H-labeled cDNA in the hybrid is related to the mass ratio of driver DNA to ^{3}H-labeled globin cDNA by the equation

$$F(R) = A + \frac{B}{1 + RC}$$

where $F(R)$, fraction [^{3}H]cDNA not able to react in vast excess of whole-cell driver DNA (data not shown); B, fraction of cDNA able to react; C, frequency of globin-gene sequence in the driver DNA expressed as fractional mass; and R, mass ratio of driver to tracer DNAs. Data were analyzed by nonlinear least-squares fitting as described by Pearson et al. (1977) (Fig. 1). The concentration of template-active DNA has been corrected for its twofold lower ability to react as compared with whole-cell DNA. Template active DNA's lower ability to react has been noted previously (Gottesfeld et al. 1974) and appears to be random with respect to sequence.

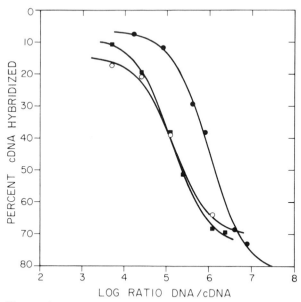

Figure 1
Titration of the globin-gene sequences in DNA from template-active fractions of Friend-cell chromatin and DNA from whole cells. (●——●) Whole-cell DNA; (□——□) template-active DNA from uninduced FSD-3 cells; (o——o) template-active DNA from uninduced FSD-3 cells.

Analysis of Chromatin from Friend Cells Refractory to Induction

Because these experiments were done with an inducible cell line, we reasoned that a good check on the accuracy of the fractionation procedure would be an analysis of the chromatin from Friend cell lines refractory to induction. We used the F4$^+$ cell line (Keppel et al. 1977) for this experiment. These cells are resistant not only to DMSO but also to hexamethylene bisacetamide (H. Eisen, pers. comm.; W. Ostertag, pers. comm.), a potent inducer of hemoglobin synthesis in Friend cells (Reuben et al. 1976).

Isolation and fractionation of chromatin and DNA hybridization analyses were carried out as in the experiment with FSD-3 cells. Template-active and -inactive DNAs were isolated as described except that the DNAs were further purified by binding to HAP in 0.12 M phosphate buffer (Britten et al. 1974); DNAs treated in this way exhibited up to 90% of the reaction of whole-cell DNA. Saturation hybridization experiments were performed as follows: 50 ng of DNA in 0.41 M phosphate buffer, pH 6.8, was mixed with increasing amounts of ^{3}H-labeled globin cDNA up to 1.0 ng. The driver DNA concentration was maintained at 5 mg/ml. The samples were denatured by boiling for 5 minutes and then were allowed to renature for 40 hours. (C_0t 10,000). Hybridization of ^{3}H-labeled globin cDNA was detected by HAP chromatography (Britten et al. 1974). Results obtained with F4$^+$ template-active DNA and F4$^+$ whole-cell DNA are shown in Figure 2. For comparison, data obtained with the FSD-3 template-active chromatin have also been plotted in this figure.

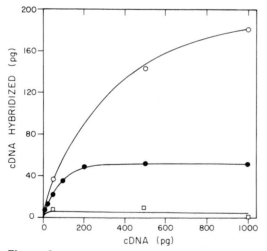

Figure 2
Saturation hybridization of ^{3}H-labeled globin cDNA to template-active DNA from uninduced FSD-3 cell chromatin and from F4$^+$ cell chromatin, and to whole-cell DNA. The 10% HAP binding of the cDNA in the absence of driver has been subtracted from each point. (■——■) Template-active DNA from FSD-3 cells; (o——o) template-active DNA from F4$^+$ cells; (●——●) template-inactive DNA from F4$^+$ cells.

RESULTS

The data in Figure 1 show that the DNA of the template-active fraction of induced FSD3 cell chromatin was enriched seven times over whole-cell DNA for the globin gene sequence. Similar results were obtained when DNA from the template-active fraction of the uninduced FSD-3 cell chromatin was analyzed. The template-active DNA of uninduced cells was also enriched for globin-gene sequence and this enrichment was essentially identical to that observed in the template-active DNA of induced cells (Fig. 1).

The data in Figure 2 show that $F4^+$ template-active DNA contained barely detectable levels of globin sequence. The data also show that the $F4^+$ cells did not suffer a deletion of the globin gene because the template-inactive DNA showed normal saturation hybridization with globin cDNA. Thus, chromatin from a cell line that does not have the potential to transcribe globin sequences does not have the globin gene in a conformation that would allow its fractionation into the template-active chromatin fraction.

CONCLUSION

Our study indicates that the globin gene in Friend cells is in a transcriptionally active configuration, both in uninduced and induced cells. Indeed, induction does not appear to involve a major change in the configuration of the globin gene. Why then was a dramatic difference found in the globin mRNA content of induced and uninduced cells? The answer is not obvious. However, in light of these data, the following argument could be made. The target cell for transformation by FVC is a cell in which the globin gene has already been switched on. The transcription of the globin gene in this cell normally may involve activation by a cytoplasmic or nuclear signal. However, upon transformation, this signal is inactivated. The inducers are able to activate this signal and to allow active transcription of the globin gene. The signal may be a small molecule or a factor affecting mRNA stability.

Acknowledgments

This work was supported by U.S. Public Health Service Grants GM 13762 and GM 26297. We thank Drs. W. Ostertag and H. Eisen for communicating to us their results on the resistance of $F4^+$ cells to HMBA induction and Dr. I. Pragnal for a gift of globin cDNA. RBW is a Fellow of the Medical Research Council of Canada. SKD was supported as a Sherman Fairchild Distinguished Scholar at the California Institute of Technology and by a lectureship at Harvard University.

REFERENCES

Axelrad, A. A., and R. A. Steeves. 1964. Assay for Friend leukemia virus: Rapid quantitative method based on enumeration of macroscopic spleen foci in mice. *Virology* **24**: 513.

Billing, R. J., and J. Bonner. 1972. The structure of chromatin as revealed by deoxyribonuclease digestion studies. *Biochim. Biophys.* Acta **281**: 453.

Britten, R. J., D. F. Graham, and B. R. Neufield. 1974. Analysis of repeating DNA sequences by reassociation. *Methods Enzymol.* **29**: 363.

Dube, S. K., G. Gaedicke, N. Kluge, B. J. Weimann, H. Melderis, G. Steinheider, T. Crozier, H. Beckmann, and W. Ostertag. 1973. Hemoglobin synthesizing mouse and human erythroleukemic cell lines as model systems for the study of differentiation and control of gene expression. *Proc. Int. Symp. Princess Takamatsu Cancer Res. Fund,* p. 99.

Ebert, P. S., I. Wars, and D. N. Buell. 1976. Erythroid differentiation in cultured Friend leukemia cells treated with metabolic inhibitors. *Cancer Res.* **36**: 1809.

Friend, C. 1957. Cell-free transmission in adult swiss mice of a disease having the character of a leukemia. *J. Exp. Med.* **105**: 307.

Friend, C., W. Scher, J. G. Holland, and T. Sato. 1971. Hemoglobin synthesis in murine virus-induced leukemic cells *in vitro:* Stimulation of erythroid differentiation by dimethyl sulfoxide. *Proc. Natl. Acad. Sci. U.S.A.* **68**: 378.

Garel, A., and R. Axel. 1976. Selective digestion of transcriptionally active ovalbumin genes from oviduct nuclei. *Proc. Natl. Acad. Sci. U.S.A.* **73**: 3966.

Gilmour, R. S., P. R. Harrison, J. M. Windass, N. Affara, and J. Paul. 1974. Globin mRNA synthesis and processing during hemoglobin induction in Friend cells. I. Evidence for transcriptional control in clone M2. *Cell Differ.* **3**: 9.

Gottesfeld, J. M., W. T. Garrard, G. Bagi, R. F. Wilson, and J. Bonner. 1974. Partial purification of the template-active fraction of chromatin: A preliminary report. *Proc. Natl. Acad. Sci. U.S.A.* **71**: 2193.

Gottesfeld, J. M., G. Bagi, B. Berg, and J. Bonner. 1976. Sequence composition of the template-active fraction of rat liver chromatin. *Biochemistry* **15**: 2472.

Gusella, J. F., and D. Housman. 1976. Induction of erythroid differentiation in vitro by purines and purine analogues. *Cell* **8**: 263.

Harrison, P. R., G. D. Birnie, A. Hell, S. Humphries, B. D. Young, and J. Paul. 1974. Kinetic studies of gene frequency. I. Use of a DNA copy of reticulocyte 9 S RNA to estimate globin gene dosage in mouse tissues. *J. Mol. Biol.* **84**: 539.

Keppel, F., B. Allet, and H. Eisen. 1977. Appearance of a chromatin protein during the erythroid differentiation of Friend virus-transformed cells. *Proc. Natl. Acad. Sci. U.S.A.* **74**: 653.

Leder, A., and P. Leder. 1975. Butyric acid, a potent inducer of erythroid differentiation in cultured erythroleukemic cells. *Cell* **5**: 319.

Liao, S.-K., and A. A. Axelrad. 1975. Erythropoietin-independent erythroid colony formation *in vitro* by hemopoietic cells of mice infected with Friend virus. *Int. J. Cancer* **15**: 467.

Marks, P. A., R. A. Rifkind, A. Bank, M. Terada, R. Reuben, E. Fibach, U. Nudel, J. Salmon, and Y. Gazitt. 1978. Induction of differentiation of murine erythroleukemia cells. In *Cell differentiation and neoplasia.* Proceedings of the 30th Annual Symposium on Fundamental Cancer Research, M. D. Anderson Hospital, Houston, Texas. (In press.)

Marmur, J. 1961. A procedure for the isolation of deoxyribonucleic acid from microorganisms. *J. Mol Biol.* **3**: 208.

Marushige, K., and J. Bonner. 1971. Fractionation of liver chromatin. *Proc. Natl. Acad. Sci. U.S.A.* **68**: 2941.

Mirand, E. A., R. A. Steeves, R. D. Lange, and J. T. Grace, Jr. 1968. Virus-induced polycythemia in mice: Erythropoiesis without erythropoietin. *Proc. Soc. Exp. Biol. Med.* **128**: 844.

Ostertag, W., H. Melderis, G. Steinheider, N. Kluge, and S. Dube. 1972. Synthesis of mouse haemoglobin and globin mRNA in leukaemic cell cultures. *Nat. New Biol.* **239**: 231.

Pearson, W. R., E. H. Davidson, and R. J. Britten. 1977. A program for least squares analysis of reassociation and hybridization data. *Nucleic Acids Res.* **4**: 1727.

Reuben, R. C., R. L. Wife, R. Breslow, R. A. Rifkind, and P. A. Marks. 1976. A new group of potent inducers of differentiation in murine erythroleukemia cells. *Proc. Natl. Acad. Sci. U.S.A.* **73**: 862.

Ross, J., Y. Ikawa, and P. Leder. 1972. Globin messenger-RNA induction during erythroid differentiation of cultured leukemia cells. *Proc. Natl. Acad. Sci. U.S.A.* **69**: 3620.

Sassa, S., F. Takaku, and K. Nakao. 1968. Regulation of erythropoiesis in the Friend leukemia mouse. *Blood* **31**: 738.

Scher, W., H. D. Preisler, and C. Friend. 1973. Hemoglobin synthesis in murine virus-induced leukemic cells in vitro. *J. Cell. Physiol.* **81**: 63.

Singer, D. 1975. Ph.D. thesis, Columbia University, New York.

Tambourin, P., and F. Wendling. 1975. Target cell for oncogenic action of polycythaemia-inducing Friend virus. *Nature* **256**: 320.

Tanaka, M., J. Levy, M. Terada, R. Breslow, R. A. Rifkind, and P. A. Marks. 1975. Induction of erythroid differentiation in murine virus infected erythroleukemia cells by highly polar compounds. *Proc. Natl. Acad. Sci. U.S.A.* **72**: 1003.

Wallace, R. B., S. K. Dube, and J. Bonner. 1977. Localization of the globin gene in the template active fraction of chromatin of Friend leukemia cells. *Science* **198**: 1166.

Weintraub, H., and M. Groudine. 1976. Chromosomal subunits in active genes have an altered conformation. *Science* **193**: 848.

Sequential Gene Expression during Induced Differentiation of Cultured Friend Erythroleukemia Cells

M. Obinata, R. Kameji, Y. Uchiyama, and Y. Ikawa

Laboratory of Viral Oncology, Cancer Institute
Tokyo 170, Japan

Friend erythroleukemia cells were obtained by culturing cells of an ascites strain originally derived from fragments of spleens of mice infected with the Friend virus complex (Friend et al. 1966; Ostertag et al. 1972; Ikawa et al. 1973). The Friend virus appears to have specific effects on committed erythroid stem cells, rendering them capable of proliferation and differentiation independently of erythropoietin (Hawkins and Krantz 1975; Liao and Axelrad 1975). The Friend leukemia cell line established in culture thus appears to comprise selected transformed erythroid precursor cells that differentiate under certain conditions and are arrested at the proerythroblastic stage of development (Friend et al. 1966). These cells are not responsive to erythropoietin and, in this respect, differ from normal erythroid precursor cells. However, when treated with dimethylsulfoxide (DMSO), these cells undergo a series of changes, many of which are characteristic of normal erythroid cell differentiation.

Other chemical reagents have been found to have similar effects on these cells (Leder and Leder 1975; Reuben et al. 1976). When such inducers are added, these cells become altered morphologically, condense their chromatin (Friend et al. 1971), develop detectable erythrocyte-specific membrane antigens (Furusawa et al. 1971), accumulate heme, increase production of several enzymes related to heme synthesis (Ebert and Ikawa 1974; Kabat et al. 1975; Sassa 1976), and synthesize adult-type globin (Ostertag et al. 1972; Boyer et al. 1972) following accumulation of globin mRNA (Ross et al. 1972; Gilmour et al. 1974).

This in vitro Friend leukemia system provides an extremely useful model for investigating a uniform population of cells before and after the induced changes have occurred that result in sequentially programmed gene expression.

EXPERIMENTAL PROCEDURES
Cells and Culture

The isolation and characterization of clonal cell lines originated from T3-Cl-2 have been described elsewhere (Ikawa et al. 1976). C-10-6 cells are most frequently used as highly differentiation-inducible cells. Friend erythroleukemia cells were grown in 250-ml plastic flasks (Falcon Plastics) containing 50 ml of Ham's F12 medium (Gibco Labotaroties) supplemented with 10% heat-inactivated calf serum. Dimethylsulfoxide 1.5% v/v and butyric acid (1 mM) were respectively added to the cultures of Friend erythroleukemia cells at 3.5×10^4 cells/ml. The cells were harvested on day 5. Hemoglobin-positive cells were counted after staining by the modified wet benzidine technique as described previously (Ikawa et al. 1976).

Preparation of RNA

Cytoplasmic RNA from Friend erythroleukemia cells and mouse reticulocyte RNA were prepared by the successive extraction with phenol-chloroform and chloroform-isoamyl alcohol as described previously (Kameji et al. 1977). Poly(A)-containing RNA was obtained through a poly(U)-sepharose column.

Nuclear RNA was prepared by a modified method of CsCl buoyant density gradient centrifugation according to Glisin et al. (1974).

Preparation of cDNA

Poly(A)-containing RNAs from induced and uninduced C-10-6 cells were used as templates for the avian myeloblastosis virus (AMV) reverse transcriptase (phosphocellulose step).

RNA (5-10 μg) was preincubated with 0.1 A_{260} unit of oligo(dT)12-18 (P. L. Biochemicals) at 37°C for 15 minutes in 0.1 M NaCl. This mixture was added to the reaction mixture (250 μl) containing 0.1 M Tris-HCl (pH 7.4); 0.004 M KCl; 6 mM Mg acetate; 2 mM dithiothreitol; 0.2 mM dATP, dGTP, and TTP; 0.07 mM [^{3}H]dCTP (24.56 Ci/mmole; New England Nuclear); 10 μg of reverse transcriptase; and 60 μg actinomycin D. After incubation at 37°C for 2 hours, the reaction mixture was processed as described previously (Kameji et al. 1977). Specific radioactivity of [^{3}H]cDNA was 2.3×10^7 dpm/μg.

cDNA-RNA Hybridization

Hybridization was carried out in a final volume of 5, 10, or 50 μl in capillary tubes (Clay-Adams), and the mixture consisted of [^{3}H]cDNA (about 500 or 1000 cpm/assay), 0.6 M NaCl, 0.013 M Tris-HCl (pH 7.4), 1.3×10^{-4} M EDTA, 50% (v/v) formamide, and sample RNA. Incubation was done at 41°C and the mixture was digested with S1 nuclease as described by Ross et al. (1974).

RESULTS AND DISCUSSION

Regulatory Step of Globin Induction in Cultured Friend Erythroleukemia Cells

Highly differentiated cells may be characterized by specialization in large-scale production of specific proteins (Palmiter 1975). Over 90% of the total proteins synthesized during erythroid differentiation is globin. The mechanisms regulating high rate of hemoglobin production operate at multiple levels. Amplification of globin genes during the course of differentiation, which can lead to synthesis of a large quantity of globin, does not take place in the induced differentiation of Friend erythroleukemia cells or in normal erythroid differentiation (Ross et al. 1974; Bishop 1974). Therefore, transcriptional control (especially the rate of RNA chain initiation), posttranscriptional control (transport of nuclear precursor into cytoplasm or modification and processing of precursor molecules), and translational control are most probable as a regulatory step for globin induction.

The synthesis of globin mRNA in induced differentiation of Friend erythroleukemia cells has been studied extensively using T3-Cl-2 cell line, and it was shown that the induction of globin was based on the transcriptional activation of globin genes (Ross et al. 1972, 1974).

Recently, several phenotypic variants of T3-Cl-2 cells were isolated by cloning—including some variants resistant to DMSO-induced differentiation. To clarify the regulatory step primarily involved in globin induction among these variants of cultured Friend erythroleukemia cells, we quantitated globin mRNA in these cells by means of nucleic acid hybridization. In addition to T3-Cl-2 variants, cell lines derived from DBA/2 mice were also used for this analysis. Cell lines were classified into four groups with respect to their responsiveness to two different inducers, as shown in Table 1. The extent of hemoglobin production is expressed by benzidine positivity as previously described (Ikawa et al. 1976).

Table 1
The Hemoglobin Inducibility of Phenotypic Variants of Cultured Friend Erythroleukemia Cells

Cell line	Percent of benzidine + cells	
	+ butyric acid	+ DMSO
C-10-6[a]	50	58
C-10-16[a]	34	41
B8/3	28	39
C-9-6[a]	18	0.3
A-10-10[a]	11	0.5
707	3.2	82
GM 86	3	79
K-1	3	0.07

[a] Clonal variant originated from T3-Cl-2 cells.

The total number of cytoplasmic globin mRNA molecules was calculated from the $C_rt_{1/2}$ values of hybridization kinetics of cytoplasmic RNA with globin cDNA. The results, summarized in Table 2, clearly indicate that the extent of benzidine positivity was well correlated with the cytoplasmic globin mRNA concentration in each variant. It is also obvious from Table 2 that none of variants of Friend erythroleukemia cells accumulated globin mRNA in the cytoplasm in an uninduced condition. The proportionality of the relative rate of specific protein synthesis and the relative concentration of its mRNA has been observed in the synthesis of ovalbumin in the hormone-stimulated chick oviduct (Palmiter 1975), in the histone synthesis in the S phase of HeLa cells (Borum et al. 1967), and in the hemoglobin synthesis by frog oocytes in response to injected globin mRNA (Gurdon et al. 1971).

Since we measured total cytoplasmic globin mRNA, the possibility can be excluded that the absence of hemoglobin synthesis in resistant variants or uninduced cells is independent of the masking of mRNA as previously observed for histone mRNA in unfertilized sea urchin eggs (Skoultchi and Gross 1973) and myosin mRNA in rat myoblasts (Buckingham et al. 1976). The translational control does not seem to play a role in the induction of erythrodifferentiation of Friend leukemia cells.

In differentiating chick embryonic erythrocytes, high retention of globin mRNAs in the nuclei and successive transport into the cytoplasm have been detected (Chan 1976). Harrison et al. (1974b) have shown that in one of the clones of Friend erythroleukemia cells, globin mRNA sequences were retained within the nuclei of uninduced cells. The existence of globin

Table 2
Estimation of Globin mRNA Content in Phenotypic Variants

Cell	Addition	Benzidine + cells (%)	$C_rt_{1/2}$	Globin content ($\% \times 10^4$)	Globin mRNA (molecules/cell)
707	none	0.5	2000	0.5	22
	butyric acid	1.2	800	1.2	40
	DMSO	82	6	167.0	4467
A-10-10	none	0	10^4	0.2	10
	butyric acid	10	170	5.9	177
	DMSO	0.7	1400	0.7	37
C-9	none	0	10^4	0.2	8
	butyric acid	8	200	5.0	167
	DMSO	0.1	5500	0.2	5
B8/3	none	0	10^4	0.2	9
	butyric acid	28	100	10.0	300
	DMSO	17	600	1.7	57
C-10-6	none	0	10^4	0.2	6
	butyric acid	53	14	71.4	2380
	DMSO	58	11	91.0	3330
GM 86	none	1.0	1200	0.8	40
	butyric acid	1.5	600	1.7	50
	DMSO	75	7	143.0	3800

mRNA sequences in the nuclei of nonerythroid cells was also observed by Humphries et al. (1976). These observations led to the consideration of post-transcriptional regulation of globin gene expression.

We have examined the possibility of post-transcriptional regulation by titration of nuclear globin mRNA. Some of the results are summarized in Table 3. Without addition of inducer, none of the cells tested retained enough globin mRNA sequences within the nuclei to be detectable under our hybridization conditions which can titrate at least one molecule per nuclei. If the stability of the nuclear globin mRNA is the same throughout the course of induction, the absence of nuclear globin mRNA sequences in uninduced Friend erythroleukemia cells excludes the regulatory process in which the inducer, in direct or indirect manner, facilitates the transport of the preexisting nuclear globin mRNA into cytoplasm. Experimental evidence suggested that globin mRNAs began to accumulate during or shortly after the proerythroblast-basophilic erythroblast transition (Harrison et al. 1974a). Cultured Friend erythroleukemia cells were expected to be transformed at a proerythroblast stage. If we could assume that these cells still retain their original characteristics of proerythroblasts even after transformation, the absence of nuclear globin mRNA in uninduced cells may suggest that proerythroblasts are still dormant in the transcription of globin genes. Histochemically, however, several variations are observable in the proerythroblastic stage.

With addition of inducer, the amount of nuclear globin mRNA was well correlated with that of cytoplasmic globin mRNA and resultant hemoglobin production or benzidine positivity. Especially, the absence or small accumulation of nuclear globin mRNA in the resistant clones indicated that the lack of hemoglobin induction was not due to the defect of transport of preexisting nuclear globin mRNA into the cytoplasm. The proportionality of nuclear and cytoplasmic globin mRNAs among the phenotypic variants indicates that the post-transcriptional regulation is not involved in these

Table 3
Estimation of Globin mRNA Content in Nucleus

Cells	Benzidine + cells (%)	Content (% × 10^4)	Molecules/cell	
			nucleus	cytoplasm
Induced				
C-10-6	14	5.3	8	2,400
A-10-10	1	0.16	1	100
707	89	83	160	11,000
B8/3	3	0.3	0.7	290
K-1	0	0.6	3.3	N.D.[a]
Uninduced				
C-10-6	0	0.31	1	
A-10-10	0	0.2	1	
707	0	0.2	1	
B8/3	0	0.2	1	

[a] N.D. indicates not done.

cells. The observed discrepancy of globin mRNA contents in our 707 clones with those of Harrison et al. (1974a) may be explained by the difference in the phenotypic properties that occurred throughout the passages in two different laboratories rather than by the difference in the quantitation of mRNA.

One serious question about our results is the assumption of a constant rate of globin mRNA degradation. This cannot be demonstrated because we lack adequate methodology for the analysis of degradation in the absence of steady-state conditions. If we postulate that globin mRNA is transcribed constitutively but is rapidly degraded in the absence of an inducer, then the key step for induction might be the inactivation of a specific nuclease for globin mRNA. This theoretical regulation, however, is not attractive because of its apparent wastefulness and because it is difficult to examine. Accordingly, in the induction of hemoglobin synthesis in Friend erythroleukemia cells, the major focus has been placed on the importance of the transcriptional control.

Complexity of Message Sequences in Cultured Friend Erythroleukemia Cells

It has been demonstrated that the persistence of high concentration of globin mRNA is necessary for the hemoglobin induction of cultured Friend erythroleukemia cells and that transcriptional regulation plays an important role. In contrast to such specific mRNAs, little information exists for the total messenger RNA population.

The fraction of transcribed genome has been determined by DNA-RNA hybridization using excess RNA and purified unique sequences of DNA (Davidson and Hough 1971). It has been shown that the population of RNA transcribed differs among tissues (Grouse et al. 1972) and developmental stages (Firtel 1972).

A most sensitive approach to the analysis of kinetics of annealing reaction of cDNA, a complementary DNA of mRNA prepared by reverse transcriptase with template mRNA, permits an estimation of the complexity of the mRNA population and the relative numbers of various messengers in it (Bishop et al. 1974). This approach might raise questions as to how many different messenger RNAs exist or how the mRNA population is modulated due to genetic regulation during the course of induced-erythrodifferentiation of Friend erythroleukemia cells.

Using avian myeloblastosis virus reverse transcriptase, cDNA was made from poly(A)-containing RNA isolated from either DMSO-induced C-10-6 cells or uninduced cells, respectively. When the cDNA was incubated with a large excess of poly(A)-containing RNA, hybridization took place with the kinetics suggesting heterogeneity in its population, as shown in Figure 1. This can be explained by the difference in the frequency distributions of message populations. By calibrating the hybridization kinetics with that of mouse globin mRNA with its cDNA, it is possible to elucidate the frequency distribution within the mRNA population (Bishop et al. 1974). Three frequency classes observed in induced and uninduced Friend erythroleukemia cells were calculated. As shown in Table 4, the first class, repre-

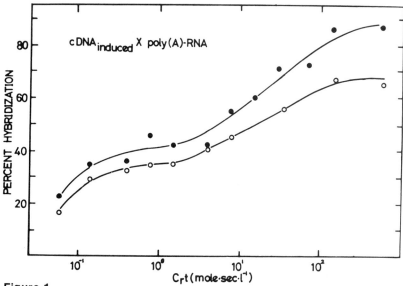

Figure 1
Hybridization of poly(A)-containing RNA with its cDNA. cDNA was made for mRNA from induced cells. (●) mRNA from induced cells; (o) mRNA from uninduced cells.

senting 26.4% of all the hybridizable messenger sequences, includes approximately five different sequences in uninduced cells; the second (29.4%) includes about 400, and the last (44.2%), about 9680 different RNA sequences of average size. Essentially similar values were obtained in induced cells. Approximately 9000 different RNA sequences are expressed in the latter cells.

Assuming an average messenger molecular weight of 6×10^5 daltons, the above RNA sequences would represent 6×10^9 daltons, or about 0.58% of total unique sentences of the whole mouse genome. These estimated

Table 4
Different Frequency Classes in Cultured Friend Erythroleukemia Cell Messages

$C_r t_{1/2}$[a]	Hybridized[b] (%)	$C_r t_{1/2}$[b]	Base-sequence complexity ($\times 10^6$)	No. of sequences	No. of RNA molecules/cell
			Uninduced cells		
0.11	26.4	0.029	3	5	13,500
7	29.4	2.1	247	412	195
75	44.2	33	5,804	9,680	12
			Induced cells		
0.15	14.3	0.021	1.2	2	18,500
2.5	28.5	0.714	81	135	550
40	57.2	22.8	5,217	8,695	17

[a] Observed value.
[b] Corrected value.

complexities correspond closely to those obtained by Birnie et al. (1974) for Friend erythroleukemia cells, by Ryffel and McCarthy (1975) for L cells, and by Williams and Penman (1975) for 3T6 cells.

Alteration of Message Sequences during Induced Differentiation

The hybridization of induced-cell cDNA to uninduced-cell poly(A)-containing RNA demonstrated the existence of a substantial amount of cDNA (approximately 20%) which does not hybridize with uninduced-cell mRNA at a higher $C_r t$ value, representing the scarce class of messages. When the hybridization of uninduced-cell cDNA to induced-cell poly(A)-containing RNA was performed, an analogous result was obtained.

The differences in the extent of hybridization of cDNA with homologous and heterologous cytoplasmic RNA were quite similar to that observed in poly(A)-containing RNA (data not shown). It is evident that these cDNA probes reflect the message population in the cytoplasmic RNA. The analysis demonstrating that cDNA also has three abundance classes indicated that these cDNAs are faithful copies of messages (data not shown). The observed sequence heterogeneity might therefore be due to the alteration in the message sequences during induced differentiation. To measure the complexity of these specific sequences present in either induced or uninduced cells, the cDNAs representing these sequences were purified by hybridization with heterologous template at a saturation level, followed by separation of hybridized and unhybridized cDNA on hydroxylapatite (Ryffel and McCarthy 1975). After this procedure, cDNA was separated into three portions. The major portion (80%) of cDNA reflects the messages present in both induced and uninduced cells. These messages are assumed to be involved in the common cellular functions. The second portion (20%) reflects the messages present only in induced cells. These messages are expected to relate to the erythrodifferentiation. The third portion (20%) reflects the messages present only in uninduced cells. These are expected to be suppressed, depending on the induced differentiation. These three cDNA probes are called, respectively, common cDNA, inducible cDNA, and suppressible cDNA.

The hybridization experiments of these cDNA probes with induced- or uninduced-cell messages are shown in Figure 2. Common cDNA can react to more than 80% of both messages with the same annealing kinetics. The hybridization data show that these messages have three abundance classes. The first class (42%) includes approximately 18 different sequences, the second class (22%) includes 64, and the third (36%), about 5300 different RNA sequences of average size, if we assume that 80% of total messages are under reaction.

In the case of the inducible cDNA, this probe can hybridize only with the messages from induced cells, and not with those from uninduced cells (Fig. 2B). The hybridization data show that these sequences belong to the scarce-class component. For the induced-cell messages, the third class comprises 57.2% of total cDNA (Table 3). If 20% of the sequences in this class were present in the induced cells, the numbers of these sequences could be $8695 \times 20 \div 57.2 = 3040$ sequences of average size. Therefore, the

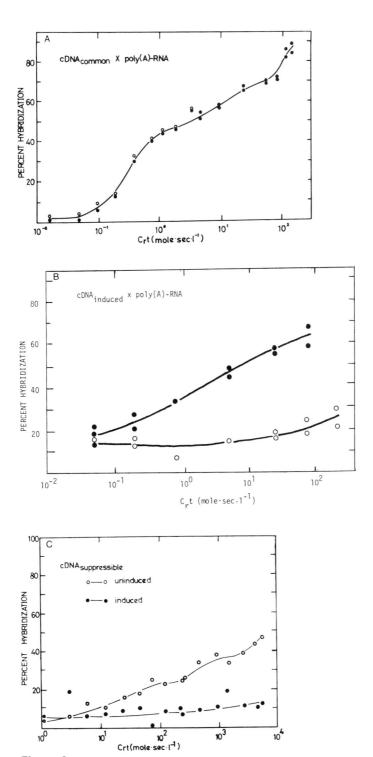

Figure 2
Hybridization of separated cDNA with messages. cDNA separated by hydroxylapatite was hybridized with induced cell messages (●) or uninduced cell messages (○). In A and B, poly(A)-containing RNA was used; in C, cytoplasmic RNA was used. (A) cDNA for common sequences; (B) cDNA for inducible sequences; (C) cDNA for suppressible sequences.

remaining (8832 − 3040 = 5792) sequences are assumed to be common sequences. The number of common sequences (5792) estimated from the hybridization data of inducible cDNA is essentially similar to that (5400) from the previous estimation.

In contrast to inducible cDNA, suppressible cDNA can hybridize only with the messages from uninduced cells. A similar estimate on the number of suppressible messages gave 4380 sequences of average size. Common sequences are 10097 − 4380 = 5717, again showing good coincidence with the previous estimation.

These results indicated that some 5400 to 5700 genes are constitutively expressed throughout the course of differentiation, approximately 3000 genes become expressed, and another 4400 genes become inactive during the induced differentiation. Assuming an average messenger molecular weight of 6×10^5 daltons, 0.31–0.34% of total mouse unique-sequence DNA is expressed constitutively, 0.17% becomes turned-on, and 0.26% becomes turned-off during the course of induced differentiation of Friend erythroleukemia cells. However, it should be stressed that this is only a rough approximation.

With addition of DMSO, the kinetics of globin mRNA accumulation in the cytoplasm as well as in the nuclei shows an exponential increase. In comparison to globin mRNA induction, the relationship of these mRNA sequences to the induced erythrodifferentiation was investigated. Cytoplasmic RNA was extracted from the C-10-6 cells with or without inducer at daily intervals. The hybridization was performed with four cDNA probes for globin, inducible, suppressible, and common mRNAs, respectively. As shown in Figure 3A, globin mRNA accumulated in accordance with the time after addition of DMSO with a certain lag phase that occurred because of underestimation of mRNA content under these conditions. Without inducer, however, no increase was observed. With the common-type sequences (Fig. 3D), the pattern of appearance of these messages was different from those of globin mRNA. The extent of hybridization stayed almost the same throughout the course of cultivation with or without inducer. These results are consistent with the previous observation that these messages are expressed constitutively in the cells.

When the inducible cDNA was used for the titration, a notable increase in the extent of hybridization was observed with time course of induction (Fig. 3C). It seems that appearances of these messages are not completely coordinated, but are somewhat related to the globin mRNA time course of accumulation. However, it is not certain whether the same messages are under the reaction in the course of induction. It is obvious that the appearance of these messages depends on the addition of inducer. The experiments on the suppressible messages demonstrated that these messages decreased with the addition of inducer. This decrease, which is observed only in induced cells, seems to show the first-order reaction on their degradation, suggesting a stochastic process. The data shown in Fig. 3E indicated that alteration of mRNA sequences is dependent on the induction of differentiation. It seems probable that alteration of messages is based on sequentially controlled gene expression.

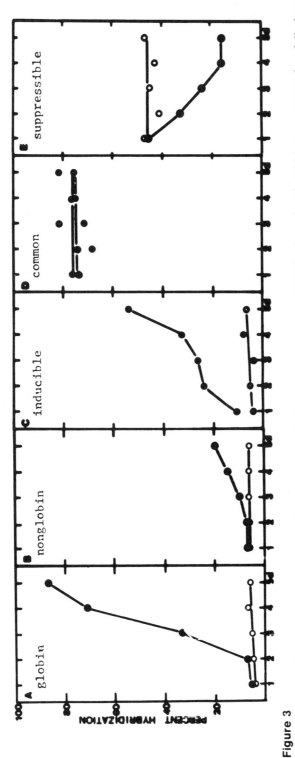

Figure 3
Titration of specific sequences during the differentiation of C-10-6 cells induced by DMSO. Cytoplasmic RNA was extracted at daily intervals from either cells induced with 1.5% (v/v) DMSO or uninduced cells. Titration was made with D_0t of 0.25 and RNA/cDNA of 10^5. (*A*) globin mRNA; (*B*) reticulocyte nonglobin mRNA; (*C*) inducible mRNA; (*D*) common mRNA; (*E*) suppressible mRNA. (●) RNA from induced cells; (○) RNA from uninduced cells.

Comparison of mRNAs of Induced Friend Erythroleukemia Cells with Those of Reticulocytes

Cultured Friend erythroleukemia cells were established from the proerythroblasts transformed by Friend virus infection. Although several of the biochemical and morphological markers examined suggest the similarity of induced differentiation to normal erythroid differentiation, these descriptions are fragmentary. Questions still remain as to whether this induced in vitro differentiation is completely comparable to the normal differentiation.

To examine these points, we compared the message sequences of differentiated Friend erythroleukemia cells with those of mouse reticulocytes which might be in a fully differentiated stage of erythroblastoid cells. In reticulocytes, most of the poly(A)-containing RNA is globin mRNA. When cDNA was made for mouse reticulocyte poly(A)-containing RNA and annealed with either the same template or globin mRNA, approximately 10% of total message consisted of nonglobin sequences. After separation of these nonglobin sequences as cDNA, the complexity of reticulocyte nonglobin mRNA was demonstrated (Fig. 4). Compared with that of globin mRNA, $C_r t_{1/2}$ of nonglobin mRNA is approximately five times slower, and therefore four species of mRNAs of average size other than globin mRNA are present in reticulocytes. This result agrees with the previous observation by Lodish and coworkers that reticulocyte lysates synthesize in vitro

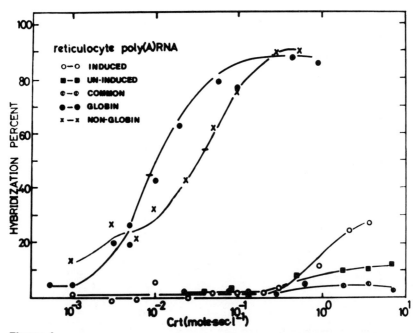

Figure 4
Hybridization of cDNA with mouse reticulocyte poly(A)-containing RNA. (o) cDNA for induced Friend-cell messages; (■) cDNA for uninduced Friend-cell messages; (Φ) cDNA for common messages; (●) cDNA for mouse globin messages; (X) cDNA for reticulocyte nonglobin messages.

six predominant proteins as well as globin (Lodish and Desalu 1973; Lodish and Small 1975). The mouse reticulocyte messages are less complex and different from those obtained in such mouse tissues as liver, brain, and kidney.

Hybridization of Friend-cell cDNA with reticulocyte messages showed that 20% of induced Friend erythroleukemia-cell messages are detectable in the reticulocyte in small quantity, whereas no messages of uninduced cells are detectable. The common cDNA from Friend cells could not hybridize with reticulocyte messages. These data all suggest that only inducible messages of Friend erythroleukemia cells remain in the reticulocyte, whereas common messages disappear during maturation of erythroid cells. This supports the idea that the patterns of gene expression in the induced erythrodifferentiation of Friend erythroleukemia cells resemble those observed in normal erythrodifferentiation.

When the cDNA for reticulocyte nonglobin messages was used for the quantitation of these sequences in induced Friend erythroleukemia cells, it was demonstrated that these reticulocyte messages are not expressed in these cells (Fig. 3B). This implies that the expression of globin and nonglobin messages present in reticulocytes is not coordinated in this induced differentiation. It cannot be clarified at present whether this is due to the programmed gene expression involved in the sequential process of cellular differentiation or to the unusual feature of this induced differentiation.

Recently, Orkin et al. (1975) showed that during the induction of Friend erythroleukemia cells the ratio of α- to β-globin mRNAs decreases from about 3.5 to about 1.0, the value characteristic of the normal reticulocyte. However, it seems that this differential timing of expression of α and β-globin genes is not a characteristic of normal erythroid-cell differentiation.

Although the normal erythropoietic systems are limited due to difficulties in obtaining large numbers of cells at specific stages of differentiation, they are required for comparison of the patterns of gene expression in both induced and normal erythrodifferentiation processes. Our results on the patterns of message sequences are summarized in Figure 5. The comparison of induced and uninduced Friend erythroleukemia cells suggests that substantial changes in gene expression accompany the transition of cells toward reticulocytes. These changes include the suppression of previously expressed genes as well as the induction of expression of certain repressed genes. The coordinated activation and inactivation of several thousands of genes occur only in response to the addition of inducer. How the inducer directly or indirectly affects the regulation of the coordinated expression of genes remains an important question in our understanding of the mechanism of induction of differentiation.

The comparison of Friend erythroleukemia cells with reticulocytes demonstrated a remarkable difference of messages. Reticulocytes are at too late a stage for comparison with Friend leukemia cells, since this stage might be a final step in terms of message expression through nuclear inactivation and enucleation. To reach the reticulocytic stage, the induced Friend erythroleukemia cells require further expression of certain additional genes following the complete suppression of common messages. More attention must be given to the relative stability of differentiation-related messages to in-

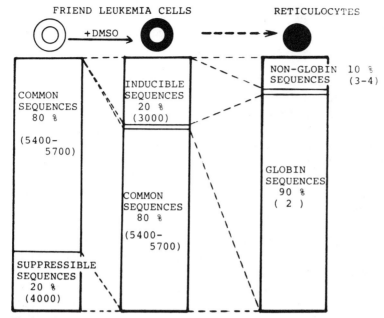

Figure 5
Comparison of poly(A)-containing mRNA composition in Friend erythroleukemia cells and mouse reticulocytes.

crease our understanding of further processes of erythrodifferentiation as well as sequentially controlled gene expression.

Acknowledgments

Thanks are due to Miss M. Aida and Mr. T. Matsugi, respectively, for the harvest of cells and the animal care. This study was partly supported by grants for cancer research from the Ministry of Education, Science and Culture, Japan; by a grant from the Foundation for Promotion of Cancer Research; and by a grant for research in natural science from the Mitsubishi Foundation. AMV reverse transcriptase was a gift from Dr. J. W. Beard, Life Science Research Laboratory, St. Petersburg, Florida, within the Virus Cancer Program, National Cancer Institute, U.S.A.

REFERENCES

Birnie, G. D., E. Macphail, B. D. Young, M. J. Getz, and J. Paul. 1974. The diversity of the messenger RNA population in growing Friend cells. *Cell Differ.* **3:** 221.

Bishop, J. O., J. G. Morton, M. Rosbash, and M. Richardson. 1974. Three abundance classes in HeLa cell messenger RNA. *Nature* **250:** 199.

Borum, T. W., M. D. Scharff, and E. Robbins. 1967. Rapidly labeled polysome-associated RNA having the properties of histone messenger. *Proc. Natl. Acad. Sci. U.S.A.* **58:** 1977.

Boyer, S. H., K. D. Wuu, A. N. Noyes, R. Young, W. Scher, C. Friend, H. D. Preisler, and A. Bank. 1972. Hemoglobin biosynthesis in murine virus-induced leukemic cells in vitro: Structure and amounts of globin chains produced. *Blood* **40:** 823.

Buckingham, M. E., A. Cohen, and F. Gros. 1976. Cytoplasmic distribution of pulse-labeled poly(A)-containing RNA, particularly 26 S RNA, during myoblast growth and differentiation. *J. Mol. Biol.* **103:** 611.

Chan, L.-N. L. 1976. Transport of globin mRNA from nucleus into cytoplasm in differentiating embryonic red blood cells. *Nature* **261:** 157.

Davidson, E. H., and B. R. Hough. 1971. Genetic information in oocyte RNA. *J. Mol. Biol.* **56:** 491.

Ebert, P. and Y. Ikawa. 1974. Induction of δ-aminolevulinic acid synthetase during erythroid differentiation of cultured leukemia cells. *Proc. Soc. Exp. Biol. Med.* **146:** 601.

Firtel, R. A. 1972. Changes in the expression of single-copy DNA during development of the cellular slime mold *Dictyostelium discoideum*. *J. Mol. Biol.* **66:** 363.

Friend, C., M. C. Patuleia and E. de Harven. 1966. Erythrocytic maturation in vitro of murine (Friend) virus-induced leukemic cells. *Natl. Cancer Inst. Monogr.* **22:** 505.

Friend, C., W. Scher, J. G. Holland, and T. Sato. 1971. Hemoglobin synthesis in murine virus-induced leukemic cells *in vitro:* Stimulation of erythroid differentiation by dimethyl sulfoxide. *Proc. Natl. Acad. Sci. U.S.A.* **68:** 378.

Furusawa, M., Y. Ikawa, and H. Sugano. 1971. Development of erythrocyte membrane-specific antigen(s) in clonal cultured cells of Friend virus-induced tumor. *Proc. Jpn. Acad.* **47:** 220.

Gilmour, R. S., P. R. Harrison, J. M. Windass, N. A. Affara, and J. Paul. 1974. Globin messenger RNA synthesis and processing during haemoglobin induction in Friend cells. I. Evidence for transcriptional control in clone M2. *Cell Diff.* **3:** 9.

Glisin, V., R. Crkvenjakov, and C. Byus. 1974. Ribonucleic acid isolated by cesium chloride centrifugation. *Biochemistry* **13:** 2633.

Grouse, L., M. D. Chilton, and B. J. McCarthy. 1972. Hybridization of ribonucleic acid with unique sequences of mouse deoxyribo nucleic acid. *Biochemistry* **11:** 798.

Gurdon, J. B., C. Lane, H. Woodland, and G. Marbaix. 1971. Use of frog eggs and oocytes for the study of messenger RNA and its translation in living cells. *Nature* **233:** 177.

Harrison, P. R., D. Conkie, N. Affara, and J. Paul. 1974a. In situ localization of globin messenger RNA formation. I. During mouse fetal liver development. *J. Cell Biol.* **63:** 402.

Harrison, P. R., R. S. Gilmour, N. A. Affara, D. Conkie, and J. Paul. 1974b. Globin messenger RNA synthesis and processing during haemoglobin induction in Friend cells. II. Evidence for post-transcriptional control in clone 707. *Cell Diff.* **3:** 23.

Hawkins, W. D. and S. B. Krantz. 1975. In vitro expression of erythroid differentiation induced by Friend polycythaemic virus. *Nature* **253:** 73.

Humphries, S., J. Windass, and R. Williamson. 1976. Mouse globin gene expression in erythroid and non-erythroid tissues. *Cell* **7:** 267.

Ikawa, Y., M. Aida, and Y. Inoue. 1976. Isolation and characterization of high and low differentiation-inducible Friend leukemia lines. *Gann* **67:** 767.

Ikawa, Y., H. Sugano, and M. Furusawa. 1973. Erythrocyte membrane specific antigens in Friend virus induced leukemia cells. *Bibl. Haematol.* **39:** 955.

Kabat, D., C. C. Sherton, L. H. Evans, R. Bigley, and R. D. Koler. 1975. Synthesis of erythrocyte-specific poteins in cultured Friend leukemia cells. *Cell* **5**: 331.

Kameji, R., M. Obinata, Y. Natori, and Y. Ikawa. 1977. Induction of globin gene expression in cultured erythroleukemia cells by butyric acid. *J. Biochem.* **81**: 1901.

Leder, A. and P. Leder. 1975. Butyric acid, a potent inducer of erythroid differentiation in cultured erythroleukemia cells. *Cell* **5**: 319.

Liao, S.-K. and A. A. Axelrad. 1975. Erythropoietin-independent erythroid colony formation *in vitro* by hemopoietic cells of mice infected with Friend virus. *Int. J. Cancer* **15**: 467.

Lodish, H. F. and O. Desalu. 1973. Regulation of synthesis of non-globin proteins in cell-free extracts of rabbit reticulocytes. *J. Biol. Chem.* **248**: 3520.

Lodish, H. F. and B. Small. 1975. Membrane proteins synthesized by rabbit reticulocytes. *J. Cell Biol.* **65**: 51.

Orkin, S. H., D. Swan, and P. Leder. 1975. Differential expression of α- and β-globin genes during differentiation of cultured erythroluekemia cells. *J. Biol. Chem.* **250**: 8753.

Ostertag, W., H. Melderis, G. Steinheider, N. Kluge, and S. Dube. 1972. Synthesis of mouse haemoglobin and globin mRNA in leukaemic cell cultures. *Nat. New Biol.* **239**: 231.

Palmiter, R. D. 1975. Quantitation of parameters that determine the rate of ovalbumin synthesis. *Cell* **4**: 189.

Reuben, R. C., R. L. Wife, R. Breslow, R. A. Rifkind, and P. A. Marks. 1976. A new group of potent inducers of differentiation in murine erythroleukemia cells. *Proc. Natl. Acad. Sci. U.S.A.* **73**: 862.

Ross, J., Y. Ikawa, and P. Leder. 1972. Globin messenger-RNA induction during erythroid differentiation of cultured leukemia cells. *Proc. Natl. Acad. Sci. U.S.A.* **69**: 3620.

Ross, J., J. Gielen, S. Packman, Y. Ikawa, and P. Leder. 1974. Globin gene expression in cultured erythroleukemia cells. *J. Mol. Biol.* **87**: 697.

Ryffel, G. U. and B. J. McCarthy, 1975. Complexity of cytoplasmic RNA in different mouse tissues measured by hybridization of polyadenylated RNA to complementary DNA. *Biochemistry* **14**: 1379.

Sassa, S. 1976. Sequential induction of heme pathway enzymes during erythroid differentiation of mouse Friend virus-infected cells. *J. Exp. Med.* **143**: 305.

Skoultchi, A., and P. R. Gross. 1973. Maternal histone messenger RNA: Detection by molecular hybridization. *Proc. Natl. Acad. Sci. U.S.A.* **70**: 2840.

Williams, J. G. and S. Penman, 1975. The messenger RNA sequences in growing and resting mouse fibroblasts. *Cell* **6**: 197.

Section 3

GRANULOCYTE AND MONOCYTE
DIFFERENTIATION AND REGULATION

Introduction

M. A. S. Moore

Memorial Sloan-Kettering Cancer Center
New York, New York 10021

Since the advent of in vitro cloning techniques for detection of normal or leukemic granulocyte-macrophage progenitor cells (colony-forming units—culture, CFU-C), an extensive literature has evolved indicating heterogeneity of the responsive cell population and of the stimulatory or inhibitory factors influencing CFU-C proliferation and differentiation. Biophysical separation procedures and, more recently, cell-surface antigenic characterization, have revealed CFU-C subpopulations committed to neutrophil, granulocyte, mixed granulocyte-macrophage, macrophage, or eosinophil granulocyte differentiation. The interrelationships of these subpopulations of CFU-C, their differential response to various species of stimulatory factors, and their kinetics in steady-state or perturbed hematopoiesis await further investigation.

In vitro proliferation of CFU-C is absolutely dependent on provision of a suitable source of activities variously termed colony-stimulating factor (CSF), colony-stimulating activity (CSA), or macrophage-granulocyte inducer (MGI). In the mouse, CSF is contained in or elaborated by a wide variety of tissues. Sources subjected to the most extensive functional and biochemical characterization have been endotoxin-treated mouse-lung conditioned medium, L-cell conditioned medium, and pokeweed-mitogen-stimulated spleen conditioned medium. Due to species specificity, human CFU-C require a human source of CSF, and that source in most frequent use is one elaborated by human leukocytes, specifically, by the monocyte fraction. Other sources of human CSF that are being subjected to extensive purification are phytohemagglutinin-stimulated lymphocyte conditioned medium, placental conditioned medium, embryonic-kidney conditioned medium, and human urine. Interestingly, the latter source lacks the capacity to stimulate human colony formation but provides a potent stimulus for mouse colony formation. The necessity of purifying and characterizing CSFs from a variety of sources is due to the documented functional and biochemical heterogeneity of CSF. Various species of CSF have been shown to stimulate predominantly macrophage colony formation (M-CSF), whereas others stimulate mixed granulocyte-macrophage (GM-CSF), neutrophil-granulocyte (NG-CSF), or eosinophil (Eo-CSF) colony formation.

Recognition that monocytes and macrophages provide a major source of CSF and that CSF promotes the production of monocytes and can induce the proliferation of macrophages implies the existence of a positive feedback control. The existence of negative feedback mechanisms regulating granulopoiesis and monocyte-macrophage production has been suggested both by in vivo and in vitro studies. There are inherent limitations to in vitro assessment of inhibitory activities of physiological significance, but there is increasing evidence that products of mature granulocytes may in some way mediate negative feedback and that defined activities such as interferon and prostaglandin E may play an important role in hematopoietic regulation.

The development of in vitro systems supporting the proliferation and differentiation of various types of hematopoietic cells has made it possible to study the regulation of leukemic cells. In vitro analysis of human myeloid leukemic cells has clearly documented that these neoplastic cells are dependent rather than autonomous with respect to their requirement for CSF for in vitro cloning. More controversial is the question of induction of differentiation of myeloid leukemic cells following exposure to appropriate stimulatory or inductive agents. As will be discussed in this section, the leukemic cell phenotype is not irreversibly frozen at a particular stage of maturation but can be induced to express more differentiated functions. Further insight into the regulatory responsiveness of myeloid leukemic cells may provide alternative therapeutic approaches for selective suppression or ablation of leukemic clones.

Regulation of Hematopoietic Differentiation and Proliferation by Colony-Stimulating Factors

A. W. Burgess, D. Metcalf, and S. Russell

Cancer Research Institute, Walter and Eliza Hall Institute of Medical Research
Melbourne, Victoria 3050, Australia

The use of semisolid cultures for the analysis of granulocyte-macrophage (GM) (Bradley and Metcalf 1966; Pluznik and Sachs 1966) and erythroid (E) progenitor cells (Stephenson et al. 1971; McLeod et al. 1974) has accelerated our understanding of many of the cellular aspects of hematopoiesis. Even from the earliest observations (Bradley and Metcalf 1966), it was apparent that clonal proliferation and differentiation of the GM-progenitors in vitro were directly dependent on the presence of a specific glycoprotein which was called the colony-stimulating factor (CSF). This glycoprotein has been detected in many biological fluids such as human urine (Stanley and Metcalf 1969) and endotoxin mouse serum (Metcalf 1971), in tissue extracts (Sheridan and Stanley 1971), and in culture fluids (Sheridan and Metcalf 1973). There is a dearth of information on the mode of action of GM-CSF. A major difficulty in studying some molecular aspects of hematopoietic differentiation has been the lack of purified regulator molecules. The purification both of GM-CSF and erythropoietin (which stimulates erythropoiesis both in vivo and in vitro) has taken a considerable time. Apart from the low concentrations of these proteins in biological fluids, purification studies have been hindered because of large losses of biological activity. However the observation that detergents (Burgess et al. 1977) such as Triton X-100 or polymers (Stanley et al. 1975) such as polyethylene glycol (Stanley et al. 1976) prevent the loss of purified GM-CSF has allowed the purification of GM-CSF from mouse-lung conditioned medium (MLCM) (Burgess et al. 1977) and mouse L-cell conditioned medium (Stanley and Heard 1977). Erythropoietin has also been purified recently from human urine (Miyake et al. 1977) and from plasma taken from anemic sheep (Goldwasser and Kung 1971).

Although the functional and structural characteristics of these molecules remain to be investigated in detail, it has already emerged that there is some heterogeneity with regard to molecular weight and charge for GM-CSF (Stanley et al. 1976; Tsuneoka and Shikita 1977) and for erythropoietin

(Goldwasser 1975). Both erythropoietin and asialo-erythropoietin appear to stimulate erythropoiesis in vitro (Goldwasser et al. 1974). Affinity chromatography of GM-CSF from MLCM and mouse L-cell conditioned medium on concanavalin A-Sepharose has shown directly that there is some heterogeneity with respect to carbohydrate moieties (Stanley et al. 1976; Burgess et al. 1977).

Analysis of the mechanism by which these molecules stimulate their specific progenitor cells (or mature cells [Burgess and Metcalf, 1977b]) requires a structural analysis of the molecule to determine the functional amino acids and modification of the regulators to produce radiolabeled derivatives. Previous studies have reported considerable losses of biological activity when an attempt was made to modify both GM-CSF (Stanley et al. 1975; Burgess and Metcalf 1977a, 1977c) and erythropoietin (Goldwasser and Kung 1971; Goldwasser 1975). This report describes some analytical experiments that characterize the different molecular species of GM-CSF present in MLCM, the microheterogeneity associated with a particular species of GM-CSF purified from MLCM, and the difficulties associated with the production of radiolabeled GM-CSF.

Characterization of the molecules that regulate the proliferation and differentiation of the progenitors for other hematopoietic cells such as megakaryocytes (Metcalf et al. 1975) and eosinophils (Metcalf et al. 1974) has not progressed to the same extent as has that for GM-CSF or erythropoietin. There are glycoproteins present in pokeweed-mitogen-stimulated spleen conditioned medium (SCM) (Metcalf et al. 1974) which stimulate GM, megakaryocyte (MEG), eosinophil (Eo), and E colony formation in vitro. These molecules appear to have similar molecular weights (Metcalf et al. 1978) and all contain terminal glucose or mannose carbohydrate moieties (Metcalf et al. 1978). However, the work described in this report indicates that those CSFs specific for a particular cell type have different charges. The possibility of purifying CSFs specific for the stimulation of particular classes of progenitor cells is discussed with reference to the microheterogeneity and the availability of materials from which these molecules may be prepared.

MATERIALS AND METHODS

Conditioned media were prepared as described previously: MLCM (Sheridan and Metcalf 1973), yolk-sac conditioned medium (YSCM) (Johnson and Metcalf 1978; G. R. Johnson and A. W. Burgess, 1978), and pokeweed-mitogen-stimulated SCM (Metcalf et al. 1974).

GM-CSF was purified from MLCM as described by Burgess et al. (1977). Solutions containing purified GM-CSF contained polyethylene glycol (0.005%, w/v) and azide (0.02%, w/v). Purified GM-CSF from MLCM migrated as a single band on polyacrylamide-gel (15%) electrophoresis in the presence and absence of sodium dodecyl sulfate (SDS) (Burgess et al. 1977) and had a specific activity of 7×10^7 colonies/mg of protein. The protein fraction passing through concanavalin A-Sepharose was processed further by gel filtration on Sephadex G-100 (S) (2.5×10^2

cm), equilibrated with 0.01 M sodium phosphate buffer, pH 7.3, containing 0.15 M sodium chloride. The buffer flow rate was maintained at 5 ml/hour. Fractions (5 ml) were collected and aliquots (0.1 ml) assayed for GM-CSF by the semisolid agar technique (Metcalf 1970). The molecular weights of the different molecular species of GM-CSF were calculated from the logarithmic relationship between the molecular weights (Andrews 1964) of bovine serum albumin, 69,000; hemoglobin, 64,000; ovalbumin, 45,000; α-chymotrypsinogen, 23,500; myoglobin, 17,800; and cytochrome C, 12,500; and of their elution volumes.

Preparative polyacrylamide-gel electrophoresis (Davis 1964) at pH 8.6 was performed with 1.5×10 cm running gels containing 15% acrylamide monomer overlaid with a stacking gel equal in volume to the sample volume. Electrophoretic mobilities were measured relative to bromophenol blue or xylene cyanol FF tracking dyes.

Pure GM-CSF was iodinated in the presence of dimethylsulfoxide (DMSO) by modification of the chloramine-T procedure suggested by Stagg et al. (1970) and the iodine-monochloride procedure (Roholt and Pressman 1972).

Biosynthetically labeled GM-CSF was separated from the labeled amino acids using a Sephadex G-25 column (0.9×8 cm) equilibrated with 0.2 M sodium acetate buffer, pH 5.0. ^{125}I-labeled GM-CSF was separated from ^{125}I$^-$ in the same way. The radioactive proteins eluted at the void volume and were applied to a column containing concanavalin A-Sepharose (Pharmacia) (0.9×8 cm) equilibrated with the same buffer. The radiolabeled glycoproteins bound to the concanavalin A-Sepharose and were recovered by addition of α-methyl-D-glucopyranoside (0.05 M) to the elution buffer. After concentration (to 0.5 ml) over an Amicon YM-10 ultrafiltration membrane, the radiolabeled GM-CSF was chromatographed on Sephadex G-100 (S) (1.5×18 cm), equilibrated with 0.01 M sodium phosphate buffer, pH 7.3, containing 0.15 M sodium chloride. The ^{125}I-labeled GM-CSF eluted from the Sephadex G-100 (S) between 25 and 30 ml.

Analysis of amino acids was performed on purified GM-CSF (15 μg) from MLCM. The final electrophoresis was performed with a Tris-borate buffer system (Neville 1971) instead of the usual Tris-glycine buffer system reported by Davis (1964). GM-CSF was hydrolyzed at 115°C for 24 hours with p-toluene sulphonic acid (4 M) (Liu and Chang 1971).

SCM was concentrated 20 times by ultrafiltration with an H1PD-10 hollow-fiber membrane. The concentrate was washed with 5 volumes of 0.01 M sodium phosphate buffer, pH 7.3, containing 0.15 M sodium chloride. A portion of this concentrate (3 ml) was chromatographed on wheat-germ agglutinin (WGA)-Sepharose (0.9×8 cm) equilibrated with the same buffer. After the initial protein had passed through the column, N-acetyl-D-glucosamine (0.1 M) was added to the buffer to elute the glycoproteins containing terminal N-acetyl-D-glucosamine carbohydrate moieties.

Isoelectric focusing was performed in thin-layer granulated gel beds (12×30 cm) according to the method of Radola (1973). Ampholines were purchased from LKB and used at a final concentration of 2% (v/v). Samples (5 ml) were applied to 70-ml beds of Sephadex G-75 or Ultrodex (LKB) evaporated to within 25% of their crackpoints. The system was focused overnight at 400 v and for a further 3 hours at 1 kv. After the system had been focused, the pH gradient was determined with an Ingold microsurface electrode calibrated with potassium hydrogen phthalate buffer, pH 4.01, and a Radiometer standard phosphate buffer (pH 6.8). The gel was sliced into 30 fractions for elution with 0.01 M sodium phosphate buffer, pH 7.3, containing 0.15 M sodium chloride. Before assay for GM-, E- or MEG-CSF, each fraction was deionized with a mixed-bed ion-exchange resin AG-501-X8 (D) (Baumann and Chrambach 1975).

CFU were assayed in 1-ml cultures containing 75,000 C57BL marrow cells in Dulbecco's modified Eagle's medium (DMEM) containing fetal calf serum (20% v/v) and 0.3% agar, or in cultures containing 20,000 12-day CBA fetal liver cells in DMEM with heat-inactivated human plasma (20%) and 0.3% agar (Johnson and Metcalf 1977). Colony counts were made after 7 days of incubation at 37°C in a fully humidified atmosphere containing 10% CO_2 (Metcalf 1970).

RESULTS

Different Molecular Species of GM-CSF

GM-CSF from MLCM eluted as a broad peak from DEAE-Sepharose (Burgess et al. 1977), suggesting that there may be some charge heterogeneity associated with the molecule. However, the presence of high concentrations of albumin and other contaminating proteins were contributing factors to the shape of the colony-stimulating activity (CSA) profile. During the next step in the purification scheme (concanavalin A-Sepharose) approximately 70 to 80% of the GM-CSF bound to the affinity column and could be specifically eluted with α-methyl-D-glucopyranoside. The specifically eluted GM-CSF had a high specific activity ($> 2 \times 10^6$ colonies mg of protein) compared to the GM-CSF passing through the column (specific activity, 2×10^5 colonies mg of protein). When the GM-CSF passing through the concanavalin A-Sepharose column was chromatographed on Sephadex G-100 (S), two peaks of biological activity were observed which corresponded to apparent molecular weights of 65,000 and 29,000 (Fig. 1). The position of lower molecular weight peak was coincident with the gel filtration elution volume of the GM-CSF which bound to concanavalin A-Sepharose (Fig. 1). Thus, there appeared to be at least three distinct molecular species present in MLCM: (1) GM-CSF which bound to concanavalin A-Sepharose and had an apparent molecular weight of 29,000 of gel filtration; (2) GM-CSF with the same apparent molecular weight but which did not bind to concanavalin A-Sepharose, and (3) a higher molecular weight species (65,000) of GM-CSF which did not bind to concanavalin A-Sepharose.

There appeared to be microheterogeneity within the GM-CSF which bound to concanavalin A-Sepharose. Although the molecular weight distribution of this GM-CSF was narrow, some heterogeneity was apparent during polyacrylamide-gel electrophoresis at pH 8.6. GM-CSF after the penultimate step during the purification (Burgess et al. 1977) (gel filtration on Ultrogel AcA44) yielded a broad band of CSA. Even on 15% gels,

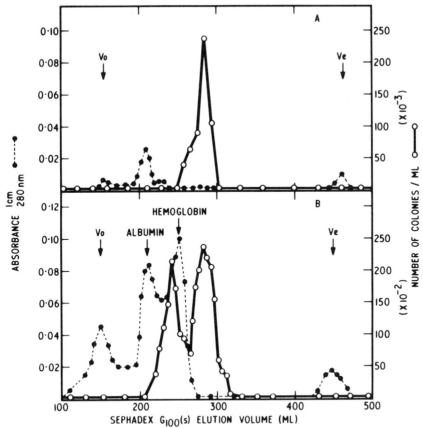

Figure 1
Sephadex G-100 (S) (2.5 × 100 cm) gel filtration in 0.01 M sodium phosphate, pH 7.3, containing 0.15 M sodium chloride of (*A*) GM-CSF (from MLCM) eluted from concanavalin A-Sepharose with α-methylglucoside (0.05 M); and (*B*) GM-CSF from MLCM which passed through concanavalin A-Sepharose. (o——o) GM-CSF. (●-----●) Protein absorbance at 280 mm.

activity was detected between electrophoretic mobilities 0.4 and 0.6 relative to xylene cyanol FF (Fig. 2). Contaminating proteins such as albumin migrated as sharp zones usually less than 0.03 relative mobility units in width. The spread of GM-CSF on electrophoresis was not due to the presence of Triton X-100 (which increased the recovery of GM-CSF from the gel two to three times), as identical experiments in the absence of Triton X-100 also indicated that GM-CSF from MLCM migrated as a broad band on electrophoresis (Fig. 2). Fractions from three positions (A, trailing edge; B, center; and C, leading edge) across the electrophoretic profile for GM-CSF shown in Figure 2 were pooled and rerun separately under the same conditions as prevailed in the original electrophoresis (Fig. 3). The electrophoretic mobilities and relative yield of GM-CSF from each of the pools (A, B, and C, Fig. 3) corresponded to their positions in the initial electrophoresis. To purify GM-CSF, we had to discard the CSA corresponding to pool A of Figure 3, as analytical gel electrophoresis in

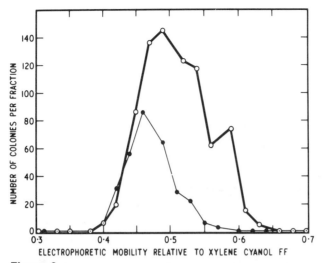

Figure 2
Polyacrylamide-gel electrophoresis of GM-CSF from the penultimate step of the purification from MLCM. A stacking gel (3 ml) was used over a 2 × 10 cm polyacrylamide gel (15%) polymerized in the presence of (o——o) and in the absence of (●——●) Triton X-100 (0.01%, v/v). Slices (1.5 mm) were crushed into 0.01 M sodium phosphate buffer (2 ml) containing fetal calf serum (5%, v/v) and assayed for GM-CSF using the semisolid agar culture system (Metcalf 1970).

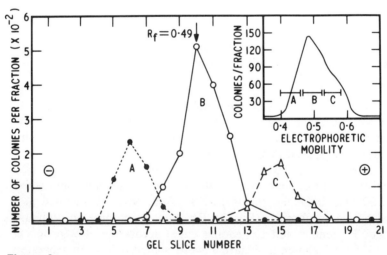

Figure 3
Electrophoresis of fractions from three different positions across the GM-CSF polyacrylamide-gel electrophoresis profile in the presence of Triton X-100 (0.01%, v/v) shown in Figure 2. The polyacrylamide-gel electrophoresis profile from each pool (A, B, C of insert) was determined separately. (A) Trailing edge (●----●); center (o——o); leading edge (△-----△).

the presence of SDS indicated that contaminating proteins were present. Pure GM-CSF was obtained by concentrating the fractions corresponding to pools B and C of Figure 3. When analyzed by electrophoresis in the presence of SDS, a single protein and activity band was observed which migrated with an apparent molecular weight of 23,000.

Amino Acid Composition

To determine the amino acid composition of GM-CSF, we modified the final purification step slightly. Glycine in the electrophoretic buffer system was not easily removed before analysis and a Tris-borate buffer system (Neville 1971) was used. The electrophoretic mobility of the GM-CSF did not appear to change. The amino-acid composition obtained after a 24-hour hydrolysis in the presence of p-toluene sulphonic acid (4 M) (Table 1) indicated that there were twice as many acidic residues as basic ones. The content of glycine appeared to be unusually high. Although tryptophan was not detected, the presence of Tris buffer interfered with the analysis of this amino acid, so that further modifications to the electrophoretic system will be necessary before this amino acid can be determined quantitatively. The only carbohydrate moieties analyzed were the hexosamines of which glucocamine, mannosamine, and galactosamine were detected. Further work is needed to define the exact ratio of carbohydrate to protein but these initial results suggest that it will be less than 1:10.

Radiolabeling of GM-CSF

Initially 100–200 μg of GM-CSF were purified from 4 liters of MLCM obtained from 800 C57BL mice (Burgess et al. 1977). This amount of protein was insufficient for a detailed structural analysis of the molecule. Even with the most sensitive amino-acid sequencing techniques, we could not have sequenced a protein of molecular weight 23,000 with less than 500 μg of protein. However, GM-CSF from MLCM was prepared in serum-

Table 1
Preliminary Amino Acid Analysis of GM-CSF from MLCM

Amino acid	Residues per 100[a]	Amino acid	Residues per 100
Lysine	4.8	Alanine	6.1
Histidine	2.2	1/2 Cystine	1.0
Arginine	3.4	Valine	4.9
Aspartate	7.3	Methionine	0.9
Threonine	6.0	Isoleucine	2.8
Serine	9.0	Leucine	7.0
Glutamate	10.3	Tyrosine	2.8
Proline	5.9	Phenylalanine	3.9
Glycine	21.7	Tryptophan	not detected
		Carbohydrate:Hexosamines	2.0

[a] As determined by 24-hour hydrolysis at 115°C in vacuo with p-toluene sulphonic acid (4 M).

free medium and its production was dependent on protein synthesis (Table 2). When DMEM was used to culture the lungs, only a small amount of ^{3}H was incorporated from amino acids into protein; however, when the amino acids were omitted from the medium, almost 2% of the initial radioactivity was incorporated into protein (Table 2). Only a small proportion of the ^{3}H-labeled protein appeared to be associated with glycoproteins (as defined by binding to concanavalin A-Sepharose) such as GM-CSF (Table 1). It was reported recently that lithium chloride increased the concentration of GM-CSF in MLCM (Harker et al. 1977). Although our data confirmed this observation, we found no stimulation of ^{3}H-labeled amino-acid incorporation into the acid-precipitable proteins (Table 3) or glycoproteins of MLCM. The level of incorporation of ^{3}H-labeled amino acids into "GM-CSF-like" proteins was rather low (Table 4) but was detectable even in highly purified samples of GM-CSF (Table 4). This level of labeling would need to be increased at least tenfold to produce ^{3}H-labeled GM-CSF suitable for sequencing.

Considerably higher levels of radioactivity must be incorporated into GM-CSF before its biological distribution or binding to its target cells can be studied. This may be achieved by modification of purified GM-CSF with methods that incorporate ^{125}I. Although the chloramine-T iodination method (Hunter and Greenwood 1962) destroys all of the biological activity associated with GM-CSF (Burgess and Metcalf 1977c), the inclusion of DMSO into the buffers used for the iodination (Stagg et al. 1970) allowed some biological activity to be retained (Table 5). This analysis of the ^{125}I products (after chloramine-T iodination) (Stagg et al. 1970) by polyacrylamide-gel electrophoresis in the presence of SDS indicated that considerable breakdown and aggregation of the pure molecule had occurred as a result of the iodination. It was necessary to reprocess the ^{125}I-labeled products by affinity chromatography with concanavalin A-Sepharose (Fig. 4) before most

Table 2

Incorporation of Tritium-Labeled Amino Acids into the Proteins of MLCM

Medium	Additions	Biological activity[a] (colonies/mg of protein) ($\times 10^{-4}$)	Incorporation of ^{3}H-labeled amino acids[c] into	
			total protein (%)	glycoprotein[b] (%)
Serum-free Dulbecco's modified Eagle's	no free amino acids	3.0 ± 1.5	1.9 ± 0.5	0.020 ± 0.001
	essential amino acids (1 mM)	3.7 ± 1.2	0.2 ± 0.1	0.001 ± 0.001
	essential amino acids (10 μM)	3.1 ± 1.6	1.9 ± 0.3	0.017 ± 0.003
	puromycin (100 μg/ml)	0.4 ± 0.3	0.7 ± 0.1	0

[a] The biological activity was estimated using 0.1 ml of a 1:20 dilution of the conditioned medium in the standard 1-ml semisolid agar system.

[b] As determined by ^{3}H binding and elution from concanavalin A-Sepharose.

[c] ^{3}H amino acids were added to 1 μCi/ml using a tritiated amino-acid mixture (TRK 440, The Radiochemical Centre).

Table 3
Effect of Lithium Chloride on the Incorporation of ^{3}H-Labeled Amino Acids into MLCM Proteins

Lithium chloride[a] concentration (mM)	Biological activity[b] (colonies/mg of protein) ($\times 10^{-4}$)	Acid precipitable[c] ^{3}H (cpm/ml) ($\times 10^{-4}$)
0	3.6 ± 1.3	2.3 ± 0.3
10	4.6 ± 1.1	2.2 ± 0.4
30	8.2 ± 2.0	2.3 ± 0.4
50	5.5 ± 2.6	2.2 ± 0.4

[a] Lithium chloride was added to serum-free Dulbecco's modified Eagle's medium (free of amino acids).
[b] The biological activity was estimated by assaying 0.1 ml of a 1:20 dilution of the conditioned medium in the standard 1-ml semisolid clonal assay.
[c] Conditioned medium (1 ml) was chilled on ice and an equal volume of cold trichloroacetic acid (TCA, 20% w/v) added. After 1 hour on ice, the precipitate was harvested on a glass filter and washed with 10% TCA and 70% ethanol. The ^{3}H was determined by liquid scintillation count.

of the ^{125}I counts appeared to be associated with biologically active GM-CSF. Although most of the ^{125}I and at least one-half of the biological activity failed to bind to concanavalin A-Sepharose, the protein eluted with α-methylglucoside (Fig. 4) contained only two radioactive species when it was analyzed by polyacrylamide-gel electrophoresis in the presence of SDS; the majority of the ^{125}I radioactivity (70%) was associated with protein migrating at 23,000 mol. wt., but a significant proportion (30%) was also associated with a higher molecular-weight species. The lower molecular-weight species of ^{125}I–GM-CSF was prepared by gel filtration from the α-methylglucoside eluate from concanavalin A-Sepharose with Sephadex G-100 (S) (Table 5).

Iodine monochloride was also used to radiolabel GM-CSF with ^{125}I. There was a significant increase in the recovery of biological activity, but the level of incorporation of ^{125}I into protein was also lower (Table 5). Polyacrylamide-gel electrophoresis indicated that there were fewer break-

Table 4
Incorporation of ^{3}H-Labeled Amino Acids into Protein at Different Stages of the Purification of GM-CSF from MLCM

Stage	Protein fraction	Incorporation of ^{3}H-labeled amino acids into protein (%)
I – Conditioned medium	Total	1.9
II – Concanavalin A-Sepharose	α-methylglucoside eluate	0.02
III – Sephadex G-100(S)	20–30,000 mol. wt.	0.01
IV – Polyacrylamide-gel electrophoresis	Mobility relative to bromophenol blue 0.24–0.34	0.001

Table 5
Incorporation of ^{125}I into Purified GM-CSF from MLCM

Purification step	Percentage recovery of biological activity		Percentage incorporation of ^{125}I into GM-CSF	
	Chloramine T[a]	Iodine-monochloride[b]	Chloramine T[a]	Iodine-monochloride[b]
Sephadex G 25:void volume	18	29	8	2.5
Concanavalin A-Sepharose:α-methyl-glucoside eluate	9	17	0.6	0.4
Sephadex G-100 (S): 20–30,000 mol. wt.	6	13	0.3	0.15

[a] The procedure of Stagg et al. (1970) was followed.
[b] The procedure of Roholt and Pressman (1972).

down and aggregation products, but again ^{125}I was associated with two molecular-weight species on polyacrylamide-gel electrophoresis in the presence of SDS. ^{125}I-labeled GM-CSF specifically eluted from concanavalin A-Sepharose was purified by gel filtration on Sephadex G-100 (S) to yield ^{125}I-labeled GM-CSF which migrated as a single molecular weight species (23,000) on analytical gel electrophoresis in the presence of SDS.

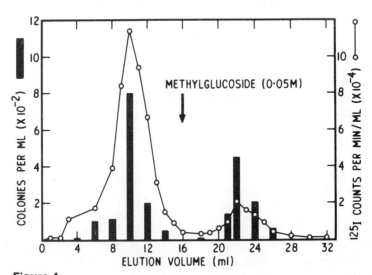

Figure 4
Affinity chromatography of ^{125}I-labeled GM-CSF using concanavalin A-Sepharose. Pure GM-CSF was radioiodinated with iodine monochloride and the ^{125}I associated with protein isolated after gel filtration on Sephadex G-25. (■) GM-CSF was determined by bioassay using semisolid agar cultures (1 ml) of C57BL mouse bone-marrow cells (75,000). (o———o) ^{125}I counts per minute per ml.

Isoelectric Focusing

Pure GM-CSF from MLCM (Burgess et al. 1977), GM-CSF from MLCM which did not bind to concanavalin A-Sepharose, and ^{125}I-labeled GM-CSF eluted from concanavalin A-Sepharose with α-methylglycopyranoside were analyzed separately by isoelectric focusing in thin-layer granulated gels (Fig. 5). Both the pure GM-CSF and ^{125}I-labeled GM-CSF focused sharply in the range pH 4.7–5.0, with the peak of biological activity and radioactivity close to pH 4.9. The pure GM-CSF focused as a single band of biological activity; however, there was a small amount (<10%) of ^{125}I-labeled GM-CSF detectable at pH 5.4 (Fig. 5). The GM-CSF passing through concanavalin A-Sepharose led to a broader distribution of biological activity (pH 4.9–5.6), with a peak at pH 5.1. The recovery of GM-CSF from isoelectric focusing was usually less than 20%, and the ampholines used to create the pH gradient inhibited colony growth even at dilutions of 1:100. CSF could only be assayed after removal of the ampholines on small columns of mixed-bed ion-exchange resins (Baumann and Chrambach 1975).

CSFs with other cellular specificities were analyzed with isoelectric focusing in thin-layer granulated gels (Metcalf et al. 1977). The hematopoietic regulators focused in the slightly acidic pH range but each ap-

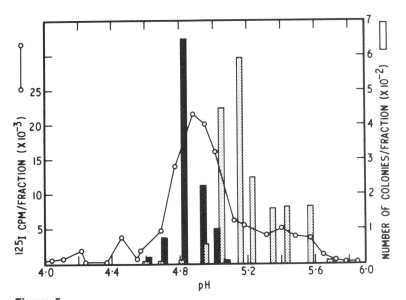

Figure 5

Isoelectric focusing in a thin-layer granulated gel (15 × 30 cm) using ampholines (2%) to generate the pH gradient. Focusing was performed overnight at 4°C with a constant voltage (400 V) and for a further 3 hours at 1000 V. The pH gradient was determined with a surface electrode, and 30 equal fractions were sliced for elution with 0.01 M sodium phosphate buffer (5 ml), pH 7.3, containing FCS. Fractions were deionized to remove ampholines and assayed for biological activity and ^{125}I. (■) Pure GM-CSF from MLCM. (▦) GM-CSF from MLCM which did not bind to concanavalin A-Sepharose. (o——o) ^{125}I counts per minute per fraction.

Table 6
Isoelectric Points of CSFs with Different Cellular Specificities

CSF	Conditioned medium	Isoelectric point[a] (pH)
Granulocyte-macrophage	mouse lung	4.9
Granulocyte-macrophage	spleen	4.8
Macrophage	yolk sac	5.1
Eosinophil	spleen	5.4–6.5
Megakaryocyte	spleen	4.4–5.9

[a] Determined by isoelectric focusing in the presence of ampholines (2%) using a thin-layer granulated gel bed (Radola 1973).

peared to have a distinctive isoelectric properties (Table 6). M-CSF from YSCM focused at a pH slightly higher than did GM-CSF from MLCM. Pokeweed-mitogen-stimulated SCM contains at least three colony-stimulating activities—GM-CSF, Eo-CSF, and MEG-CSF. Although most of the GM-CSF present in SCM focused at 4.8, the Eo-CSF focused between pH 5.4 and 6.5 (Table 6). MEG-CSF did not focus well in this system, but

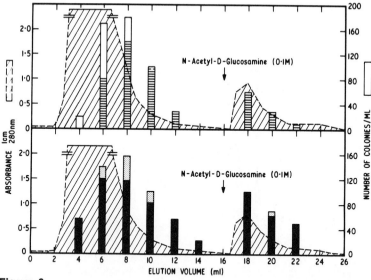

Figure 6
Affinity chromatography of concentrated pokeweed-mitogen SCM using WGA-Sepharose. (▨) Protein absorbance at 280 mm. Colony-stimulating activities for (■) GM-CSF ($\times 10^{-1}$); (▦) Eo-CSF; (▤) MEG-CSF; and (□) E-CSF are expressed independently. The number of colonies for each type is represented by the length of each bar. The start of the specific elution with N-acetyl-D-glucosamine is shown by the arrow. Note that the colony scale for GM colony stimulation should be multiplied by 10.

the distribution of activity indicated that it is distinct both from GM- and Eo-CSFs.

WGA-Sepharose Chromatography

When GM-CSF from MLCM or M-CSF from YSCM were chromatographed on WGA-Sepharose, no biological activity appeared to be eluted from the column with N-acetyl-D-glucosamine. Indeed both GM-CSF and M-CSF appeared to be recovered quantitatively in the protein fraction which had failed to bind to the column. When GM-, Eo-, MEG-, and E-CSFs from SCM were chromatographed on WGA-Sepharose, all of the factors (except E-CSF) were detected both in the unbound and specifically eluted proteins (Fig. 6). Almost 30% of the GM-CSF, 20% of the MEG-CSF and <5% of the total protein bound to WGA-Sepharose were recovered in the specifically eluted fractions (Fig. 6). Less than 10% of the Eo-CSF and none of the E-CSF were detected in the specifically eluted fractions.

DISCUSSION

Different Molecular Species of GM-CSF

There are many reports indicating that CSFs from different tissues vary considerably in molecular weight (Sheridan and Stanley 1971). Mouse L-cell conditioned medium was reported to contain several molecular weight species (Stanley et al. 1976), although this finding may have been complicated by partial interaction between a single molecular species of GM-CSF and the contaminating proteins. MLCM also appears to contain at least two molecular-weight species of GM-CSF, but most of the activity (i.e., that binding to concanavalin A-Sepharose) eluted from Sephadex G-100 as a single peak (mol. wt. 29,000). GM-CSF passing through concanavalin A-Sepharose eluted as two peaks from Sephadex G-100, one at the same elution volume as GM-CSF specifically eluted from concanavalin A-Sepharose and the other eluting at higher molecular weight (between albumin and hemoglobin). If such a column is developed in the absence of a detergent such as Triton X-100 or polyethylene glycol, the low molecular weight GM-CSF probably would not be detected since, at low protein concentrations, GM-CSF is adsorbed onto glass, plastic, and cellulose surfaces). Similarly, during polyacrylamide-gel electrophoresis, Triton X-100 (0.01%, v/v) must be included when the electrophoretic mobility distribution of GM-CSF is being determined. Microheterogeneity was present even in highly purified GM-CSF and could be detected by electrophoresis. The heterogeneity appeared to be due to a charge distribution on the molecule rather than reversible aggregation of subunits. When fractions from the leading, center, and trailing edges of the CSA electrophoretic profile were reelectrophoresed they migrated with different electrophoretic mobilities, corresponding to their mobilities on the original electrophoresis. Such charge heterogeneity has been observed for other leucocyte-produced glycoproteins such as interferon (Bose et al. 1976) and for another hematopoietic regulator, erythropoietin (Goldwasser 1975).

Amino Acid Composition

GM-CSF contains a relatively small number of basic amino acids, so that the use either of automated or manual techniques should allow production of tryptic peptides of suitable size for sequence analysis. The presence of at least two and possibly three methionine residues per molecule should also allow the production of cyanogen bromide fragments that would be suitable for structural analysis. Considerable difficulties were experienced in implementing an electrophoretic buffer system (for the final step in the purification) which did not include Tris (which interferes to some extent with the amino-acid analysis). The hexosamine content in this initial analysis was only five to six residues per molecule; however, a more complete sugar analysis must be made before the total carbohydrate content can be assessed.

Radiolabeling of GM-CSF

GM-CSF is actively synthesized in MLCM; however, many other proteins must also be synthesized, as the level of incorporation of ^{3}H-labeled amino acids into glycoproteins is only 0.01 of the ^{3}H-labeled amino-acid incorporation into the total protein. Such a small proportion of the initial label appeared to enter "CSF-like" molecules that it was unlikely that pure, uniformly radiolabeled GM-CSF would be available for sequencing studies. However, it should have been possible to introduce one or two labeled residues into specific positions to aid the sequence analysis. Although lithium chloride stimulated the production of GM-CSF in MLCM (Harker et al. 1977), there was no increase in the incorporation of ^{3}H amino acids into total protein or glycoprotein.

It appears to be possible to introduce ^{125}I into GM-CSF, but with rather low efficiency and with a considerable loss of biological activity. Although all of the pure GM-CSF bound to concanavalin A-Sepharose after iodination with the chloramine-T method or iodine monochloride, only 50% of the biological activity was still able to bind to the lectin column. Similarly, although the pure GM-CSF migrated as a single band on electrophoresis in the presence of SDS, after the iodination several radioactive molecular species were produced. The ^{125}I-labeled GM-CSF eluted from concanavalin A-Sepharose contained ^{125}I-labeled molecules which had migrated with the same mobility as had GM-CSF on polyacrylamide-gel electrophoresis and eluted in the same place from Sephadex G-100 (S). Although the ^{125}I and biological activities were coincident, it was not possible to distinguish between biologically active ^{125}I-labeled GM-CSF and a mixture of inactive ^{125}I-labeled GM-CSF and unmodified GM-CSF. In the case of erythropoietin, the biological activity appeared to be destroyed when the molecule was iodinated (Goldwasser 1975).

Isoelectric Focusing

Analysis of the molecular properties of hormones which are in low concentration in plasma or conditioned media is often made extremely difficult by the high concentrations of contaminating proteins. These proteins lead to artifacts because of their specific binding proteins and differential non-specific carrier effects during protein recovery, and it is not possible to apply sufficient levels of biological activity to techniques with high resolving power such as polyacrylamide-gel electrophoresis. The introduction of thin-layer granulated gel bed isoelectric focusing (Radola 1973) allowed analysis of charge heterogeneity for several CSFs, even at protein loads up to 200 mg. We analyzed pure GM-CSF by this technique and the isoelectric point appeared to be close to pH 4.9. The isoelectric point distribution both for ^{125}I-labeled GM-CSF and pure GM-CSF appeared to be the same. GM-CSF which did not bind to concanavalin A-Sepharose had a broader isoelectric point distribution, with maximum activity near pH 5.1.

The chemistry of a single cell-specific GM-CSF appeared to be complicated by a small amount of heterogeneity with respect to molecular weight and charge. However, in spite of this heterogeneity, it appeared possible to isolate particular molecular species for detailed study, especially when the level of activity in a biological fluid was reasonably high. Progress was made with the purification of at least two types of GM-CSF and erythropoietin (Table 7). The level of GM-CSF in the starting conditioned media was considerably elevated above biological levels and, unless large quantities (e.g., 2500 liters) of starting fluid could be processed, the amount of pure regulator which could be obtained was too small for chemical characterization. Levels of MEG-CSF, Eo-CSF, and E-CSF in SCM made the purification of these factors from mice rather difficult. A report indicating that there is a reasonable level of MEG-CSF in human urine (McLeod et al. 1976) indicates that this might be a suitable starting material for the purification of this factor, but, as yet, there is no information to suggest that this MEG-CSF will stimulate MEG-progenitor cells from its own species.

Chromatography using insolubilized lectins can be used effectively both for the preparation of a particular molecular species of a given CSF (e.g., GM-CSF from MLCM) (Burgess et al. 1977) or for the analysis of the heterogeneity of particular carbohydrate moieties of a given CSF. Although all of the GM-, Eo-, MEG- and E-CSF in SCM bound to concanavalin A-Sepharose, the CSFs, when examined with WGA-Sepharose, showed different levels of binding. Further molecular analysis of the CSFs specifically eluted from WGA-Sepharose with *N*-acetyl-D-glucosamine was considerably easier than analysis of the CSFs present in the unbound fraction (the protein content was much lower in the former). However, the specifically eluted CSF often represented less than 30% of the total CSA (Johnson and Burgess 1978), so that these procedures were not always effective as preparative techniques.

Even if the complete purification of these factors proves difficult, isoelectric focusing indicates that the charge distribution on different CSFs is sufficiently different to allow the separation of cellular specific CSFs. Until reagents such as these and the macrophage-specific CSF from YSCM are

Table 7
Sources of Colony-Stimulating Factors for Different Mouse Hematopoietic Progenitor Cells

Cellular specificity	Source of activity	Protein concentration (mg/ml)	No. colonies per mg protein per 10^5 cells	No. progenitor[a] cells per 10^5 cells	Normalized[b] specific activity (colonies per mg protein)	Reference
Granulocyte-Macrophage	human urine	0.14	200	300	70	Stanley and Metcalf 1969
	mouse lung CM	0.7	36,000	300	12,000	Sheridan and Metcalf 1973
	mouse L-cell CM	0.07	50,000	300	17,000	Austin et al. 1971
	yolk sac CM	0.18	10,000	300	3,300	Johnson and Burgess 1978
Megakaryocyte	spleen cell CM	4.0	90	20	450	Metcalf et al. 1975
	mouse L-cell CM	8.0	13	10	130	Nakeff and Daniels-McQueen 1976
	sheep plasma	80	0.04	12	0.33	McLeod et al. 1976
	human urine	0.14	60	12	500	McLeod et al. 1976
Eosinophil	spleen cell CM	4.0	90	20	450	Metcalf et al. 1974
Erythroid	spleen cell CM	4.0	650	250[c]	260	Johnson and Metcalf 1977
	sheep plasma	80	0.14	38	0.37	Goldwasser and Kung 1971
	human urine	0.14	5.4	38	14	Espada and Gutnisky 1970

[a] Progenitor cell numbers maximal figures taken from semisolid assays with mouse bone-marrow cells.
[b] The specific activities have been normalized to 100 progenitor cells per 10^5 target cells.
[c] Semisolid agar assays were performed with liver cells from fetuses at 12 days of gestation.

used in competition titration experiments (Johnson and Burgess 1978), the relationships between the progenitor cells of the hematopoietic differentiation states will remain a matter of conjecture.

Acknowledgments

This work was generously supported by the Anti-Cancer Council of Victoria, the National Health and Medical Research Council of Australia, and the National Cancer Institute, Washington, Contract NOI-CB-33854. We are grateful to the expert technical assistance provided by Mrs. S. Delves, Mrs. Sue Webb, and Ms. Lynette Wadeson. Dr. G. R. Johnson kindly provided the yolk-sac conditioned medium used in these studies and the amino-acid analysis was performed by Dr. Charles Roxborough, C.S.I.R.O. Division of Protein Chemistry, Parkville, Victoria, Australia.

REFERENCES

Andrews, P. 1964. Estimation of the molecular weights of proteins by Sephadex gel-filtration. *Biochem. J.* **91**: 222.

Austin, P. E., E. A. McCulloch, and J. E. Till. 1971. Characterization of the factor in L-cell conditioned medium capable of stimulating colony formation by mouse marrow cells in culture. *J. Cell. Physiol.* **77**: 121.

Baumann, G., and A. Chrambach. 1975. Quantitative removal of carrier ampholytes from protein fractions derived from isoelectric focusing. *Anal. Bioch.* **69**: 649.

Bose, S., D. Gurari-Rotman, U. Th. Ruegg, L. Corley, and C. B. Anfinsen. 1976. Apparent dispensability of the carbohydrate moiety of human interferon for antiviral activity. *J. Biol. Chem.* **251**: 1659.

Bradley, T. R., and D. Metcalf. 1966. The growth of mouse bone marrow cells in vitro. *Aust. J. Exp. Biol. Med. Sci.* **44**: 287.

Burgess, A. W., and D. Metcalf. 1977a. Serum half-life and organ distribution of radiolabeled colony stimulating factor in mice. *Exp. Hematol.* (Copenh.) **5**: 456.

―――. 1977b. The effect of colony stimulating factor on the synthesis of ribonucleic acid by mouse bone marrow cells in vitro. *J. Cell. Physiol.* **90**: 471.

―――. 1977c. Colony stimulating factor and the differentiation of granulocytes and macrophages. In *Experimental hematology today* (ed. S. J. Baum), p. 135. Springer-Verlag, New York.

Burgess, A. W., J. Camakaris, and D. Metcalf. 1977. Purification and properties of colony-stimulating factor from mouse lung-conditioned medium. *J. Biol. Chem.* **252**: 1998.

Davis, B. J. 1964. Disc electrophoresis. II. Method and application to human serum proteins. *Ann. N.Y. Acad. Sci.* **121**: 404.

Espada, J., and A. Gutnisky. 1970. Purificacion de eritropoyetina urinaria humana. *Acta Physiol. Lat. Am.* **20**: 122.

Goldwasser, E. 1975. Erythropoietin and the differentiation of red blood cells. *Fed. Proc.* **34**: 2285.

Goldwasser, E., and C. K. H. Kung. 1971. Purification of erythropoietin. *Proc. Natl. Acad. Sci. U.S.A.* **68**: 697.

Goldwasser, E., and C. K.-H. Kung, and J. Eliason. 1974. On the mechanism of erythropoietin-induced differentiation. XIII. The role of sialic acid in erythropoietin action. *J. Biol. Chem.* **249**: 4202.

Harker, W. G., G. Rothstein, D. Clarkson, J. W. Athens, and J. L. Macfarlane. 1977. Enhancement of colony-stimulating activity production by lithium. *Blood* **49:** 263.

Hunter, W. M., and F. C. Greenwood. 1962. Preparation of iodine-131 labelled human growth hormones of high specific activity. *Nature* **194:** 495.

Johnson, G. R., and A. W. Burgess. 1978. Molecular and biological properties of a macrophage colony-stimulating factor from mouse yolk sacs. *J. Cell Biol.* **77:** 35.

Johnson, G. R., and D. Metcalf. 1978. Sources and nature of granulocyte-macrophage colony stimulating factor in fetal mice. *Exp. Hematol.* (Copenh.). (In press.)

Johnson, G. R., and D. Metcalf. 1977. Pure and mixed erythroid colony formation in vitro stimulated by spleen conditioned medium with no detectable erythropoietin. *Proc. Natl. Acad. Sci. U.S.A.* **74:** 3879.

Liu, T.-Y., and Y. H. Chang. 1971. Hydrolysis of proteins with p-toluenesulfonic acid: Determination of tryptophan. *J. Biol. Chem.* **246:** 2842.

McLeod, D. L., M. M. Shreeve, and A. A. Axelrad. 1974. Improved plasma culture system for production of erythrocytic colonies in vitro: Quantitative assay method for CFU-E. *Blood* **44:** 517.

———. 1976. Induction of megakaryocyte colonies with platelet formation *in vitro*. *Nature* **261:** 492.

Metcalf, D. 1970. Studies on colony formation *in vitro* by mouse bone marrow cells. II. Action of colony stimulating factor. *J. Cell. Physiol.* **70:** 89.

———. 1971. Acute antigen-induced elevation of serum colony stimulating factor (CSF) levels. *Immunology* **21:** 427.

Metcalf, D., S. Russell, and A. W. Burgess. 1978. Production of hematopoietic stimulating factors by pokeweed mitogen-stimulated spleen cells. *Transplant. Proc.* (Copen). (In press.)

Metcalf, D., H. R. MacDonald, N. Odartchenko, and B. Sordat. 1975. Growth of mouse megakaryocyte colonies *in vitro*. *Proc. Natl. Acad. Sci. U.S.A.* **72:** 1744.

Metcalf, D., J. Parker, H. M. Chester, and P. W. Kincade. 1974. Formation of eosinophilic-like granulocytic colonies by mouse bone marrow cells *in vitro*. *J. Cell. Physiol.* **84:** 275.

Miyake, T., C. K.-H. Kung, and E. Goldwasser. 1977. The purification of human erythropoietin. *J. Biol. Chem.* **252:** 5558.

Nakeff, A., and S. Daniels-McQueen. 1976. *In vitro* colony assay for a new class of megakaryocyte precursor colony-forming unit megakaryocyte (CFU-M). *Proc. Soc. Exp. Biol. Med.* **151:** 587.

Neville, D. M. 1971. Molecular weight determination of protein-dodecyl sulfate complexes by gel electrophoresis in a discontinuous buffer system. *J. Biol. Chem.* **246:** 6328.

Pluznik, D. H., and L. Sachs. 1966. The induction of clones of normal mast cells by a substance from conditioned medium. *Exp. Cell Res.* **43:** 553.

Radola, B. J. 1973. Isoelectric focusing in layers of granulated gels. I. Thin layer isoelectric focusing of proteins. *Biochim. Biophys. Acta* **295:** 412.

Roholt, O. A., and D. Pressman. 1972. Iodination-isolation of peptides from the active site. In *Methods in enzymology* (eds. C. H. W. Hirs and S. N. Timasheff), Vol. XXV, part B, p. 438. Academic Press, New York.

Sheridan, J. W., and D. Metcalf. 1973. A low molecular weight factor in lung conditioned medium stimulating granulocyte and monocyte colony formation *in vitro*. *J. Cell. Physiol.* **81:** 11.

Sheridan, J., and E. R. Stanley. 1971. Tissue sources of bone marrow colony stimulating factor. *J. Cell. Physiol.* **78:** 451.

Stagg, B. H., J. M. Temperley, H. Rochman, and J. S. Morley. 1970. Iodination of the biological activity of gastrin. *Nature* **228:** 58.

Stanley, E. R., and P. M. Heard. 1977. Factors regulating macrophage production and growth: Purification and some properties of the colony stimulating factor from medium conditioned by mouse L cells. *J. Biol. Chem.* **252:** 4305.

Stanley, E. R., and D. Metcalf. 1969. Partial purification and some properties of the factor in normal and leukaemic human urine stimulating mouse bone marrow colony growth *in vitro*. *Aust. J. Exp. Biol. Med. Sci.* **47:** 467.

Stanley, E. R., M. Cifone, P. M. Heard, and V. Defendi. 1976. Factors regulating macrophage production and growth: Identity of colony-stimulating factor and macrophage growth factor. *J. Exp. Med.* **143:** 631.

Stanley, E. R., G. Hansen, J. Woodcock, and D. Metcalf. 1975. Colony-stimulating factor and the regulation of granulopoiesis and macrophage production. *Fed. Proc.* **34:** 2272.

Stephenson, J. R., A. A. Axelrad, D. L. McLeod, and M. M. Shreeve. 1971. Induction of colonies of hemoglobin-synthesizing cells by erythropoietin *in vitro*. *Proc. Natl. Acad. Sci. U.S.A.* **68:** 1542.

Tsuneoka, K., and M. Shikita. 1977. A sialoglycoprotein-stimulating proliferation of granulocyte-macrophage progenitors in mouse bone marrow cell cultures. *Febs (Fed. Eur. Biochem. Soc.) Lett.* **77:** 243.

Inhibitors of Hematopoiesis Obtained from Mature Neutrophils, Established Cell Lines, and Patients with Bone-Marrow-Failure Syndromes

M. J. Cline, T. Olofsson, M. Frölich, S. Herman, and D. W. Golde
Division of Hematology-Oncology, Department of Medicine
University of California School of Medicine, Los Angeles, California 90024

A variety of inhibitors of proliferating hematopoietic cells have been described in experimental model systems and in studies of human disease. Some of these inhibitors undoubtedly play important roles in the regulation of hematopoiesis; others are of uncertain significance. Two broad categories can be distinguished: those inhibitors that are thought to be part of the normal regulatory mechanisms by which the levels of circulating blood cells are governed and those inhibitors that arise as a consequence of disease.

In the studies covered by this report, we sought to obtain information about three types of inhibitors of myelopoiesis: first, the identification of normally occurring products of mature neutrophils that specifically inhibit proliferating myeloid precursors; second, the identification of inhibitors of cell proliferation produced by established lines of hematopoietic and nonhematopoietic cells; and third, the identification of inhibitors of myelopoiesis and erythropoiesis in the serum of patients with bone-marrow-failure syndromes.

The studies concerned with inhibitory products of mature neutrophils were stimulated by a large and controversial literature on granulocyte chalones (Rytömaa and Kiviniemi 1968a,b; Paukovitz 1971, 1973; Benestad et al. 1973; Lajtha 1973; Bateman 1974; Lord et al. 1974). These are putative, tissue-specific, species-nonspecific negative regulators of granulopoiesis. We utilized purified populations of human neutrophils to condition medium and then examined this conditioned medium for ability to inhibit the proliferation of various hematopoietic cell populations.

If physiological inhibitors of myelopoiesis occur naturally, then one might have a reasonable expectation of identifying similar inhibitors produced by some continuously growing hematopoietic cell lines. This expectation formed the basis of our second approach.

Finally, the concept of immune inhibitors of hematopoiesis occurring in human disease has had considerable display in the recent medical literature

(Krantz 1974; Ascensao et al. 1976; Hoffman et al. 1976, 1977) and it seemed to us appropriate to reexamine this area with newer technology.

METHODOLOGY

Samples Assayed for Inhibitors

Heparinized blood from normal subjects was separated into polymorphonuclear neutrophil (PMN), mononuclear leukocyte (MNL), and erythrocyte fractions of $91 \pm 4\%$, $96 \pm 4\%$, and 100% purity, respectively, as described previously by Herman et al. (1978). Purified cell populations were incubated in sterile saline with antibiotics at concentrations of $1-100 \times 10^6$/ml from 2 to 36 hours. The cell-free medium was collected by centrifugation and stored frozen until assay.

Established cell lines (Table 1) were grown under standard tissue culture conditions and cell-free conditioned medium obtained from late, logarithmic-phase cultures. Medium was stored at $-20°C$ until assayed for inhibitory activity. Serum was obtained from 80 patients with various hematologic disorders including bone-marrow-failure syndromes, 30 normal subjects, and 22 patients with various nonhematopoietic disorders. Serum was stored frozen until examined for inhibitory activity.

Inhibitor Assays

Cloning Assays

Human bone-marrow myeloid progenitors (colony-forming units—culture, CFU-C) and erythroid progenitors (colony-forming units—erythroid,

Table 1

Human Cell Lines Tested for Inhibitory Activity against Human and Mouse CFU-C

Cell lines	Origin
K562[a]	myeloid leukemia
253J, J82, T24, 647V	bladder cancer
Chago,[a] SKMES	lung cancer
HeLa	uterine cervix
Colo 16	squamous cancer of skin
FL	amniotic cells
MT	osteosarcoma
LAN-N1, -N2, IMR-32, SK-N-MC, SK-N-SH	neuroblastoma
LA-96, 4265, Raji, LA85, LA-259, LAIM-8, LA109, WiL2, LA237	B lymphocytes
Reis, Prit, Moor	hairy cell leukemia
MOLT-4	T cell

[a] Inhibitory activity found.

CFU-E) were cloned in agar and methylcellulose, respectively, as previously described (Golde and Cline 1972; Golde et al. 1976). Leukocyte-conditioned medium served as a source of colony stimulating activity (CSA).

Potential inhibitors were added directly to the suspending matrix or else the cells were preincubated with inhibitor and subsequently suspended either in agar or methylcellulose and cultured in vitro. Human and mouse myeloid colonies were enumerated at 12 and 7 days, respectively. Human and mouse erythroid colonies were enumerated at 7 and 2 days, respectively.

^{3}H-Labeled Thymidine (TdR) Incorporation

Swiss Webster mouse or normal human hematopoietic cells were incubated in 0.44 ml of 10% aged fetal calf serum (FCS) and normal saline containing 1–100 μl of inhibitors and 0.04 μCi ^{3}H-labeled TdR (6.7 Ci per mM, 0.35 μg/ml). After 2 hours' incubation at 37°C, isotope incorporation was stopped with iced buffered saline solution (BSS); cells were pelleted and resuspended in 7.5% cold trichloroacetic acid, collected, and washed with acid and ethanol on Millipore filters (no. AP250500). Tritium incorporation was determined in a liquid-scintillation spectrometer.

Mitotic Index

Cells freshly isolated from normal mouse-bone marrow or phenylhydrazine-treated mouse spleens (Glass et al. 1975) were incubated in complete tissue culture medium at 1×10^6/ml with colcemide, 0.1 μg/ml in the presence or absence of 0.1 volume granulocyte-conditioned medium. At intervals of up to 7 hours, aliquots were removed and stained with Giemsa, and 5000 cells were counted to determine the mitotic index (mitoses/100 cells $\times$ 100).

Cytotoxicity Assay

Complement-dependent antibodies active against CFU-C were assayed by a modification of the method of Terasaki and McClelland (1964). Normal bone-marrow cells at 10×10^6 ml were suspended in 15% heat-inactivated (56°C for 30 minutes) FCS plus McCoy's 5A medium. To 0.1 ml of cells was added 0.1 ml of test serum and the combination mixed and incubated at room temperature for 30 minutes. To the mixture, 0.1 ml of fresh-frozen (-70°C) normal rabbit serum (NRS) was added and incubation continued for an additional 60 minutes. The NRS served as a source of complement and was not in itself cytotoxic for human CFU-C.

At the completion of incubation, sufficient agar was added to establish a concentration of 0.3% and the cell suspension plated in duplicate for the CFU-C assay. Controls included normal human AB serum in place of test serum (negative control), a complement-dependent rabbit antiserum cytotoxic for human CFU-C (Billing et al. 1976; Cline and Billing 1977) (positive control), and bone marrow plated in the absence of the feeder layers. As a further control, one tube received heat-inactivated FCS in lieu of a complement source.

RESULTS

Inhibitory Products from Human Neutrophils

Human neutrophils, mononuclear leukocytes, or erythrocytes were used to condition media, and these media were subsequently assayed for their ability to inhibit ^{3}H-labeled TdR incorporation in proliferating hematopoietic cell populations of varying compositions. It has been established (Herman et al. 1978) that the production of inhibitors of ^{3}H-labeled TdR incorporation by PMN are not influenced or are minimally influenced by the following: (1) a concentration of PMN between 2 and 50×10^6/ml in the conditioned medium; (2) conditioning times between 16 and 36 hours; (3) mouse bone-marrow target cell numbers between 8×10^4 and 20×10^5; and (4) ^{3}H-labeled TdR additions between 30 and 300 nmoles. Therefore, using standard conditions of conditioned medium obtained from 50×10^6 human PMN/ml incubated for 16 hours and a target cell number of 4×10^5 hematopoietic cells, we obtained the results shown in Table 2. Neutrophil-conditioned medium actively inhibited ^{3}H-labeled TdR incorporation in normal mouse-bone marrow and inhibited it less potently in human marrow. It had little effect on mitogen-stimulated human lymphocytes. Inhibitory activity was not specific for proliferating myeloid precursors but was at least equipotent for mouse populations enriched in erythroid progenitors such as those obtained from phenylhydrazine-treated mouse spleens. Mouse marrow populations enriched in myeloid precursors were not especially sensitive to the inhibitor. These observations were consistent with the effects of PMN-conditioned medium observed in clonogenic and mitogenic inhibition assays (Table 3). In brief, inhibitory activity was demonstrable only against CFU-E and mitotically active erythroid populations. We concluded that we could find no evidence for products of mature neutrophils which directly and specifically affected proliferating myeloid precursors. Clearly, however, neutrophils did produce a factor or factors that inhibited proliferation of some hematopoietic cells under in vitro conditions. These neutrophil factors have been partially characterized (Herman et al. 1978). Medium conditioned by erythrocytes lacked inhibitory activity, and that conditioned by nonadherent peripheral blood mononuclear leukocytes had 0–21% of the PMN activity. Monocyte-enriched, surface-adherent MNL had 22–50% of the neutrophil activity. These levels of inhibitory activity could possibly be explained by a neutrophil contamination of these populations of between 2 and 7%.

Inhibitors Produced by Established Cell Lines

We next examined a continually propagating cell line (K562) established from a patient with chronic myelocytic leukemia in blast crisis (Lozzio and Lozzio 1975; Klein et al. 1976) for the presence of possible myelopoietic inhibitors. These cells retained a Ph^1 chromosome and had surface antigenic markers of differentiated granulocytes (Drew et al. 1977a,b). The medium conditioned by these cells contained a potent inhibitor of human, but not mouse, myeloid stem cells (Table 4). Stimulated by these observations, we examined 29 human cell lines (Table 1) for the presence of inhibitors. We

Table 2
Effect of Neutrophil-Conditioned Medium on Inhibition of ^{3}H-Labeled TdR in Various Target Cells

Target cells	% Myeloid cells[a]	% Inhibition of ^{3}H-labeled TdR at various concentrations of neutrophil-conditioned medium		
		10 µl	50 µl	100 µl
Normal mouse bone marrow	54 ± 6%	27 ± 8 (10)	49 ± 15 (9)	61 (2)
Hypertransfused mouse marrow	81 ± 5%	13 ± 9 (4)	28 ± 6 (3)	
Phenylhydrazine-treated mouse spleen	10 ± 5%†	22 ± 11 (3)	42 ± 7 (3)	50 ± 12 (3)
Human bone marrow	53 ± 20%	0 (3)	14 ± 4 (3)	30 ± 9 (3)
Human lymphocytes (PHA-stimulated)	95% lymphoid	<1 (3)	8 ± 6 (3)	17 (2)

Medium conditioned by 50×10^6 PMN/ml for 16 hours was added to 4×10^5 target cells in 0.44 ml of medium containing 0.4 µCi ^{3}H-labeled TdR. After 2 hours, ^{3}H-labeled TdR incorporation was determined. Results are expressed as mean ± S.D. of the percentage of inhibition. Numbers of experiments are given in parentheses.
[a] All identifiable granulocytic cells; 80 ± 4% of these cells were erythroid.

Table 3
Effect of Neutrophil-Conditioned Medium on Inhibition of CFU-C, CFU-E, and Mitotic Indices

Target cells	Inhibition of		
	CFU-C (%)	CFU-E (%)	Mitotic Index[a] (%)
Mouse bone marrow	5 ± 5	69 ± 9	0 ± 2
Phenylhydrazine mouse spleen (erythroid)	–	–	40 ± 4

[a] Mitotic index was measured at 7 hours with 1×10^6 cells/ml incubated with colcemide 0.1 μg/ml in McCoy's 5A medium containing 10% aged FCS. Results are expressed as mean ± S.D. of the percentage inhibition.

identified one of the cell lines, Chago (Rabson et al. 1973), established from a carcinoma of the lung, which produced a potent inhibitor of human CFU-C (Table 4).

The characteristics of the inhibitors found in the medium of these two cell lines are summarized in Table 5. Although these inhibitors have not been characterized completely, it is clear that they are potently active against a variety of proliferating hematopoietic cells. Activity may be species-specific in that mouse hematopoietic cells are minimally retarded in their growth by medium from these cell lines. In addition to inhibiting CFU-C, both K562 and Chago retarded the growth of mitogen-stimulated lymphocytes either in clonogenic or radioisotopic assays (Fig. 1). Both cell lines also inhibited the growth in agar of myeloid cells from a patient with chronic myelocytic leukemia but did not inhibit proliferation of cells from three patients with acute leukemia.

Thus far, it appears that the inhibitor Chago-conditioned medium has

Table 4
Inhibition of Human Myeloid and Erythroid Progenitors by Medium of Two Established Cell Lines

Cell lines	Final dilution	% Inhibition of human cell lines		% Inhibition of mouse cell lines	
		CFU-C	CFU-E	CFU-C	CFU-E
K562, Myeloid leukemia	1:10	100	64	74	30
	1:20	100	58	44	30
	1:40	–	15	–	–
	1:100	98	6	23	–
	1:200	94	–	–	–
Chago, Carcinoma of lung	1:10	100	73	88	47
	1:20	–	64	77	47
	1:40	–	58	–	43
	1:100	71	27	34	3
	1:1000	49	–	–	–

Table 5
Characteristics of Inhibitors Produced by Two Established Human Cell Lines

	Cell line	
Inhibitory activity[a]	K562	Chago
Against human CFU-C	strong	strong
Against human CFU-E	weak	weak
Against human CFU-L[b]	strong	strong
Against epithelial bladder cell lines (3)	no	no
Against human diploid fibroblasts	no	no
Against cloning of CML cells (1)	yes	yes
Against human acute leukemia cells (3)	no	no
Against K562 cells	no	yes
Against mouse CFU-C, CFU-E	weak	weak
Against mouse Friend leukemia cells	no	no
Stable at 56°C, 30 minutes	no	yes
Stable at 70°C, 30 minutes	no	no
Stable at −20°C, 2 weeks	yes	yes
Approximate molecular weight	>100,000	>100,000
Adheres to Millipore filters	yes	no

[a] Numbers of cell lines tested are given in parentheses.
[b] Colony-forming units for T lymphocytes.

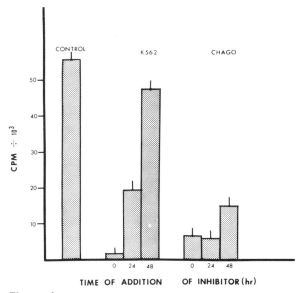

Figure 1
Conditioned medium from continuously growing myeloid (K562) or lung carcinoma (Chago) cells was added to phytohemagglutinin-stimulated human lymphocytes at initiation of culture or after 24 or 48 hours. ^{3}H-labeled TdR was added at 72 hours and cultures were harvested 16 hours later. The results of acid-precipitable radioactive counts incorporated into lymphocytes (mean ± S.D.) are shown.

different physical characteristics from those produced by K562. The inhibition by K562 appears restricted to some human hematopoietic cells.

Inhibitors of Myelopoiesis in Patients with Bone-Marrow Failure

We recently applied a modification of the microcytotoxicity assay of Terasaki and McClelland (1964) to look for inhibitors of myelopoiesis in the serum of patients with a variety of disorders. Many of these patients had the common feature of granulocytopenia and abnormal myeloid development in the bone marrow. In this assay system, a serum containing presumed inhibitor was incubated with normal human marrow cells in the presence or absence of complement. The CFU-C content of the treated marrow was then determined. Sera from 30 normal subjects and 102 patients were screened. These included congenital neutropenia (10), acquired neutropenia (20), aplastic anemia (29), systemic lupus erythematosus (11), Felty's syndrome (3), and idiopathic thrombocytopenic purpura (4), in addition to miscellaneous disorders. Table 6 summarizes the results of these studies. Inhibitor was not found in any normal sera. Of the 102 patients, 12 had sera containing detectable cytotoxins against human CFU-C. In nine patients, cytotoxic activity was complement-dependent, and in three it was not. In one of the three patients with complement-independent inhibitor, the activity was localized in the immunoglobulin fraction. The diseases in which humoral antibody against CFU-C were detected included systemic lupus erythematosus, autoimmune panleukopenia, atypical cases of aplastic anemia, and acquired neutropenia, as well as miscellaneous disorders such as acute leukemia, carcinoma of the breast, and renal transplant rejection. Inhibitor was not detected in seven multiply transfused patients with non-

Table 6
Serum Inhibitory Activity against Human Bone Marrow CFU-C

Disease	Number of patients	Cytotoxic activity against CFU-C		
		C'-dependent	C'-independent	Absent
Congenital neutropenia	10	0	0	10
Idiopathic acquired neutropenia	17	0	2	15
Drug-induced neutropenia	3	0	0	3
Aplastic anemia	29	3	1	25
Systemic lupus erythematosus[a]	11	3	0	8
Felty's syndrome	3	0	0	3
Multiply transfused patients	7	0	0	7
Other disease	22	3	0	19
Total	102	9	3	90

[a] Four of these patients were neutropenic.

hematologic disease and 25 multiple-transfused patients with aplastic anemia.

DISCUSSION

In 1960, Bullough and Lawrence introduced the concept of tissue-specific products of differentiated cells that serve as mitotic inhibitors of early cells of the same line. They termed such inhibitors "chalones" from the Greek word meaning to loosen or make slack. Recently, various investigators have described evidence of the existence of granulocyte chalones (Rytömaa and Kiviniemi 1968a,b; Paukovitz 1971, 1973; Benestad et al. 1973; Bateman 1974; Beran 1974; Lord et al. 1974). However, the diversity of model systems, starting materials, target cells, and methodologic limitations make interpretations difficult and the subject has remained controversial. Most previous assays of granulocyte "chalone" have utilized inhibition of ^{3}H-labeled TdR incorporation into hematopoietic cell populations as the primary assay. Assays of chalonelike material based on this system are susceptible to several artifacts, including dilution of isotope with cold thymidine, changes in nucleotide pool size, interference with phosphorylation and transport of thymidine, and cell death.

Despite these limitations, the concept of negative feedback regulation of hematopoiesis is important and deserves careful study. We undertook detailed examination of the effects of products of mature granulocytes on hematopoietic cell proliferation. We observed definite inhibitory activity against erythroid precursors in a variety of assay systems but no unequivocal inhibitory activity against normal neutrophil progenitors. We could not, therefore, detect a granulocyte chalone. Our failure to find such inhibitory materials acting directly and specifically on proliferating granulocyte precursors does not exclude the possibility of other negative feedback controls acting indirectly through cells producing CSA. Furthermore, our failure to confirm the findings of earlier investigators in this field may reflect differences in starting materials and methodology. For the moment, however, the existence of specific granulocyte chalones must remain tentative.

We examined a continuously growing human myeloid leukemia line (K562) with the hope that this might furnish a source of granulocyte-specific inhibitory factor. Indeed, conditioned medium from this cell potently inhibited human CFU-C but had little activity against erythroid progenitors, fibroblasts, or continuously growing epithelial cells. It did inhibit the proliferation of mitogen-stimulated lymphocytes in a time-dependent fashion. The inhibitory product of this cell line is now being characterized. It is of high molecular weight and heat labile. The critical question relates to whether it represents a physiologically important negative feedback regulator of hematopoietic cells or merely an interesting, but trivial, phenomenon.

In surveying a spectrum of established cell lines for the occurrence of other inhibitors, we identified one—the Chago cell line from a carcinoma derived from the lung. The inhibitor produced by this cell line is different from that of K562 both in its physical characteristics and its broader spectrum of antiproliferative activity. Whether such carcinoma-derived inhibitors have relevance either for normal physiology or human disease is still unknown.

Immunologic destruction of mature blood cells has long been recognized. In contrast, little information exists about immune destruction of the proliferating precursors of immature blood cells. Antibody-induced suppression of erythropoiesis in the syndrome of pure red-cell aplasia has been known for almost a decade (Krantz 1974). Only recently has it been documented that immunologic suppression can involve other immature hematopoietic cells. Several recent investigations suggest that cellular cytotoxic mechanisms as well as humoral antibody may be important in the suppression of hematopoiesis (Ascensao 1976; Hoffman et al. 1976, 1977; Litwin and Zanjani 1977). In this study, we examined serum from 102 patients with a variety of disorders for the presence of both complement-dependent and complement-independent inhibitors of granulopoiesis. Target cells were normal bone-marrow stem cells (CFU-C) which were grown in agar to form discrete colonies of granulocytes and macrophages. Inhibitors of normal CFU-C were found in the serum of 12 patients with unusual forms of leukopenia, aplastic anemia, systemic lupus erythematosus, and occasionally in other disorders. In 9 of the 12, the inhibitor was complement-dependent and in one other it was characterized as an immunoglobulin. Thus, in at least 10 of the 12 patients, the inhibitor probably was an antibody. In one of the remaining patients, the inhibitor was heat labile, was labile in storage at $-20\,°C$, and was probably not an immunoglobulin. It was found in a patient who had developed aplastic anemia in the setting of infectious mononucleosis.

SUMMARY

We have investigated three classes of potential inhibitors of myelopoiesis: products of mature neutrophils which specifically inhibit myeloid progenitors (chalones); inhibitory products of established cell lines from hematopoietic and nonhematopoietic tissues; circulating inhibitors in patients with a variety of bone-marrow-failure syndromes.

We failed to find granulocyte-specific chalones using purified human granulocytes as a tissue source and a variety of hematopoietic cell populations as targets. Granulocyte products inhibited erythropoiesis in a variety of assays but had little effect on myelopoiesis. We concluded that the concept of granulocyte chalones is still unsubstantiated.

We identified inhibitors of hematopoiesis in the conditioned medium of two cell lines: one established from chronic myelocytic leukemia in blast crisis (K562) and the other from carcinoma of the lung (Chago). Both inhibitors were of high molecular weight and had no detectable activity against a variety of nonhematopoietic cells. Screening of 27 other cell lines failed to reveal other inhibitors. Chago and K562 inhibitors had different physical characteristics.

We examined the serum from 102 patients with bone-marrow-failure syndromes and other disorders for complement-dependent and -independent inhibitors of myeloid stem cells (CFU-C). Circulating inhibitors were found in 12 patients; in nine, these were complement dependent. In another patient, the inhibitor was an immunoglobulin that did not require complement for cytotoxicity. Diseases in which antibody-mediated inhibition of myelopoiesis occurs include aplastic anemia, acquired neutropenia, and systemic lupus erythematosus.

Acknowledgment

Human urinary erythropoietin was a gift from the National Heart and Lung Institute. This work was supported by U.S. Public Health Service Grants CA 15688, CA 15614, and RR 00865.

REFERENCES

Ascensao, J., W. Kagan, M. Moore, R. Pahwa, J. Hansen, and R. Good. 1976. Aplastic anaemia: Evidence for an immunological mechanism. *Lancet* **i**: 699.

Bateman, A. E. 1974. Cell specificity of chalone-type inhibitors of DNA synthesis released by blood leucocytes and erythrocytes. *Cell Tissue Kinet.* **7**: 451.

Benestad, H. B., T. Rytömaa, and K. Kiviniemi. 1973. The cell specific effect of the granulocyte chalone demonstrated with the diffusion chamber technique. *Cell Tissue Kinet.* **6**: 147.

Beran, M. 1974. Serum inhibitors of in vitro colony formation: Relation to haemopoietic tissue in vivo. *Exp. Hematol.* **2**: 58.

Billing, R. J., B. Rafizadeh, I. Drew, G. Hartman, R. Gale, and P. Terasaki. 1976. Human B-lymphocyte antigens expressed by lymphocytic and myelocytic leukemia cells. I. Detection by rabbit antisera. *J. Exp. Med.* **144**: 167.

Bullough, W. S., and F. B. Lawrence. 1960. The control of epidermal mitotic activity in the mouse. *Proc. R. Soc. Lond., B, Biol. Sci.* **151**: 517.

Cline, M. J., and R. J. Billing. 1977. Antigens expressed by human B lymphocytes and myeloid stem cells. *J. Exp. Med.* **146**: 1143.

Drew, S. I., O. Bergh, J. D. McClelland, R. Mickey, and P. I. Terasaki. 1977a. Antigenic specificities detected on papainized human granulocytes by microgranulocytotoxicity. *Transplant. Proc.* **9**: 639.

Drew, S. I., P. I. Terasaki, R. J. Billing, O. J. Bergh, J. Minowada, and E. Klein. 1977b. Group-specific human granulocyte antigens on a chronic myelogenous leukemia cell line with a Philadelphia chromosome marker. *Blood* **49**: 715.

Glass, J., L. M. Lavidor, and S. H. Robinson. 1975. Studies of murine erythroid cell development: Synthesis of heme and hemoglobin. *J. Cell Biol.* **65**: 298.

Golde, D. W., and M. J. Cline. 1972. Identification of the colony-stimulating cell in human peripheral blood. *J. Clin. Invest.* **51**: 2981.

Golde, D. W., N. Bersch, and M. J. Cline. 1976. Potentiation of in vitro erythropoiesis by dexamethasone. *J. Clin. Invest.* **57**: 57.

Herman, S. P., D. W. Golde, and M. J. Cline. 1978. Neutrophil products that inhibit cell proliferation. Relation to granulocytic "chalone." *Blood* **51**: 207.

Hoffman, R., E. D. Zanjani, J. D. Lutton, R. Zalusky, and L. R. Wasserman. 1977. Suppression of erythroid-colony formation by lymphocytes from patients with aplastic anemia. *N. Engl. J. Med.* **296**: 10.

Hoffman, R., E. D. Zanjani, J. Vila, R. Zalusky, J. D. Lutton, and L. R. Wasserman. 1976. Diamond-Blackfan syndrome: Lymphocyte-mediated suppression of erythropoiesis. *Science* **193**: 899.

Klein, E., H. Ben-Bassat, H. Neumann, P. Ralph, J. Zeuthen, A. Polliack, and F. Vánsky. 1976. Properties of the K562 cell line derived from a patient with chronic myeloid leukemia. *Int. J. Cancer* **18:** 421.

Krantz, S. B. 1974. Pure red-cell aplasia. *N. Engl. J. Med.* **291:** 345.

Lajtha, L. G. 1973. Commentary on "Chalone of the granulocyte system," by T. Rytömaa, and "Granulopoiesis-inhibiting factor," by W. R. Paukovitz. *Natl. Cancer Inst. Monogr.* **38:** 157.

Litwin, S. D., and E. D. Zanjani. 1977. Lymphocytes suppressing both immunoglobulin production and erythroid differentiation in hypogammaglobulinaemia. *Nature* **266:** 57.

Lord, B. I., L. Cercek, B. Cercek, G. P. Shah, T. M. Dexter, and L. G. Lajtha. 1974. Inhibitors of haemopoietic cell proliferation: Specificity of action within the haemopoietic system. *Br. J. Cancer* **29:** 168.

Lozzio, C. B., and B. B. Lozzio. 1975. Human chronic myelogenous leukemia cell-line with positive Philadelphia chromosome. *Blood* **45:** 321.

Paukovitz, W. R. 1971. Control of granulocyte production: Separation and chemical identification of a specific inhibitor (chalone). *Cell Tissue Kinet.* **4:** 539.

―――. 1973. Granulopoiesis-inhibiting factor: Demonstration and preliminary chemical and biological characterization of a specific polypeptide (chalone). *Natl. Cancer Inst. Monogr.* **38:** 147.

Rabson, A. S., S. W. Rosen, A. H. Tashjian, Jr., and B. D. Weintraub. 1973. Production of human chorionic gonadotropin in vitro by a cell line derived from a carcinoma of the lung. *J. Natl. Cancer Inst.* **50:** 669.

Rytömaa, T., and K. Kiviniemi. 1968a. Control of granulocyte production. I. Chalone and antichalone, two specific humoral regulators. *Cell Tissue Kinet.* **1:** 329.

―――. 1968b. Control of granulocyte production. II. Mode of action of chalone and antichalone. *Cell Tissue Kinet.* **1:** 341.

Terasaki, P. I., and J. D. McClelland. 1964. Microdroplet assay of human serum cytotoxins. *Nature* **204:** 998.

Use of Molecular Probes for Detection of Human Hematopoietic Progenitors

G. B. Price and R. L. Krogsrud

Division of Biological Research, Ontario Cancer Institute
Toronto, Ontario M4X 1K9, Canada

We are embarking upon a new era of investigation of cell growth and differentiation. In vitro culture assays for all the recognized committed progenitors of hematopoiesis now exist. Johnson and Metcalf (this volume) have described an assay for progenitors that are present in day-11 fetal liver and that exhibit pluripotentiality (as single cells) by giving rise to multiple lineages of hematopoiesis—erythrocytes, megakaryocytes, and granulocytes. And so, even pluripotent stem cells of hematopoiesis may succumb in the near future to in vitro analysis.

Although fruitful information may be obtained by analysis of minority populations of progenitors and stem cells with partially purified or purified probes, we believe from experience with purification and characterization of regulators of hematopoiesis (Price et al. 1973, 1974, 1975, 1976) that the data still present problems of interpretation; that is, are the observations a result of primary stimulatory effects or merely a consequence of a secondary effect, or cascade, involving molecular and cellular interactions. Effective understanding of regulatory factors of progenitors and stem cells, hematopoietic stem cells in particular, requires that essential factors and target cells be available either in homogeneity or in a state of isolation from possible secondary effectors. Although physical separation procedures provide significant enrichment, it is unlikely that any single procedure will be sufficient. Our approach is to break down further the heterogeneity still present in physically enriched populations of progenitors and stem cells (1) by attempting to recognize qualitative or quantitative discriminatory surface antigens on progenitor cells; and (2) by using the technique of fluorescence polarization with activated cell-sorting capacity to detect appropriately responsive progenitors or stem cells, e.g., the erythroid progenitors—the colony-forming units—erythroid (CFU-E) or burst-forming units-erythroid (BFU-E)—that are responsive to erythropoietin.

CURRENT STATUS OF ANTIGEN DISCRIMINATION

The development of an automated device for sorting cells on the basis of fluorescence or light-scattering characteristics (Bonner et al. 1972; Hulett et al. 1973) has opened the door to purification of minority cell populations such as pluripotent stem cells. Fluorescence resulting from labeled antibodies has already been used to purify subpopulations of lymphocytes (Strober 1976; Cantor et al. 1975) and leukemic cells (Janossy et al. 1976). The sorting of subpopulations of bone marrow on the basis of combined small-angle (2–10°) and large-angle (80–100°) light scattering (Visser et al. 1977) and by small-angle light scattering alone (data not shown) has demonstrated that colony-forming units—spleen (CFU-S) are heterogeneous in this characteristic; hence, there is need for more specific markers.

The search for a unique surface marker for stem cells began with the observations by Golub (1972) and van den Engh and Golub (1974) that antimouse brain serum was cytotoxic for CFU-S, the pluripotent stem cell. We found that the human brain also exhibited these antigens and that rabbit antisera to human brain crossreacted with mouse CFU-S (Krogsrud et al. 1977). This antiserum was found to be cytotoxic in the human system as well, reducing the number of granulocyte progenitors, CFU-C, and erythroid progenitors BFU-E, by up to 50%. More interestingly, our assay revealed that the amount of ΔC, a probable precursor to CFU-C, was reduced by 50 to 100% by treatment with this antiserum. The greater effectiveness toward this precursor may have indicated a higher density of antigen, implying a quantitative difference between progenitors that could be exploited on the cell sorter.

What other antigens occur on stem cells? Presumably histocompatibility-related antigens may be found which maintain the self-identity of the organism. Evidence has been presented (van den Engh et al. 1977) that H2 (K and D) antigens but not Ia antigens occur on CFU-S. If one considers the similarities of characteristics of the hematopoietic stem cell and embryonic cells with regard to pluripotentiality, then another possible marker comes to mind. The T/t locus of the mouse produces a surface antigen that is found on early embryonic cells (Bennett 1975). Some mutations at this locus cause embryonic death, probably by impaired intracellular recognition (Bennett et al. 1976).

In the mouse, such antigens are found also on sperm. Our investigation of a parallel situation in the human has led to the observation that autoimmune antisperm sera do indeed recognize specificities on the early hematopoietic progenitors in human marrow and also crossreact with mouse CFU-S (Krogsrud et al. 1977).

SORTING BY FLUORESCENCE

Using unseparated bone marrow and the rabbit antihuman brain serum, we attempted to sort progenitor cells; to do this we employed an automated cell sorter that had been built in our laboratory. Table 1 gives the results of three such experiments. The sorted cells were those that met both criteria of fluorescence by antibody-labeled cells and light scattering indicative of live cells. Significant enrichment was obtained both in CFU-S (14.5 times)

Table 1
Progenitor Cells Sorted with Rabbit Antihuman Brain Serum Label

		Colonies per 2×10^5 cells			Enrichment[d] ($\times$)	Recovery[e] (%)
Serum[a]	Progenitor	sorted	unsorted[b]	straight through[c]		
RAHuBr	M CFU-S	318	20	–	14.5	96
NRS	M CFU-S	13	20	22	–	95
RAHBr	M CFU-E	1003	125	–	20.5	100[f]
NRS	M CFU-E	0	123	49	–	100
RAHBr F(ab')$_2$	M CFU-S	168	1	39	3.4	40
NRS F(ab')$_2$	M CFU-S	60	54	49	1.2	100
RAHBr F(ab')$_2$	M CFU-C	178	3	168	1.2	6
NRS F(ab')$_2$	M CFU-C	99	244	152	–	100
RAHuBr	Hu ΔC[g]	100%	0%	150%	–	–
RAHuBr	Hu CFU-C	81	23	18	4.5	100
RAHuBr	Hu CFU-E	20	4	3	6.6	100

Abbreviations: M, mouse; Hu, human.

[a] Mouse bone marrow was incubated with rabbit antihuman brain serum (mouse liver, kidney, and thymus adsorbed) and then with fluorescein-isothiocyanate–conjugated goat anti-rabbit IgG. Human bone marrow was incubated in the same manner but with serum adsorbed with type-AB erythrocytes.

[b] This represents those cells that were not deflected.

[c] Sorting function was disabled while a number of cells equal to the total observed during sorting were passaged straight through the machine.

[d] Based on the number of CFU/10^5 cells in the sorted population divided by the number of CFU/10^5 cells in the sample run straight through without sorting.

[e] Total number of CFU recovered both in sorted and unsorted populations.

[f] In actuality, greater than 100% recovery was observed. See text.

[g] Expressed as percentage net increase of CFU-C in liquid culture: $\left(\frac{\text{CFU-C day 7} - \text{CFU-C day 0}}{\text{CFU-C day 0}}\right) \times 100$. Progenitors responsible for ΔC effect were detected only in sorted fractions. Because no direct clonal assay was available, no enrichment could be calculated.

and CFU-E (20.5 times) for the first experiment. Enrichment was based on comparison with cells that were merely passed through the machine. When recovery of activity was calculated, we found that more than 100% of the CFU-E were recovered, suggesting that depletion of an inhibitory cell population from the sorted population may have occurred.

Using the F(ab')$_2$ fragment of antibody for labeling, we accomplished a low recovery and only a modest enrichment of CFU-S (3.4 times).

In the case of the human bone marrow, significant enrichment was achieved both for BFU-E (6.6 times) and CFU-C (4.5 times) and the ΔC activity was found entirely within the sorted fraction.

With these preliminary results for encouragement, we next sought to enrich the stem cells overall by combining the ficoll-hypaque gradient separation with enhanced fluorescence labeling techniques for sorting.

CURRENT STATUS OF MOLECULAR PROBES

Our attempt to characterize molecular probes began with a study of the observations of Austin et al. (1971). We then proceeded to investigate factors of human-leukocyte conditioned medium essential for growth of human granulocyte progenitors, CFU-C, ΔC, and T-lymphocyte progenitors. Three high-molecular-weight protein species that stimulate granulocyte colony growth (colony-stimulating activities, CSA) have been described (Price et al. 1975). The CFU-C targets of these stimulators are essentially identical since the velocity sedimentation profiles show that similar profiles of CFU-C responding to any of the species of CSA can be obtained (Price and McCulloch 1977). A cross-reaction of these factors with β_2-microglobulin (Price et al. 1976) prompted us to test human urinary β_2-microglobulin for activity. Our preliminary tests resulted in a very modest number of clusters which we decided to ignore for the moment. Subsequently, samples of purified β_2-microglobulin were prepared from the urine of patients with renal dysfunction. Preliminary tests of the preparation showed biochemical and immunochemical homogeneity with β_2-microglobulin. In the two experiments shown in Figure 1, β_2-microglobulin stimulated the formation of granulocyte colonies. (After 1 month's storage at $-20°$ C, potency was significantly reduced.) The difference in the previous preparation and that available in the experiments shown in Figure 1 may have been a result of storage method rather than degree of purification; however, new samples are being obtained for future work at which time homogeneity of all preparations will be checked either by N-terminal analysis or finger-

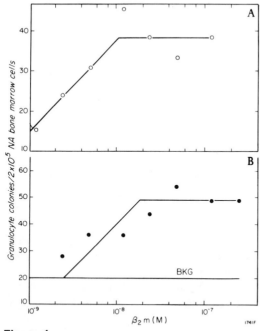

Figure 1
Two experiments showing stimulation of human granulocytic colonies by human urinary β_2-microglobulin (β_2 m). (NA indicates nonadherent.) (*A*) Experiment in which no background of granulocyte colony formation was measured. (*B*) Experiment in which an average of 20 granulocyte colonies was observed as a background in the absence of exogenous sources of CSA.

printing of peptide digests. However, with regard to the CSA of conditioned medium, N-terminal amino acid analysis of 15,000 mol. wt. CSA did not result in amino acid residues compatible with the N terminus of β_2-microglobulin.

Factors exhibiting CSA have been obtained from membranes of peripheral leukocytes; therefore, it was thought that a potential candidate for these factors might be the histocompatibility antigens. Tests of the ability of appropriate HL-A antisera to block CSA, as anti–β_2-microglobulin did, indicated a likelihood that these factors may have been HL-A–associated antigens (Till et al. 1977). Successful pursuance of these observations should provide additional data in support of histocompatibility antigens possessing tissue or organ-organizing and stimulating effects. Of course, the best-documented analogy is that of the H-Y gene-product-associated organizing ability of testicular development (Ohno et al. 1976; Ohno 1977; Wachtel 1977).

A low-molecular-weight factor that showed CSA has been described (Price et al. 1973,1974). It has the interesting chemical properties of a peptide which partitions into diethyl ether and then redissolves in aqueous solvents. This material is a poor CSA of CFU-C; however, we found it essential for stimulation of ΔC (Niho et al. 1975; Till et al. 1977). Preliminary experiments by J. Graves (presently at the Department of Physiology, Medical University of South Carolina, Charleston) in which the effect of low-molecular-weight CSA factors on synthetic lipid bilayer membranes was tested suggested that some ion-transporting potential may exist within this class of stimulators. However, further experiments will be necessary to confirm this possible ionophoric activity and to identify the ions which may be transported by low-molecular-weight CSA factors. Indirect supportive evidence was obtained from the observation that the Ca^{++} ionophore A23187 could successfully replace the factor from leukocyte conditioned medium (Price and McCulloch 1978).

Because of these experiments, P. Ottensmeyer et al. (in prep.) investigated the molecular structure of low-molecular-weight factors that show CSA and found them to be cyclic compounds. One class was distinguished by its biological reactivity (Price et al. 1974); it was observed only in some preparations of a low-molecular-weight CSA factor from leukemic peripheral blood and was examined after exposure to 10^{-3} M $CaCl_2$ and extraction into diethyl ether. A Ca^{++} ion appeared to be visible in one of the two looped structures, further suggesting the potential and mechanism whereby this stimulator may act. Representatives of the other classes (obtained from normal and most leukemic peripheral blood) were structurally different and appeared to be coiled in one loop. We have not as yet attempted to charge these molecules with ions.

A fourth factor present only in mitogen-stimulated leukocyte conditioned medium was found to be essential for growth of T lymphocyte colonies from their progenitors. This activity was observed usually as an approximately 40,000 mol. wt. protein, although in leukemia-peripheral-leukocyte conditioned medium (stimulated with phytohemagglutinin) (Price and McCulloch 1977) it was observed occasionally as an approximately 25,000 mol. wt. protein. Preliminary experiments with reduction of this 40,000 mol. wt.

TABLE 2
Effect of Histamine on Cell-Cycle State of Hematopoietic Progenitors

| | *Percentage loss due to hydroxyurea* | | | |
| | *Mouse* | | *Human* | |
Stimulus	CFU-S	CFU-C	ΔC	CFU-C
Control	9	34	9	38
2-Methylhistamine				
10^{-8} M	8	34	7	42
10^{-7} M	9	32	16	35
4-Methylhistamine				
10^{-8} M	23	32	78	41
10^{-7} M	47	33	85	36
4-Methylhistamine + 10 M metiamide				
10^{-8} M	3	30	0	36
10^{-7} M	61	33	9	35

Day 0, 14 colonies; day 7, 66 colonies; $\Delta C = \frac{66-14}{14} = 3.7$ times.

factor by 2-mercaptoethanol yielded two molecules of approximately 25,000 and 15,000 mol. wt. The 40,000 mol. wt. species possessed no detectable CSA for CFU-C, although the 15,000 mol. wt. subunit moiety did. (The 15,000 mol. wt. moiety may be similar or identical to the protein originally isolated as a high-molecular-weight CSA factor.)

In addition to these biological probes, J. Byron (1975, 1976) has shown that a number of other molecular probes may be used successfully to manipulate the pluripotent stem cell of mouse bone marrow. In particular, 4-methylhistamine (Byron 1976) appeared to be a potentially specific probe for hematopoietic stem cells. The possibility that this reagent might be useful in subsequent analysis by flow microfluorometry of early mouse and hematopoietic progenitors prompted us to repeat the experiments of Byron (1976) and extend them to the human system. Byron's assay depended upon the activation of progenitor cells into cell cycle by specific drugs. He determined this activation by detecting the increased number of progenitors in S phase using the cycle-specific drug hydroxyurea to kill such progenitors. As shown in Table 2, 4-methylhistamine but not 2-methylhistamine activated CFU-S and human progenitors ΔC but not mouse or human CFU-C. This activation could also be blocked by metiamide, the specific inhibitor of 4-methylhistaminebinding. Thus, 4-methylhistamine provided us with a potentially specific probe for some of the earliest progenitors of hematopoiesis; it should provide an interesting complement to the CSA probes for CFU-C.

CURRENT STATUS OF FLUORESCENCE POLARIZATION MEASUREMENTS OF CELL RESPONSES TO PROBES

The detection of a biological response to molecular regulators has always represented a biochemical problem until the last decade when biophysical characterizations became possible, for example, by velocity sedimentation (Miller and Phillips 1969; Miller et al. 1975) or by buoyant density (Leif and Vinograd 1964; Shortman 1969; Gorczynski et al. 1970). More recently, an association between cellular microviscosities or "structuredness" and either cell-cycle state or response has been uncovered (Cercek and Cercek 1972; Inbar and Shinitzky 1974; Cercek et al. 1975; Arndt-Jovin et al. 1976; Cercek and Cercek 1976; Price et al. 1977). The fluorescence polarization methods used to make such biophysical assessments of cell character and response were pioneered by Weber (1972), who studied biological macromolecular structure; these methods were subsequently applied to the study of membranes (Radda and Vanderkooi 1972; Arndt-Jovin and Jovin 1976).

Similarly, it is now possible to measure cytoplasmic fluorescence depolarization of single cells in a flow system (Price et al. 1977). The probes, low-molecular-weight CSA factors (Price et al. 1977) and A23187, together with erythropoietin and pokeweed-mitogen-stimulated mouse spleen conditioned medium, have been used successfully to produce changes in cytoplasmic fluorescence polarization of mouse and human bone marrow (data

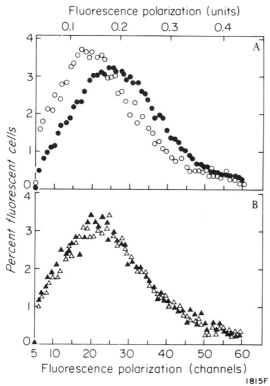

Figure 2

Fluorescence polarization changes in mouse bone marrow. (*A*) 4-Methylhistamine versus control. (o) Response to stimulation by 10^{-8} M 4-methylhistamine for 1 hour's exposure at 37°C; (●) control, untreated. (*B*) Metiamide versus 4-methylhistamine + metiamide. (△) 10^{-5} M metiamide; (▲) metiamide + 4-methylhistamine.

TABLE 3
Effect of Histamine Upon Marrow Cells as Measured by Fluorescence Polarization

Stimulus[a]	P^b mean	Fiducial limits[c]
No control	0.190	0.188–0.192
Metiamide	0.174	0.170–0.178
4-Methylhistamine	0.158	0.156–0.165
4-Methylhistamine + metiamide	0.178	0.175–0.182

[a] Metiamide, 10^{-5} M; 4-methylhistamine, 10^{-8} M.

[b] Polarization $= \dfrac{I_{\|} - I_{\perp}}{I_{\|} + I_{\perp}}$, where $I_{\|}$ is intensity of fluorescence in the parallel plane and $I_{\perp}$ is intensity of fluorescence in the perpendicular plane of the polarized excitation light.

[c] Fiducial limits are 98%.

not shown). Such changes in fluorescence polarization in response to appropriate probes may prove to be early indicators of cell response.

The interesting probes of early hematopoietic progenitors, 4-methylhistamine and metiamide, produce fluorescence polarization changes which may be examined for correlation with biological stimulation; specific blockage of the stimulus was also observed (Fig. 2; Table 3). When these experiments were carried out on mouse bone marrow, fluorescence polarization changes approximated the observed biological response of cell activation, as suggested in Table 2; 4-methylhistamine stimulated or caused a decrease of polarization (0.158 versus 0.190 or 0.174). Furthermore, blockage by metiamide of the effects of 4-methylhistamine as measured by polarization (0.178 versus 0.174) was also observed. Therefore, analytical measurement of molecular probes appeared to correlate with cytoplasmic fluorescence depolarization. However, satisfactory proof of the true correspondence of these fluorescence polarization changes to biological response remains to be found. The implementation of activated sorting on the basis of fluorescence polarization has begun, and preliminary results suggest that progenitor populations may be detected and enriched in an appropriately selected and sorted fraction.

SUMMARY

A biophysical approach to the characterization of hematopoietic progenitor cell is possible. Well-known techniques of physical enrichment of hematopoietic stem cells used in conjunction with the modern technology of fluorescence-activated cell sorting provide one the opportunity to evaluate the direct effects of potential regulators of hematopoiesis. Preparation of antigenic probes, purification and description of biological molecular probes, and the apparent analytical success of fluorescence polarization measurements have been refined to the point that some definition of the processes of the differentiation of and commitment by hematopoietic progenitors appears possible.

Acknowledgments

We wish to express our gratitude to Ms. K. Benzing and Ms. P. Thompson for expert technical assistance. Dr. R. G. Miller, Mr. W. Taylor, and Mr. M. McCutcheon were responsible primarily for the construction and maintenance of the fluorescence flow microfluorimetry apparatus. Preparations of purified human β_2-microglobulin were obtained from Cedarlane Laboratories, London, Ontario. Research was supported by the Medical Research Council of Canada, the National Cancer Institute of Canada, and the Ontario Cancer Research and Treatment Foundation.

REFERENCES

Arndt-Jovin, D. J., and T. M. Jovin. 1976. Cell separation by using fluorescence anisotropy. *Prog. Clin. Biol. Res.* **9**: 123.

Arndt-Jovin, D. J., W. Ostertag, H. Eisen, F. Klimek, and T. M. Jovin. 1976. Studies of cellular differentiation by automated cell separation. Two model systems: Friend virus-transformed cells and *Hydra attenuata*. *J. Histochem. Cytochem.* **24**: 332.

Austin, P. E., E. A. McCulloch, and J. E. Till. 1971. Characterization of the factor in L-cell conditioned medium capable of stimulating colony formation by mouse marrow cells in culture. *J. Cell. Physiol.* **177**: 121.

Bennett, D. 1975. The T-locus of the mouse. *Cell* **6**: 441.

Bennett, D., E. A. Boyse, and L. J. Old. 1976. Cell interactions. In *3rd Lepetit Colloquim* (ed. L. G. Silvestri), p. 247. North Holland Publishing Co., Amsterdam.

Bonner, W. A., H. R. Hulett, R. G. Sweet, and L. A. Herzenberg. 1972. Fluorescence activated cell sorting. *Rev. Sci. Instrum.* **43**: 404.

Byron, J. W. 1975. Manipulation of the cell cycle of the hemopoietic stem cell. *Exp. Hematol.* (Copenh.) **3**: 44.

———. 1976. Bone marrow toxicity of metiamide. *Lancet* **ii**: 1350.

Cantor, H., E. Simpson, V. L. Sato, C. G. Fathman, and L. A. Herzenberg. 1975. Characterization of subpopulations of T lymphocytes. I. Separation and functional studies of peripheral T cells binding different amounts of fluorescent anti-Thy 1.2 (theta) antibody using a fluorescence cell sorter (FACS). *Cell. Immunol.* **15**: 180.

Cercek, L., and B. Cercek. 1972. Studies on the structuredness of cytoplasm and rates of enzymatic hydrolysis in growing yeast cells. I. Changes induced by ionizing radiation. *Int. J. Radiat. Biol.* **21**: 445.

———. 1976. Changes in the structuredness of cytoplasmic matrix (SCM) in human lymphocytes induced by PHA and cancer basic protein as measured in single cells. *Br. J. Cancer* **33**: 539.

Cercek, L., P. Milenkovic, B. Cercek, and L. G. Lajtha. 1975. Induction of PHA response in mouse bone marrow cells by thymic extracts as studied by changes in the structuredness of cytoplasmic matrix. *Immunology* **29**: 885.

Golub, E. S. 1972. The distribution of brain associated θ antigen cross reactive with mouse in the brain of other species. *J. Immunol.* **109**: 168.

Gorczynski, R. M., R. G. Miller, and R. A. Phillips. 1970. Homogeneity of antibody-producing cells as analysed by their buoyant density in gradients of Ficoll. *Immunology* **19**: 817.

Hulett, H. R., W. A. Bonner, R. G. Sweet, and L. A. Herzenberg. 1973. Development and application of a rapid cell sorter. *Clin. Chem.* **19**: 813.

Inbar, M., and M. Shinitzky. 1974. Cholesterol as a bioregulator in the development and inhibition of leukemia. *Proc. Natl. Acad. Sci. U.S.A.* **71**: 4229.

Janossy, G., M. F. Greaves, T. Revesz, T. Alister, M. Roberts, J. Durrant, B. Kirile, D. Catovsky, and M. E. J. Beard. 1976. Blast crisis of chronic myeloid leukemia (CML). II. Cell surface marker analysis of "lymphoid" and myeloid cases. *Br. J. Haematol.* **34**: 179.

Krogsrud, R. L., J. Bain, and G. B. Price. 1977. Serologic identification of hemopoietic progenitor cell antigens common to mouse and man. *J. Immunol.* **119**: 1486.

Leif, R. C., and J. Vinograd. 1964. The distribution of buoyant density of human erythrocytes in bovine albumin solutions. *Proc. Natl. Acad. Sci. U.S.A.* **51**: 520.

Miller, R. G., and R. A. Phillips. 1969. Separation of cells by velocity sedimentation. *J. Cell. Physiol.* **73**: 191.

Miller, R. G., R. M. Gorczynski, L. Lafleur, H. R. MacDonald, and R. A. Phillips. 1975. Cell separation analysis of B and T lymphocyte differentiation. *Transplant. Rev.* **25**: 59.

Niho, Y., J. E. Till, and E. A. McCulloch. 1975. Granulopoietic progenitors in suspension culture: A comparison of stimulatory cells and conditioned media. *Blood* **45**: 811.

Ohno, S. 1977. The original function of MHC antigens as the general plasma membrane anchorage site of organogenesis-directing proteins. *Immunol. Rev.* **33**: 59.

Ohno, S., L. C. Christian, S. S. Wachtel, and G. C. Koo. 1976. Hormone-like role of H–Y antigen in bovine freemartin gonad. *Nature* **261**: 597.

Price, G. B., and E. A. McCulloch. 1978. Cell surfaces and the regulation of hemopoiesis. *Semin. Hematol.* (In press.)

Price, G. B., E. A. McCulloch, and J. E. Till. 1973. A new human low molecular weight granulocyte colony stimulating activity. *Blood* **42**: 341.

———. 1976. Cross reactivity of β_2-microglobulin with human granulocyte colony-stimulating activity. *J. Immunol.* **117**: 416.

Price, G. B., M. McCutcheon, W. B. Taylor, and R. G. Miller. 1977. Measurement of cytoplasmic fluorescence depolarization of single cells in a flow system. *J. Histochem. Cytochem.* **25**: 597.

Price, G. B., J. S. Senn, E. A. McCulloch, and J. E. Till. 1974. Heterogeneity of molecules with low molecular weight isolated from media conditioned by human leukocytes and capable of stimulating human granulopoietic progenitor cells. *J. Cell. Physiol.* **84**: 383.

———. 1975. The isolation and properties of granulocyte colony stimulating activities. *Biochem. J.* **148**: 209.

Radda, G. K., and J. Vanderkooi. 1972. Can fluorescence probes tell us anything about membranes. *Biochim. Biophys. Acta* **265**: 509.

Shortman, K. 1969. Equilibrium density gradient separation and analysis of lymphocyte populations. In *Progress in separation and purification* (ed. T. Gerritsen), vol. 2, p. 167. John Wiley & Sons, New York.

Strober, S. 1976. Maturation of B lymphocytes in rats. III. Two subpopulations of memory B cells in the thoracic duct lymph differ by size, turnover rate, and surface immunoglobulin. *J. Immunol.* **177**: 1288.

Till, J. E., G. B. Price, J. S. Senn, and E. A. McCulloch. 1977. Cell interactions in the control of hemopoiesis. In *Growth kinetics and biochemical regulation of normal and malignant cells* (eds. B. Drewinko and R. M. Humphrey), p. 223. Williams & Wilkins Co., Baltimore.

van den Engh, G. J., and E. S. Golub. 1974. Antigenic differences between hemopoietic stem cells and myeloid progenitors. *J. Exp. Med.* **139**: 1621.

van den Engh, G. J., J. Russell, and D. de Cicco. 1977. Expression of H-2 antigens on the hemopoietic stem cell. *Exp. Hematol.* (Copenh.) **5** (suppl. 2): 4.

Visser, J. W. M., S. Bol, G. J. van den Engh, and A. Jongeling. 1977. Analysis and sorting of early hemopoietic cells by flow system light scattering measurements. *Exp. Hematol.* (Copenh.) **5** (suppl. 2): 39.

Wachtel, S. S. 1977. H-Y antigen: Genetics and serology. *Immunol. Rev.* **33**: 33.

Weber, G. 1972. Use of fluorescence in biophysics: Some recent developments. *Annu. Rev. Biophys. Bioeng.* **1**: 553.

The Maturation and Activation of Mononuclear Phagocytes

Z. A. Cohn

Laboratory of Cellular Physiology and Immunology
The Rockefeller University, New York, New York 10021

PRECURSORS AND DEVELOPMENT

It is now generally conceded that the very heterogenous population of mononuclear phagocytes found in the tissues are derived from common precursors in the marrow. The earliest recognizable forms are the replicating monoblasts and promonocytes. Although generally assumed that these arise from multipotential stem cells, formal proof of the sort employed for the granulocyte series is still lacking. Nevertheless, considerable information exists on the progeny of committed stem cells that produce macrophage colonies in soft agar systems, and this will be reviewed in other chapters of this volume.

It is important to realize that the majority of the information concerning this series of phagocytes derives from the mouse. In this species, it has been estimated that the total population of monoblasts and promonocytes represents approximately $2-3 \times 10^5$ cells and that the bone-marrow content of monocytes is in the neighborhood of 10^6 cells. Monocytes numbering approximately 10^6 cells also are constantly in the circulation and leave this compartment randomly under steady-state conditions (Table 1).

When responding to an inflammatory event in the periphery, the monocyte system has rather limited potential as compared with that of the granulocytes. Much of this information comes from the studies of van Furth and his colleagues (van Furth 1976). First, intravascular monocytes can be focused into local lesions, attaching to the endothelial cell surface in postcapillary venules and emigrating into the tissues. It has been estimated that 70% of the total daily production of monocytes can be relegated to a site such as the peritoneal cavity. This raises the important problem of the supply from other tissues with their steady-state number of cells as well as of other infectious or tumor-bearing sites. A second response occurs in the bone marrow and is characterized by a decrease in the generation time of monocyte precursors leading, perhaps, to a twofold increase in the production rate of monocytes. When one considers that the intramarrow pool of adult granulo-

Table 1
Life History of Mononuclear Phagocytes

Bone marrow	Blood	Tissues
Monoblasts, 2×10^5 Promonocytes, 5×10^5 Monocytes, 2×10^6	→ monocytes, 1×10^6 →	macrophages $> 10^8$
Production rate 1–2×10^6/day	halving time ~ 24 hours	lifespan > 60 days

Data taken from van Furth 1976.

cytes ready to be released under similar circumstances is 100 times greater than the circulating level, the distinction between the two cell series is evident. A third response is the release of colony-forming precursors from the bone marrow (Lin and Stewart 1974; Stewart et al. 1975). It has been known for some time that conditioned medium from fibroblasts stimulates the limited division of mononuclear phagocytes.

This response occurred only with cell populations that had been evoked with an inflammatory stimulant. Similar populations induced with thioglycolate medium contained about 5% of cells which form colonies in liquid systems when cultured at low density ($\pm 10^{-4}$). Although little is known about the fate of these colony-forming units in vivo, they are unlikely to add more than a limited number of cells to the local environment.

EMIGRATION AND CHEMOTAXIS

One of the more critical steps in the life history of the mononuclear phagocytes is that of their delivery to tissue sites both within and without the vasculature. Macrophages intercalate with endothelial cells to form an important barrier in the liver on the walls of sinusoids. Similarly, large numbers of macrophages are found in the spleen, lung, adrenal and serous cavities, and outside of the walls of all small blood vessels. This long-lived population carries out many of the basic functions of macrophages involved with the surveillance of the *milieu interieur*. In a sense, it is this population, not the granulocytes, that responds initially to tissue insults.

The process of emigration is poorly understood; however, it does involve at least two phases. First is the adherence of circulating monocytes to the surface of endothelial cells and the subsequent spreading reaction whereby they conform to the shape of the substrate. Second is the translational motility necessary to pass between endothelial cells, penetrate the basement membrane, and enter the tissues. These steps occur regularly during the steady state and random dissemination of monocytes to the tissues and at a much accelerated and focused pace during inflammation.

We recently became interested in these phenomena and, in the past few years, developed in vitro assays that mimic early phases of the reaction. In brief, we examined the ability of freshly explanted cells to spread on glass or plastic substrates, a process in which the apparent surface area of the cell may increase from six to eight times in a short time. Two distinct and dissociable systems present in plasma and serum were originally described by Bianco et al. (1976). The first was related to the contact phase of blood coagulation. This was activated by glass and not by plastic surfaces, was depleted by kaolin adsorption, and was inhibited by soybean trypsin inhibitors. Presumably, in the presence of Hageman factor, a trypsinlike protease was generated which led to the rapid spreading of macrophages. This finding was in keeping with the prior observations of Rabinovitch (1975) on the role of proteases in rapid spreading. The second factor could be produced in kaolin-adsorbed plasma and was associated with the activation either of the classical or alternative pathways of complement activation.

THE NATURE OF THE COMPLEMENT-DEPENDENT FACTOR

Further analysis of the complement-associated factor was carried out (O. Götze et al., in prep.). It was first ascertained that individual components of the classical pathway failed to induce spreading, even though activation of the pathway generated a spreading factor. This led to the finding that mixtures of factors B, $\overline{D}$ and C_{3b} resulted in a potent spreading activity and implicated the alternative, or properdin, system. Further analysis revealed that the active factor was neither C_{3b} nor Ba but, instead, was the large split product of proteolytic cleavage of factor B, namely Bb. Purification of this 64,000-dalton fragment yielded a product that produced spreading at a concentration of 1.4 μg. Treatment of Bb with diisopropyl fluorophosphate (DFP) resulted in a loss both of enzymatic and spreading activities and suggested that the intact catalytic site was required. Inasmuch as it had been shown that macrophages synthesize factor B, the possibility was raised that cleavage of B by proteases either extracellularly or on the plasma membrane could lead to spreading and other parameters of activation.

THE ROLE OF FACTOR Bb AS AN INHIBITOR OF THE TRANSLATIONAL MOTILITY OF MACROPHAGES

It was soon suspected that a factor that induced spreading and a high degree of polarization of the macrophage cytoplasm would also alter the motility of these cells. To examine this in more detail (C. Bianco et al., in prep.), we utilized a capillary tube assay in which concentrated macrophages crawl out onto a cover slip and form a face of emigrating cells. This procedure, which is commonly employed to assay the macrophage-inhibitory-factor activity of lymphokines, gives a semiquantitative index of the motility of cell populations. Initial experiments revealed that factors generated during the contact phase of blood coagulation and during complement activation markedly inhibited the migration of macrophages. This result was in keeping with prior studies on spreading. The analogy was extended further when purified factor Bb yielded marked inhibition at microgram concentrations. It was of interest that a natural antagonist of Bb was generated during the activation of the classical pathway. Peptides produced during the cleavage of C_5 enhanced motility and stimulated migration out of the capillary tubes. These findings suggest that products of the classical and alternative pathways operating upon the mononuclear phagocytes of the tissues in an antagonistic fashion might control both the extent of cell migration as well as the number of cells entering a local lesion. Such speculations, however, will require experiments under in vivo conditions.

THE ACTIVATION OF MACROPHAGES

Tissue macrophages exist in many functional states, depending upon their location and exposure to particulates. Under steady-state conditions, macrophages of organs such as liver and spleen are constantly involved in the physiological destruction of formed elements such as erythrocytes, platelets, and leukocytes. Such cells are influenced by the uptake of digestible particles, as we shall describe in a subsequent section, and function in a state of activation. Activation in its simplest terms represents an elevation of function from steady-state resting values and may be concerned with a wide variety of cellular parameters.

Historically, activation has been associated with the alterations of the vacuolar apparatus. The word stems from the in vitro studies of Cohn and colleagues (Cohn and Weiner 1963; Gordon and Cohn 1973) and from in vivo modifications described by Mackaness (1964) related to the antimicrobial properties of the cell which evolve during the course of infection. More recently, the antitumor or cytotoxic role of mononuclear phagocytes has been included in the definition. It is apparent, however, that activation should be considered in a much broader context—one based upon our current knowledge of the physiology and biochemistry of the cell. I shall briefly review, therefore, some of the information concerning these properties, properties which have occurred both under inflammatory situations in which the macrophage was exposed to humoral and cellular elements and when it was exposed to lymphocyte-derived products (lymphokines).

PLASMA MEMBRANE COMPONENTS

Striking changes occur in the exteriorly disposed plasma-membrane polypeptides, as revealed by lactoperoxidase-mediated iodination. A comparison of inflammatory macrophages and resident mouse peritoneal cells has been made by Yin et al. (in prep.). Cells were iodinated in suspension (for method, see Hubbard and Cohn 1972) and then allowed to adhere to the surface of a dish. Normal cells exhibited more than 15 labeled bands with molecular weights from 12,000 to 200,000 daltons. The inflammatory cell, however, had a much reduced number of bands and an increase in one component constituting a glycoprotein with a molecular weight slightly over 100,000 daltons. This component was quite sensitive to proteases but had a long half-life when the iodinated cells were cultured in vitro.

Plasma-membrane-associated enzymes also demonstrated a characteristic series of alterations. The ectoenzyme 5'-nucleotidase had 80% of its activity available to substrate on the cell surface of resident cells (Table 2). The

Table 2
Properties of Resident and Inflammatory Macrophages
Obtained from the Mouse Peritoneal Cavity

Property	Resident cell	Inflammatory cell
Cell protein (μg/10^6 cells)	80	130
Superoxide anion (nmole/mg protein/90′)	45	520
5′-Nucleotidase (U/mg protein)	59	0.7
Alkaline phosphodiesterase I (mU/mg protein)	1.43	3.91
Spreading (% cells/hour)	4	>90
Pinocytosis (nl/10^6/hour)	46	247
Phagocytosis		
E-IgG	600	1600
E-IgM-C′	40	1000
Secretory enzymes		
Lysozyme (μg/mg protein)	56	47
Plasminogen activator (U/mg protein)	1	800
Collagenase (U/10^7 cells)	1	15
Elastase (U/10^7 cells)	1.8	68
Lysosomal hydrolases		
acid phosphatase	4	55
β-glucoronidase	15	75
cathepsin	21	65

inflammatory macrophage, however, exhibited less than 5% of normal activity, which was associated with a marked decrease in the half-life of the enzyme (Edelson and Cohn 1976a,b). Calculations suggest that inflammatory cells were inactivating the enzyme at twice the normal rate. Another ectoenzyme, alkaline phosphodiesterase I, was present in greater amounts on inflammatory cells where its activity was increased 2.5 times (Edelson and Erbs 1978). As in the case of 5′-nucleotidase, it was inactivated at a more rapid rate and had a half-life that was 50% of that occurring in resident cells. These observations suggest, but do not prove, that the increased pinocytic rate of the inflammatory cell is a determinant of plasma-membrane turnover.

Functional properties of the plasma membrane were also modified in macrophages activated during an inflammatory process. For example, the rate of fluid phase pinocytosis increased from three to five times (Table 2) that exhibited by resident cells. This form of endocytosis enclosed many solutes that did not bind specifically to the plasma membrane, and interiorized them within membrane-derived vesicles (Silverstein et al. 1977). It was a process which in the macrophage (Steinman et al. 1976) resulted in the interiorization of huge amounts of plasma membrane and in resident cells led to the internalization of the entire surface area of the cell within approximately 30 minutes. The increase in solute uptake could be related to an even more rapid membrane internalization; however, this value would be tempered by the larger surface area of the inflammatory cell which is not now precisely known (Edelson et al. 1975). In any event, it seems likely that both membrane internalization and recycling of membrane back to the cell surface is occurring at a more rapid rate in the inflammatory cell.

Other forms of endocytosis were also accelerated in the inflammatory macrophage (Table 2). Phagocytosis, which is accomplished via the Fc receptor, was increased. Both the rate of uptake and the number of erythrocytes ingested per cell were enhanced. A portion of this increment was probably related to the greater surface area exposed to the particles. These quantitative differences with the Fc receptor were mirrored by more qualitative increments in the case of the complement receptor and the uptake of E-IgM-C'. Although resident cells had easily demonstrable complement receptor and bound many erythrocytes, few, if any, were actually ingested. In contrast, the inflammatory cell ingested large numbers (Bianco et al. 1975). Although this dissociation between the attachment and ingestion phases of endocytosis is poorly understood, it could relate to the integration of cytosol contractile elements with membrane receptors in the activated state.

THE VACUOLAR APPARATUS

Considerable data are available to indicate that the inflammatory macrophage has higher levels of lysosomal hydrolases (Cohn and Benson 1965). These enzymes accumulated in cells that had taken up macromolecular solutes and particulates of a digestible nature, whereas relatively inert components did not lead to this response (Axline and Cohn 1970). There is no known substrate-specific induction of these enzymes.

METABOLIC EVENTS

Prior studies had reported differences in the glucose oxidation of resident macrophages and in those from infected animals (Nathan et al. 1970). More recently, interest has focused on oxygen intermediates which in granulocytes served roles in the myeloperoxidase-mediated bactericidal mechanism. Studies (R. Johnston and Z. Cohn, in prep.) have revealed that inflammatory macrophages (Table 2) are a rich source of superoxide anion (O_2^-), whereas resident cells produce only traces of this compound. Neither population produced O_2^- in the resting state and had to be stimulated either by phagocytosis or by the application of nanogram concentrations of phorbol myristate (PMA). In the macrophage, O_2^- served as the precursor of H_2O_2 in the presence of the dismutase.

Direct measurement of the production of H_2O_2 has been reported by Nathan and Root (1977) employing the scopeletin fluorometric assay.

Macrophages obtained from animals infected with bacillus Calmette-Guérin (BCG) produced as much H_2O_2 (1.6 nmole/10^6 cells/min) as did granulocytes, whereas resident cells were inactive. Again, triggering agents such as PMA, ionophore A23187, or phagocytosis were required. The formation of H_2O_2 by inflammatory cells has not been investigated in detail.

SECRETORY PRODUCTS

Within the last few years it has become apparent that the mononuclear phagocytes are active secretory cells that release enzymes, inhibitors, complement components, and factors into their environment (Gordon 1977). In many instances, more than 95% of a newly synthesized product has been found extracellularly and a small fraction has been found in the cell-associated compartment. This is in contrast to the release of small amounts of acid hydrolases during the ingestion of large particulates—a phenomenon probably related to the imperfect closure of the phagocytic vacuole. It is generally assumed, although not proven, that the secretory pathway is similar to that found in pancreas and liver; namely, that secretion occurs via a vesicular mechanism, the membrane being derived from the Golgi complex.

A CONSTITUTIVE ENZYME

The major bulk secretory product of the macrophage is the enzyme lysozyme. This aminopolysaccharidase, in contrast to the neutral proteinases, was produced and secreted at constant rates by monocytes and by resident and inflammatory macrophages (Gordon et al. 1974a). Modification in the serum concentration length of in vitro cultivation or state of in vivo activation had relatively small effects on the rate of secretion. Finally, colchicine and particle ingestion, both of which stimulate protease secretion (Gordon et al. 1974b; Gordon and Werb 1976) did not influence its output. It seemed likely that a significant proportion of lysozyme production in vivo occurred via mononuclear phagocytes although its function remained uncertain. Clearly it lysed the cell walls of certain nonpathogenic microorganisms but whether there was a natural mammalian substrate is uncertain.

THE INDUCTION OF SECRETORY ENZYMES

A number of neutral proteinases, in contrast to lysozyme, were largely the products of activated macrophages, whereas their normal or resident counterparts secreted only traces of the enzymes. The three proteinases which have been best characterized to date are a plasminogen activator, collagenase, and elastase (Werb and Gordon 1975a,b). The proteolytic attack on substrates such as proteoglycans and casein may have represented the presence of other distinct enzymes in macrophage conditioned medium.

The requirements for enzyme secretion have received extensive study.

First, cells obtained from inflammatory environments, such as induced by thioglycolate broth, secreted large amounts of enzyme and maintain this secretion for many days in vitro. A second stimulus related to the uptake of particulate material by the macrophage. Ingestion of polystyrene latex particles, immune complexes, or microorganisms stimulate the secretion of neutral proteinases. Of interest is the finding that the uptake of digestible or degradable particles led to a short-lived burst of secretion, whereas the phagocytosis of nondegradable particles resulted in long-term production— probably for the life span of the cell in vitro. The controls that regulate this process are unknown.

A third system concerns the ability of lymphokines to stimulate secretion and represents a more immunologically significant mechanism. Studies by Nogueira et al. (1977a,b) have clearly demonstrated that the addition of specific antigen to cultures of sensitized spleen cells led to the release of a factor which induced enzyme secretion in resident macrophages. The production of this factor requires thymus-derived lymphocytes, and cultures depleted of T cells with anti-θ antiserum were deficient in this activity. Thus far, the microbial antigens of *Trypanosoma cruzi* and BCG, as well as those involved in the mixed lymphocyte reaction, have been employed.

MICROBICIDAL AND CYTOCIDAL ACTIVITY

One of the most striking gaps in our knowledge of the mononuclear phagocytes is in the area of microbicidal activity. It is clear that macrophages activated in vivo as a result of infection with a number of facultative intracellular bacteria or with certain protozoa became highly microbicidal. Such cells also demonstrated cytocidal activity against a variety of tumor targets. In the first case, the microbicidal event presumably was intracellular, whereas in the latter it was extracellular. Neither mechanism is understood; yet, it is interesting that the induction of microbicidal activity was an event associated with in vivo infection. None of the inflammatory stimuli associated with the functional properties shown in Table 2 gave rise to this population of cells. This could indicate either a qualitative or quantitative difference in the activated state. My bias at this time is toward the latter difference, first, because cells expressing microbicidal activity often lose this property when cultured in vitro, and, second, because the recent studies of N. Nogueira and Z. Cohn (in prep.) have indicated that the exposure either of normal or inflammatory cells to lymphokines results in the in vitro induction of microbicidal activity. Similarly, when cells expressing microbicidal activity were exposed to similar products in vitro, they maintained a high degree of activation. When completed, this work should allow a more detailed analysis of the induction phase as well as the underlying cytocidal event.

Acknowledgments

This work was supported in part by U.S. Public Health Service Grants AI07012 and AI12975 and by a grant from The Rockefeller Foundation.

REFERENCES

Axline, S. G., and Z. A. Cohn. 1970. In vitro induction of lysosomal enzymes by phagocytosis. *J. Exp. Med.* **131:** 1239.

Bianco, C., A. Eden, and Z. A. Cohn. 1976. The induction of macrophage spreading: Role of coagulation factors and the complement system. *J. Exp. Med.* **144:** 1531.

Bianco, C., F. M. Griffin, Jr., and S. C. Silverstein. 1975. Studies of the macrophage complement receptor: Alteration of receptor function upon macrophage activation. *J. Exp. Med.* **141:** 1278.

Cohn, Z. A., and B. Benson. 1965. The differentiation of mononuclear phagocytes: Morphology, cytochemistry, and biochemistry. *J. Exp. Med.* **121:** 153.

Cohn, Z. A., and E. Wiener. 1963. The particulate hydrolases of macrophages. I. Comparative enzymology, isolation, and properties. *J. Exp. Med.* **118:** 991.

Edelson, P. J., and Z. Cohn. 1976a. 5'-Nucleotidase activity of mouse peritoneal macrophages. I. Synthesis and degradation in resident and inflammatory populations. *J. Exp. Med.* **141:** 1581.

―――――. 1976b. 5'-Nucleotidase activity of mouse peritoneal macrophages. II. Cellular distribution and effects of endocytosis. *J. Exp. Med.* **144:** 1596.

Edelson, P. J., and C. Erbs. 1978. Plasma membrane localization and metabolism of alkaline phosphodiesterase I in mouse peritoneal macrophages. *J. Exp. Med.* **147:** 77.

Edelson, P. J., R. Zweibel, and Z. A. Cohn. 1975. The pinocytic rate of activated macrophages. *J. Exp. Med.* **142:** 1150.

Gordon, S. 1977. Macrophage neutral proteinases and defense of the lung. *Fed. Proc.* **36:** 2707.

Gordon, S., and Z. A. Cohn. 1973. The macrophage. *Int. Rev. Cytol.* **36:** 171.

Gordon, S., and Z. Werb. 1976. Secretion of macrophage neutral proteinase is enhanced by colchicine. *Proc. Natl. Acad. Sci. U.S.A.* **73:** 872.

Gordon, S., J. Todd, and Z. A. Cohn. 1974a. In vitro synthesis and secretion of lysozyme by mononuclear phagocytes. *J. Exp. Med.* **139:** 1228.

Gordon, S., J. C. Unkeless, and Z. A. Cohn. 1974b. Induction of macrophage plasminogen activator by endotoxin stimulation and phagocytosis: Evidence for a two-stage process. *J. Exp. Med.* **140:** 995.

Hubbard, A. L., and Z. A. Cohn. 1972. The enzymatic iodination of the red cell membrane. *J. Cell Biol.* **55:** 390.

Lin, H.-S. and C. C. Stewart. 1974. Peritoneal excudate cells. I. Growth requirement of cells capable of forming colonies in soft agar. *J. Cell Physiol.* **83:** 369.

Mackaness, G. B. 1964. The immunological basis of acquired cellular resistance. *J. Exp. Med.* **120:** 105.

Nathan, C. F., and R. K. Root. 1977. Hydrogen peroxide release from mouse peritoneal macrophages: Dependence on sequential activation and triggering. *J. Exp. Med.* **146:** 1648.

Nathan, C. F., M. L. Karnovsky, and J. R. David. 1970. Alterations of macrophage functions by mediators from lymphocytes. *J. Exp. Med.* **133:** 1356.

Nogueira, N., S. Gordon, and Z. Cohn. 1977a. *Trypanosoma cruzi:* Modification of macrophage function during infection. *J. Exp. Med.* **146:** 157.

―――――. 1977b. *Trypanosoma cruzi:* The immunological induction of macrophage plasminogen activator required thymus-derived lymphocytes. *J. Exp. Med.* **146:** 172.

Rabinovitch, M. 1975. Macrophage spreading *in vitro*. In *Mononuclear phagocytes in infection, immunity and pathology* (ed. R. van Furth), p. 369. Blackwell, Oxford.

Silverstein, S. C., R. M. Steinman, and Z. Cohn. 1977. Endocytosis. *Annu. Rev. Biochem.* **46:** 669.

Steinman, R. M., S. E. Brodie, and Z. A. Cohn. 1976. Membrane flow during pinocytosis: A stereologic analysis. *J. Cell. Biol.* **68:** 665.

Stewart, C. C., H.-S. Lin, and C. Adles. 1975. Proliferation and colony-forming ability of peritoneal exudate cells in liquid culture. *J. Exp. Med.* **141:** 1114.

van Furth, R. 1976. Origin and kinetics of mononuclear phagocytes. *Ann. N.Y. Acad. Sci.* **278:** 161.

Werb, Z., and S. Gordon. 1975a. Elastase secretion by stimulated macrophages: Characterization and regulation. *J. Exp. Med.* **142:** 361.

———. 1975b. Secretion of specific collagenase by stimulated macrophages. *J. Exp. Med.* **142:** 346.

Regulatory Interactions in Normal and Leukemic Myelopoiesis

M. A. S. Moore, J. Kurland, and H. E. Broxmeyer

Sloan-Kettering Institute for Cancer Research
New York, New York 10021

The development of a semisolid agar culture system for cloning murine granulocyte-macrophage committed stem cells (colony-forming unit—culture, CFU-C) (Bradley and Metcalf 1966) and its subsequent adaptation for human studies (Pike and Robinson 1970) have resulted in the accumulation of much information on this committed stem-cell population in health and disease.

Granulopoiesis in vitro is dependent on diffusible activities termed colony-stimulating factors (CSF). Addition of CSF to cultures at concentrations as low as 10^{-12} to 10^{-14} M will promote colony formation. The factor is not a simple inducing agent; its presence is required throughout the growth of the colony for promoting CFU-C survival, proliferation, and differentiation (Metcalf and Moore 1971). Recently, purified preparations of CSF have been shown also to act as macrophage growth-stimulating or mitogenic factors, stimulating extensive peritoneal macrophage or blood monocyte proliferation in agar culture (Stanley et al. 1976; Lin and Freeman 1977). Extensive functional and biochemical studies on CSF obtained from a number of sources have revealed considerable heterogeneity of material with colony-stimulating activity.

Recognition that CSF are produced by monocytes and macrophages (Moore and Williams 1972; Golde and Cline 1972; Moore et al. 1974) and that they act to promote increased monocyte production and macrophage proliferation introduces the problem of what mechanisms act to counterbalance this positive feedback drive. A number of mechanisms have been revealed in in vitro studies and many, if not all, may be of physiological significance in vivo. The functional heterogeneity of the phagocytic mononuclear cell population must firstly be considered, since marked variation in CSF producing capacity exists. "Virgin" macrophages developing in agar culture from CFU-C and macrophages generated in continuous marrow culture are not constitutive producers of CSF; however, exposure of these cells to macrophage-activating agents such as lipopolysaccharide

(LPS) or bacillis Calmette-Guérin (BCG) rapidly induces CSF synthesis and secretion (M. A. S. Moore, unpubl.). In this sense, CSF recruitment of additional monocytes and macrophages would not, ipso facto, lead to increased CSF production in the absence of an exogenous source of stimulation such as endotoxemia due to gram-negative bacterial infection. Neoplastic monocyte or macrophage cell lines also retain the capacity to produce CSF; however, in some cases the leukemic cell lines are constitutive producers and, in others, CSF production is observed only after LPS stimulation, suggesting retention of a degree of normal responsiveness by the transformed cells (Ralph et al. 1977).

A second feature of monocyte macrophage CSF production is found in the functional heterogeneity of the activities. Medium conditioned by human monocytes contains two species of CSF, one of apparent molecular weight 30,000, is a true human CSF, stimulating human granulocyte-macrophage colony formation. A high molecular-weight (150,000) factor is also produced that stimulates mouse-marrow, but not human-marrow, colony formation. (Shah et al. 1977). This latter CSF may be identical to the species of CSF purified from human urine (Stanley et al. 1975). The physiological significance of these two species of CSF resides in the temporal course of their production. Human active CSF is produced early in monocyte cultures but production ceases in 1–2 weeks, whereas mouse active CSF are produced continuously for many weeks.

There is increasing evidence that mature granulocytes and their products participate as one of the regulators of myelopoiesis. Negative feedback control of granulopoiesis has been reported in various systems, and the concept of a granulocyte chalone specifically inhibitory to CFU-C or to more differentiated myeloid cells has received some experimental support (Rytömaa 1973; Lord et al. 1974). The studies of Broxmeyer et al. (1977a, b) have indicated a more indirect mechanism of granulocyte negative feedback that recognizes the implications of the bipotentiality of the granulocyte-monocyte committed stem cell. In the marrow cultures of all species that have been investigated so far, "spontaneous" colony formation in the absence of an exogenous source of CSF is observed when the cells are cultured at a sufficiently high concentration (Moore and Williams 1972). This spontaneous colony formation is due to endogenous elaboration of CSF by marrow monocytes and macrophages and is considerably enhanced by removal of mature granulocytes from the cultured cell population. Addition of mature granulocytes, granulocyte extracts, or medium conditioned by incubation with granulocytes markedly inhibits spontaneous colony formation (Broxmeyer et al. 1977b). This granulocyte-derived colony-inhibiting activity (CIA) acts in a nonspecies-specific manner to suppress CSF production by monocytes and macrophages. CIA does not inhibit monocyte-macrophage proliferation and is clearly distinct from the granulocyte chalone, since no inhibition of granulocytic colony formation is observed in the presence of an exogenous source of CSF.

Broxmeyer et al. (1977a) reported that extracts of PMN from patients with chronic myelocytic leukemia (CML) at all stages of the disease were quantitatively deficient in CIA. An additional regulatory defect in CML was the reduced responsiveness of CSF-producing cells to inhibitory activity derived from normal polymorphonuclear (PMN) neutrophils. The combination of these two defects may, in part, play a role in the profound granulocytic hyperplasia associated with CML.

The regulatory interactions involving diffusible stimulatory and inhibitory activities elaborated by granulocytes, lymphocytes, and phagocytic mononuclear cells clearly can involve specific macromolecules or, alternatively, nonspecific modulating activities. Pharmacological studies have shown that prostaglandins of the E series and other agents capable of elevating intracellular levels of cyclic adenosine monophosphate (cAMP) profoundly inhibit granulopoiesis and macrophage proliferation in vitro (Kurland and Moore 1977a). Just as CSF promotes continued replication of the CFU-C and its progeny, PGE limits this effect by an opposing action on the responsiveness of the myeloid stem cell and its proliferative progeny to stimulation by CSF. Kurland and Moore (1977 a, b) have shown that prostaglandin synthesized by phagocytic mononuclear cells may be of central importance in the modulation of hematopoiesis. Measurement of production of prostaglandin E by murine macrophages and human monocytes, performed with a sensitive radioimmunoassay, has shown a linear relationship between the number of phagocytic mononuclear cells and the concentration of prostaglandin E in the conditioned medium (Kurland et al. 1978). This observation explains the lack of correlation between the numbers of monocytes and macrophages used to stimulate granulocyte-macrophage colony formation and the incidence of colonies. Titration of various numbers of adherent macrophages or blood monocytes as a source of stimulus for human or murine marrow CFU-C has shown clearly that colony formation is stimulated by low numbers of phagocytic mononuclear cells ($0.05-2 \times 10^5$) and is inhibited if higher concentrations are used. Parallel studies in which monocytes or macrophages treated with indomethacin, a potent inhibitor of prostaglandin synthesis, have revealed a linear relationship between the number of colonies stimulated and the number of phagocytic mononuclear cells used as the source of CSF (Kurland et al. 1978). These observations point to the unique ability of the macrophage to control the proliferation of its own progenitor cell by elaboration of opposing regulatory influences. Recently developed methodologies now permit prolonged maintenance of normal hematopoiesis in an in vitro continuous culture system of mouse bone-marrow cells (Dexter et al, this volume). Regulatory interactions between an adherent population of bone-marrow-derived cells and pluripotential stem cells allow prolonged in vitro stem-cell replication, differentiation into progenitor cells of the granulocytic, monocytic, megakaryocytic, and erythroid series, and terminal differentiation of a number of hematopoietic lineages. This system offers considerable potential for further analysis of regulatory interactions in normal hematopoiesis and has been used to investigate both chemical and viral leukemogenesis in vitro (Dexter and Lajtha 1976; Dexter et al. 1977). This chapter describes further studies on functional and regulatory characterization of leukemic cell lines derived from in vivo tumors or from tumors that developed spontaneously in in vitro continuous bone-marrow culture.

MATERIALS AND METHODS

Production of CSF by Normal or Neoplastic Macrophages

Peritoneal cells were harvested from 2- to 3-month-old C3HeB/FeJ and C3H/HeJ mice (Jackson Laboratories) by intraperitoneal lavage with 8-ml ice-cold serum-free McCoy's 5A modified medium containing 1 U heparin/ml. After washing in McCoy's medium containing 15% fetal calf serum (FCS), varying numbers of PEC were allowed to adhere to 35-mm plastic culture dishes (Lux Scientific Corp.) for 1.5 hours at 37°C and the nonadherent peritoneal cells removed. Cultures containing 1×10^5 adherent macrophages, as determined by morphology and neutral red and α-napthyl-acetate esterase positivity, were incubated for 48 hours in McCoy's medium containing 15% FCS. The supernatant media were harvested after 48 hours and a sensitive radioimmunoassay was used to assay for prostaglandin E. The CSF content of the conditioned medium was determined by titration of the dialyzed medium in standard agar cultures containing 7.5×10^4 BDF_1 bone marrow cells with colonies scored after 7 days' incubation.

In vitro generated macrophages were obtained by culturing 1×10^6 BDF_1 bone-marrow cells in 15-ml plastic tubes containing 5 ml of 0.8% methylcellulose in McCoy's 5A modified medium, 15% FCS, and 10% conditioned medium from a murine myelomonocytic leukemic cell line (WEHI-3) as a source of CSF. CFU-C clonal proliferation under these conditions resulted in the generation of a mixed population of granulocytes and macrophages by 7 days of incubation, and by 14 days 5×10^5 to 1×10^6 cells per tube were recovered by centrifugation; these were $> 95\%$ macrophages. These cells were allowed to adhere and condition medium under conditions identical to those used for the peritoneal macrophages.

Production of CSF and prostaglandin was also determined in media conditioned for 48 hours by 5×10^5 viable cells of the following neoplastic murine cell lines: WEHI-3, a myelomonocytic leukemia; J774 and RAW-264, macrophage tumors, the latter induced by Abelson leukemia virus; SK2.2, a spontaneous monocytic cell line generated in vitro; EL4, a T-cell lymphoma; RBL-3, a spontaneous B-cell lymphoma; and $HSDM_1C_1$, a fibrosarcoma. Conditioned media from cultures of normal macrophages or of the neoplastic cell lines were also obtained following 48 hours' incubation in the presence of 1 µg of *Salmonella typhosa* LPS (WO901, Difco Laboratories) or 10% of a CSF source obtained by tenfold concentration of serum-free medium conditioned by the WEHI-3 cell line.

Establishment of Continuous Hematopoietic Cell Cultures

To establish an adherent microenvironment capable of supporting continuous stem-cell replication, we flushed the contents of a single femur either from CBA or $B_6D_2F_1$ mice into 25 cm² plastic culture flasks (Corning Laboratories) containing 10 ml of Fisher's medium (Gibco Laboratories) supplemented with 25% horse serum (Flow Laboratories) and antibiotics as described by Dexter et al. (this volume). Sixteen cultures were established and maintained at 33°C in an atmosphere of 7.5% CO_2 in air. After 1 week, all the growth medium and suspension cells were removed and fresh medium was added. To four cultures containing either a $B_6D_2F_1$ or CBA adherent layer were added 5×10^6 washed $B_6D_2F_1$ bone-marrow cells. Also used as a source of stem cells were 5×10^6 CBA fetal liver cells; these were inoculated either into four $B_6D_2F_1$ or four CBA-adherent marrow cultures. The cultures were subjected twice weekly to removal of one-half the growth medium and suspension cells, at which time fresh medium was added. The nonadherent cells that had been removed were counted, centrifuged, and resuspended in Fisher's medium. Granulocyte-macrophage progenitor cells were assayed in semisolid agar and McCoy's medium in the presence or absence of WEHI-3 conditioned medium as a source of CSF.

RESULTS

Inducible and Constitutive Production of CSF by Macrophages and Neoplastic Cell Lines

Constitutive production of CSF characterized the cultures of murine peritoneal macrophages (Table 1) and, confirming earlier observations, 1 μg of LPS augmented CSF production threefold in cultures of C3HeB/FeJ macrophages. In contrast, the LPS-unresponsive C3H/HeJ mouse macrophages showed no increase in CSF production in the presence of 1–10 μg of LPS. Macrophages generated from bone-marrow-derived CFU-C in methylcellulose culture exhibited very low constitutive CSF production but showed a marked increase in CSF elaboration following LPS treatment. Three of the four neoplastic monocyte or macrophage cell lines were not

Table 1

Endogenous and Exogenous CSF Requirement for Production of Prostaglandin E by Normal or Transformed Macrophages

Cell source	CSF U/0.1 ml		Production of prostaglandin E (pg)		
	control	+LPS	control	+LPS	+CSF
Peritoneal macrophage					
C_3HeB/FeJ	26 ± 6	74 ± 9	200	1905	6445
C_3H/HeJ	49 ± 5	51 ± 4	469	198	1949
CFU-C derived macrophage	3 ± 1	41 ± 3	33	164	N.D.
Monocyte Macrophage cell lines					
RAW 264	0	111 ± 1	0	305	393
SK 2.2	0	40 ± 2	295	3400	N.D.
J 774	0	6 ± 1	0	1050	N.D.
WEHI-3	121 ± 7	130 ± 6	175	1100	N.D.
T-cell lymphoma					
EL 4	0	0	0	0	0
B-cell lymphoma					
RBL-3	152 ± 12	154 ± 10	0	0	0
Fibrosarcoma					
$HSDM_1C_1$	0	0	5050	4252	4575

N.D., not determined.

constitutive producers of CSF but all were capable of production following LPS stimulation. The apparent low level of CSF activity elaborated by the cell line J774 following LPS treatment may have been due to the associated induction of a nondialyzible CFU-C inhibitory activity. The myelomonocytic leukemic cell line WEHI-3 elaborated large quantities of CSF in the absence of LPS stimulation. WEHI-3 differed also from the preceding monocyte-macrophage cell lines and normal macrophages in the spectrum of biological activities elaborated. The CSF species elaborated by the latter cell types stimulated predominantly macrophage and mixed neutrophil-granulocyte-macrophage colony formation (GM-CSF), whereas WEHI-3 produced, in addition, activities stimulating eosinophil (Eo-CSF) and megakaryocyte (Meg-CSF) colony formation. CSF production was not demonstrable in cultures of the T-cell lymphoma EL4 and was not induced by LPS treatment; however, exposure of these cells to the T-cell mitogen concanavalin A resulted in rapid induction of CSF. The B-cell lymphoma RBL-3, like the WEHI-3 cell line, produced large amounts of CSF and no increased production occurred after addition of LPS.

A radioimmunoassay used to determine production of prostaglandin E revealed that normal peritoneal macrophages and, to a lesser extent, in vitro generated macrophages produced significant quantities of prostaglandin E (Table 1). In vitro activation of these cells with LPS produced a tenfold increase in PGE production by C3HeB/FeJ macrophages and a fivefold increase by in vitro derived macrophages; C3H/HeJ macrophages were unresponsive. These results suggested a link between LPS induction of increased CSF and of prostaglandin-E synthesis. In order to test the possibility that prostaglandin synthesis was activated by increased CSF levels in the culture medium rather than by LPS directly, we incubated the normal macrophages in the presence of a concentrated source of CSF from WEHI-3 conditioned medium. As can be seen in Table 1, CSF treatment increased prostaglandin-E production by 32 times in normal C3HeB/FeJ macrophages, a level three times higher than that observed following LPS activation. That this CSF effect was not due to contamination by LPS was indicated by the ability of CSF to increase markedly the production of prostaglandin E by the LPS-unresponsive C3H/HeJ macrophages.

Retention of a similar linkage between induction of CSF and subsequent synthesis of prostaglandin E was observed in the monocyte-macrophage cell lines. RAW-264 and J774 showed no basal production of CSF or of prostaglandin E; however, both activities were induced by LPS. CSF treatment alone induced significant PGE production in the RAW-264 cell line. Low basal levels of PGE production characterized SK2.2 and WEHI-3, but LPS treatment increased this by 7 to 10 times.

Normal T and B lymphocytes are recognized as being capable of elaborating CSF following mitogen or antigen stimulation. The neoplastic T-cell line EL4 revealed retention of this capacity following concanavalin-A activation, and the B-cell lymphoma, RBL-3, is a constitutive CSF-producing cell line. Production of prostaglandin E by normal or neoplastic B or T lymphocytes was not observed, suggesting that the link between CSF production and prostaglandin-E synthesis represents a phenomenon restricted to normal or neoplastic monocytes and macrophages. Likewise, the fibrosarcoma cell line $HSDM_1C_1$ was a high constitutive producer of prostaglandin E but did not elaborate detectable CSF.

Hematopoiesis and Leukemogenesis in Continuous Hemopoietic Cell Cultures

The continuous allogeneic or syngeneic co-cultures of bone marrow or bone marrow plus fetal liver showed a normal pattern of sustained granulopoiesis and continuous generation of CFU-C for 4 to 5 weeks; the exception was that of the CBA bone marrow plus fetal liver cultures, where CFU-C production was not sustained beyond 2 weeks. During this time, colony formation in agar was totally dependent upon the addition of CSF, and normal colony granulocyte-macrophage differentiation was observed. The syngeneic CBA-marrow–fetal-liver cultures showed a rapid loss of CFU-C production, converting to a predominantly macrophage morphology within 2 to 4 weeks. On the 35th day of culture, one of four cultures showed a marked increase in cellularity, with reappearance of CSF-dependent agar colony formation (Fig. 1). The remaining cultures in this group showed no evidence of transformation over a 25-week period of observation. The transformed culture was characterized initially by a total dependence on CSF for its cloning in agar with a 10–20% cloning efficiency. The cells in continuous culture had a monoblastic morphology, and the agar colonies comprised monoblasts and monocytes with no evidence of macrophage differentiation. The majority of transformed cells were Fc and C' receptor positive and lysozyme producing. CSF-independent colony formation was first evident after 84 days of continuous culture, with progression to autonomous proliferation being subsequently observed.

Individual agar colonies developing from the transformed culture 42 days after its initiation were inoculated in culture flasks at 33° and 37°C in the presence or absence of CSF. Continuous cell lines were produced that were dependent upon CSF (SK2.2$^+$) or were autonomous (SK2.2$^-$) and which proliferated optimally either at 33° or 37°C. The original cell line (SK2.2$^+$) had a modal chromosome number of 65 with 12 metacentric marker chromosomes. The properties of this cell line paralleled those of the original continuous culture, showing retention of CSF-dependent agar cloning and induction of CSF and prostaglandin production following exposure to LPS (Table 1). Subcutaneous or intravenous injection of the cell line into the syngeneic CBA recipients has not, to date, resulted in leukemia or lymphoma development; however, subcutaneous injection into nude mice resulted in development of tumors within 2 to 3 weeks.

A second pattern of spontaneous transformation was observed in the allogeneic combination of CBA and BDF_1 bone marrow. A normal pattern of sustained granulopoiesis and CFU-C production was observed for 4 to 5 weeks and, by the end of the 5th week, the suspension cells in culture were exclusively macrophage and cell production declined. On the 42nd day it was apparent that one of the cultures had a markedly increased cellularity, and 10% of the suspension cells had a blast-cell morphology. Agar cultures established at this time showed a return of colony formation in the presence or absence of CSF, and colony morphology was exclusively undifferentiated

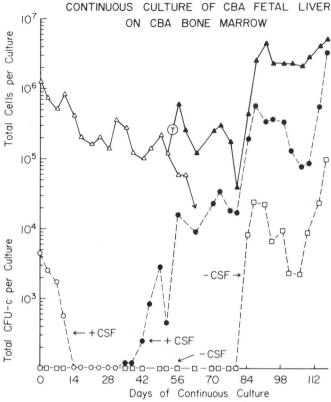

Figure 1
Spontaneous CSF-dependent monocytic transformation in continuous cultures of CBA marrow plus fetal liver. (△——△) Total cells per culture pretransformation; and (▲——▲) posttransformation. (o-----o) Total CSF-dependent granulocyte-macrophage CFU-C; (●-----●) total CSF-dependent transformed monocyte CFU-C; and (□-----□) total CSF-independent transformed monocyte CFU-C. Cultures were subjected to twice weekly 50% depopulation of suspension cells.

blast cell. The remaining cultures underwent a similar transformation at 70, 77, and 84 days, respectively (Table 2). The blast cells in all four transformed cultures were negative for Fc or C' receptors, surface and cytoplasmic immunoglobulin, lysozyme, and terminal transferase. The cloning efficiency of blast cells in the four cultures remained constant over 10–15 sequential depopulations and ranged from 2.7% in culture T-3 to 25.6% in culture T-2 (Table 2). Interestingly, although these transformed cultures were characterized by spontaneous colony formation in agar, the addition of CSF invariably increased cloning efficiency by 50–300%. Cytogenetic analysis revealed marked aneuploidy with modal chromosome numbers ranging from 62 to 115 with a unique spectrum of metacentric, long telocentric, and minute marker chromosomes characterizing each culture. Although lacking a number of features that would allow cell lineage characterization, the transformed blast cells all produced CSF and prostaglandin-E and showed

Table 2
Spontaneous Transformation of Continuous Allogeneic
Bone-Marrow Cultures (CBA + BDF$_1$)

Phenotypic characteristics	T_1	T_2	T_3	T_4
Time of transformation (days)	42	70	77	84
Cloning efficiency % + CSF	8.0	25.6	2.7	4.3
Cloning efficiency % − CSF	5.1	16.2	1.7	1.6
CSF production-U/0.1 ml	267	32	459	44
CSF production + 1 µg LPS	357	608	565	80

some augmentation of production of these activities following LPS stimulation.

Continuous cell lines were established with individual agar colonies from each culture and were adapted to growth at 37°C in the absence of the adherent marrow microenvironment. These cell lines, as well as cells from the primary cultures, produced rapidly growing tumors following subcutaneous injection in nude mice. Transplantation of these tumors into BDF$_1$ and CBA mice (syngeneic either with the second source of marrow or the marrow used to establish the adherent layer) resulted in rapidly growing tumors in BDF$_1$ mice, indicating that the transformed cells had originated from the second addition of washed marrow cells rather than from a cell type in the adherent marrow microenvironment.

DISCUSSION

The capacity of phagocytic mononuclear cells to regulate myelopoiesis is suggested by production of macrophage CSF and prostaglandin E. That production of these opposing activities is an inducible rather than a constitutive property of macrophages is indicated by the very low levels of CSF and prostaglandin E elaborated by macrophages generated in vitro from CFU-C, in contrast to the greater levels produced by peritoneal macrophages from conventional mice and the marked induction of CSF and prostaglandin E following in vitro activation of the cells by LPS. The apparent paradox of associated induction of two biologically active factors with opposing activity can be resolved if one considers the temporal sequence of induction. Increased synthesis and secretion of CSF can be detected within minutes of in vivo or in vitro activation of macrophages with LPS, with the maximum rate of synthesis being observed after 2 to 3 hours. In contrast, induction of prostaglandin-E synthesis does not occur until 18 to 24 hours after LPS stimulation. Our present observations have implicated CSF directly in the induction of macrophage prostaglandin-E synthesis and have shown a correlation between the concentration of CSF in the macrophage mileu and the subsequent levels of prostaglandin E induced. This suggests a unique surveillance role for macrophages in the regulation of myelopoiesis since, under steady-state conditions, elaboration of CSF is associated with recruitment of monocytes and granulocytes by an action on the marrow CFU-C and CSF-dependent local macrophage proliferation, both proliferative responses being limited by prostaglandin E elaborated in response to increasing levels of CSF. Additional homeostatic control may be provided by the modulation of monocyte-macrophage production of CSF by CIA derived from mature granulocytes (Broxmeyer et al. 1977 a, b). Inhibition of production of CSF by granulocyte-derived CIA provides further evidence for the concept that prostaglandin E is elaborated in proportion to variations in levels of CSF, inasmuch as production of basal prostaglandin E by human monocytes is also reduced following treatment with CIA (Kurland et al. 1978). Perturbations in myelopoiesis following infection or inflammation are associated with a marked neutrophil leukocytosis and local macrophage proliferation, responses which may be directly correlated with the acute induction of increased production of macrophage CSF in response to macrophage activating agents such as LPS. A genetic defect in this LPS response is provided by the C_3H HeJ mouse whose peritoneal macrophages showed no increase in induction of CSF or prostaglandin E following in vitro LPS activation. Unlike the response to LPS, that of C_3H HeJ macrophages was a normal tenfold increase in production of prostaglandin E following treatment with CSF, further indicating the secondary nature of prostaglandin-E induction in LPS-stimulated cells and providing a control for the possibility that the macrophage response to preparations of CSF was due to the latter's contamination by endotoxin.

The neoplastic monocyte-macrophage cell lines retained a number of responses exhibited by their normal counterparts. Capacity to produce CSF was evident in all cases and in most lines this was an inducible function associated with LPS activation. The link between activation and production of prostaglandin E was clearly retained by the neoplastic cells since, in all cases, LPS induced prostaglandin-E and CSF production. Furthermore, prostaglandin-E could be induced directly by treatment of the cells with CSF. The inducibility of CSF and of prostaglandin E appears to be a property restricted to normal or neoplastic monocytes and macrophages, since production of CSF by normal B or T lymphocytes is observed following treatment with mitogen (Parker and Metcalf 1974; Ruscetti and Chervenick 1975), but it is not associated with production of prostaglandin E. Our present results show that the T-cell line EL-4 produced neither CSF nor prostaglandin E constitutively or following LPS treatment; however, in other studies, the T-cell mitogen concanavalin A induced production by this cell line of CSF but not of prostaglandin E (P. Ralph et al., in prep.). The high constitutive levels of CSF produced by a B-cell lymphoma in the absence of detectable prostaglandin E, and the high levels of prostaglandin E produced by a fibrosarcoma without associated production of CSF further support the view that induction of these opposing activities is inherently linked only in cells of the monocyte-macrophage lineage.

It is important to determine which, if any, of the preceding phenotypic characteristics of the leukemic cell lines are significant to the natural development and progression of leukemia. Studies on the properties of leukemic cells established as continuous cell lines may be misleading, since phenotypic changes may be of secondary derivation. It was of interest, therefore, to monitor certain parameters of regulatory responsiveness associated with leukemogenesis occurring in vitro in continuous marrow culture. The spontaneous transformations observed in allogeneic continuous bone-marrow cultures were characterized by an increased rate of cell production, development of agar cloning capacity in the absence of CSF, and failure of the cells to mature. These events were paralleled by marked aneuploidy, capacity of the cells to grow continuously in the absence of an adherent marrow cell population, and tumorigenic potential in nude mice or normal mice syngeneic with the second marrow cell population (BDF_1). The transformed cells both in vivo and in vitro were null by a variety of functional, enzymatic, or surface-marker studies; however, the augmentation of agar cloning in the presence of CSF, the capacity of the cells to produce CSF or prostaglandin E, and the augmentation of these activities by in vitro LPS treatment suggest that the transformed target cells were early in the monocyte-myeloid lineage. The transformation observed in the syngeneic cultures of CBA bone marrow plus fetal liver cells progressed in three phases, with initial evidence of transformation being the reappearance of CSF-dependent cloning, with maturation within the colonies being exclusively monocytic without progression to typical macrophages. This morphological transformation was associated with aneuploidy and continuous replication of the agar cloned cells in suspension culture in the absence of the adherent marrow microenvironment but in the presence of a source of CSF. This phase of CSF dependence was followed by progression to partial autonomy of the cells with some spontaneous cloning capacity in agar and still later to fully autonomous cloning and CSF-independent continuous proliferation in suspension culture. This progression from CSF dependence to autonomy appeared to be paralleled by an increased rate of in vivo growth of the cells in nude mice. Interestingly, in both the dependent and autonomous phase of development of this neoplastic monocytic cell line, constitutive production of CSF or of prostaglandin E was not observed unless the cells were activated by in vitro exposure to LPS.

It is possible that these models of "spontaneous" in vitro leukemogenesis may more accurately reflect events associated with the natural development of leukemia, unlike the more artefactual transformation systems involving infection with murine leukemia viruses such as Friend or Abelson. However, a viral etiology of the spontaneous transformations is certainly not excluded, since activation of endogenous viruses may be associated with allogeneic hematopoietic cell interactions or with fetal hematopoietic cell culture.

Acknowledgments

This work was supported by the Viral Oncology Program of the National Cancer Institute, Contract CP-V0-71008-60 and by American Cancer Society Grant CH-3.

REFERENCES

Bradley, T. R., and D. Metcalf. 1966. The growth of mouse bone marrow cells *in vitro*. *Aust. J. Exp. Biol. Med. Sci.* **44**: 287.

Broxmeyer, H. E., N. Mendelsohn, and M. A. S. Moore. 1977a. Abnormal granulocyte feedback regulation of colony stimulating activity-producing cells from patients with chronic myelogenous leukemia. *Leukemia Res.* **1:** 3.

Broxmeyer, H. E., M. A. S. Moore, and P. Ralph. 1977b. Cell free granulocyte colony inhibiting activity derived from human PMN. *Exp. Hematol.* **5:** 87.

Dexter, T. M., and L. G. Lajtha. 1976. Proliferation of haemopoietic stem cells and development of potentially leukaemic cells *in vitro*. *Bibl. Haematol.* **43:** 1.

Dexter, T. M., D. Scott, and N. M. Teich. 1977. Infection of bone marrow cells *in vitro* with FLV: Effects on stem cell proliferation, differentiation and leukemogenic capacity. *Cell* **12:** 355.

Golde, D. W., and M. J. Cline. 1972. Identification of colony stimulating cells in human peripheral blood. *J. Clin. Invest.* **51:** 2981.

Kurland, J., and M. A. S. Moore. 1977a. Modulation of hemopoiesis by prostaglandins. *Exp. Hematol.* **5:** 357.

———. 1977b. The regulatory role of the macrophage in normal and neoplastic hemopoiesis. In *Experimental hematology today* (eds. S. Baum and G. Ledney), p. 51. Springer-Verlag, Berlin.

Kurland, J. I., R. S. Bockman, H. E. Broxmeyer, and M. A. S. Moore. 1978. Limitation of excessive myelopoiesis by the intrinsic modulation of macrophage-derived prostaglandin E. *Science* **199:** 552.

Lin, H.-S., and P. G. Freeman. 1977. Peritoneal exudate cells. IV. Characterization of colony forming cells. *J. Cell. Physiol.* **90:** 407.

Lord, B. I., L. Cercek, B. Cercek, G. P. Shah, T. M. Dexter, and L. G. Lajtha. 1974. Inhibitors of haemopoietic cell proliferation?: Specificity of action within the haemopoietic system. *Br. J. Cancer* **29:** 168.

Metcalf, D., and M. A. S. Moore. 1971. *Haemopoietic cells*. North Holland Research Monographs. *Front. Biol.* **24.**

Moore, M. A. S., and N. Williams. 1972. Physical separation of colony stimulating cells from *in vitro* colony forming cells in hemopoietic tissue. *J. Cell. Physiol.* **80:** 195.

Moore, M. A. S., G. Spitzer, D. Metcalf, and D. G. Penington. 1974. Monocyte production of colony stimulating factor in familial cyclic neutropenia. *Br. J. Haematol.* **27:** 47.

Parker, J. W., and D. Metcalf. 1974. Production of colony-stimulating factor in mitogen-stimulated lymphocyte cultures. *J. Immunol.* **112:** 502.

Pike, B. L., and W. A. Robinson. 1970. Human bone marrow colony growth in agar-gel. *J. Cell. Physiol.* **76:** 77.

Ralph, P., H. E. Broxmeyer, and I. Nakoinz. 1977. Immunostimulators induce granulocyte/macrophage colony-stimulating activity and block proliferation in a monocyte tumor cell line. *J. Exp. Med.* **146:** 611.

Ruscetti, F. W. and P. A. Chervenick. 1975. Release of colony stimulating activity from thymus-derived lymphocytes. *J. Clin. Invest.* **55:** 520.

Rytömaa, T. 1973. Annotation: Role of chalone in granulopoiesis. *Br. J. Haematol.* **24:** 141.

Shah, R. G., L. H. Caporale, and M. A. S. Moore. 1977. Characterization of colony-stimulating activity produced by human monocytes and phytohemagglutinin-stimulated lymphocytes. *Blood* **50:** 811.

Stanley, E. R., M. Cifone, P. M. Heard, and V. Defendi. 1976. Factors regulating macrophage production and growth: Identity of colony-stimulating factor and macrophage growth factor. *J. Exp. Med.* **143:** 631.

Stanley, E. R., G. Hensen, J. Woodcock, and D. Metcalf. 1975. Colony stimulating factor and the regulation of granulopoiesis and macrophage production. *Fed. Proc.* **34:** 2272.

Microenvironmental Influences on Granulopoiesis in Acute Myeloid Leukemia

P. Greenberg and B. Mara

Department of Medicine, Veterans Administration Hospital, Palo Alto, California 94304 and Stanford University School of Medicine, Stanford, California 94305

Microenvironmental influences within bone marrow and spleen have been shown to be critical for hematopoietic stem-cell proliferation and differentiation in experimental animals (McCulloch et al. 1965; Gallagher et al. 1971; Knospe and Crosby 1971; Trentin 1971; Chamberlin et al. 1974; Matioli and Rife 1976; Cline et al. 1977). Histologic and functional studies by these investigators have demonstrated that locally active cell-derived factors provide stromal influences that contribute to the support of hematopoiesis.

In vitro marrow culture techniques have permitted analysis of factors involved in the regulation of granulopoiesis by evaluation of the ability of granulocytic progenitor cells (colony-forming units—culture, CFU-C) to form granulocyte-macrophage colonies in agar under the necessary influence of the humoral stimulatory substance termed colony-stimulating activity (CSA) (Rickard et al. 1970; Metcalf 1973). Human marrow cells require cellular sources of CSA for their in vitro proliferation, whereas murine marrow is stimulated as well by CSA present in serum and urine (Foster et al. 1968; Metcalf and Stanley 1969; Pike and Robinson 1970; Metcalf and Moore 1975). Recent studies in mice have shown that proliferation of marrow CFU-C is related predominantly to intramedullary CSA elaboration by cells firmly adherent to the inner surface of hematopoietic bone (Chan and Metcalf 1972, 1973). Thus, local production of CSA within the marrow plays a major role in influencing granulopoiesis.

Cellular sources of CSA are also present within human marrow and can be selectively harvested by their adherence and density characteristics (Haskill et al. 1972; Messner et al. 1973; Moore et al. 1973; Senn et al. 1974). We have employed these physical separation techniques to evaluate marrow-cell-derived CSA levels in normal subjects and in patients with acute myeloid leukemia (AML) in order to determine the possible role of human marrow CSA provision as a microenvironmental stimulus for granulopoiesis.

METHODS

The methodology for these studies was recently described in detail (P. Greenberg et al. 1978). The buoyant component of aspirated human marrow cells was obtained by ficoll-hypaque density centrifugation. These cells were then permitted to adhere to plastic tissue culture dishes (Messner et al. 1973). The nonadherent cells were rinsed off and the remaining adherent cells were incubated in modified McCoy's medium containing 15% fetal calf serum and 0.5 mM 2-mercaptoethanol for 7 days at 37°C. After this incubation, the conditioned medium was harvested and stored at −20°C until use. Target normal human marrow cells were obtained from buoyant cells less dense than 1.068 g/cm^3; the bovine serum albumin neutral density (density cut) procedure (Greenberg et al. 1976; Heller and Greenberg 1977) was utilized. These cells were then permitted to adhere to tissue culture dishes and the nonadherent buoyant cell population was harvested and used as target cells. Continuous albumin density gradients were performed in selected experiments. Cells and test-conditioned media were incubated for 7 to 10 days in agar culture at 37°C and colonies were counted as previously described (Greenberg, et al. 1971; Greenberg et al. 1976). Colonies consisted of more than 50 cells with granulocytic-monocytic differentiation. Quantitative estimates of effective concentrations of CSA were obtained from titration curves of conditioned mediums. For standardization, the numbers of colonies produced by these concentrations were compared to those stimulated by a stable leukocyte conditioned medium CSA source. The data were analyzed by curve-fitting computer programs (Greenberg et al. 1974).

RESULTS

A sigmoid-shaped dose-response curve of CSA values was obtained with increasing numbers of adherent and total marrow cells. Plateau levels of CSA occurred at approximately $3–15 \times 10^5$ adherent cells. Most normal specimens provided this number of adherent cells, with approximately 9% of the marrow cells being adherent. All of the CSA was provided by the adherent cell population, and, specifically, no CSA was provided by the nonadherent target marrow cells. Plasma from normal marrow or peripheral blood had no demonstrable CSA when nonadherent target cells were used, whereas the presence of the adherent cells permitted colony formation to occur. This indicated that substances present in normal serum enhanced CSA production by endogenous CSA-producing cells rather than by providing CSA itself. Control plates lacking a CSA source had no colony formation. Density distribution profiles showed that the CSA-producing cells represented a subpopulation of the adherent cells, with a peak density of 1.066 g/cm^3.

The morphologic, cytochemical, and phagocytic characteristics of the adherent marrow CSA-producing cells were assessed. It was found that 84–87% of these cells were α-naphthyl acetate esterase positive, resembled monocytes morphologically, and were capable of phagocytosing latex particles. These data suggest that middensity monocytes and macrophages contribute a major portion of the marrow CSA.

With these studies providing a background for quantitating and characterizing the normal marrow CSA-producing cells, we turned our attention to patients with AML. All AML patients received the same chemotherapeutic induction regimen (daunomycin, cytosine arabinoside, and 6-thioguanine) and maintenance program (monthly cytosine arabinoside and 6-thioguanine) as previously reported (Embury et al. 1977). In comparison with 16 control subjects, 13 of 21 patients with AML were found to have significantly low marrow CSA levels at diagnosis or relapse. Only four of the 13 patients (33%) with low marrow CSA entered complete remission, whereas seven of eight patients (88%) with normal CSA did achieve complete remission ($P < 0.01$). These data have been described more completely (P. Greenberg et al. 1978). All four patients in partial remission and 42 of 46 in complete remission had normal marrow CSA values. In comparison with control subjects, low CSA values were found particularly in patients with acute myeloblastic as opposed to myelomonocytic leukemia. These two entities were distinguished by previously defined morphological criteria (Hayhoe and Cawley 1972). It should be emphasized, however, that this method of categorization is less sensitive and specific than are current cytochemical techniques (Bennett et al. 1976). Other clinical parameters and patterns of marrow colony formation were evaluated and showed no significant correlation with complete remission rates or marrow CSA levels. Sequential monthly studies of three patients with AML in stable remission demonstrated persisting normal marrow CSA and CFU-C values. In contrast, in six patients in remission when these studies were begun who subsequently relapsed, marrow CSA values decreased within 2 to 3 months of relapse. Marrow CSA values paralleled marrow CFU-C during remission until relapse, with normal CSA values persisting longer than CFU-C.

DISCUSSION

In these studies we quantitated marrow-cell CSA provision, characterized the marrow CSA-producing cells, and assessed alterations of marrow-cell CSA levels in normal subjects and patients with AML. Alteration of marrow CSA levels in AML correlated well with the patients' clinical status and prognoses. Patients at diagnosis or relapse, particularly those patients with AML and those failing remission induction, had significantly decreased marrow CSA values. Low marrow CSA was a significant negative prognostic indicator, since only 33% of patients with this finding entered drug-induced complete remission, whereas complete remission occurred in 88% of patients with normal CSA.

Sequential investigations showed that marrow CSA provision was generally normal in patients during stable remission. In contrast, low or progressively decreasing marrow CSA values occurred in patients with impending relapse. Marrow CFU-C and CSA correlated well in these patients during remission, with a decrease in marrow CFU-C occurring earlier than low marrow CSA in impending relapse. This prolonged persistence of marrow CSA may relate to the relatively longer life-span of monocytes and macrophages in comparison with granulocytic progenitor cells (VanFurth 1970). The sequential marrow CFU-C patterns found during remission of AML were similar to those previously reported from this and other laboratories (Greenberg et al. 1971; Bull et al. 1973; Heller and Greenberg 1977).

That normal marrow CSA values were associated both with achievement and persistence of complete remission suggests that adequate marrow CSA provision may be essential for sustaining normal granulopoiesis in AML following induction chemotherapy. The absence of CSA in plasma from marrow or peripheral blood further implicated local cellular sources as being important providers of CSA. Thus, marrow CSA production may provide a measure of the intramedullary influences involved in regulating granulopoiesis. Histologic examination of human marrow has indicated that granulopoiesis occurs within the marrow parenchyma (Weiss 1970) where granulocytic precursors and cells of the monocyte-macrophage series are in proximity. Thus, it is possible that short-range interactions could occur in this microenvironment between adherent-cell elaborated stimulatory substances and granulocytic precursors. In this model, long-range influences could participate in the response to such major perturbations as infection or antigenic challenge.

The monitoring of marrow CSA provision appears useful for evaluating microenvironmental influences and intramedullary cellular interactions on granulopoiesis and for assessing prognosis and clinical status in AML.

REFERENCES

Bennett, J. M., D. Catovsky, M. T. Daniel, G. Flandrin, D. A. G. Galton, H. R. Gralnick, and C. Sultan. 1976. Proposals for the classification of the acute leukaemias. *Br. J. Haematol.* **33**: 451.

Bull, J. M., M. J. Duttera, E. D. Stashick, J. Northup, E. Henderson, and P. P. Carbone. 1973. Serial in vitro marrow culture in acute myelocytic leukemia. *Blood* **42**: 679.

Chamberlin, W., J. Barone, A. Kedo, and W. Fried. 1974. Lack of recovery of murine hematopoietic stromal cells after irradiation-induced damage. *Blood* **44**: 385.

Chan, S. H., and D. Metcalf. 1972. Local production of colony-stimulating factor within the bone marrow: Role of nonhemopoietic cells. *Blood* **40**: 646.

––––––. 1973. Local and systemic control of granulocytic and macrophagic progenitor cell regeneration after irradiation. *Cell Tissue Kinet.* **6**: 185.

Cline, M. J., C. Le Fevre, and D. W. Golde. 1977. Organ interactions in the regulation of hematopoiesis: In vitro interactions of bone, thymus, and spleen with bone marrow stem cells in normal, Sl/Sl^d, and W/W^v mice. *J. Cell. Physiol.* **90**: 105.

Embury, S. H., L. Elias, P. H. Heller, C. E. Hood, P. L. Greenberg, and S. L. Schrier. 1977. Remission maintenance therapy in acute myelogenous leukemia. *West. J. Med.* **136**: 267.

Foster, R., D. Metcalf, W. A. Robinson, and T. R. Bradley. 1968. Bone marrow colony stimulating activity in human sera: Results of two independent surveys in Buffalo and Melbourne. *Br. J. Haematol.* **15**: 147.

Gallagher, M. T., M. P. McGarry, and J. Trentin. 1971. Defect of splenic stroma (hemopoietic inductive microenvironments) in the genetic anemia of Sl/Sl^d mice. *Fed. Proc.* **30**: 684.

Greenberg, P. L., B. Mara, and P. Heller. 1978. Marrow adherent cell colony stimulating activity production in acute myeloid leukemia. *Blood.* (In press.)

Greenberg, P. L., W. C. Nichols, and S. L. Schrier. 1971. Granulopoiesis in acute myeloid leukemia and preleukemia. *N. Engl. J. Med.* **284**: 1225.

Greenberg, P., I. Bax, B. Mara, and S. Schrier. 1974. Alterations of granulopoiesis following chemotherapy. *Blood* **44**: 375.

Greenberg, P. L., B. Mara, I. Bax, R. Brossel, and S. L. Schrier. 1976. The myeloproliferative disorders: Correlation between clinical evolution and alterations of granulopoiesis. *Am. J. Med.* **61**: 878.

Haskill, J. S., R. D. McKnight, and P. R. Galbraith. 1972. Cell-cell interaction in vitro: Studied by density separation of colony-forming, stimulating, and inhibiting cells from human bone marrow. *Blood* **40**: 394.

Hayhoe, F. G. J., and J. C. Cawley. 1972. Acute leukemia: Cellular morphology, cytochemistry and fine structure. *Clin. Haematol.* **1**: 49.

Heller, P., and P. L. Greenberg. 1977. Marrow colony forming cell density distribution patterns during remission of acute myeloid leukemia. *J. Natl. Cancer Inst.* **59**: 313.

Knospe, W., and W. Crosby. 1971. Aplastic anemia: A disorder of the bone marrow sinusoidal microcirculation rather than stem cell failure? *Lancet* **i**: 20.

Matioli, G., and L. L. Rife. 1976. Hemopoietic stem cell kinetics in 4000 r irradiated spleens. *J. Reticuloendothel. Soc.* **20**: 429.

McCulloch, E. A., L. Siminovich, J. Till, E. S. Russell, and S. E. Bernstein. 1965. The cellular basis of the genetically determined hemopoietic defect in anemic mice of genotype Sl/Sl^d. *Blood* **26**: 399.

Messner, H. A., J. E. Till, and E. A. McCulloch. 1973. Interacting cell populations affecting granulopoietic colony formation by normal and leukemic human marrow cells. *Blood* **42**: 701.

Metcalf, D. 1973. Regulation of granulocyte and monocyte-macrophage proliferation by colony stimulatory factor (CSF): A review. *Exp. Hematol.* (Copenh.) **1**: 185.

Metcalf, D., and M. Moore. 1975. Growth and responsiveness of human granulocytic leukemic cells in vitro. *Bibl. Haematol.* **40**: 235.

Metcalf, D., and E. R. Stanley. 1969. Quantitative studies on the stimulation of mouse bone marrow colony growth in vitro by normal human urine. *Aust. J. Exp. Biol. Med. Sci.* **47**: 453.

Moore, M. A. S., N. Williams, and D. Metcalf. 1973. In vitro colony formation by normal and leukemic human hemopoietic cells: Interaction between colony-forming and colony-stimulating cells. *J. Natl. Cancer Inst.* **50**: 591.

Pike, B. L., and W. A. Robinson. 1970. Human bone marrow colony growth in agar-gel. *J. Cell. Physiol.* **76**: 77.

Rickard, D. A., R. K. Shadduck, A. Marley, and F. Stohlman, Jr. 1970. In vitro and in vivo colony technique in the study of granulopoiesis. In *Hemopoietic cellular proliferation* (ed. F. Stohlman, Jr.), p. 238. Grune & Stratton, New York.

Senn, J. S., H. A. Messner, and E. R. Stanley. 1974. Analysis of interacting cell populations in cultures of marrow from patients with neutropenia. *Blood* **44**: 33.

Trentin, J. 1971. Determination of bone marrow stem cell differentiation by stromal hemopoietic inductive microenvironments (HIM). *Am. J. Pathol.* **65**: 621.

VanFurth, R. 1970. Origin and kinetics of monocytes and macrophages. *Semin. Hematol.* **7**: 125.

Weiss, L. 1970. The histology of the bone marrow. In *Regulation of hematopoiesis* (ed. A. S. Gordon), vol. 1, p. 79. Appleton-Century-Crofts, New York.

Regulation of Normal Cell Differentiation and Malignancy in Myeloid Leukemia

L. Sachs

Department of Genetics, The Weizmann Institute of Science, Rehovot, Israel

THE DIFFERENTIATION-INDUCING PROTEIN MGI

The development of experimental systems for the culture and cloning of normal hematopoietic cells (Ginsburg and Sachs 1963, 1965; Sachs 1964, 1974 a, b; Pluznik and Sachs 1965, 1966; Bradley and Metcalf 1966; Ichikawa et al. 1966; Paran et al. 1970) has made possible the study of controls that regulate hematopoietic cell differentiation and the blocks that can occur in leukemia. All the main types of mammalian hematopoietic cells can be cloned in culture (Ginsburg and Sachs 1963; Pluznik and Sachs 1965; Bradley and Metcalf 1966; Stephenson et al. 1971; Metcalf et al. 1975; Fibach et al. 1976; Gerassi and Sachs 1976; Sredni et al. 1976). Normal myeloid precursors can be induced to differentiate to mature macrophages and granulocytes by the protein inducer (Pluznik and Sachs 1965, 1966; Ichikawa et al. 1976) that we now call MGI (macrophage and granulocyte inducer) (Landau and Sachs 1971; Sachs 1974 a, b). This inducer, which is secreted by various types of cells including fibroblasts and macrophages, can also be found in human serum (Mintz and Sachs 1973). Unless otherwise stated, all the experiments described in this paper and summarized elsewhere (Sachs 1978a) were carried out with cells from mice. MGI, which has also been referred to as mashran gm (Ichikawa et al. 1967), colony-stimulating factor (Metcalf 1969), or colony-stimulating activity (Austin et al. 1971), is specific for the induction of macrophages and granulocytes.

Purified MGI from fibroblasts has a molecular weight of about 68,000 (Landau and Sachs 1971; Guez and Sachs 1973; Stanley and Heard 1977) and purified MGI from lungs, a molecular weight of about 23,000 (Burgess et al. 1977); the lower molecular weight may be derived from the higher molecular weight MGI. Purified MGI can induce the formation of macrophages and granulocytes. However, there also may be molecular forms that induce only macrophages or granulocytes, and different cofactors (Landau and Sachs 1971; Sachs 1974b) that influence the differentiation to one or the other cell type. Incubation of the precursor cells with MGI for different periods of time indicated that MGI had to be present until the process of differentiation had been completed (Pluznik and Sachs 1966; Paran and Sachs 1968). Normal myeloid precursors require MGI for cell viability, growth, and differentiation. It remains to be determined whether these properties can all be induced by MGI molecules with the same chemical composition.

INDUCTION OF NORMAL DIFFERENTIATION OF MYELOID LEUKEMIC CELLS BY MGI

The finding of the differentiation-inducing protein MGI raised the question as to whether this protein can also induce the normal differentiation of myeloid leukemic cells. To test this possibility, we studied leukemic cells from different myeloid leukemias. Our results indicated that there is one type of myeloid leukemic cell in humans (Paran et al. 1970) and mice (Sachs 1974a,b, 1978b) that can be induced to differentiate normally both in vitro and in vivo (J. Lotem and L. Sachs, unpubl.). The experiments with mice showed that these leukemic cells can be induced with purified MGI to dif-

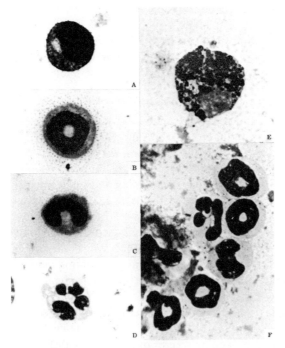

Figure 1
Differentiation of MGI^+D^+ cells to mature macrophages and granulocytes by MGI. (*A*) Undifferentiated blast cell; (*B–D*) stages in the differentiation to mature granulocytes; (*E*) macrophage; and (*F*) group of granulocytes in different stages of differentiation.

ferentiate to mature macrophages and granulocytes (Fibach et al. 1972) (Fig. 1). This type of leukemia has been cloned in culture (Ichikawa 1969; Fibach et al. 1972, 1973) and will be referred to as MGI$^+$D$^+$ (D$^+$ for differentiation to mature cells) (Fibach et al. 1973). Like normal macrophages and granulocytes, the mature cells induced from these leukemic cells were no longer malignant in vivo and no longer multiplied in vitro (Fibach and Sachs 1975). Unlike Friend erythroleukemic cells (Friend et al. 1971) that have lost the ability to respond to the normal erythroid-inducing protein erythropoietin (Kluge et al. 1974), these MGI$^+$D$^+$ myeloid leukemic cells could still respond to the normal myeloid-inducing protein MGI.

Isolation of the myeloid precursor cells from normal bone marrow (Lotem and Sachs 1977a) made possible a comparison of the sequence of differentiation in normal myeloid and MGI$^+$D$^+$ leukemic cells. The results indicated that both normal cells and these leukemic cells were induced by MGI to produce the same mature cells, macrophages and granulocytes, and that the process of differentiation occurred in the same sequence. The sequence of differentiation in both cases was induction of C3 and Fc rosettes (Lotem and Sachs 1974), C3 and Fc immune phagocytosis (Lotem and Sachs 1977b), synthesis and secretion of lysozyme (Krystosek and Sachs 1976), and formation of mature macrophages and granulocytes (Paran et al. 1970; Fibach et al. 1972, 1973; Lotem and Sachs 1977b).

ORIGIN OF MYELOID LEUKEMIA

Although both the normal and MGI$^+$D$^+$ leukemic cells could be induced to differentiate normally by MGI, the normal cells differed from the leukemic ones in that the leukemic cells were viable and could grow in the absence of MGI, whereas the normal cells required MGI for cell viability and growth (Fibach and Sachs 1976). Therefore, these leukemic cells were malignant, not because they could not be induced to differentiate by the normal inducing protein, but because they no longer required this protein for viability and growth. The absence of an adequate supply of MGI in vivo would limit the growth of the normal cells but would not affect the growth of the leukemic cells. Thus, a change in the cells resulting in a partial or complete loss of the requirement of this protein for viability and growth may be the cause of myeloid leukemia.

The MGI$^+$D$^+$ leukemic cells studied did not have a normal dipoid karyotype, and the change in the requirement for MGI for cell viability and growth was associated with a chromosome abnormality (Hayashi et al. 1974; Azumi and Sachs 1977). This suggests a genetic origin, one that is due to a chromosomal change, for the leukemia that arises from a loss in the requirement for MGI for viability and growth.

GENETIC DISSECTION OF THE CONTROL OF DIFFERENTIATION

Studies with different types of myeloid leukemic cells have indicated that, once the change has occurred which allows the leukemic cells to grow in the absence of MGI, other genetic changes may follow that produce blocks in the induction of differentiation by MGI. The isolation and study of such mutants (Fibach et al. 1973; Lotem and Sachs 1974, 1975a, 1976, 1977b; Krystosek and Sachs 1976) has shown that there can be blocks at different stages of differentiation and that there are separate controls for the induction of Fc rosettes, C3 rosettes, Fc immune phagocytosis, C3 immune phagocytosis, synthesis and secretion of lysozyme, formation of mature macrophages, and formation of mature granulocytes (Lotem and Sachs 1977b). Mutants have also been isolated that differed in the time of induction and in the sequence of differentiation. In one such mutant, MGI induced the synthesis and secretion of lysozyme without going through the stage of Fc and C3 immune phagocytosis (Lotem and Sachs 1977b). The isolation of other mutants should make possible further identification not only of the degree of independence of the control for each marker but also of the extent to which other stages in the sequence can be omitted, or sequences reversed, and the cells still proceed to the next stage. It will also be interesting to find mutants that still require MGI for viability and growth but are blocked in differentiation, and to determine whether such cells are malignant.

In addition to the induction of Fc and C3 rosettes (Lotem and Sachs 1974), Fc and C3 immune phagocytosis (Lotem and Sachs 1977b), lysozyme synthesis and secretion (Krystosek and Sachs 1976), and the formation of morphologically mature macrophages and granulocytes (Fibach et al. 1972, 1973), we also found that induction of MGI^+D^+ cells by MGI resulted in an induction of many other markers. These included cell migration (Fig. 2), nonimmune phagocytosis (Fibach and Sachs 1975), and chemotaxis (G. Symonds and L. Sachs, unpubl.), synthesis of actin (B. Hoffman and L. Sachs, unpubl.), increase in β_2-adrenergic receptors (Simantov and Sachs 1978), agglutinability and stimulation of the pentose cycle by concanavalin A (Vlodavsky et al. 1976), glycolytic production of ATP (Vlodavsky et al. 1976), and a regain of the normal requirement for MGI for cell viability and growth (Fibach and Sachs 1976). An analysis of protein synthesis by two-dimensional gel electrophoresis has shown 73 protein

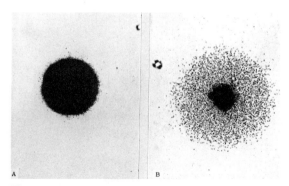

Figure 2
Colonies of MGI^+D^+ cells in agar (*A*) without MGI and (*B*) with MGI. The colony in B shows migration of the cells induced by MGI.

changes after induction of MGI⁺D⁺ cells by MGI (B. Hoffman and L. Sachs, unpubl.). Thus, there are many other markers that can be used for the genetic dissection of the controls for normal differentiation in these cells.

CELL COMPETENCE FOR THE INDUCTION OF NORMAL DIFFERENTIATION

We have found that differences in competence for induction of normal differentiation in myeloid leukemic cells by MGI were associated with membrane differences between MGI⁺D⁺ and the less competent cells. These included differences in concanavalin-A-induced cell-to-cell binding between a fixed cell and an unfixed cell (Rutishauser and Sachs 1974, 1975a); cell binding to nylon fibers coated with different densities of concanavalin A (Rutishauser and Sachs 1975b); concanavalin-A agglutin-ability after ATP depletion (Vlodavsky et al. 1976); the frequency of free or specifically anchored concanavalin-A surface receptors as measured by capping (Lotem et al. 1976); cap formation by H-2 antigens (U. Bushkin and L. Sachs, unpubl.) and murine leukemia virus gs antigens (Liebermann and Sachs 1977); differences in desensitization of functional β_2-adrenergic receptors, such differences possibly being due to an alteration in the uncoupling system between adrenergic receptors and adenylate cyclase (Simantov and Sachs 1978); and differences in the amount of surface ecto-ATPase activity (Weiss and Sachs 1977). This suggests that certain membrane properties may be associated with an appropriate arrangement of surface receptors for MGI or its internalization, such arrangement possibly being required for the induction of normal cell differentiation by this protein. Cell competence for differentiation by MGI was also associated with differences in the production and inducibility of type-C RNA virus; the highest degree of competence was associated with the highest degree of virus production (Liebermann and Sachs 1977). Thus, there may be an association of viral sequences with regulatory sites for differentiation. The increased virus production may have indicated a different state of the regulatory sites for differentiation in the competent cells, and the higher virus production may either be a by-product of these regulatory sites or may influence them directly. Therefore, this appears to be a favorable system for the investigation of the possible role of type-C virus in modifying cell competence for differentiation by a normal regulator.

Isolation of segregants from the MGI⁺D⁺ to a less competent phenotype and vice versa has shown that these segregants could differ in the degree of stability of the altered phenotype. In many segregants, the altered phenotype was unstable and, after multiplying, the cells soon reverted to the parental phenotype, whereas in other segregants the new phenotype was stable for many cell generations. Chromosome studies have indicated that only the stable, but not the unstable, segregants have specific chromosome changes that can be observed with chromosome banding (Hayashi et al. 1974). Therefore, the stable segregants were mutants with chromosome changes that could be identified with the light microscope. One possible explanation which can be experimentally tested in this system is that the cell's competence to be induced for the phenotypic changes associated with differentiation may have been due to transposable genetic elements (Nevers and Saedler 1977) whose stabilization requires these specific chromosome changes.

Karyotype analysis of cells with different degrees of competence for the induction of differentiation by MGI has shown that the genes that control cell competence are located on mouse chromosomes 2 and 12 and that inducibility by MGI is controlled by the balance between these genes (Azumi and Sachs 1977). We have suggested that these chromosomes also carry genes that control the malignancy of these cells. Results with cell hybridization have shown that hybrids between myeloid leukemic cells and normal macrophages suppressed the malignancy of the leukemic cells (Shkolnik and Sachs 1978), presumably because of a change in the balance of these genes (Sachs 1974a).

INDUCTION OF SOME NORMAL CELL MARKERS BY TREATMENT WITH OTHER COMPOUNDS

Studies with various compounds, including those used in cancer therapy, have shown that some of the stages of differentiation can be induced in appropriate clones of myeloid leukemic cells by certain steroids such as dexamethasone, prednisolone, and estradiol (Lotem and Sachs 1974, 1975a); by such presumably surface-acting agents as the lectins concanavalin A, phytohemmaglutinin, pokeweed mitogen (J. Lotem and L. Sachs, unpubl.), lipopolysaccharides (LPS), lipid A (Weiss and Sachs 1978), and dimethylsulfoxide (DMSO) (Krystosek and Sachs 1976; Maeda and Sachs 1978); by actinomycin D and other compounds that can interact with DNA such as cytosine arabinoside, mitomycin C, 5-bromodeoxyuridine (Lotem and Sachs 1974, 1975b), nitrosoguanidine (A. Falk and L. Sachs, unpubl.); by X-irradiation (A. Falk and L. Sachs, unpubl.); and by some carcinogenic hydrocarbons (Z. Schwarzbard and L. Sachs, unpubl.). Compounds that induced some normal cell markers in an appropriate clone of MGI^+D^+ leukemic cells are shown in Table 1. Not all the compounds induced the same markers or induced changes in the same clones. MGI was the only compound that induced all the changes to mature macrophages and granulocytes.

An example of the dissection of the controls for induction by such steroid inducers (SI) as dexamethasone and MGI is shown in Figure 3. The data show that MGI and SI did not induce the same markers even in clones (SI^+MGI^+) that responded to both compounds and that it was possible to isolate mutants that were SI^+MG^-, SI^-MG^- (Lotem and Sachs 1976), or SI^-MGI^+ (L. Cohen and L. Sachs, unpubl.). The lack of response to dexamethasone in the SI clones was not due to any detectable defect in the number, nuclear transport, or association with DNA-containing structures of the steroid receptors (Krystosek and Sachs 1977). The results indicated that there are different cellular sites for MGI and a steroid inducer such as dexamethasone. DMSO induced C3 but not Fc rosettes and macrophages but not granulocytes in a $MGI^+SI^+DMSO^+$ clone. We also isolated

Table 1
Inducers and Noninducers for Normal Cell Markers in MGI$^+$D$^+$ Myeloid Leukemic Cells

Type of compound	Inducers	Noninducers
Peptide hormones	MGI	erythropoietin
		nerve growth factor
		insulin
		ubiquitin
		thymopoietin
		interferon
Steroids	dexamethasone	progesterone
	prednisolone	testosterone
	hydrocortisone	epitestosterone
	estradiol	androstenedione
		cortisone
Lectins	concanavalin A	
	phytohemagglutinin	
	pokeweed mitogen	
Polycyclic hydrocarbons	benzo(a)pyrene	benz(a)anthracene
	dimethylbenz(a)-anthracene	dibenz(a,c)anthracene
		dibenz(a,h)anthracene
		phenanthrene
Other compounds	lipopolysaccharide	colchicine
	lipid A	vinblastine
	mitomycin C	Na butyrate
	dimethylsulfoxide	cycloheximide
	cytosine arabinoside	db cyclic AMP
	hydroxyurea	db cyclic GMP
	thymidine	cordycepin
	5-iododeoxyuridine	deoxyglucose
	5-bromodeoxyuridine	ouabain
	5-fluorodeoxyuridine	ionophore 23187
	nitrosoguanidine	
	actinomycin D	
	adriamycin	
	daunomycin	
	X-irradiation	

The different inducers were not all active on the same clone and did not all induce the same markers.

MGI$^+$SI$^+$DMSO$^-$ clones (Krystosek and Sachs 1976; Maeda and Sachs 1978), so that there appeared to be different cellular sites for DMSO, SI, and MGI. Friend erythroleukemia cells could be induced to differentiate partially by various compounds; there also appeared to be different cellular sites for different compounds (Nudel et al. 1977; Ohta et al. 1976). However, in contrast to these erythroleukemic cells that could be induced to differentiate partially by a compound such as DMSO but which had lost the ability to respond to erythropoietin, the MGI$^+$DMSO$^+$ myeloid leukemic cells could still respond to MGI.

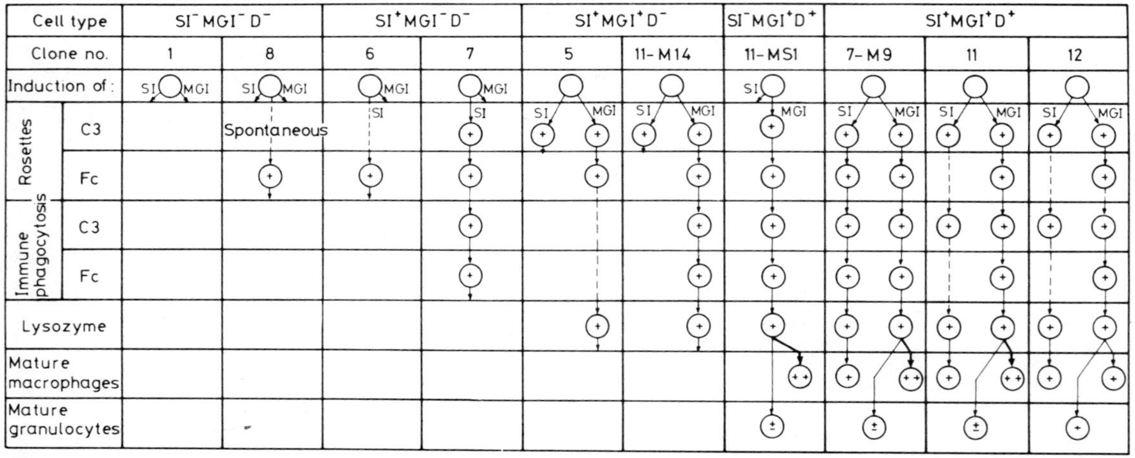

Figure 3
Inducibility for differentiation-associated markers in different clones of myeloid leukemic cells by the SI dexamethasone and the normal regulatory protein MGI. Clones that could be induced to differentiate to mature cells are referred to as D^+.

INDUCTION OF MGI IN MGI⁺D⁺ LEUKEMIC CELLS

Among the compounds that induced differentiation to macrophages in some MGI⁺D⁺ clones were LPS from different bacteria. LPS induced a high frequency of Fc and C3 rosettes, lysozyme, and macrophages in certain clones of MGI⁺D⁺ cells but not in MGI⁺D⁻ clones. Similar results were obtained with lipid A, so that the active part of the LPS molecule appeared to be lipid A. The results also showed that treatment of the responding MGI⁺D⁺ clones with LPS or lipid A induced an activity in the conditioned medium that behaved like MGI (Weiss and Sachs 1978). This indicates that appropriate clones of leukemic cells can be induced to produce their own normal differentiation inducer.

The activity of LPS and lipid A could be distinguished from MGI in that MGI, but not LPS or lipid A, induced the formation of colonies with macrophages or granulocytes from normal bone-marrow cells, induced some stages of differentiation in MGI⁺D⁻ clones, and induced differentiation in a MGI⁺D⁺ clone that had been selected for resistance to induction by LPS. The inducing activity of the protein MGI, but not of lipid A, was also sensitive to trypsin. These tests made it possible to assay for MGI in medium that still may have contained residual LPS or lipid A. Studies on the time of induction of detectable MGI after treatment of MGI⁺D⁺ cells with LPS have indicated that induction of MGI was detected before the induction of rosettes or lysozyme. These results showed that the lipid-A portion of LPS indirectly induced differentiation of MGI⁺D⁺ myeloid leukemic cells by inducing in these cells production of the differentiation-inducing protein MGI (Weiss and Sachs 1978).

It will be of interest to determine which of the other compounds that are able to induce differentiation-associated properties in MGI⁺D⁺ leukemic cells act directly and which, like lipid A, may induce differentiation indirectly by inducing the production of MGI. The induction of regulatory proteins that can induce specific cell differentiation may represent a more general mechanism for the induction of differentiation by various compounds in different cell types.

POSSIBILITIES FOR THERAPY

Our results obtained with these myeloid leukemic cells suggest some novel possibilities for treatment of these leukemias (Paran et al. 1970; Fibach et al. 1972; Sachs 1974a, 1976), treatment which may also be applicable to other types of tumors. The finding of MGI⁺D⁺ myeloid leukemic cells that could be induced to differentiate normally by MGI suggests MGI injection, grafting of MGI-producing cells, or stimulation of the in vivo production of MGI to induce the normal differentiation of these leukemic cells. This would be a form of tumor therapy that would not be based on the search for cytotoxic agents that kill tumor cells more readily than normal cells. The membrane differences found between cells which differed in their degree of competence to be induced to differentiate by MGI may represent useful markers for prediction of the response of the leukemic cells to MGI in vivo.

MGI$^+$D$^+$ leukemic cells could be induced by MGI again to require this protein for cell viability and growth. This suggests that induction of differentiation of the leukemic cells to this stage, followed by the withdrawal of MGI, may also result in the loss of viability and growth of the induced MGI$^+$D$^+$ leukemic cells in vivo.

The induction of normal macrophage and granulocyte differentiation by MGI also suggests that injection of MGI or stimulation of its production in vivo may produce a rapid recovery of the normal macrophage and granulocyte populations following the cytotoxic therapy that seriously depletes the normal population required to combat infection.

The present results may also explain the response of some, but not of all, patients to chemical and irradiation cytotoxic therapy. We have shown that chemicals and irradiation used in therapy can induce some stages of differentiation in clones of myeloid leukemic cells with the appropriate genotype, and that clonal differences in inducibility for normal differentiation-associated properties are not necessarily associated with differences in the response of these clones to the cytotoxic effect of these compounds. Cells with induced Fc and C3 receptors, phagocytosis, and other macrophagelike properties may be expected to behave differently in the body in their reponse to a variety of factors, including antibodies, than would cells without these properties. The in vivo growth of leukemic cells with the appropriate genotype may thus be controlled by therapeutic agents used not only because of their cytotoxic effect but also because they induce these differentiation-associated properties. Differences in the cell's competence to be induced by these agents may explain the differences in response to therapy found in different individuals. The possible induction of MGI by these compounds may also play a role in the therapeutic effects obtained in vivo.

The results obtained with these myeloid leukemic cells suggest possible forms of therapy based on the use of a normal regulatory protein such as MGI for induction of normal differentiation in malignant cells and a more rapid recovery of the normal cell population after the present forms of therapy have been utilized. They also suggest the use of other compounds that can induce the normal regulatory protein or can affect mutant malignant cells at differentiation sites that are no longer susceptible to the normal regulator.

REFERENCES

Austin, P. E., E. A. McCulloch, and J. E. Till. 1971. Characterization of the factor in L-cell conditioned medium capable of stimulating colony formation by mouse marrow cells in culture. *J. Cell. Physiol.* **77:** 121.

Azumi, J.-I., and L. Sachs. 1977. Chromosome mapping of the genes that control differentiation and malignancy in myeloid leukemic cells. *Proc. Natl. Acad. Sci. U.S.A.* **74:** 253.

Bradley, T. R., and D. Metcalf. 1966. The growth of mouse bone marrow cells *in vitro*. *Aust. J. Exp. Biol. Med. Sci.* **44:** 287.

Burgess, A. W., J. Camakaris, and D. Metcalf. 1977. Purification and properties of colony-stimulating factor from mouse lung conditioned medium. *J. Biol. Chem.* **252:** 1998.

Fibach, E., and L. Sachs. 1975. Control of normal differentiation of myeloid leukemic cells. VIII. Induction of differentiation to mature granulocytes in mass culture. *J. Cell. Physiol.* **86:** 221.

———. 1976. Control of normal differentiation of myeloid leukemic cells. XI. Induction of a specific requirement for cell viability and growth during the differentiation of myeloid leukemic cells. *J. Cell. Physiol.* **89:** 259.

Fibach, E., E. Gerassi, and L. Sachs. 1976. Induction of colony formation *in vitro* by human lymphocytes. *Nature* **259:** 127.

Fibach, E., M. Hayashi, and L. Sachs. 1973. Control of normal differentiation of myeloid leukemic cells to macrophages and granulocytes. *Proc. Natl. Acad. Sci. U.S.A.* **70:** 343.

Fibach, E., T. Landau, and L. Sachs. 1972. Normal differentiation of myeloid leukaemic cells induced by a differentiation-inducing protein. *Nat. New Biol.* **237:** 276.

Friend, C., V. Scher, J. G. Holland, and T. Sato. 1971. Hemoglobin synthesis in murine virus-induced leukemic cells *in vitro:* Stimulation of erythroid differentiation by dimethyl sulfoxide. *Proc. Natl. Acad. Sci. U.S.A.* **68:** 378.

Gerassi, E., and L. Sachs. 1976. Regulation of the induction of colonies *in vitro* by normal human lymphocytes. *Proc. Natl. Acad. Sci. U.S.A.* **73:** 4546.

Ginsburg, H., and L. Sachs. 1963. Formation of pure suspensions of mast cells in tissue culture by differentiation of lymphoid cells from the mouse thymus. *J. Natl. Cancer Inst.* **31:** 1.

———. 1965. Destruction of mouse and rat embryo cells in tissue culture by lymph node cells from unsensitized rats. *J. Cell. Comp. Physiol.* **66:** 199.

Guez, M., and L. Sachs. 1973. Purification of the protein that induces cell differentiation to macrophages and granulocytes. *FEBS* (Fed. Eur. Biochem. Soc.) *Lett.* **37:** 149.

Hayashi, M., E. Fibach, and L. Sachs. 1974. Control of normal differentiation of myeloid leukemic cells. V. Normal differentiation in aneuploid leukemic cells and the chromosome banding pattern of D^+ and D^- clones. *Int. J. Cancer* **14:** 40.

Ichikawa, Y. 1969. Differentiation of a cell line of myeloid leukemia. *J. Cell. Physiol.* **74:** 223.

Ichikawa, Y., D. H. Pluznik, and L. Sachs. 1966. In vitro control of the development of macrophage and granulocyte colonies. *Proc. Natl. Acad. Sci. U.S.A.* **56:** 488.

———. 1967. Feedback inhibition of the development of macrophage and granulocyte colonies. I. Inhibition by macrophages. *Proc. Natl. Acad. Sci. U.S.A.* **58:** 1480.

Kluge, N., G. Gaedicke, G. Steinheider, S. Dube, and W. Ostertag. 1974. Globin synthesis in Friend-erythroleukemia mouse cells in protein- and lipid-free medium. *Exp. Cell Res.* **88:** 257.

Krystosek, A., and L. Sachs. 1976. Control of lysozyme induction in the differentiation of myeloid leukemic cells. *Cell* **9:** 675.

———. 1977. Steroid hormone receptors and the differentiation of myeloid leukemic cells. *J. Cell. Physiol.* **92:** 345.

Landau, T., and L. Sachs. 1971. Characterization of the inducer required for the development of macrophage and granulocyte colonies. *Proc. Natl. Acad. Sci. U.S.A.* **68:** 2540.

Liebermann, D., and L. Sachs. 1977. Type C RNA virus production and cell competence for normal differentiation in myeloid leukaemic cells. *Nature* **269:** 173.

Lotem, J., and L. Sachs. 1974. Different blocks in the differentiation of myeloid leukemic cells. *Proc. Natl. Acad. Sci. U.S.A.* **71:** 3507.

———. 1975a. Induction of specific changes in the surface membrane of myeloid leukemic cells by steroid hormones. *Int. J. Cancer* **15:** 731.

———. 1975b. Control of normal differentiation of myeloid leukemic cells. VI. Inhibition of cell multiplication and the formation of macrophages. *J. Cell. Physiol.* **85:** 587.

———. 1976. Control of Fc and C3 receptors on myeloid leukemic cells. *J. Immunol.* **117:** 580.

———. 1977a. Control of normal differentiation of myeloid leukemic cells. XII. Isolation of normal myeloid colony-forming cells from bone marrow and the sequence of differentiation to mature granulocytes in normal and D$^+$ myeloid leukemic cells. *J. Cell. Physiol.* **92:** 97.

———. 1977b. Genetic dissection of the control of normal differentiation in myeloid leukemic cells. *Proc. Natl. Acad. Sci. U.S.A.* **74:** 5554.

Lotem, J., I. Vlodavsky, and L. Sachs. 1976. Regulation of cap formation by concanavalin A and the differentiation of myeloid leukemic cells: Relationship to free and anchored surface receptors. *Exp. Cell Res.* **101:** 323.

Maeda, S., and L. Sachs. 1978. Control of normal differentiation of myeloid leukemic cells. XIII. Inducibility for some stages of differentiation by dimethylsulfoxide and its dissociation from inducibility by MGI. *J. Cell. Physiol.* **94:** 181.

Metcalf, D. 1969. Studies on colony formation *in vitro* by mouse bone marrow cells. I. Continuous cluster formation and relation of clusters to colonies. *J. Cell. Physiol.* **74:** 323.

Metcalf, D., G. J. V. Nossal, N. L. Warner, J. F. A. P. Miller, T. E. Mandel, J. E. Layton, and G. A. Gutman. 1975. Growth of B-lymphocyte colonies in vitro. *J. Exp. Med.* **142:** 1534.

Mintz, U., and L. Sachs. 1973. Differences in inducing activity for human bone marrow colonies in normal serum and serum from patients with leukemia. *Blood* **42:** 331.

Nevers, P., and H. Saedler. 1977. Transposable genetic elements as agents of gene instability and chromosomal rearrangements. *Nature* **268:** 109.

Nudel, U., J. E. Salmon, M. Terada, A. Bank, R. A. Rifkind, and P. A. Marks. 1977. Differential effects of chemical inducers on expression of β globin genes in murine erythroleukemia cells. *Proc. Natl. Acad. Sci. U.S.A.* **74:** 1100.

Ohta, Y., M. Tanaka, M. Terada, O. J. Miller, A. Bank, P. A. Marks, and R. A. Rifkind. 1976. Erythroid cell differentiation: Murine erythroleukemia cell variant with unique pattern of induction by polar compounds. *Proc. Natl. Acad. Sci. U.S.A.* **73:** 1232.

Paran, M., and L. Sachs. 1968. The continued requirement for inducer for the development of macrophage and granulocyte colonies. *J. Cell. Physiol.* **72:** 247.

Paran, M., L. Sachs, Y. Barak, and P. Resnitzky. 1970. *In vitro* induction of granulocyte differentiation in hematopoietic cells from leukemic and non-leukemic patients. *Proc. Natl. Acad. Sci. U.S.A.* **67:** 1542.

Pluznik, D. H., and L. Sachs. 1965. The cloning of normal "mast" cells in tissue culture. *J. Cell. Comp. Physiol.* **66:** 319.

———. 1966. The induction of clones of normal mast cells by a substance from conditioned medium. *Exp. Cell Res.* **43:** 553.

Rutishauser, U., and L. Sachs. 1974. Receptor mobility and the mechanism of cell-cell binding induced by concanavalin A. *Proc. Natl. Acad. Sci. U.S.A.* **71:** 2456.

———. 1975a. Cell to cell binding induced by different lectins. *J. Cell Biol.* **65:** 247.

———. 1975b. Receptor mobility and the binding of cells to lectin-coated fibers. *J. Cell Biol.* **66:** 76.

Sachs, L. 1964. The analysis of regulatory mechanisms in cell differentiation. In *New perspectives in biology* (ed. M. Sela), p. 246. Elsevier Co., Amsterdam.

———. 1974a. Regulation of membrane changes, differentiation and malignancy in carcinogenesis. *Harvey Lect.* **68:** 1.

―――. 1974b. Control of growth and differentiation in normal hematopoietic and leukemic cells. In *Control of proliferation in animal cells* (eds. B. Clarkson and R. Baserga), p. 915. Cold Spring Harbor Laboratory, Cold Spring Harbor, New York.

―――. 1976. Control of normal cell differentiation in leukemic cells. *Bibl. Haematol.* **43:** 6.

Sachs, L. 1978a. Control of normal cell differentiation and the phenotype of malignancy in myeloid leukemia. *Nature*. (In press.)

Sachs, L. 1978b. Control of normal cell differentiation in leukemic white blood cells. In *M. D. Anderson Symposium on Cell Differentiation and Neoplasia* (ed. L. Wildrick). Williams and Wilkins, Baltimore. (In press.)

Shkolnik, T., and L. Sachs. 1978. Suppression of the *in vivo* malignancy and *in vitro* cell multiplication of myeloid leukemic cells by hybridization with normal macrophages. *Exp. Cell Res.* **113:** 197.

Simantov, R., and L. Sachs. 1978. Differential desensitization of functional adrenergic receptors in normal and malignant myeloid cells. Relationship to receptor mediated hormone cytotoxicity. *Proc. Natl. Acad. Sci. U.S.A.* (In press.)

Sredni, B., Y. Kalechman, H. Michlin, and L. A. Rozenszajn. 1976. Development of colonies *in vitro* of mitogen-stimulated mouse T lymphocytes *Nature* **259:** 130.

Stanley, E. R., and P. M. Heard. 1977. Factors regulating macrophage production and growth, purification and some properties of the colony stimulating factor from medium conditioned by mouse L cells. *J. Biol. Chem.* **252:** 4305.

Stephenson, J. R., A. A. Axelrad, D. L. McLeod, and M. M. Shreeve. 1971. Induction of colonies of hemoglobin-synthesizing cells by erythropoietin *in vitro*. *Proc. Natl. Acad. Sci. U.S.A.* **68:** 1542.

Vlodavsky, I., E. Fibach, and L. Sachs. 1976. Control of normal differentiation of myeloid leukemic cells. X. Glucose utilization, cellular ATP and associated membrane changes in D^+ and D^- cells. *J. Cell. Physiol.* **87:** 167.

Weiss, B., and L. Sachs. 1977. Differences in surface membrane ecto-ATPase and ecto-AMPase in normal and malignant cells. I. Decrease in ecto-ATPase in myeloid leukemic cells and the independent regulation of ecto-ATPase and ecto-AMPase. *J. Cell. Physiol.* **93:** 183.

―――. 1978. Indirect induction of differentiation in myeloid leukemic cells by lipid A. *Proc. Natl. Acad. Sci. U.S.A.* **75:** 1374.

Section 4

LYMPHOCYTE DIFFERENTIATION AND REGULATION

Introduction

H. Cantor

Harvard Medical School, Dana Cancer Center
Boston, Massachusetts 02115

E. A. Boyse

Memorial Sloan-Kettering Cancer Center
New York, New York 10021

Differentiation in higher organisms refers to the generation of diverse, individually specialized sets of cellular progeny from a single stem cell. This process connotes a series of absolute commitments that step by step dictate the production of a mature organism from the zygote. One may approach the study of the molecular basis of this process by focusing on the ontogeny of a single cell type.

A good case can be made for the use of lymphocyte populations of the mouse for this purpose. First, the surface components of lymphocytes have been more extensively characterized than have those of any other cell type. The majority of these components have been serologically identified and the genes coding for them have been precisely mapped. In most cases, these surface structures (1) are controlled by genes that are expressed exclusively during lymphocyte differentiation and (2) mediate important specialized functions carried out by lymphocytes and no other cells.

Many of these gene products are expressed exclusively on subsets of T or B lymphocytes, and so they can serve as identification markers of cells at different stages of differentiation. In addition, many assays of immune function in vitro are now available, so one can determine the possession or lack of a variety of immune functions within a given set of cells. Thus, according to this strategy, the cell surface "antigen profile" of a subpopulation of lymphocytes can be easily correlated with its functional properties. This approach has led to an increasingly accurate delineation of each step or branch in the lymphocyte differentiation path.

Two major lines of lymphocyte differentiation have been identified so far: T cells, which differentiate in the thymus and are concerned mainly with cell-mediated immunity, and B cells, which differentiate elsewhere. Both types of lymphocytes carry receptors for antigen. B cells, when stimulated by antigen, develop into plasma cells which secrete these receptors in the form of free antibodies. T cells do not. These diversified surface receptors can be considered as markers for individual lymphocytes or lymphocyte clones.

B LYMPHOCYTES

The myeloma cell has so far received the greatest attention of any of the B-lymphocyte sets. These are malignant B cells (or plasma cells) that secrete free antibody; they have been exploited to elucidate the fine structure of antibody or immunoglobulin. However, myeloma cells constitute a terminal cell set with no further commitments to make. Attention has now been directed toward the inductive signals required for expression of the B-lymphocyte genetic program responsible for the formation and secretion of free antibody. These studies have resulted in the characterization of several surface structures which may play important roles in the triggering of this cell type. A special class of immunoglobulin, IgD, is thought to play an important role both in antigen recognition and in the antigen-dependent triggering of B lymphocytes. Characterization of this cell-surface structure and its role in the differentiation of B cells has been extensively investigated (Uhr et al., this volume). Attention has also been given to B-cell surface molecules which may act as receptors for molecules that selectively induce B-cell division. This approach is based to a large extent on the following model to explain B-cell activation: (1) the relevant antigen selects a B-cell clone expressing an appropriate immunoglobulin receptor; (2) this interaction "focuses" a second molecule attached to the antigen onto the B cell. It is this second molecule which binds to a "mitogen receptor" that induces B-cell proliferation and differentiation. Stimulation of the mitogen receptor by substances such as lipopolysaccharides in microculture has revealed a great deal about the kinetics of individual clones of B cells. These studies have indicated, for example, that approximately 10% of the B cells that are induced to secrete the IgM class of immunoglobulin will "switch" to IgG secretion after several divisions (Melchers, this volume).

The underlying assumption of this approach is based on the fact that antigen alone is not sufficient to make B cells produce antibody. Induction of this B-cell program requires, in addition, a signal produced by T lymphocytes. Thus, the product of one cell, the T lymphocyte, can induce another cell, the B lymphocyte, to express its final differentiative program. In addition to the surface molecules discussed above, there are a number of candidate molecules on the surface of B cells which may act as receptors for T-cell signals. For example, Huber et al. (1977) have serologically identified a cell surface component expressed on a subset of B cells that provisionally has been termed Lyb3. Administration of anti-Lyb3 antibody, together with purified B cells, allows these cells to produce antibody to thymus-dependent antigen; that is, the effect of the antiserum is to replace partially the missing T-cell signal. These results have suggested that the Lyb3 surface component may act as a receptor on the B cell for the T-cell signal which, in concert with antigen, triggers B cells. Attachment of anti-Lyb3 to this B-cell surface component is thought to trigger the B cell by mimicry of the T-cell signal. Isolation of the Lyb3 component by conventional immunoprecipitation techniques has revealed that anti-Lyb3 serum binds a single molecular species having a molecular weight of approximately 68,000 daltons (R. Cone et al., in prep.). The polypeptides recognized by anti-Lyb3 are not composed of disulfide-linked subunits and bear no antigenic relationship with known B-cell membrane immunoglobulins including IgM and IgD. Most importantly, absorption of anti-Lyb3 serum with the isolated 68,000-dalton polypeptide removes the ability of the anti-Lyb3 serum to augment the in vivo immune response of B cells. This latter observation provides direct evidence that the 68,000-dalton polypeptide isolated after anti-Lyb3 immunoprecipitation is, in fact, the component on the B-cell membrane that is responsible for triggering of the B cell by anti-Lyb3.

T-LYMPHOCYTES

There is good evidence that stem cells migrate from yolk sac and liver in the embryo and from spleen and bone marrow in the adult into the thymus, where they differentiate to thymocytes. Most available data indicate that thymus stem cells cannot also arise in situ, although reports to the contrary in Amphibia must be considered.

Experiments with adult bone-marrow cells of the mouse indicate that a subpopulation of marrow cells is committed to migrate to the thymus and to differentiate to thymocytes ("prothymocyte"), and may be distinguished from other hematopoietic precursors. For example, El-Arini and Osoba (1973) studied the capacity of bone-marrow cells of different buoyant densities to transfer adoptively either hematopoietic function or T-cell function to irradiated hosts. Their results imply that progenitors of thymocytes and T cells are distinguishable from multipotential stem cells on the basis of the cells' different buoyant densities.

A direct approach to determining the developmental relationship between cells programed to give rise to thymocytes and those committed to generate hematopoietic cells is based on the following assumption. Cells transformed at a stage in which a recognizable set of options are retained may maintain their special differentiative potentials. In other words, a variety of murine leukemias may be "arrested" at an intermediate phase of their differentiation history. Studies of the surface phenotypes and function of such leukemias may lead to a clearer understanding of the relationships among pluripotential stem cells, prothymocytes, and cells committed to give rise only to hematopoietic and granulocytic cells (Silverstone et al., this volume). A similar strategy was used by Fialkow (1974) who, using genetic rather than cell surface markers, sought to elucidate the developmental relationship between stem cells committed to the lymphocyte lineage, termed "prolymphocytes," and other hematopoietic stem cells. One may also focus on the following question. How does the thymocyte program come to be expressed? As mentioned above, the cell that is to become a thymocyte resides in the bone marrow and spleen during adult life. This cell, the prothymocyte, migrates to the thymus, where it is induced by thymopoietin, a thymic hormone or "inducer," to express the thymocyte phenotype (Goldstein, this volume).

Three points are worth noting. (1) Migration of cells to a destination where the next step in differentiation is induced is a typical feature of ontogeny. (2) Analysis of prothymocyte induction in vitro shows that the phenotype can be induced not only by thymopoietin but also by cAMP and by agents that react with the cell surface but are physiologically irrelevant. Therefore the prothymocyte is committed before it reaches the thymus. Prothymocyte induction illustrates that initiation of differentiation by a specific inducer under physiological circumstances is not incompatible with nonspecific induction under abnormal conditions (a situation seen frequently in studies of embryogenesis). Goldstein (this volume) has found that induction of prothymocytes in vitro requires transcription and translation but not synthesis of DNA, and it entails the manifestation of five or more unlinked genes within an hour or so. (3) Manifestation of such a complex phenotype as the thymocyte without cell division seems remarkable. But this could well be a feature of any induced cell set whose surface phenotype is known in detail.

This T-cell induction system, preferably used in combination with a cloned, inducible cell line of prothymocytes, is obviously an attractive model for studying the sequence of events that begins with an interaction between a chemically defined (and sequenced) inducer molecule and its cell-surface receptor, and ends with the expression of a new genetic program by the target cell.

T-CELL SUBLINES

At the present time, the most effective technique for identifying and separating functional subpopulations of thymocytes and peripheral T cells has come from studies of the cell-surface components that become expressed on cells undergoing thymus-dependent differentiation. This type of classification is based upon the description of alloantisera that define several cell-surface differentiation components, Lyt1, Lyt2, and Lyt3, that are all expressed on mouse thymocytes but only on some peripheral T lymphocytes (Cantor and Boyse 1976). Since Lyt components have not been detected on cells of other tissues (brain, epidermal cells, kidney cells, liver cells, B lymphocytes, or bone-marrow cells), they are evidently specified by genes expressed exclusively during thymus-dependent differentiation. Studies of the reactivity of these Lyt antisera have revealed that T cells are divisible into three subpopulations on the basis of the following phenotypes: TL^-Ly123^+, TL^-Ly1^+, and TL^-Ly23^+. Thus, according to their distinctive expression of Lyt components, these cells obey different sets of genetic instructions. Do these instructions include commitment to one or another T-cell function?

Mature T cells have several different immunological tasks. Some must induce B cells to make antibody; some must kill cells that they recognize as being antigenically foreign; some regulate these activities by suppressing antibody production by B cells and immune responses mediated by other T cells. These three functions are the elements of an exquisitely balanced network of cellular interactions, and all three can be analyzed by stringent assays in vitro.

Each of these T-cell functions are mediated by cells expressing different Ly phenotypes. The Ly1 set provides help. The Ly23 set is cytotoxic and suppressive; separate surface phenotypes have not yet been definitely established for these two functions. Thus, the programs for the Ly1 and Ly23 sets combine information for phenotype and function, and are denoted Ly1:H and Ly23:CS. Evidence to date indicates that cells expressing the Ly123 surface phenotype may represent an intermediate set that has acquired receptors for antigen but is not yet committed to H or CS functions. Thus, it is probably the reservoir from which H and CS cells are supplied.

An additional question has been asked. Do the different sets derived from prolymphocytes carry out their respective functions independently of one another? They do not. The functions of cell sets derived from the prolymphocyte are precisely interlocked. As indicated above, the Ly1:H cell cooperates with the B cell, enabling it to produce antibody. Secondly, a collaborating Ly1:H cell enables the Ly2:CS cell to generate killer progeny more effectively. This latter interaction is discussed in detail by Sundharadas et al. (this volume).

WHAT DO T CELLS "SEE"

There is increasing evidence that a major occupation of T cells involves recognition and reaction to "altered" or polymorphic variants of HL-A (in man) and H-2 (in mice) gene products. These genes are collectively termed the major histocompatibility complex because they induce the strongest homograft reactions. In the mouse, two major types of MHC antigens have been identified: K/D and I. T cells screen the mouse for cells that have been antigenically modified, as by a virus. Efficient screening may require that the T cells have the same H-2 and I types as the target cells. The mechanism of their screening may be summarized as follows. The Ly1:H cell sees the unfamiliar antigen in conjunction with I. A message is passed to an Ly2:CS cell, whose progeny now kill cells which they see in conjunction with D or K. (An additional circuit involves passage of the Ly1:H message to a B cell, whose progeny now produce antibody against the foreign determinant.)

What is the fine structure of MHC gene products? Do some have binding sites that allow close association with foreign antigens such as viruses? The biochemical structure of MHC molecules of man is known in greater detail than those of the mouse, in large part from analyses of HL-A molecules by Strominger et al. (this volume). These studies, combined with parallel studies of the T cell receptor, may provide a definitive insight into the molecular basis of the fine specificity of T-cell recognition of antigens.

REFERENCES

Cantor, H., and F. A. Boyse. 1976. Regulation of cellular and humoral immune responses by T-cell subclasses. *Cold Spring Harbor Symp. Quant. Biol.* **41:** 23.

El-Arini, M. O. and D. Osoba. 1973. Differentiation of thymus-derived cells from precursors in mouse bone marrow. *J. Exp. Med.* **137:** 821.

Fialkow, P. J. 1974. The origin and development of human tumors studied with cell markers. *N. Engl. J. Med.* **291:** 26.

Huber, B., R. K. Gershon, and H. Cantor. 1977. Identification of a B-cell surface structure involved in antigen-dependent triggering: Absence of this structure on B cells from CBA/N mutant mice. *J. Exp. Med.* **145:** 10.

Correlating Terminal Deoxynucleotidyl Transferase and Cell-Surface Markers in the Pathway of Lymphocyte Ontogeny

A. E. Silverstone,* N. Rosenberg,[†] and D. Baltimore

Center for Cancer Research and Department of Biology
Massachusetts Institute of Technology, Cambridge, Massachusetts 02139

V. L. Sato

The Biological Laboratories, Harvard University, Cambridge, Massachusetts 02138

M. P. Scheid and E. A. Boyse

Memorial Sloan-Kettering Cancer Center, New York, New York 10021

Tumor cells usually maintain a well-defined phenotype over many generations. Although there is always the possibility that a given phenotype has no in vivo homologue, the constellation of properties of the tumor cell often has at least a superficial relationship to a normal cell counterpart. In such a situation, the tumor cell can be viewed as an indefinitely proliferating clone of cells fixed at a single stage of cellular differentiation. Such clones are extremely favorable models for characterizing particular compartments of cellular differentiation.

Tumor cell lines derived from leukemias and lymphomas are potential models of distinct stages in B-lymphocyte and T-lymphocyte development. We have been attempting to fit such tumor cells into the pathways of lymphocyte development, especially during the less defined early phases of these cell lineages. In this process, our discovery of new tumor cell phenotypes has suggested experiments for the isolation of normal developmental equivalents. Conversely, in vivo results have led to new methods of characterizing tumor cells.

Our main focus in this work has been the exceptional DNA polymerase called terminal deoxynucleotidyl transferase (TdT). This enzyme can be assayed unambiguously by standard methods of enzymology (Bollum 1974), by immunofluorescence (Baltimore et al. 1976), and by radioimmunoassay (Kung et al. 1976). In spite of extensive attempts to find TdT in other cell types (Chang 1971; Kung et al. 1975, 1976), it has been found thus far only in lymphoid cells of the thymus and bone marrow. Importantly, only the lymphoid cells of these two tissues are positive; circulating lymphoid cells, lymph node cells, and normal spleen cells contain no detectable activity (McCaffrey et al. 1973, 1975; Coleman et al. 1974). Therefore, TdT can be considered to be a marker of immature lymphoid cells. Many lymphoid

---

*Present address: Memorial Sloan-Kettering Cancer Center, New York, New York 10021
[†] Present address: Cancer Research Center and Department of Pathology, Tufts University School of Medicine, Boston, Massachusetts 02111.

cell tumors are TdT-positive and, in all cases, the cells appear to represent immature lymphoid cells (Baltimore et al. 1976).

In addition to its value as a marker for compartments in lymphoid differentiation, TdT is of great interest for the possible functional significance that derives from its enzymatic properties (Baltimore 1974). It is now clear from recent work (Tonegawa et al. 1976; Leder et al. 1976) that the production of immunoglobulin—and, therefore, probably of the other specific recognition molecules of the immune system such as those on the surface of T cells (Binz and Wigzell 1975; Eichman and Rajewsky 1975; Marchalonis 1975)—requires somatic alterations of DNA in the course of ontogeny. These alterations include the joining of genetic regions coding for the variable (V) portion of the immunoglobulin molecules with that which encodes the constant (C) portion, and the generation of diverse V regions from a limited number of germ-line genes. These changes must occur at particular steps during lymphoid-cell ontogeny and, therefore, are presumably mediated by enzymes programed to be expressed in particular ontogenetic sets. TdT is the only enzyme known thus far that has as its substrate DNA and that fulfills the developmental characteristics mentioned above. Therefore, it is a good candidate for a fundamental role in the DNA alterations in lymphoid ontogeny.

Our approach has been to relate the expression of TdT in particular sets of lymphoid cells—defined on the basis of surface antigen phenotype—to physical properties such as size and to other criteria such as sensitivity to cortisone (Kung et al. 1975). The occurrence of TdT in immature, but not mature, lymphoid cells implies that during lymphocyte differentiation there must occur a stage of induction of TdT synthesis followed by repression of its synthesis. This approach should ultimately allow us to isolate the cells in which the functional role of TdT can be investigated. With the advent of newer markers, many tumor lines whose phenotype was formerly regarded as "null" can now be identified as analogues of discrete cell sets. We summarize here the current state of evidence that TdT is found in cells developmentally intermediate between the colony-forming units—spleen (CFU-S) and the mature T and B effector cells. We also describe a number of new cell-surface markers that will be valuable in further elaboration of these stages.

MURINE TUMOR MODELS: THYMOCYTES AND T CELLS

Because TdT is normally found in thymocytes but not in peripheral cells, T-cell tumor lines bearing the enzyme should be related to more immature T cells (thymocytes). By contrast, cells lacking the enzyme should be related to more developed, peripheral T cells. The cell surface markers Thy-1 and TL have been used to define T-cell tumors (Boyse and Old 1969; Old and Boyse 1972). The Thy-1 antigen is found normally on all sets of T cells as well as in many other tissues, such as brain and epidermis (Reif and Allen 1966; Scheid et al. 1972; Trowbridge et al. 1975). It is not found on B cells. TL, the thymus-leukemia antigen, is detectable only on thymocytes and certain leukemia cells but not on peripheral T cells (Boyse et al. 1965; Boyse and Old 1969). Both antigens are lacking in bone-marrow cells but can become expressed on the surface of bone-marrow cells committed to the T pathway of differentiation (prothymocytes) when they receive an inductive signal such as the hormone thymopoietin (Komuro and Boyse 1973; Scheid et al. 1975). The strict demarcation of the period during which TL is expressed in T-cell ontogeny makes it an especially valuable marker to relate to the absence or presence of TdT.

Almost all those murine lymphoid tumors that have been positive for the TL alloantigen have had relatively high levels of TdT, ranging from 0.14 to 5.4 U of enzyme per 10^9 cells (Table 1). RADA1 is the only TL$^+$ tumor in which we have found no TdT.

Table 1
TL$^+$ Tumor Lines

Line	Description	Thy-1	Lyt-1	Lyt-2	TdT U/10^9 cells	Source or reference
ERLD	radiation-induced leukemia in C57BL/6	+	+	+	1.1	a
RL♂1	radiation-induced leukemia in BALB/c	+	+	+	1.9	a
EA.RAD.1	radiation-induced leukemia in B6 × A	+			2.6	a
ASL1	spontaneous leukemia in A	+	+	+	0.14	a
S49.1	mineral-oil-induced leukemia in BALB/c	+			0.56	b
L251A	Gross virus-induced leukemia in C57BR	+			0.87	c
R1	spontaneous leukemia in C58	+			2.2	b
B241	cloned line from Moloney MuLV-induced thymoma in BALB/c	+			3.6	d
B246	cloned line from Moloney MuLV-induced thymoma in BALB/c	+			3.8	d
L691	radiation-induced lymphoma in C57L mice	+			0.62	e
RADA1	radiation-induced leukemia in A	+	+	+	<0.002	a

Thy-1 and TL antisera were prepared and used as described by Shiku et al. (1975). Sera and procedures as described by Shen et al. (1975) were used to determine the Lyt phenotype. TdT was determined biochemically as described in Silverstone et al. (1976). Lines were obtained from the laboratories indicated in the following footnotes and were described in the indicated publications.

[a] E. A. Boyse and M. P. Scheid, Memorial Sloan-Kettering Cancer Center, New York, New York.
[b] R. Acton, University of Alabama, Birmingham; Hyman et al. 1972; Ralph 1973.
[c] R. Acton, University of Alabama, Birmingham; Imagawa et al. 1968.
[d] N. E. Rosenberg, Center for Cancer Research, Massachusetts Institute of Technology, Cambridge.
[e] P. Arnstein, California State Department of Public Health, Berkeley; Arnstein et al. 1976.

Table 2
TL⁻ T-Cell Tumor Lines

Line	Derivation	Thy-1	Lyt-1	Lyt-2	TdT U/10^9 cells
BALENTL 13	ethyl-nitrosurea–induced leukemia in BALB/c	+	+	−	<0.01
EL-4	DMBA-induced leukemia in C57BL/6	+	+	+	0.03
E♂ G2	Gross virus-induced leukemia in C57BL/6	+	+	−	<0.002
HRS-ST23	spontaneous leukemia in hr/hr mice	+	+	−	<0.01
HRS-ST25	spontaneous leukemia in hr/hr mice	+	+	−	<0.008
HRS-ST32	spontaneous leukemia in hr/hr mice	+	+	−	<0.002
HRS-ST37	spontaneous leukemia in hr/hr mice	+	+	−	<0.004

Surface phenotypes and enzyme measurements were made as explained in Table 1. BALENTL 13 was supplied by Dr. M. Potter, National Institute of Allergy and Infectious Diseases, Bethesda, Maryland. Its phenotype was determined by Dr. B. J. Mathieson et al. (1978), and the TdT level was determined on cells harvested from ascites tumors from passage in pristane-primed mice. All other lines were supplied and typed by Dr. M. P. Scheid. Assays were done on ascites and solid tumor-derived cells. DMBA is dimethyl benzanthracene.

A series of Thy-1⁺ TL⁻ tumor lines have also been examined (Table 2). The majority of these lines had no detectable TdT, although one (EL4) did have a low level of enzyme. It may be relevant that the surface phenotype of EL4 is Lyt-1⁺ Lyt-2⁺, whereas functional T lymphocytes generally express one or the other of these antigens but not both (Cantor and Boyse 1975). TL⁻ TdT⁺ lines, therefore, may represent a type of cell intermediate between the immature TL⁺ and the mature TL⁻ sets, perhaps corresponding to the intermediary Lyt-123: ARC set proposed by Cantor (Boyse and Cantor 1978).

As shown in Figure 1, there are tumor homologues for each of the cell sets thus far characterized in the normal population. Therefore, the presence of TdT in developing thymocytes, but not in circulating T cells, is mirrored in its presence or absence in tumor populations of corresponding phenotypes. The more differentiated a T cell becomes, the less TdT it has; the more differentiated the tumor cell line is with respect to its surface antigens, the less TdT is evident.

MURINE TUMOR MODELS: B CELLS

If TdT plays an obligatory role in generating the somatic changes in DNA that are required for the production of immunoglobulin-secreting cells,

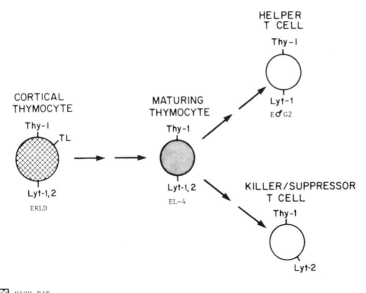

Figure 1
Thymocytes and T cells. The pathway of thymocyte development as suggested in Boyse and Cantor (1977) is shown. Tumor homologues for most types are indicated below each phenotype.

then we should expect to find TdT in some compartment of early B-cell differentiation. Plasmacytomas did not contain TdT (Table 3, lines 1–3), which is reasonable because the alterations in DNA should terminate once a cell is capable of producing a particular immunoglobulin (unless clonal selection theory is modified to allow a degree of diversification in single immunoglobulin-producing clones [Cunningham 1976]). A more disturbing result has been the inability to detect TdT in the lymphocytes from the bursas of Fabricius of 3- to 5-week-old chickens (Baltimore et al. 1976). Such a result might suggest that TdT plays no role in B-cell development except that, as documented below, many other experiments suggest that in certain stages of B-cell development TdT is present. We believe, therefore, that the TdT-positive stages of B-lymphocyte differentiation are not represented among the bursal lymphocytes we have studied.

While searching for tumor models that might correspond phenotypically to early B cells, we examined many tumors and cell lines that were induced in vivo and in vitro by the Abelson murine leukemia virus (A-MuLV) (Baltimore et al. 1976). Abelson virus induces a thymus-independent leukemia thought, for two main reasons, to be of B-cell origin: (1) the virus potentiates pristane-mediated induction of plasmacytomas in BALB/c mice (Potter et al. 1973), and (2) certain lines of cells grown from tumors induced by Abelson virus appear to have surface and/or intracellular immunoglobulin (Ig) (Premkumar et al. 1975; Pratt et al. 1977). Other po-

Table 3
B-Cell and Stem-Cell Tumor Lines

Line	Derivation		Ig	Lyb-2	Lyt-4	Sc-1	TdT U/10^9 cells	Source or reference
MOPC-70A	mineral-oil-induced plasmacytoma in BALB/c		+	–	+		<0.002	a
MOPC 315	mineral-oil-induced plasmacytoma in BALB/c		+	–	+		<0.002	b
MPC 11	mineral-oil-induced plasmacytoma in BALB/c		+	+			<0.002	b
I.29	spontaneous ascites in I strain		–	+	+		<0.002	a
L1210	methyl-cholanthrene-induced leukemia in DBA/2		–	+	+		<0.002	e
J-16-55	DMBA-induced leukemia in neonatally thymectomized SJL mice		–	–	+		0.05	c
J-16-78	DMBA-induced leukemia in neonatally thymectomized SJL mice		–	+	+		0.11	c
Fr-1-92	DMBA-induced leukemia in neonatally thymectomized SJL mice		–	–	+		0.08	c
BM-18-4		BALB/c	–	+	+	+	0.04	d
BM-18-8	single focus bone-marrow lymphoid	BALB/c	μ		+		0.01	d
93–4	colony derived by in vitro	C57BL/6	–			+	0.02	d
L-1	infection with Abelson murine	C57L	–	+	+	+	0.03	d
BR-48	leukemia virus from	C57BR	–			+	0.05	d
SWR-4		SWR/J	–		+	+	0.02	d
L-6		C57L	–	Low	+	+	<0.002	d

| J-15-46 | DMBA-induced leukemia in neonatally thymectomized SJL mice | − | − | + | <0.002 | c |

Immunoglobulin was determined either directly by the method of Hammerling et al. (1975) or by direct immunofluorescence with the FACS (Cantor et al. 1975). BM-18-8 has been described as having IgM heavy chain protein as determined by immunoprecipitation of radioactively labeled cell extracts (Pratt et al. 1977). Results suggesting the correctness of this observation have been obtained by Dr. E. J. Siden (Massachusetts Institute of Technology, Center for Cancer Research, pers. comm.). However, such a protein has not been found in BM-18-4, L-1, or L-6 (E. Siden, pers. comm.). Lyb-2 and Lyt-4 were determined by absorption (Boyse et al. 1970), with sera described by Sato and Boyse (1976) or Komuro et al. (1975b), respectively. Sc-1 was determined by indirect immunofluorescence with the thymus-absorbed RAMB-II antiserum (RAMB-Sc) described in the text. TdT was determined biochemically (Silverstone et al. 1976). Lines were obtained from the laboratories indicated in the following footnotes and were described in the indicated publications.

[a] E. A. Boyse and M. P. Scheid, Memorial Sloan-Kettering Cancer Center, New York, New York.
[b] M. Gefter, Department of Biology, Massachusetts Institute of Technology, Cambridge.
[c] N. Haran-Ghera, Weizmann Institute, Rehovot; Haran-Ghera and Peled 1973; Haimovich et al. 1977.
[d] N. E. Rosenberg, Center for Cancer Research, Massachusetts Institute of Technology, Cambridge; Rosenberg and Baltimore 1976.
[e] P. R. Ralph, Memorial Sloan-Kettering Institute, Rye, New York; Freund et al. 1976.

tentially interesting types of tumors are those chemically induced in neonatally thymectomized SJL mice (Haran-Ghera and Peled 1973). These tumors lack conventional T-cell markers (Table 3).

Most of the Abelson virus-induced cell lines and tumors we have examined, as well as the non-T-cell chemically induced tumors, contained low levels of TdT (0.01–0.1 U/10^9 cells) (Table 3). However, in none of these lines could we detect Ig on the cell surface by immunofluorescence (V. Sato and N. Rosenberg, unpubl.). We could not detect Ig in the cytoplasm of BM-18-4, L-1, BR-48, SWR-4, L-6, and 93-4 by immunofluorescence (P. Burrows et al., unpubl.), nor could we detect surface Ig after induction with dimethylsulfoxide in BM-18-4 by immunofluorescence (V. Sato and N. Rosenberg, unpubl.). Some apparent μ chain can be found, however, by biosynthetic labeling of one Abelson virus-induced tumor cell line, BM-18-8 (Pratt et al. 1977; E. Siden et al., unpubl.). In general these lines do not appear to express the conventional T- or B-cell surface markers. However, several new surface markers have been discovered which have made it possible for us to begin to understand the relationship of these "null" tumors to lymphoid ontogeny.

Lyb-2 is an alloantigen that has been described as a B-cell marker by Sato and Boyse (1976). It is found on B cells in spleen and lymph nodes and also on 30–35% of normal bone-marrow cells. Ontogenetic studies indicate that the expression of Lyb-2 precedes detectable expression of surface Ig on fetal-liver and bone-marrow cells (M. P. Scheid, unpubl.). The "null" tumors we have tested that contained TdT were Lyb-2$^+$ (Table 3). Therefore, if Lyb-2 is the hallmark of a cell set that carries an option for the B-cell pathway, we can tentatively propose a link between TdT and B-cell differentiation.

Another recently described cell-surface alloantigen that we have used in these studies is Lyt-4 (previously Ly-5 [Boyse et al. 1977]). It has been detected on Thy-1$^+$ cells (Komuro et al. 1975b) and recent results have placed it on some sets of Thy-1$^-$ cells in spleen, fetal liver, and bone marrow of normal and nu/nu mice. It is also found on several murine plasmacytomas (M. P. Scheid, unpubl.); further characterization of Lyt-4$^+$ cells is in progress.

The "null" cell lines that expressed TdT and Lyb-2 also had Lyt-4 (Table 3) and so cannot be said to relate solely to B-cell differentiation. One tumor of spontaneous origin (L1210) has been found to have Lyb-2 and Lyt-4, but it lacks detectable TdT (Table 3). Another tumor, I.29, is of a similar phenotype. Two tumors containing little (L-6) or no (J-15-46) Lyb-2 have been found to be Lyt-4$^+$ and TdT$^-$ (Table 3, last two lines).

These studies of new cell-surface markers have yet to provide a clear delineation of the relationship of the "null" tumors to mature B and T cells but rather have defined three new cell phenotypes. These are: (1) Lyt-4$^+$ Lyb-2 (low or −) TdT$^-$, (2) Lyt-4$^+$ Lyb-2$^+$ TdT$^+$, and (3) Lyt-4$^+$ Lyb-2$^+$ TdT$^-$. To characterize these tumors further, we turned to a different type of cell-surface marker.

Xenogeneic sera can be used to study surface antigens. One type of serum is the hyperimmune rabbit antiserum to mouse brain (called RAMB-II). Such a serum will react with pluripotent stem cells of bone marrow and spleen (Golub 1972; Van den Engh et al. 1974). After absorption with thymocytes, our serum will still eliminate 95% of CFU-S as defined by the spleen focus-forming assay of Till and McCulloch (1961) (V. L. Sato, unpubl.). For the sake of discussion, we have called the serum, after absorption, RAMB-Sc and have named the antigen(s) detected by the absorbed serum Sc-1. Using indirect immunofluorescence and the fluorescence-activated cell sorter (FACS), we find that Sc-1 is not detectable on thymocytes but is present on a small population of bone-marrow cells (<5%).

All of the Abelson virus-induced cell lines we have studied using FACS analysis have had easily detectable levels of Sc-1 (Table 3). Two of these lines were tested for their ability to absorb the anti-CFU-S activity of the serum and both could remove it. We believe, therefore, that the Abelson tumors have a proximal developmental relationship to stem cells.

A possible sequence of development for B and T cells in which we can place the various phenotypes described above is shown in Figures 2 and 3. We postulate that there is a prolymphocyte with stem-cell antigens, differentiation antigens of both T and B cells and TdT (Fig. 2). Such a cell would be typified by tumors like L-1. Maturation along the B-cell pathway would then generate cells like L1210—cells that have Lyb-2, and have lost TdT. Such cells would further mature through stages of cytoplasmic Ig, surface Ig, and secretion of Ig as documented by others (Hammerling et al. 1975; Cooper et al. 1976; Melchers et al. 1976).

The exact relationship of these postulated prolymphocytes to the pathways of T- and B-lymphocyte development are not clear. Figure 3 shows alternative schemes between which we are unable to choose. The TdT$^+$ prolymphocyte must have a TdT$^-$ progenitor and the characteristics of tumor L-6 are consistent with its being a model for such a cell. Expressing Lyt-4, a low level of Lyb-2, a high level of Sc-1, and no TdT, it could represent a stage just before the TdT$^+$ prolymphocyte.

Comparing the schemes of B- and T-cell development (Figs. 1–3), we found one prominent and crucial cell type in the T pathway but not in the B pathway: the high TdT cell. Although we have not found a murine tumor model for such cells, evidence from studies on human tumors as well as on normal murine bone marrow have identified high TdT cells that appear to be related to B cells.

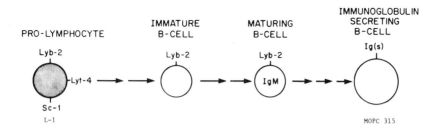

Figure 2
B-cell pathway. A partial pathway showing possible early stages of B-cell development is shown with tumor homologues for each stage indicated.

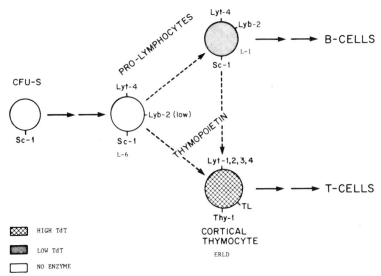

Figure 3
Lymphocyte stem cells. A suggested set of ontogenetic relations for the phenotypes of lymphocyte stem cells is shown. Tumor homologues for these phenotypes are indicated below each set.

HUMAN TUMORS

The same type of analysis we have applied to mouse tumors is applicable to human tumors. Our initial interest in TdT was, in fact, first sparked by the discovery of the high level of enzyme in cells from most patients with acute lymphoblastic leukemia (ALL) (McCaffrey et al. 1973; Kung et al. 1978). Recently it has become evident that there are subtypes of ALL as determined by surface phenotypes. One antigen, $P_{23,30}$, is similar to Lyb-2 in that it is found on human B cells but not T cells (Chess et al. 1976; Friedman et al. 1977). It is also found on "null" cells and on bone-marrow lymphocytes (Greaves et al., this volume). It is apparently the product of the HLA-D locus and is probably an Ia antigen (Springer et al. 1976). The

Table 4
Marker Studies on Childhood Human Lymphoblastic Leukemia Cells

Patient subgroups (% total number of patients)	E rosette	BK	$P_{23,30}$	TdT
75	−	−	+	+
20	+	+	−	+
2–3	+	+	+	+
1–2	−	−	−	−

BK and $P_{23,30}$ were determined as described in Schlossman et al. (1976). TdT assays were performed by McCaffrey et al. (1975). TdT+ values ranged from 3 to 100 U/10^9 cells both in E-rosette negative and E-rosette positive tumor lines.

other known antigen is BK which, like mouse TL, is found on thymocytes but not on circulating T cells (Schlossman et al. 1976).

ALL cells are usually either $P^+_{23,30}$ (75%) or BK$^+$, indicating apparent B- and T-cell relationships (Schlossman et al. 1976). Correlation of results of McCaffrey and his colleagues with those of Schlossman and his colleagues, both groups having studied the same patients, has shown that cells of both the two major ALL surface phenotypes have high levels of TdT (Table 4). There are a few cases with other sets of characteristics. The majority class of human ALL represents a type of cell we have not yet found in mice: cells with high TdT and a potential relationship to B cells. The occurrence of these human cells encourages a belief that there may be a stage of pre-B-cells with a high level of TdT.

TdT IN NORMAL MURINE CELL POPULATIONS

We have previously described the development of an antibody to TdT and an indirect immunofluorescence assay for the enzyme (Kung et al. 1976; Baltimore et al. 1976). We have found it possible to stain murine lymphoid cells with this methodology and to analyze by FACS analysis the quantity of TdT in individual lymphoid cells (P. C. Kung and A. E. Silverstone, unpubl.). As shown in Figure 4a, murine bone marrow contains a small percentage of cells that have very high levels of TdT. In fact, these cells contain more TdT than does any equivalent population in the thymus (Fig. 4b). Further evidence for a Thy-1$^-$ bone-marrow population that contains high levels of TdT comes from the work of Pazmino et al. (1977) in which they found a small population of bone-marrow cells having TdT activity at least equal to that of the average thymocyte.

PHENOTYPE OF NORMAL MURINE TdT$^+$ CELLS

A method has been developed to determine the phenotype of bone-marrow cells containing TdT (Fig. 5). The methodology involves eliminating specific sets of cells with antiserum to relevant markers in the presence of complement (mass cytolysis). The lysed cells are removed by discontinuous density gradients or with trypsin and DNase. The amount of TdT in the remaining cells indicates whether cells with a chosen marker contained a significant fraction of the TdT in the bone-marrow cell population. Thus, correlations between cell surface and TdT phenotypes can be established for sets of normal cells (Silverstone et al. 1976). Table 5 shows that about 50% of bone-marrow TdT occurs in Lyb-2$^+$ cells; i.e., elimination of the Lyb-2$^-$ population, which comprises less than 35% of bone-marrow cells, removes about 50% of bone-marrow TdT. Elimination of the Thy-1$^+$ population (about 1% of bone-marrow cells) does not demonstrably reduce the yield of TdT.

We also examined normal bone marrow for the presence of TdT$^+$ Sc-1$^+$ cell types. We used both a standard rabbit antimouse brain serum (RAMB-I) (Golub 1971) and the hyperimmune serum described above (RAMB-II).

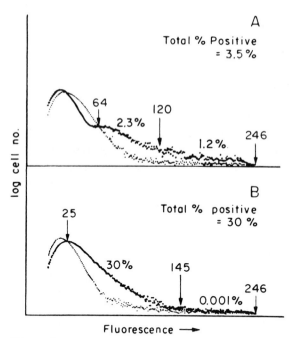

Figure 4
Quantitation of TdT by fluorescence in bone marrow and thymus lymphocytes using FACS. (*A*) FACS profile of acetone-fixed murine bone-marrow cells from C57BL/6 mice. Cells were indirectly stained with rabbit antibody to calf TdT (a gift of Dr. P. C. Kung). The logarithm of the number of cells in each channel is shown. Of the bone-marrow cells, 1.2% stained strongly with the antibody as described previously (Baltimore et al. 1976). Increasing channel number is directly related to increasing fluorescence. (*B*) FACS profile of acetone-fixed C57BL/thymocytes. The same gain settings were used as were used for the analysis in (*A*).

Such xenogeneic sera react not only with thymocytes and T cells but also with some prothymocytes (Sato et al. 1976; Stout et al. 1976). There was no significant loss of TdT after mass cytolysis with RAMB-I; but RAMB-II serum, with its marked activity against stem cells, removed more than 60% of the TdT of bone marrow (Table 5). Thus, TdT in normal bone marrow is associated with Lyb-2$^+$ and Sc-1$^+$ cell sets (which may be different or overlapping).

TdT IN PROTHYMOCYTES

Komuro and Boyse (1973) and Komuro et al. (1975a) had shown that some "null" cells in the murine bone marrow could be induced in vitro to express such T-cell surface components as Thy-1 and TL, and we have already reported that some portion of these inducible cells contain TdT (Silverstone et al. 1976). The evidence is illustrated in Table 6. Thus, induction by exposure of bone-marrow cells to thymopoietin (in vitro) does not change the

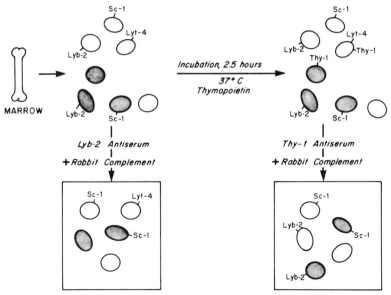

Figure 5

Analysis of surface phenotypes of TdT containing bone-marrow cells. Bone-marrow cells were treated as shown and as described in the text. Cells of a particular phenotype were eliminated by complement-mediated lysis. Using these antisera, we could eliminate no more than 40% of the cells. The loss of a significant amount of TdT from the population remaining was seen as a decline in units of enzyme per 10^9 remaining cells.

Table 5

Phenotype of TdT-Positive Cells in Marrows of $nu/+$ or C57L Mice

Antiserum used for cytotoxic elimination	TdT remaining ($U/10^9$ cells)
None	0.20
Normal rabbit serum, normal mouse serum	0.20
Anti-Thy-1	0.19
Anti-Lyb-2	0.11
RAMB-I	0.18
RAMB-II	0.07

Bone-marrow lymphocytes were incubated for 30 minutes with the indicated antiserum at a dilution calculated to give optimal lysis. The cells were then washed and resuspended in a crude preparation of rabbit complement optimized for the particular antiserum. Incubation was continued for 30–40 minutes at 37°C. Cells were then washed to remove dead cells. In some cases, dead cells were removed by discontinuous density gradient centrifugation. RAMB-I is the conventional rabbit antimouse brain serum described by Golub (1971). RAMB-II is the hyperimmune rabbit antimouse brain serum described in the text. Cells remaining after the elimination (>90% of controls) were assayed biochemically for TdT as previously described (Silverstone et al. 1976).

Table 6

Induction of Antigens on TdT-Containing Bone-Marrow Cells in C57L Mice

Inductive treatment	Postinductive treatment (+ complement)	TdT remaining ($U/10^9$ cells)
None	normal mouse serum	0.20
Thymopoietin	normal mouse serum	0.20
Thymopoietin	anti-Thy-1	0.09
Thymopoietin	anti-Lyb-2	0.10
Thymopoietin	anti-Thy-1 + anti-Lyb-2	0.03

Cells were induced for 2.5 hours at 37°C in RPMI-1640, 2% fetal calf serum with 500 ngm/ml thymopoietin, and 5% CO_2. Antisera and complement treatment were as described in Table 5, and enzyme levels were determined biochemically.

quantity of demonstrable TdT, but elimination of that cell set which can be induced to express Thy-1 reduced TdT by 50% (Table 6, line 3). On the other hand, induction by thymopoietin does not measurably increase or decrease the amount of TdT in cells expressing Lyb-2 (compare Table 6, line 4 with Table 5, line 4). Thus, it does not appear that Lyb-2$^+$ TdT$^+$ cells become Lyb-2$^-$ or that Lyb-2$^-$ TdT$^+$ cells become Lyb-2$^+$. However, Table 6, line 5, suggests that there are at least two distinct TdT$^+$ cell sets in bone marrow, because Thy-1 induction with elimination, coupled with Lyb-2 elimination, are additive in regard to reduction of TdT. One of these sets is the classical prothymocyte population which can be induced to express Thy-1. The other set is not inducible for Thy-1 but is for Lyb-2$^+$. Elimination of both sets removes over 85% of TdT from bone marrow. Thus 85% or more of the total TdT in bone marrow resides in the 40% or less of cells classified as either prothymocytes (Thy-1 inducible) or Lyb-2$^+$.

TdT IN nu/nu BONE MARROW

We have reported (Baltimore et al. 1976) that we could not relate the TdT found in nu/nu mice marrow to the prothymocyte set or to any other cell set. As shown in Table 7 (lines 1 and 2), the amount of TdT in nu/nu mouse bone marrow is about half that of normal or of heterozygous +/nu mice (Table 5, lines 1 and 2). As with normal mice, the elimination of nu/nu bone-marrow cell sets with Thy-1 or RAMB-I antiserum (Table 7, lines 3 and 4) does not significantly reduce TdT despite the fact that nu/nu mouse marrow does contain significant numbers of Thy-1$^-$ cells that react with RAMB-I (Sato et al. 1976). But the elimination of nu/nu mouse marrow cells that react with the RAMB-II antiserum does lower TdT (Table 7, line 5), indicating that in nu/nu mice as well as in normal mice a substantial amount of TdT is in Sc-1$^+$ cells.

It should be noted that some nu/nu mouse stocks show much lower levels of bone marrow TdT than do others (Pazmino et al. 1977), and that this enzyme activity is found mostly in a cell population distinguishable from

Table 7
Phenotype of TdT-Positive Cells in Marrows of nu/nu Mice

Antiserum used for cytotoxic elimination	TdT remaining ($U/10^9$ cells)
None	0.09
Normal rabbit serum, normal mouse serum	0.08
Anti-Thy-1	0.08
RAMB-I	0.07
RAMB-II	0.04

Methods were as described in Table 5. Nu/nu mice were obtained under a National Cancer Institute contract from Dr. F. J. Farrow, Life Sciences, St. Petersburg, Florida. These mice were outbred on a National Institutes of Health-Swiss background.

the normal murine TdT$^+$ cell populations by discontinuous density gradient sedimentation. Whether the TdT$^+$ cells of nu/nu and normal mice belong to different serologically defined cell sets has yet to be determined.

DISCUSSION

The ultimate aim of this line of research will be to provide a detailed sequence of developmental relationships for the sets of cells involved in T- and B-lymphocyte ontogeny. We should like to know what biochemical events mark the various stages of maturation of these cells and to correlate surface antigen phenotypes with biochemical events. We especially want to characterize the currently obscure early stages of lymphocyte differentiation, in which the pluripotent hematopoietic stem cell becomes committed to either the B- or the T-lymphocyte pathway.

At present we are focusing on a single enzyme, TdT, because this enzyme appears and disappears during lymphocyte maturation. Also, the known enzymatic activity of TdT involves DNA metabolism, so that TdT is likely to play a role in the choice or nature of the genes that function in a given lymphocyte to produce immunoglobulin or T-cell receptors (Baltimore 1974).

Before commenting on the data presented here, we would like to add a word about more general conceptual problems. A lymphocyte tumor cell population is static in its properties and in this way differs fundamentally from most normal cell populations that retain division potential. In normal animals, generally only end-stage cells have a static phenotype, and most such cells have no further division potential (nerve, muscle, erythrocytes, plasma cells, etc.). The static phenotype of a lymphocyte tumor cell population often resembles that of some normal cell. Such a normal cell and its progeny, however, would not be static but would be transiting a particular developmental sequence during which they would alter their phenotype with time. The constant phenotype of the lymphocyte tumor cell suggests that its capacity to continue differentiation has been blocked, and that may, in fact, be an important aspect of the action of leukemia viruses and other leukemogenic stimuli.

It must be acknowledged that other types of tumor cells may be much less fixed in phenotype. For example, a multipotential stem cell that has undergone malignant transformation may still be capable of engendering most or all of the normal cell sets for which it is programed (Illmensee and Mintz 1976). It is possible, in fact, that our methods of isolation and the passage of the lymphocyte tumor cells we have studied have selected for stable phenotypes.

To rigorously examine the proposition that a given lymphocyte tumor could represent a defined set of a normal developmental pathway, we need a battery of markers with which to compare the tumor cells with classes of normal cells. In the present work, the occurrence of certain surface antigens as well as the enzyme TdT has been used as a marker. To some extent, we have been able to find homologies between tumor cells and normal cells that have suggested developmental relationships. Because the tumor cells have stable phenotypes, we can closely define their properties. It is harder to define sets in the ontogeny of normal lymphocytes and harder yet to purify them.

The limitations of the methodologies we have used are potentially severe. First, we have as yet no way to know if antigens or enzymes are being ectopically expressed in tumor cells, giving a phenotype that may have no relationship to normal cells. Given the paucity of markers for early stages in lymphocyte ontogeny, the homologies seen may be merely fortuitous. Second, the sera we have used to define cell sets have themselves been defined either genetically or by absorption protocols and, thus, their "monospecificity" is only relative. At least in the case of the xenogeneic serum, they may react with multiple proteins. Furthermore, even antisera made in congenic mice may identify more than one cell set, not only because a single component may be shared by more than one program (e.g., the Lyt-4 determinant), but also because of the possibility of tightly linked polymorphic genes which determine products that are expressed in different though related cell sets (Boyse and Cantor 1978).

Despite these reservations, certain generalities are apparent at this stage of our work. The tumor cells we have studied generally have had the properties of differentiating cells such as the Sc-1 antigen, TdT, and TL. None of these characteristics are to be found in the functional lymphocytes of normal animals. Also, the tumor cells can generally be fit into apparent developmental sequences (Figs. 1–3), although individually there may be exceptions. A third point is that TdT, both in normal cells and tumor cells, appears only in developing cells of the lymphocyte pathway and not in mature functional cells or tumor homologs of functional cells. This is illustrated in the overall sequence diagramed in Figure 6.

The validity of the use of tumor cells as models of developing cell sets will depend on future analyses. We have already developed some new markers for the earliest stages of lymphocyte ontogeny which should increase the precision of the phenotypes assigned to individual cell sets. If the genetic reorganization and the reprograming involved in normal lymphocyte development can be proved with tumors, their use will have been validated. Study of gene structure and function in such cells will be required before we can know the answer.

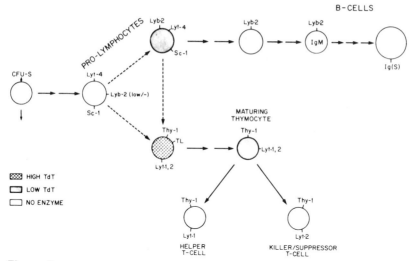

Figure 6

An overall composite scheme combining Figures 1, 2, and 3 is shown. TdT is located in an early developing lymphocyte population.

Acknowledgments

We would like especially to thank Dr. R. P. McCaffrey (Sidney Farber Cancer Center, Boston, Massachusetts) for the correlations of TdT levels with surface phenotype in human acute lymphoblastic leukemia. The generous donation of purified rabbit anticalf TdT antibody by Dr. P. C. Kung (Ortho Pharmaceutical, Roritan, New Jersey) is gratefully acknowledged. We thank Dr. E. Unanue and E. Adams of the Department of Pathology, Harvard Medical School, for the operation of the FACS. The able technical assistance of Virginia Bryan, Regan Ihde, and Dennis Triglia is greatly appreciated. This work was supported by National Institutes of Health Grants CA-230103, CA 13374, CA 22241, CA 22131, HD 08415, and CA 14051; by an American Cancer Society Grant VC-4I; by a grant from the American Cancer Society Massachusetts Division, and by a contract from the Virus Cancer Program of the National Cancer Institute. A. E. S. and N. R. were postdoctoral fellows of the American Cancer Society Massachusetts Division. E. A. B. is an American Cancer Society research professor of cell surface immunogenetics. D. B. is an American Cancer Society research professor of microbiology.

REFERENCES

Arnstein, P., J. L. Riggs, L. S. Oshiro, R. J. Huebner, and E. H. Lennette, 1976. Induction of lymphoma and associated xenotropic type C virus in C57L mice by whole-body irradiation. *J. Natl. Cancer Inst.* **57**: 1085.

Baltimore, D. 1974. Is terminal deoxynucleotidyl transferase a somatic mutagen in lymphocytes? *Nature* **248**: 409.

Baltimore, D., A. E. Silverstone, P. C. Kung, T. A. Harrison, and R. P. McCaffrey.

1976. Specialized DNA polymerases in lymphoid cells. *Cold Spring Harbor Symp. Quant. Biol.* **41**: 63.

Binz, H., and H. Wigzell. 1975. Shared idiotypic determinants on B and T lymphocytes reactive against the same antigenic determinants. I. Demonstration of similar or identical idiotypes on IgG molecules and T-cell receptors with specificity for the same alloantigens. *J. Exp. Med.* **142**: 197.

Bollum, F. J. 1974. Terminal deoxynucleotidyl transferase. In *The enzymes* (ed. P. D. Boyer), vol. 10, p. 145. Academic Press, New York.

Boyse, E. A., and H. Cantor. 1978. Immunogenetic aspects of biological communication: A hypothesis of evolution by program duplication. In *Molecular basis of cell-cell interaction* (eds. R. A. Lerner and D. Bergsma), p. 249. Alan R. Liss, New York. (In press.)

Boyse, E. A., and L. J. Old. 1969. Some aspects of normal and abnormal cell surface genetics. *Annu. Rev. Genet.* **3**: 269.

Boyse, E. A., L. J. Old, and E. Stockert. 1965. The TL (thymus leukemia) antigen: A review. In *Immunopathology 4th international symposium.* p. 23. Schwabe, New York.

Boyse, E. A., H. Cantor, F.-W. Shen, and J. F. C. McKenzie. 1977. Nomenclature for antigens demonstrable on lymphocytes. *Immunogenetics* **5**: 189.

Boyse, E. A., L. Hubbard, E. Stockert, and M. E. Lamm. 1970. Improved complementation in the cytotoxic test. *Transplantation (Baltimore)* **10**: 446.

Cantor, H., and E. A. Boyse. 1975. Functional subclasses of T lymphocytes bearing different Ly antigens. I. The generation of functionally distinct T-cell subclasses is a differentiative process independent of antigen. *J. Exp. Med.* **141**: 1376.

Cantor, H., E. Simpson, V. L. Sato, C. G. Fathman, and L. A. Herzenberg. 1975. Characterization of subpopulations of T lymphocytes. I. Separation and functional studies of peripheral T-cells binding different amounts of fluorescent anti-Thy 1.2 (theta) antibody using a fluorescence-activated cell sorter (FACS). *Cell. Immunol.* **15**: 180.

Chang, L. M. S. 1971. Development of terminal deoxynucleotidyl transferase activity in embryonic calf thymus gland. *Biochem. Biophys. Res. Commun.* **44**: 124.

Chess, L., R. Evans, R. E. Humphreys, J. L. Strominger, and S. F. Schlossman. 1976. Inhibition of antibody-dependent cellular cytotoxicity and immunoglobulin synthesis by an antiserum prepared against a human B-cell Ia-like molecule. *J. Exp. Med.* **144**: 113.

Coleman, M. S., J. J. Hutton, and F. J. Bollum. 1974. Terminal deoxynucleotidyl transferase and DNA polymerases in classes of cells from rat thymus. *Biochem. Biophys. Res. Commun.* **58**: 1104.

Cooper, M. D., J. F. Kearney, P. M. Lydyard, C. E. Grossi, and A. R. Lawton. 1976. Studies of generation of B-cell diversity in mouse, man, and chicken. *Cold Spring Harbor Symp. Quant. Biol.* **41**: 139.

Cunningham, A. J. 1976. Implications of the finding that antibody diversity develops after antigenic stimulation. In *The generation of antibody diversity: A new look* (ed. A. J. Cunningham), p. 89. Academic Press, New York.

Eichman, K., and K. Rajewsky. 1975. Induction of T and B cell immunity by anti-idiotypic antibody. *Eur. J. Immunol.* **5**: 661.

Freund, J. G., A. Ahmed, R. E. Budd, M. E. Dorf, K. W. Sell, W. E. Vannier, and R. E. Humphreys. 1976. The L1210 leukemia cell bears a B lymphocyte specific, non-H-2 linked alloantigen. *J. Immunol.* **117**: 1903.

Friedman, S. M., J. M. Breard, R. E. Humphreys, J. L. Strominger, S. F. Schlossman, and L. Chess. 1977. Inhibition of proliferative and plaque-forming cell responses by human bone-marrow-derived lymphocytes from peripheral blood by antisera to the p23,30 antigen. *Proc. Natl. Acad. Sci. U.S.A.* **74**: 711.

Golub, E. S. 1971. Brain-associated θ antigen: Reactivity of rabbit anti-mouse brain with mouse lymphoid cells. *Cell. Immunol.* **2:** 353.

———. 1972. Brain-associated stem cell antigen: An antigen shared by brain and hemopoietic stem cells. *J. Exp. Med.* **136:** 369.

Haimovich, J., Y. Bergman, M. Linker-Israeli, and N. Haran-Ghera. 1977. Cell surface components of carcinogen-induced lymphoid tumors on SJL/J mice. *Eur. J. Immunol.* **7:** 226.

Hammerling, U., A. F. Chin, J. Abbott, and M. P. Scheid. 1975. The ontogeny of murine B lymphocytes. I. Induction of phenotypic conversion of Ia$^-$ to Ia$^+$ lymphocytes. *J. Immunol.* **115:** 1425.

Haran-Ghera, N., and A. Peled. 1973. Thymus and bone marrow derived lymphatic leukaemia in mice. *Nature* **241:** 396.

Hyman, R., P. Ralph, and S. Sarkar. 1972. Cell-specific antigens and immunoglobulin synthesis of murine myeloma cells and their variants. *J. Natl. Cancer Inst.* **48:** 173.

Illmensee, K., and B. Mintz. 1976. Totipotency and normal differentiation of single teratocarcinoma cells cloned by injection into blastocysts. *Proc. Natl. Acad. Sci. U.S.A.* **73:** 549.

Imagawa, D. T., H. Issa, and M. Nakai. 1968. Cultivation of Gross virus-induced murine thymic lymphoma cells *in vitro*. *Cancer Res.* **28:** 2017.

Komuro, K., and E. A. Boyse. 1973. Induction of T lymphocytes from precursor cells in vitro by a product of the thymus. *J. Exp. Med.* **138:** 479.

Komuro, K., G. Goldstein, and E. A. Boyse. 1975 a. Thymus-repopulating capacity of cells that can be induced to differentiate to T cells in vitro. *J. Immunol.* **115:** 195.

Komuro, K., K. Itakura, E. A. Boyse, and M. John. 1975 b. Ly-5: A new T-lymphocyte antigen system. *Immunogenetics* **1:** 452.

Kung, P. C., P. D. Gottlieb, and D. Baltimore. 1976. Terminal deoxynucleotidyltransferase: Serological studies and radioimmunoassay. *J. Biol. Chem.* **251:** 2399.

Kung, P. C., A. E. Silverstone, R. P. McCaffrey, and D. Baltimore. 1975. Murine terminal deoxynucleotidyl transferase: Cellular distribution and response to cortisone. *J. Exp. Med.* **141:** 855.

Kung, P. C., J. C. Long, R. P. McCaffrey, R. L. Ratliff, T. A. Harrison, and D. Baltimore. 1978. Terminal deoxynucleotidyl transferase in the diagnosis of leukemia and malignant lymphoma. *Am. J. Med.* **64:** 788.

Leder, P., T. Honjo, J. Seidman, and D. Swan. 1976. Origin of immunoglobulin gene diversity: The evidence and a restriction-modification model. *Cold Spring Harbor Symp. Quant. Biol.* **41:** 855.

Marchalonis, J. J. 1975. Lymphocyte surface immunoglobulins. *Science* **190:** 20.

Mathieson, B. J., P. S. Campbell, M. Potter, and R. Asofsky. 1978. The expression of Ly-1, Ly-2, Thy-1 and TL differentiation antigens on mouse T-cell tumors. *J. Exp. Med.* **147:** 1267.

McCaffrey, R., D. F. Smoler, and D. Baltimore. 1973. Terminal deoxynucleotidyl transferase in a case of childhood acute lymphoblastic leukemia. *Proc. Natl. Acad. Sci. U.S.A.* **70:** 521.

McCaffrey, R., T. A. Harrison, R. Parkman, and D. Baltimore. 1975. Terminal deoxynucleotidyl transferase activity in human leukemia cells and in normal human thymocytes. *New Engl. J. Med.* **292:** 775.

Melchers, F., J. Andersson, and R. A. Phillips. 1976. Ontogeny of murine B lymphocytes: Development of Ig synthesis and of reactivities to mitogens and to anti-Ig-antibodies. *Cold Spring Harbor Symp. Quant. Biol.* **41:** 147.

Old, L. J., and E. A. Boyse. 1972. Current enigmas in cancer research. *Harvey Lect.* **67:** 273.

Pazmino, N. H., R. N. McEwan, and J. N. Ihle. 1977. Distribution of terminal deoxynucleotidyl transferase in bovine serum albumin gradient-fractionated thymocytes and bone marrow cells of normal and leukemic mice. *J. Immunol.* **119:** 494.

Potter, M., M. D. Sklar, and W. P. Rowe. 1973. Rapid viral induction of plasmacytomas in pristane-primed BALB/c mice. *Science* **182:** 592.

Pratt, D. M., J. Strominger, R. Parkman, D. Kaplan, J. Schwaber, N. Rosenberg, and C. D. Scher. 1977. Abelson virus-transformed lymphocytes: Null cells that modulate H-2. *Cell* **12:** 683.

Premkumar, E., M. Potter, P. A. Singer, and M. D. Sklar. 1975. Synthesis, surface deposition, and secretion of immunoglobulins by Abelson virus-transformed lymphosarcoma cell lines. *Cell* **6:** 149.

Ralph, P. 1973. Retention of lymphocyte characteristics by myelomas and θ^+ lymphomas: Sensitivity to cortisol and phytohemagglutinin. *J. Immunol.* **110:** 1470.

Reif, A. E., and J. M. V. Allen. 1966. Mouse nervous tissue iso-antigens. *Nature* **209:** 523.

Rosenberg, N., and D. Baltimore. 1976. A quantitative assay for the transformation of bone marrow cells by Abelson murine leukemia virus. *J. Exp. Med.* **143:** 1453.

Sato, H., and E. A. Boyse. 1976. A new alloantigen expressed selectively on B cells: The Lyb-2 system. *Immunogenetics* **3:** 565.

Sato, V. L., S. D. Waksal, and L. A. Herzenberg. 1976. Identification and separation of pre T-cells from nu/nu mice: Differentiation by preculture with thymic reticuloepithelial cells. *Cell. Immunol.* **24:** 173.

Scheid, M., E. A. Boyse, E. A. Carswell, and L. J. Old. 1972. Serologically demonstrable alloantigens of mouse epidermal cells. *J. Exp. Med.* **135:** 938.

Scheid, M., G. Goldstein, U. Hammerling, and E. A. Boyse. 1975. Induction of T and B lymphocyte differentiation *in vitro*. In *Membrane receptors of lymphocytes* (ed. M. Seligman), p. 353. Elsevier, Amsterdam.

Schlossman, S. F., L. Chess, R. E. Humphreys, and J. L. Strominger. 1976. Distribution of Ia-like molecules on the surface of normal and leukemic human cells. *Proc. Natl. Acad. Sci. U.S.A.* **73:** 1288.

Shen, F.-W., E. A. Boyse, and H. Cantor. 1975. Preparation and use of Ly antisera. *Immunogenetics* **2:** 591.

Shiku, H., P. Kisielow, M. A. Bean, T. Takahashi, and E. A. Boyse. 1975. Expression of T-cell differentiation antigens on effector cells in cell-mediated cytotoxicity in vitro: Evidence for functional heterogeneity related to surface phenotypes of T cells. *J. Exp. Med.* **141:** 227.

Silverstone, A. E., H. Cantor, G. Goldstein, and D. Baltimore. 1976. Terminal deoxynucleotidyl transferase is found in prothymocytes. *J. Exp. Med.* **144:** 543.

Springer, T. A., J. F. Kaufman, L. A. Siddoway, M. Giphart, D. L. Mann, C. Terhorst, and J. L. Strominger. 1976. Chemical and immunological characterization of HL-A–linked B-lymphocyte alloantigens. *Cold Spring Harbor Symp. Quant. Biol.* **41:** 387.

Stout, R. D., S. D. Waksal, V. L. Sato, K. Okumura, and L. A. Herzenberg. 1976. Functional studies of lymphoid cells defined and isolated by a fluorescence activated cell sorter (FACS) from normal and athymic mice. In *Proceedings of the 10th leukocyte culture conference: Leukocyte membrane determinants regulating immune reactivity* (eds. V. O. Eijsvoogel, D. Roos, and W. Zeijlemaker). Academic Press, New York.

Till, J. E., and E. A. McCulloch. 1961. A direct measurement of the radiation sensitivity of normal mouse bone marrow. *Radiat. Res.* **14:** 213.

Tonegawa, S., N. Hozumi, C. Matthyssens, and R. Schuller. 1976. Somatic changes in the content and context of immunoglobulin genes. *Cold Spring Harbor Symp. Quant. Biol.* **41:** 877.

Trowbridge, I. S., I. L. Weissman, and M. J. Bevan. 1975. Mouse T-cell surface glycoprotein recognised by heterologous anti-thymocyte sera and its relationship to Thy-1 antigen. *Nature* **256:** 652.

Van den Engh, G. J., and E. S. Golub. 1974. Antigenic differences between hemopoietic stem cells and myeloid progenetors. *J. Exp. Med.* **139:** 1621.

Polypeptides Regulating Lymphocyte Differentiation

G. Goldstein
Ortho Pharmaceutical Corporation
Raritan, New Jersey 08869

The lymphocytes of the body represent a complex mixture of cells derived from common progenitors and related by linear or branching differentiation. Traced back in their ontogeny, lymphocytes derive from stem cells common to other hematopoietic differentiations. The processes that determine commitment to certain differentiative pathways remain unknown but increasing knowledge is being gained concerning the chemical signals that induce fulfillment of precommitted programs of differentiation. For lymphocytes, a number of these regulatory substances appear to be relatively small polypeptides, and this fact has enabled structural elucidation to a point not yet possible with the "-poietins" for other hematopoietic cells.

THYMOPOIETIN

Isolation

Thymopoietin was isolated from bovine thymus (Goldstein 1974). Two closely related polypeptides were isolated and these were termed thymopoietin I and II since they were immunologically crossreactive and had indistinguishable activities.

Chemical Structure

Thymopoietin is a 49 amino-acid polypeptide chain (Schlesinger and Goldstein 1975a) whose structure is summarized in Figure 1. Thymopoietin I and II differ only by two residues. Since these substitutions did not affect the biological activity of thymopoietin, they probably represented isohormonal variations in cattle. A tridecapeptide fragment of thymopoietin was synthesized by solid-phase methodology and was shown to manifest the biological activity of entire molecule (Schlesinger et al. 1975a). More recently Fujino et al. (1977) completed a classical solution synthesis of the entire

```
                      5                    10                         15
TP I:  NH₂-Gly-Gln-Phe-Leu-Glu-Asp-Pro-Ser-Val-Leu-Thr-Lys-Glu-Lys-Leu-
TP II:      Ser-

                        20                   25                      30
         -Lys-Ser-Glu-Leu-Val-Ala-Asn-Asn-Val-Thr-Leu-Pro-Ala-Gly-Glu-

                    35                  40                   45
         -Gln-Arg-Lys-Asp-Val-Tyr-Val-Gln-Leu-Tyr-Leu-Gln-His-Leu-Thr-
                                                         -Thr-

                  49
         -Ala-Val-Lys-Arg-COOH
```

Figure 1
Amino-acid sequences of thymopoietin I (TP I) and thymopoietin II (TP II). Synthesis of a tridecapeptide (boldface) and the entire 49 amino-acid sequence of thymopoietin II produced biologically active molecules.

49 amino-acid chain of thymopoietin II and demonstrated that it had similar biological activity to native thymopoietin II in the T-cell induction assay.

Site of Production

Thymopoietin is synthesized and secreted by epithelial cells of the thymus. This is demonstrable by immunofluorescent localization with antithymopoietin antisera (Goldstein 1975). A neuromuscular assay was used to show that thymopoietin is secreted by the thymus (Goldstein and Hofmann 1969) and is present in thymus extracts but not in extracts of other tissues (Goldstein 1968). More recently, the T-cell induction assay was used to bioassay thymopoietin in the serum. Circulating thymopoietin was not detectable after thymectomy nor in nude mice (a mutant that lacks a thymus) (Twomey et al. 1977). Additionally, thymopoietin levels were low or absent in human patients with certain immunodeficiency diseases involving absence or hypoplasia of the thymus and were detectable following successful reconstitution therapy (Lewis et al. 1977).

A radioimmunoassay for thymopoietin has been developed (Goldstein 1976), but this requires further refinement before it can be applied to measurements of serum thymopoietin. Immunological or chemical identification of thymopoietin would clearly be preferable, since certain of the bioassays could be spuriously activated by other substances (Scheid et al. 1973). However, spurious induction by ubiquitin or related substances could be obviated by deliberate addition of 100 µg/ml ubiquitin to the assays; this concentration of ubiquitin inhibited ubiquitin receptors but left the cells susceptible to induction by thymopoietin (Brand et al. 1976; Twomey et al. 1977).

Actions of Thymopoietin

Induction of Prothymocyte to Thymocyte Differentiation

Thymopoietin, in concentrations down to 20 pg/ml, induced the differentiation of prothymocytes (cells committed to thymocyte differentiation but which lack the surface characteristics of thymocytes) to thymocytes (as detected by cell-surface molecules and functional characteristics) (Basch and Goldstein 1974, 1975). This differentiative event occurred within 2 hours in vitro. Induction involved only brief (less than 10 minutes) exposure to thymopoietin, elevation of cAMP within the first 30 minutes, and the intracellular processes of transcription and translation (but not DNA replication) (Goldstein et al. 1976; Storrie et al. 1976). Thus, an irreversible differentiation was initiated in a committed precursor cell, and this differentiation became manifest 2 hours after the inductive stimulus. The subsequent fate of the differentiating cell has not yet been determined and it is not presently known to what stage differentiation will proceed and what other requirements, if any, are needed for additional inductive stimuli.

Other Action of Thymopoietin in Vitro

Thymopoietin has shown a number of actions not related to its effects on prothymocyte to thymocyte differentiation. These did not appear to be linked with a cAMP second signal but rather, where studied, appeared to be linked with a cGMP second signal. Thus, thymopoietin induced enhanced mixed-lymphocyte-culture responses in fractionated lymphocytes (associated with elevation of intracellular cGMP) (G. Sunshine et al., in prep.). Thymopoietin inhibited the induction of early B-cell differentiation in vitro (an induction which also appeared to be cAMP-mediated like prothymocyte to thymocyte differentiation) and induced a late stage in B-cell differentiation (appearance of the plasma-cell marker), an induction which appeared to be mediated by cGMP (Hämmerling et al. 1976). Additionally, thymopoietin induced complement-receptor appearance on early granulocytes and enhanced phagocytosis in mature granulocytes obtained from human bone marrow (Kagan et al. 1977).

Actions of Thymopoietin in Vivo

Thymopoietin had a number of detectable effects when it was injected in vivo. An interesting feature of these actions was that a single injection of peptide, which would be expected to survive only a matter of minutes in the circulation, was effective. Thymopoietin, in doses down to 4 ng per mouse, caused a delayed impairment of neuromuscular transmission; this was the bioassay used in its isolation (Goldstein 1974). Thymopoietin also affected lymphocytes and immune reactivity. Thymopoietin injections in vivo in nude mice resulted in prothymocyte induction, with the appearance of cells with TL^+ $Thy-1^+$ phenotype in the spleen and TL^- $Thy-1^+$ phenotype in lymph nodes (Scheid et al. 1975). Following adult thymectomy in the mouse, daily injections of thymopoietin prevented the T-cell decline and subclass changes associated with adult thymectomy (J. Crowle et al., in prep.). Daily injections of thymopoietin in NZB mice from the age of 4 weeks delayed the onset and reduced the severity of autoimmune hemolytic disease (E. Gershwin and G. Goldstein, in prep.). In mice with established disease, thymopoietin injections enhanced the MLC reactivity of the lymph node cells which had been reduced by the disease (E. Gershwin and G. Goldstein, in prep.). In 2-year-old mice, which showed marked defects in T-dependent antibody responses, daily thymopoietin injections tended to restore the youthful pattern of responsiveness (M. Weksler and G. Goldstein, in prep.). It is clear that these studies are only a beginning and that much remains to be learned of the actions of thymopoietin in the body.

UBIQUITIN

Isolation

Ubiquitin was first isolated from extracts of bovine thymus during the isolation of thymopoietin (Goldstein et al. 1975). It was the major polypeptide component in thymus extracts in the molecular weight range of thymopoietin. However, ubiquitin was also found in extracts of other mammalian tissues and, with the development of a radioimmunoassay, was shown to be present in extracts of a variety of organisms, including yeasts, bacteria, and higher plants (Goldstein et al. 1975). Small amounts of ubiquitin were isolated from *Escherichia coli* and celery and were shown to have similar biological activity and immunological reactivity to mammalian ubiquitin (G. Goldstein and D. Schlesinger, unpubl.). More recently Olson et al. (1976) have described the isolation of a nuclear protein which they termed A-24. They showed that it consists of histone 2-A covalently bound to a nonhistone protein (Goldknopf and Busch 1977) that proved, by amino acid sequence, to be ubiquitin (Hunt and Dayhoff 1977). The significance of these findings remains to be elucidated.

Chemical Structure

Ubiquitin is a 74 amino-acid polypeptide chain whose structure is summarized in Figure 2. Ubiquitin isolated from both cattle and man has this identical amino-acid sequence (Schlesinger et al. 1975a; Schlesinger and

```
          1            5               10              15
NH2-Met-Gln-Ile-Phe-Val-Lys-Thr-Leu-Thr-Gly-Lys-Thr-Ile-Thr-Leu-

                       20              25              30
   -Glu-Val-Glu-Pro-Ser-Asp-Thr-Ile-Glu-Asn-Val-Lys-Ala-Lys-Ile-

                      35              40              45
   -Glu Asp Lys Glu Gly Ile  Pro Pro Asp Gln Gln Arg Leu Ile Phe-

                    50              55              60
   -Ala-Gly-Lys-Gln-Leu-Glu-Asp-Gly-Arg-Thr-Leu-Ser-Asp-Tyr-Asn-

                 65              70          74
   -Ile-Gln-Lys-Glu-Ser-Thr-Leu-His-Leu-Val-Leu-Arg-Leu-Arg-COOH
```
Figure 2
Amino-acid sequence of ubiquitin (bovine and human). Synthesis of a hexadecapeptide (boldface) produced a biologically active molecule.

Goldstein 1975b), and preliminary studies on small amounts of ubiquitin isolated from *E. coli* and celery revealed no differences for the N terminal first 20 residues (D. Schlesinger and G. Goldstein, unpubl.). The C terminal hexadecapeptide fragment of ubiquitin was synthesized by solid-phase methodology and was shown to manifest the biological activity of the entire molecule (Schlesinger and Goldstein 1975b). Certain other structural features of ubiquitin are of interest. It is very stable in solution. Thus, the proton nuclear magnetic resonance (NMR) pattern, which contained clear evidence of tertiary structure in solution, remained virtually unchanged up to 80°C (Lenkinski et al. 1977). That ubiquitin has a stable solution confirmation was confirmed by the finding that it is extremely resistant to digestion by trypsin (G. Goldstein and D. Schlesinger, unpubl.). However, when ubiquitin was maleated, trypsin readily cleaved it at arginyl residues (Schlesinger et al. 1975a), suggesting that the ϵ amino groups of lysyl residues were involved in the stable solution conformation of ubiquitin.

Site of Production

As noted above, ubiquitin has been found in all mammalian tissues; indeed, the amino-acid sequence of ubiquitin apparently has been conserved in evolution from prokaryotes to the most complex life forms of today. Ubiquitin was not detectable in serum by radioimmunoassay and its described association with histone 2-A in protein A-24 suggested that it may function as a structural or controlling element related to DNA. These findings did not implicate ubiquitin as a physiological regulatory molecule for lymphocyte differentiation and clearly established that it was not a thymic hormone.

Actions of Ubiquitin

Ubiquitin did not affect neuromuscular transmission in the assay used to isolate thymopoietin (see above). It was originally isolated in the mistaken belief that it could be an inactive precursor molecule, giving rise to thymopoietin by enzymatic cleavage. Although ubiquitin bore no structural relationship to thymopoietin and did not affect neuromuscular transmission, it did mimic the action of thymopoietin in inducing prothymocyte to thymocyte differentiation (Goldstein et al. 1975). This ubiquitin did by engaging receptors on the prothymocyte membrane that were discrete from those activated by thymopoietin. Thus, propranolol, a β-adrenoreceptor blocking drug inhibited induction by ubiquitin, whereas induction by thymopoietin was not affected by this drug. Apparently the ubiquitin receptor was linked to the same second signal as the thymopoietin receptor (putatively adenylate cyclase) and, thus, initiated the same differentiative events in these committed precursor cells. However, unlike thymopoietin, ubiquitin did not drive T-cell differentiation selectively but, additionally, induced early stages of B-cell differentiation. Furthermore, ubiquitin also induced early T- and B-cell differentiation in vitro in a chicken system, whereas bovine thymopoietin was inactive in this system (presumably because of evolutionary divergence of avian and mammalian thymopoietins, since chicken thymus extracts induce selective T-cell differentiation in this induction assay) (Brand et al. 1976).

Ubiquitin injections in vivo in nude mice resulted in prothymocyte induction with the appearance of cells with TL⁺ Thy 1⁺ phenotype in the spleen and TL⁻ Thy 1⁺ phenotype in lymph nodes (Scheid et al. 1975). Additionally, ubiquitin injections in NZB mice with autoimmune hemolytic disease partially restored the deficient MLC responses in the lymph-node lymphocytes of these mice and daily ubiquitin injections from 4 to 6 weeks of age delayed the appearance of and reduced the titer of red-cell autoantibodies (E. Gershwin and G. Goldstein, unpubl.).

NONAPEPTIDE ISOLATED BY J.-F. BACH (FTS)

Isolation

Bach et al. (1976, 1977) reported the isolation from pig serum of a nonapeptide that was originally termed thymic factor but more recently has been termed facteur thymique serique (FTS). Fractionation of the pig serum was monitored with a bioassay based on the induction of azathioprine sensitivity of murine cells forming spontaneous rosettes with sheep red-blood cells, an assay previously shown to parallel the appearance of Thy-1 sensitivity in these cells (Bach and Dardenne 1973).

Chemical Structure

The amino-acid sequence of the Bach nonapeptide is summarized in Figure 3. Although the authors originally published this sequence (Bach et al. 1976), they subsequently had some doubts as to the form of residue at positions 1 and 5; however, they felt that position 1 was pyrocarboxylic acid and position 5 was glutamine from charge considerations. Synthesis either with glutamine or pyrocarboxylic acid at the N terminus gave molecules with comparable activity to the natural nonapeptide.

Site of Production

The rosette assay registered a substance that disappeared from the serum following thymectomy (Bach et al. 1972; Bach and Dardenne 1973; Lacombe et al. 1974). It was inferred from these studies that the nonapeptide isolated from serum with this bioassay was a circulating thymic hormone. However, our studies (see below) suggest that this nonapeptide has an active site resembling ubiquitin and lacking selectivity for T cells in the induction assay. This raises the possibility that the rosette assay (which did register induction by thymopoietin or ubiquitin) (J.-F. Bach, pers. comm.;

NH_2 -Gln-Ala-Lys-Ser-Gln-Gly-Gly-Ser-Asn-COOH

Figure 3
Amino-acid sequence of FTS, the nonapeptide isolated by J.-F. Bach. Synthesis of this nonapeptide produced a biologically active molecule.

R. A. Good, pers. comm.) may have registered thymopoietin rather than the nonapeptide in the serum. This important question needs resolution by methods that measure the nonapeptide chemically or immunologically; that is, that measure the molecule rather than its biological effect, which may be mimicked by other substances. A radioimmunoassay for the nonapeptide is currently under development (J.-F. Bach, pers. comm.).

Actions of Bach Nonapeptide

The reported activities of the Bach nonapeptide included induction of Thy-1^+ cells in vitro and in vivo (Bach et al. 1975), enhancement of the generation of alloantigen-reactive cytotoxic T cells in thymectomized mice (Bach and Beaurain 1976), induction of suppressor T cells in NZB mice assayed with antibody production against polyvinylpyrrolidone (Bach and Niaudet 1976), enhancement of mitogen response in nude mice in vitro (Bach et al. 1975) and of adult thymectomized rats in vivo (Bach et al. 1976), and normalization of the abnormally high level of autologous erythrocyte-binding cells in adult thymectomized mice (Charreire and Bach 1975). Additionally, using chicken cells, we have shown that the synthetic nonapeptide is active in inducing both B- and T-cell differentiation in vitro in the induction assay (Brand et al. 1977). This inductive activity was inhibited by the β adrenergic blocking agent propanolol, and high inhibitory concentrations of the Bach nonapeptide and ubiquitin were cross-inhibitory with inductive concentrations of the other. These data suggest that the Bach nonapeptide and ubiquitin, which do not have identical amino-acid sequences, nevertheless generate active sites with similar stereochemistry.

BURSOPOIETIN

Bursopoietin is a short peptide which is being purified from chicken bursa of Fabricius, a B-cell differentiating organ in birds whose mammalian analogue remains unknown. Bursopoietin is detectable in extracts of the bursa of Fabricius in an induction assay in which fractionated cells from the bone marrow of newly hatched chickens are used (Brand et al. 1976). Spurious results due to ubiquitin in tissue extracts are obviated by the addition of 100 μg/ml ubiquitin (a concentration that inactivitates ubiquitin receptors) during the assays. This assay has identified a B-cell differentiating activity in extracts of the bursa of Fabricius, and partial purification has shown that this activity can be related to a low-molecular-weight peptide. It is hoped that these studies will lead eventually to the detection and isolation of a similar "B-poietin" in mammals.

OTHER THYMIC HORMONE PREPARATIONS

Over the years, A. L. Goldstein and colleagues have described progress in purifying a substance which they have termed thymosin (Goldstein et al. 1977). Most biological work with these extracts has been performed with a preparation termed thymosin fraction 5 that is very heterogenous. Goldstein et al. (1977) have proposed that the multiple bands in this preparation seen in polyacrylamide gel isoelectrofocusing represent a family of thymic hormones which they classify as α, β, and γ, depending on their mobilities. They have published the complete amino-acid sequence of an anionic peptide which they have termed thymosin α-1 (Goldstein et al. 1977), but the evidence that this, or indeed that any of the polypeptide bands seen in their preparation, is a thymic hormone is not stringent. Thus, there is no evidence as to the presence or absence of these materials in extracts of control tissues or as to their presence in the circulation and its relationship to the thymus. Indeed the major peptide in thymosin fraction 5, which has been termed β-1 thymosin, proved to be ubiquitin, which is clearly not a thymic hormone. Trainin et al. (1975) have described the effects of a thymus extract which they have termed thymic humoral factor and again have reported its progressive purification over the years. The structure of this material has not yet been published so that comparisons with other polypeptides are not yet possible.

CONCLUSIONS

It is now clearly established that the differentiations of hematopoietic cells, including lymphocytes, are controlled by substances that can be termed inducing agents in the classical experimental embryological sense; that is, substances that trigger differentiation in cells already committed to this differentiation. Thymopoietin fulfills these criteria in that it is secreted only by the differentiating organ, the thymus. It triggers differentiation in a committed precursor cell, the prothymocyte, and it does so in a selective manner without indiscriminate induction of other committed precursor cells. The finding that thymopoietin induction of prothymocyte-to-thymocyte differentiation is mediated by a cAMP second signal explains the finding that other agents raising intracellular cAMP can drive this differentiation in vitro.

Our studies detecting bursopoietin in the chicken indicate that an inducing agent for B-cell differentiation is also present in the body. Hopefully, this work in birds should lead to the identification of the mammalian analogue of the bursa of Fabricius and the mammalian "B-poietin."

In addition to its role as an inducing agent for prothymocyte-to-thymocyte differentiation, thymopoietin also appears to be a hormone in the classical sense; that is, it is secreted by the thymus into the circulation at concentrations that affect a number of tissues, including T cells, B cells, macrophages, granulocytes, and the neuromuscular junction. These peripheral actions, where studied, appear to be mediated by cGMP as a second signal. This dual use of the same molecule both as inducing agent and hormone is made possible by the integrative function of the cell membrane. We assume that prothymocytes in the periphery are subjected to influences such as insulin and growth hormone which depress their intracellular cAMP levels. Thus, the circulating levels of thymopoietin, which would be effective in inducing prothymocyte-to-thymocyte differentiation in vitro, do not do so in the periphery in the face of these inhibitory influences; whereas within the thymus the immigrant prothymocyte would be subjected to even higher concentrations of thymopoietin which would elevate intracellular cAMP levels beyond the required threshold-and-trigger differentiation. If one accepts this schema, variations of circulating thymopoietin concentrations could be utilized as a classic hormonal system without perturbing the delicate balance of differentiation of prothymocytes regulated by thymopoietin within the thymus.

The Bach nonapeptide remains a puzzle. It is a circulating peptide with potent actions on lymphocyte differentiation, and it will be exceedingly important to determine by a chemical or immunological means its source of production and circulating levels. Ubiquitin is unlikely to have a physiological role in lymphocyte differentiation. It is probably sequestered within cells in vivo and its demonstrable actions in vitro are probably related to its mimicry of epinephrine in producing an active site which also is similar to that of the Bach nonapeptide.

Thymopoietin, bursopoietin, and perhaps the Bach nonapeptide induce the differentiation of committed early precursor cells and perhaps regulate the activities of differentiated cells in a hormonal rather than in an inductive sense. It is clear that both T cells and B cells undergo numerous branching differentiations subsequent to these early steps. Certain differentiations are induced by encounter with antigen. Subsequent events are regulated by products of the lymphocytes themselves. Whether further inductive signals secreted by parenchymal cells are utilized to regulate lymphocyte differentiations remains an open question.

REFERENCES

Bach, J.-F., and M. Dardenne. 1973. Studies on thymus products. II. Demonstration and characterization of a circulating thymic hormone. *Immunology* **25**: 353.

Bach, J.-F., M. Dardenne, and J.-M. Pleau. 1977. Biochemical characterisation of a serum thymic factor. *Nature* **266**: 55.

Bach, J.-F., M. Dardenne, J.-M. Pleau, and M. A. Bach. 1975. Isolation, biochemical characteristics, and biological activity of a circulating thymic hormone in the mouth and in the human. *Ann. N.Y. Acad. Sci.* **249**: 186.

Bach, J.-F., M. Dardenne, J.-M. Pleau, and J. Rosa. 1976. Caractérisation biochimique du facteur thymique circulant. *C. R. Hebd. Seances Acad. Sci.*, ser. D, **283**: 1605.

Bach, J.-F., M. Dardenne, M. Papiernik, A. Barois, P. Levasseur, and H. Le Brigand. 1972. Evidence for a serum-factor secreted by the human thymus. *Lancet* **2**: (7786): 1056.

Bach, M.-A., and G. Beaurain. 1976. Cyclic AMP and T-cell differentiation. *Ann. Immunol.* (Paris) **127**: 967.

Bach, M.-A., and P. Niaudet. 1976. Thymic function in NZB mice. II. Regulatory influence of a circulating thymic factor on antibody production against polyvinylpyrrolidone in NZB mice. *J. Immunol.* **117**: 760.

Basch, R. S., and G. Goldstein. 1974. Induction of T-cell differentiation *in vitro* by thymin, a purified polypeptide hormone of the thymus. *Proc. Nat. Acad. Sci. U.S.A.* **71**: 1474.

———. 1975. Antigenic and functional evidence for *in vitro* inductive activity of thymopoietin (thymin) on thymocytes precursors. *Ann. N.Y. Acad. Sci.* **249**: 290.

Brand, A., D. G. Gilmour, and G. Goldstein. 1976. Lymphocyte-differentiating hormone of bursa of Fabricius. *Science* **193**: 319.

———. 1977. Effects of a nonapeptide FTS on lymphocyte differentiations *in vitro*. *Nature* **269**: 597.

Charreire, J., and J.-F. Bach. 1975. Binding of autologous erythrocytes to immature T-cells. *Proc. Natl. Acad. Sci. U.S.A.* **72**: 3201.

Fujino, M., T. Fukuda, H. Kawaji, S. Shinagawa, Y. Sugino, and M. Takaoki. 1977. Synthesis of the nonatetracontapeptide corresponding to the sequence proposed for thymopoietin II. *Chem. Pharm. Bull.* (Tokyo) **25**: 1486.

Goldknopf, I. L., and H. Busch. 1977. Isopeptide linkage between nonhistone and histone 2A polypeptides of chromosomal conjugate-protein A24. *Proc. Natl. Acad. Sci. U.S.A.* **74**: 864.

Goldstein, A. L., T. L. K. Low, M. McAdoo, J. McClure, G. B. Thurman, J. Rossio, C.-Y. Lai, D. Chang, S.-S. Wang, C. Harvey, A. H. Ramel, and J. Meienhofer. 1977. Thymosin α_1: Isolation and sequence analysis of an immunologically active thymic polypeptide. *Proc. Natl. Acad. Sci. U.S.A.* **74**: 725.

Goldstein, G. 1968. The thymus and neuromuscular function. A substance in thymus which causes myositis and myasthenic neuromuscular block in guinea pigs. *Lancet* **2**: 119.

———. 1974. Isolation of bovine thymin: A polypeptide hormone of the thymus. *Nature* **247**: 11.

———. 1975. Isolation of thymopoietin (thymin). *Ann. N.Y. Acad. Sci.* **249**: 177.

———. 1976. Radioimmunoassay for thymopoietin. *J. Immunol.* **117**: 690.

Goldstein, G., and W. W. Hofmann. 1969. Endocrine function of the thymus affecting neuromuscular transmission. *Clin. Exp. Immunol.* **4**: 181.

Goldstein, G., M. Scheid, E. A. Boyse, A. Brand, and D. G. Gilmour. 1976. Thymopoietin and bursopoietin: Induction signals regulating early lymphocyte differentiation. *Cold Spring Harbor Symp. Quant. Biol.* **41**: 5.

Goldstein, G., M. Scheid, U. Hämmerling, E. A. Boyse, D. H. Schlesinger, and H. D. Niall. 1975. Isolation of a polypeptide that has lymphocyte-differentiating properties and is probably represented universally in living cells. *Proc. Natl. Acad. Sci. U.S.A.* **72**: 11.

Hämmerling, U., A. F. Chin, J. Abbott, G. Goldstein, M. Sonenberg, and M. K. Hoffman. 1976. The ontogeny of B lymphocytes. III. Opposite signals transmitted to B lymphocyte precursor cells by inducing agents and hormones. *Eur. J. Immunol.* **6**: 868.

Hunt, L. T., and M. O. Dayhoff. 1977. Amino-terminal sequence identity of ubiquitin and the nonhistone component of nuclear protein A24. *Biochem. Biophys. Res. Commun.* **74**: 650.

Kagan, W. A., G. J. O'Neill, G. S. Incefy, G. Goldstein, and R. A. Good. 1977. Induction of human granulocyte differentiation *in vitro* by ubiquitin and thymopoietin. *Blood* **50**: 275.

Lacombe, M., F. Perner, M. Dardenne, and J.-F. Bach. 1974. Thymectomy in the young pig: Effects on the level of circulating thymic hormone. *Surgery* (St. Louis): **76**: 556.

Lenkinski, R. E., D. M. Chen, J. D. Glickson, and G. Goldstein. 1977. Nuclear magnetic resonance studies of the denaturation of ubiquitin. *Biochim. Biophys. Acta* (In press.)

Lewis, V., J. J. Twomey, G. Goldstein, R. O'Reilly, E. Smithwick, R. Pahwa, S. Pawha, R. A. Good, H. Schulte-Wisserman, S. Horowitz, R. Hong, J. Jones, O. Sieber, C. Kirkpatrick, S. Polmar, and P. Bealmear. 1977. Circulating thymic-hormone activity in congenital immunodeficiency. *Lancet* **2**(8036): 471.

Olson, M. O. J., I. L. Goldknopf, K. A. Guetzow, G. T. James, T. C. Hawkins,

C. J. Mays-Rothberg, and H. Busch. 1976. The NH_2- and COOH-terminal amino acid sequence of nuclear protein A24. *J. Biol. Chem.* **25**: 5901.

Scheid, M. P., M. K. Hoffman, K. Komuro, U. Hämmerling, J. Abbott, E. A. Boyse, G. H. Cohen, J. A. Hooper, R. S. Schulof, and A. L. Goldstein. 1973. Differentiation of T cells induced by preparations from thymus and nonthymic agents. *J. Exp. Med.* **138**: 1027.

Scheid, M. P., G. Goldstein, and E. A. Boyse. 1975. Differentiation of T cells in nude mice. *Science* **190**: 1211.

Schlesinger, D. H., and G. Goldstein. 1975a. The amino acid sequence of thymopoietin II. *Cell* **5**: 361.

―――. 1975b. Molecular conservation of 74 amino acid sequence of ubiquitin between cattle and man. *Nature* **255**: 423.

Schlesinger, D. H., G. Goldstein, and H. D. Niall. 1975a. The complete amino acid sequence of ubiquitin, an adenylate cyclase-stimulating polypeptide probably universal in living cells. *Biochemistry* **14**: 2214.

Schlesinger, D. H., G. Goldstein, M. P. Scheid, and E. A. Boyse. 1975b. Chemical synthesis of a peptide fragment of thymopoietin II that induces selective T cell differentiation. *Cell* **5**: 367.

Schlesinger, D. H., G. Goldstein, P. Scheid, E. A. Boyse, and G. Tregear. 1975c. Synthesis of a lymphocyte-differentiating hexadecapeptide based on the sequence of ubiquitous immunopoietic polypeptide (UBIP). *Biochemistry* **34**: 551.

Storrie, B., G. Goldstein, E. A. Boyse, and U. Hammerling. 1976. Differentiation of thymocytes: Evidence that induction of the surface phenotype requires transcription and translation. *J. Immunol.* **116**: 1358.

Trainin, N., M. Small, D. Zipori, T. Umiel, A. I. Kook, and V. Rotter. 1975. Characteristics of THF, a thymic hormone. In *Biological activity of thymic hormones* (ed. D. W. VanBekkum), p. 117. Kooyker Scientific Publications, Rotterdam.

Twomey, J. J., G. Goldstein, V. M. Lewis, P. M. Bealmear, and R. A. Good. 1977. Bioassay determinations of thymopoietin and thymic hormone levels in human plasma. *Proc. Natl. Acad. Sci. U.S.A.* **74**: 2541.

Isolation and Structure of HLA Antigens

J. L. Strominger, W. Ferguson, A. Fuks, J. Kaufman, H. Orr, P. Parham, R. Robb, and C. Terhorst

Department of Biochemistry and Molecular Biology, The Biological Laboratories
Harvard University, Cambridge, Massachusetts 02138

M. Giphart

Department of Immunohaematology, University Hospital
Leiden, The Netherlands

D. Mann

The National Cancer Institute, Immunology Branch
National Institutes of Health, Bethesda, Maryland 20014

The major histocompatibility complex is a region on the sixth human chromosome or 17th mouse chromosome that includes the genes for a number of products displaying an important role in eliciting cellular and humoral immune responses. The glycoproteins coded for by the *H2-D* and *H2-K* regions of the mouse and the *HLA-A* and *-B* regions of man are thought to be the target sites for immune attack of foreign tissue grafts. Moreover, these cell-surface antigens are involved in the recognition by sensitized thymus-derived lymphocytes of virally infected and chemically modified syngeneic cells. In addition, experiments in the mouse and guinea pig have shown the involvement of Ia antigens in antigen presentation and in T- and B-cell collaboration. In the human, the *HLA-D* locus — defined by the mixed lymphocyte reaction (MLR) — may encompass immune response genes and the genes for the Ia antigens.

A striking feature of the products of the *HLA-A, -B* and *-C* loci, detected by alloantisera, and of the *HLA-D* locus, detected both by alloantisera and the MLR, is their high degree of genetic polymorphism. It is obvious that the molecular underpinnings of this polymorphism, the natural function of the products of these genes and the reasons for their evolutionary diversification, are of considerable biological interest.

During the past few years, we have been engaged in the isolation of the products of the HLA-A, -B and -D genes and the study of their structure (Springer et al. 1977c; Strominger et al. 1977). Earlier studies of the isolation of HLA-A and -B gene products were limited by the small amount of material that was available from individual human spleens. Cultured human lymphocytes provided a source of material that was available in greater amounts and in reproducible supply. The amount of the HLA-A, -B and -D gene products on the surface of cultured human lymphocytes was on the order of 20- 40-fold the amount on normal peripheral lymphocytes. This amplification of amount facilitated the isolation of these materials in reasonable quantities.

PAPAIN-SOLUBILIZED HLA-A AND -B ANTIGENS

H-2 antigens of the mouse were first solubilized by papain, and this procedure can also be used as the first step in the preparation of HLA antigens. Following papain solubilization, the purification required only four steps. The purification to homogeneity was only about 70-fold; i.e., the total amount of HLA antigens must be 1–2% of the total membrane protein. The yields obtained with improved procedures are in the order of 50%. It is now possible to obtain about 80 mg of pure HLA antigen from 1 kg of cultured lymphocytes (Parham et al. 1977).

The first interesting observation made regarding the purified material was that it contained two polypeptides: a glycoprotein of about 34,000 daltons and a small polypeptide of 12,000 daltons (p34,12). The 12,000-dalton polypeptide soon was shown in several laboratories to be identical with β_2-microglobulin, a protein first isolated from human urine and later sequenced in several laboratories. β_2-microglobulin was shown to have sequence homology with immunoglobulins, particularly with the C_H3 domain of IgG. Thus, there were several similarities between HLA antigens and immunoglobulins: two-chain structure, limited proteolysis by papain, and sequence homology of the small subunit of HLA antigens with immunoglobulins. We found it of interest, therefore, to see if further homologies could be found in the large glycoprotein subunit of the HLA antigens. The amino-acid compositions indicated not only that the alleles within the HLA-A and HLA-B genes were very similar to each other but also that the products of the HLA-A gene were very similar in composition to the products of the HLA-B gene. NH_2-terminal sequences of the products of the HLA-A gene, i.e., HLA-A2 obtained from two different cell lines, and the products of the HLA-B gene, i.e., HLA-B7 and HLA-B7,12, indicated that these materials were nearly identical at their NH_2-terminals (Terhorst et al. 1976). The HLA-B7,12 mixture had two amino acids at only one position in the first 25 residues, and the two different HLA-A2 preparations were identical. In comparing the four products which had been obtained, we found amino acid differences at only two positions. Sequences obtained for HLA-A and HLA-B gene products in several other laboratories confirmed a striking conservation of sequence at the NH_2-terminus. Moreover, a comparison of the NH_2-terminal sequences of the HLA-A and -B gene products with those obtained in other laboratories for the H2-K and H2-D gene products showed a striking homology, although some data suggest that there has been considerably more evolutionary diversification within the H-2 gene products than within the HLA gene products. The results of a search for homologies between NH_2-terminal sequences of the HLA-A, -B gene products and immunoglobulin sequences were disappointing. A limited homology of questionable significance with some mouse V-region genes was found (Fig. 1).

Because two intrachain disulfide bridges were found in each heavy chain of all of the HLA antigen preparations which were examined, an important question was raised about the distribution of the half-cystines. Are the intrachain disulfides linearly arranged and present in domains as they are in immunoglobulins? In order to approach this problem, we found it neces-

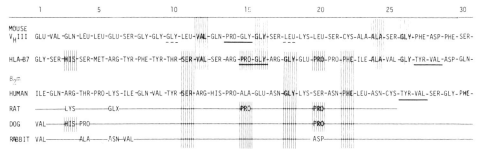

Figure 1
Comparison of NH$_2$-terminal sequence of human HLA-B7 with mouse V$_H$III sequence and β_2-microglobulin. Similar comparisons both of murine H-2 and human HLA antigen sequences were made by Capra et al. (1976). A low degree of homology between HLA sequences and β_2-microglobulin sequences was also found (bottom of figure).

sary to carry out a fragmentation of the heavy chain (Fig. 2) (Terhorst et al. 1977). This was made possible by a fortuitously located, acid-cleavable aspartyl-proline peptide linkage between the two intrachain disulfides of the HLA-B7 molecule and several (but not all) other HLA-A and -B antigens. Cleavage of HLA-B7 with acid resulted in a COOH-terminal fragment of about 13,000 daltons and an NH$_2$-terminal fragment of 23,000 daltons. Since the COOH- and NH$_2$-terminal fragments were separated from each other whether or not the disulfides were reduced, the intrachain disulfides of the heavy chain must have been linearly arranged.

The NH$_2$-terminal fragment resulting from acid cleavage could be further cleaved by CNBr. There were only two methionine residues in the heavy chain. One was the fifth amino-acid residue in the sequence and the other was fortuitously located immediately on the NH$_2$-terminal side of the first intrachain disulfide bridge. Thus, cleavage with formic acid followed by CNBr resulted in the separation of the heavy chain of the HLA antigen into three "domains." The NH$_2$-terminal region of 14,000 daltons contained a single asparagine-linked carbohydrate chain. The middle region of 9000 daltons contained an intrachain disulfide bridge as did the COOH-terminal region of 13,000 daltons. Further studies with the reagent NTCB (2-nitro-5-thiocyanatobenzoic acid) which cleaved at cysteine residues have shown that the size of the two intrachain disulfide loops is very similar to their size in the V and C regions of human κ chains and β_2-microglobulin (Fig. 2).

Thus, the distribution and spacing of the intrachain disulfide bridges suggested homology to immunoglobulins. In order to investigate that homology further, we decided to sequence the small COOH-terminal acid-cleavage fragment of HLA-B7, especially because the cleavage pattern suggested that the third cysteine residue might not be very far from the acid-cleavage site; thus, it would therefore be relatively easy for us to compare sequences around cysteine residues in immunoglobulins with those in the HLA antigen. The small acid-cleavage fragment was isolated and sequenced (Fig. 3) (Terhorst et al. 1977). The third half-cystine in the

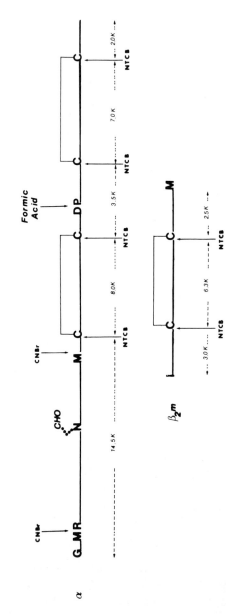

Figure 2
Fragmentation of HLA antigen compared with β_2-microglobulin. (Terhorst et al. 1977).

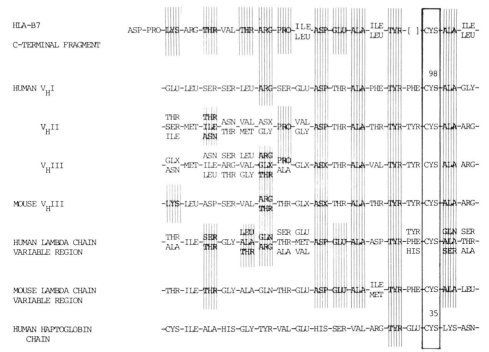

Figure 3
Comparison of the sequence of the small COOH-terminal acid cleavage fragment with various human and mouse V-region sequences. (Terhorst et al. 1977).

HLA heavy chain was located 16 residues from the aspartyl-proline cleavage site. If one deletion was inserted in the sequence, then strong homology was seen with a variety of mouse and human V-region sequences. Five residues were totally conserved in all of these molecules. Maximum homology of nine residues out of 18 was seen within a given V-region gene product, and 12 residues were found in one or another of the V-region gene products. Thus, these data indicate that there is considerable homology to immunoglobulins.

Study of the glycan residue of the HLA antigens (Parham et al. 1977) was undertaken partly because there have been various suggestions in the past that the glycan might affect the immunological specificity of the HLA antigens. A single N-linked glycan with a molecular weight of 3100 daltons was located in the NH_2-terminal region of the molecule. The tryptic glycopeptides isolated both from HLA-A2 and HLA-B7 had little or no specific inhibitory activity and, moreover, deglycosylation of both HLA-A2 and HLA-B7 antigens resulted in a product that retained full inhibitory activity although only a single acetylglucosamine residue remained on the polypeptides. It is certain, therefore, that the immunological specificity resides in the polypeptide sequence of the different HLA antigens.

The tryptic glycopeptides obtained both from HLA-A2 and HLA-B7 were sequenced (Fig. 4) (Parham et al. 1977). The HLA-B7 glycopeptide was 10 amino acids larger than that obtained from HLA-A2 because there

	Human heavy chain† variable regions				Glycopeptides		NH$_2$-terminals		NH$_2$-terminal small acid cleavage fragment
	V$_H$I	V$_H$II	V$_H$II	V$_H$III					
	EU	DAW	COR	TEI	A2	B7	A2	B7	B7
	Pro	Lys	Lys	Thr	Gly	Gly			
	Asn	Tyr	Tyr	His	Tyr	Tyr			
	Tyr	Tyr	Tyr	Tyr	Tyr	Tyr			
60†	Ala	Gly	Asx*	Ala	Asn*	Asn*			
	Gln	Ala	Thr	Val	Glx	Glx			
	Lys	Ser	Ser	Ser	Ser	Thr			
	Lys	Leu	Leu	Val	Glx	Glx			
	Phe	Glu	Glu	Gln	Ala	Ala			
65	Gly	Thr	Thr	Gly	Gly	Gly	Gly	Gly	
	Arg	Arg	Arg	Arg	Ser	Ser	Ser	Ser	
	Val	Leu	Leu	Phe	His	His	His	His	
	Thr	Ala	Thr	Thr	Thr	Ser	Ser	Ser	
	Ile	Val	Ile	Ile	Val	Val	Met	Met	
70	Thr	Ser	Ser	Ser	Glx	Glx	Arg	Arg	
	Ala	Lys	Lys	Arg	Arg	Ser	Tyr	Tyr	
	Asp	Asp	Asp	Asn		Met	Phe	Phe	
	Gln	Thr	Thr	Asp	Tyr	Phe	Tyr	Tyr	
	Ser	Ser	Ser	Ser	Gly	Thr	Thr	Thr	
75	Thr	Lys	Arg	Lys	Leu	Ser	Ser	Ser	
	Asn	Asn	Asn	Asn	Glx	Val	Val	Val	
	Thr	Gln	Gln	Thr	Gly	Ser	Ser	Ser	
	Ala	Val	Val	Leu	Gly	Arg	Arg	Arg	
	Tyr	Val	Val	Tyr	Pro	Pro	Pro	Pro	Pro
80	Met	Leu	Leu	Leu	Asp	Gly	Gly	Pro	
	Glu	Ser	Thr	Gln	Arg	Arg	Arg	Lys	
	Leu	Met	Met	Met	Gly	Gly	Arg		
	Ser	Asn	-	Leu	-	-	Thr		
	Ser	Thr	-	Ser	-	-	Val		
85	Leu	Val	-	Leu	-	-	Thr		
	Arg	Gly	Asp	Glu	Glu	Glu	Arg		
	Ser	Pro	Pro	Pro	Pro	Pro	Pro		
	Glu	Gly	Val	Glu	Arg	Pro	Phe		
	Asp	Asp	Asp	Asp	Phe	Phe	Asp		
90	Thr	Thr	Thr	Thr	Ile	Ile	Glu		
	Ala	Ala	Ala	Ala	Ala	Ala	Ala		
	Phe	Thr	Thr	Val	Val	Val	Ile/Leu§		
	Tyr	Tyr	Tyr	Tyr	Gly	Gly	Tyr		
	Phe	Tyr	Tyr	Tyr	Tyr	Tyr	-		
95	Cys	Cys	Cys	Cys			Cys		
	Ala	Ala	Ala	Ala			Ala		
	Gly	Arg	Arg	Arg			Ile/Leu§		

Figure 4

Comparison of sequences of tryptic glycopeptides, NH$_2$-terminals, and the small acid-cleavage fragment with human heavy chain variable region sequences (Parham et al. 1977).

* Site of linkage to carbohydrate.

† Ile and Leu were not distinguishable.

had been a substitution of a serine for an arginine at position 15. If one examines the first 14 amino-acid residues of the two glycopeptides, one again finds conservation of sequence in gene products that are located a great distance apart on the sixth chromosome. The differences are only at position 6, where there is a serine-threonine interchange, and at position 12, where there is the reciprocal threonine-serine interchange. The isolation of the pronase glycopeptide Tyr-Asn-Glx from both tryptic glycopeptides indicated that the carbohydrate was located on asparagine residue 4.

A comparison of these two sequences with sequences around carbohydrates in immunoglobulin molecules uncovered a sequence homology to the V_HII region of the COR immunoglobulin. The carbohydrate in the COR immunoglobulin is located at position 60, a very unusual position for a carbohydrate side chain in immunoglobulins. Sequence homology to immunoglobulins of the V_HI and V_HIII subgroups which did not contain carbohydrates at this position was also evident. The maximum homology with any one immunoglobulin was five out of 15 residues, but, when all of the subgroups were compared, then homology at nine of the 15 residues could be shown. These results have been confirmed by a computer search kindly carried out by Dr. Elvin Kabat.

By chance, the sequenced segments which were available appeared to be derived from the amino terminal regions of the putative HLA antigen "domains." Therefore, we were able to compare these sequences to see if there might be homology between HLA antigen "domains." Indeed, a suggestion of such homology was found when the NH_2-terminal sequences, the glycopeptide sequences, and the sequence of the small acid-cleavage fragment were aligned with the sequences of the V_H domains. Thus, it appears that, apart from sequence homology with immunoglobulins, the "domains" of HLA antigens may have sequence homology to each other. It seems likely, therefore, that HLA antigens and immunoglobulins evolved from a common ancestral gene.

DETERGENT-SOLUBILIZED HLA-A AND -B ANTIGENS

Several procedures were developed for the purification of HLA antigens after solubilization with detergent (Robb et al. 1976; Springer et al. 1977b). Again, the purification required to obtain pure antigen was in the order of 70-fold, and approximately 50% yields were obtained. The heavy chain of the detergent-soluble antigen was 10,000 daltons larger than that of the papain-solubilized antigen, and, moreover, the occurrence of the two chains in the detergent-solubilized product (p44,12) indicated that the two-chain structure was not an artifact due to proteolysis by papain. The heavy chain of several detergent-solubilized HLA antigens appeared to contain two additional cysteines not present in the papain-solubilized molecules which were easily reduced by dithiothreitol (in contrast to the intrachain disulfide bridges which were reduced only under denaturing conditions). The cleavage of detergent-solubilized HLA antigen by papain proceeded in two steps, the 44,000-dalton polypeptide being cleaved to an intermediate of 39,000 daltons before cleavage to the final product of about 34,000 daltons (Springer and Strominger 1976). A portion of the two additional easily reduced cysteine residues was removed at each step. The composition of the first small peptide cleaved could be deduced by the difference in the composition of p44 and that of p39. The composition of the second peptide removed could be correspondingly deduced from the compositions of p39 and p34. The first peptide was very hydrophilic; it contained a large number of charged amino-acid residues and very few nonpolar amino acids. The second peptide removed was extremely hydrophobic and contained a large number of hydrophobic amino-acid residues (Val, Leu, Ile). These observations suggested that the HLA antigens span the membrane with the

- (met) - cys - arg - arg - lys - ser - ser - gly - gly - lys

gly - gly - ser - tyr - ser - glx - ala - ala - cys - ser -

asx - ser - ala - glx - gly - ser - asx - val - ser - leu -

thr - ala

Figure 5
Sequence of the COOH-terminal hydrophilic peptide of HLA-B7 (Robb et al. 1978).

hydrophobic region inserted in the membrane and the COOH-terminal hydrophilic peptide extending inside the cell (Springer and Strominger 1976). This hydrophilic peptide has been isolated recently by another procedure and sequenced (Fig. 5). A number of striking structural similarities to the red blood cell membrane protein glycophorin which also spans the membrane were evident, an especially striking similarity being a clustering of basic amino acids near the membrane interface. A model which summarizes our present views of the structure of the HLA-A and HLA-B antigen is shown in Figure 6.

STRUCTURE OF THE HLA-D GENE PRODUCT

A polypeptide complex presumed to be the product of a gene in the HLA-D region was first detected as an impurity in the late stages of the preparation of the HLA-A and HLA-B antigens. After papain solubilization, it could be separated from the HLA-A and -B antigens by careful gel filtration and

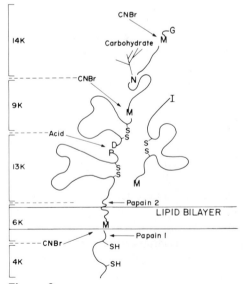

Figure 6
Model of HLA-B7.

was found to be a complex of two polypeptides of 23,000 and 30,000 daltons (p23,30) (Humphreys et al. 1976). Similar material separated from detergent-solubilized HLA antigens was a complex of polypeptides of 29,000 and 34,000 daltons (p29,34) (Springer et al. 1977a,c). Antisera prepared against p23,30 or against p29,34 were specific for peripheral human B cells or B-lymphoblastoid cells lines. In addition, they lysed a subset of peripheral null cells. They did not lyse T cells or T-lymphoblastoid cell lines. Thus, this antigen is a differentiation antigen expressed on some but not all cells of the lymphocyte lineage. This antiserum has had considerable utility in the precise classification of leukemias (Sallan et al. 1978; Schlossman et al. 1976).

A human alloantiserum prepared in Norway by planned immunization of a recipient with cells from an *HLA-A, -B, -C* identical donor who differed at the *D* locus (in this case the LD108 specificity). Highly concordant patterns were obtained when a panel of cells was tested by cytotoxicity using this B-cell-specific alloantiserum and by mixed lymphocyte reaction using homozygous typing cells. The rabbit anti-p23,30 serum and the human LD108 alloantiserum were then used in immunoprecipitation studies in which a [^{35}S]-methionine-labeled cell line derived from the immunizing donor was employed. In the double antibody technique, the rabbit hetero-antiserum and the human alloantiserum both precipitated the p29,34 polypeptide complex in comparable amounts (Giphart et al. 1977). Another HLA-D–linked human alloantiserum (BE) precipitated in small amounts from these membranes only a polypeptide of 32,000 daltons. No precipitate was obtained with normal human serum or with a one-third human HLA-D linked alloantiserum (CB). Thus, these data argue strongly that at least one of the polypeptides in the p29,34 complex (and its papain cleavage product, p23,30) is a product of a gene in the HLA-D region.

Structural studies of the HLA-D antigen have been more limited than those of the HLA-A and -B antigens. Data have been obtained that suggest that the two polypeptides of this complex may share large regions of structural homology, but the structural homology to p44, the glycoprotein subunit of the HLA-A and -B antigens, is very small, if it exists at all. NH$_2$-terminal studies of the large subunit of the p23,30 antigen revealed striking homologies to the murine IC gene products but no homology to the IA or IB gene products (Fig. 7). The NH$_2$-terminal sequence of the small subunit of the HLA-D antigen correspondingly revealed no homology to the small subunit of the murine IA antigen, but a homology to the small subunit of the guinea-pig IA antigen was evident. Obviously, further structural study of these interesting molecules will provide us with a great deal of information.

The anti-p23,30 serum has also been employed to study possible functions of the HLA-D product on human B cells (Chess et al. 1976; Friedman et al. 1977). Peripheral human B cells after separation from T cells by immunoabsorbent chromatography have undergone an initial triggering reaction which has permitted them to be stimulated by polyclonal activators such as phytohemagglutinin to become antibody-secreting, plaque-forming cells. They also have been similarly triggered by T-cell products. In either case, the induced formation of plaque-forming cells was totally abrogated

N-TERMINAL SEQUENCES OF THE LARGE SUBUNIT OF Ia ANTIGENS

	1				5				10				15				20
MAN	ILE	LYS	GLU	GLU	ARG	VAL	ILE	ILE/LEU	GLN	ALA	GLU	PHE	TYR	LEU	ASN	TYR	ASP-PHE-GLN-GLY
MOUSE $I-A^b$	-	-	-	ILE	-	ALA	-	-	VAL	-	-	-	-	-	VAL	-	
$I-A^k$	-	-	-	ILE	-	ALA	-	VAL	-	-	-	-	-	-	VAL	TYR	
$I-C^k$	ILE	-	-	-	-	ILE	-	-	ALA	-	-	TYR	LEU	-	-		

N-TERMINAL SEQUENCES OF THE SMALL SUBUNIT OF Ia ANTIGENS

	1				5				10				15				20
MAN	GLY	ASP	THR	PRO	GLU	ARG	PHE	LEU	GLU	GLN	VAL						
MOUSE $I-A^b$	-	-	-	-	ARG	-	-	VAL	TYR	-	-	-	-	-	TYR		
$I-A^k$	-	-	-	-	-	-	-	VAL	-	-	-	-	-	-	TYR		
GUINEA PIG	ILE	TYR	-	PRO	-	-	PHE	LEU	PHE	-	PHE						

Figure 7
Comparison of NH_2-terminal sequences of p29,34 (p23,30) with various mouse and guinea pig Ia-antigen sequences. Data for mouse Ia antigens were obtained by J. Silver, J. M. Cecka, M. McMillan, and L. Hood (Silver et al. 1977), and those for guinea pig Ia antigens by B. D. Schwartz, A. M. Kask, and E. M. Shevach (Schwartz et al. 1977). (See Springer et al. 1977c.)

by the addition of anti-p23,30 serum (Friedman et al. 1977). This effect was apparent when antiserum was added any time up to 4 days after addition of the polyclonal activator or the T-cell product, but little or no inhibition occurred if the antiserum was added at 5 days or after. At 8 days, at which time plaque-forming cells were measured, the antigen was still present on the cells, as could be demonstrated by their lysis in the presence both of anti-p23,30 serum and complement. However, an immunoglobulin-secreting human myeloma cell line, Simpson, had no p23,30 antigen. Thus, the p23,30 antigen may be associated with an intermediate stage of differentiation in the B-cell lineage from stem cells to antibody-forming cells, and it appears to have an important function in further differentiation of these cells to antibody-secreting cells.

It is hoped that further studies of the structure of the HLA-A, -B and -D antigens, and of the HLA-C antigen, may elucidate the interesting functions of these molecules.

Acknowledgments

These studies were supported by National Institutes of Health Grants AI 09576 and AI 01736 to The Biological Laboratories, Harvard University, and National Cancer Institute Grant CA 06516 to the Sidney Farber Cancer

Institute. H. O. was supported by an NIH Fellowship, A. F. by the Medical Research Council of Canada, and M. G. by a grant from the Dutch Organization for the Advancement of Pure Research (Z. W. O.).

REFERENCES

Capra, J. D., E. S. Vitetta, D. G. Klapper, J. W. Uhr, and J. Klein. 1976. Structural studies on protein products of murine chromosome 17: Partial amino acid sequence of an H-2K^b molecule. *Proc. Natl. Acad. Sci. U.S.A.* **73**: 3661.

Chess, L., R. Evans, R. E. Humphreys, J. L. Strominger, and S. F. Schlossman. 1976. Inhibition of antibody-dependent cellular cytotoxicity and immunoglobulin synthesis by an antiserum prepared against a human B-cell Ia-like molecule. *J. Exp. Med.* **144**: 113.

Friedman, S. M., J. M. Breard, R. E. Humphreys, J. L. Strominger, S. F. Schlossman, and L. Chess. 1977. Inhibition of proliferative and plaque-forming cell responses by human bone-marrow-derived lymphocytes from peripheral blood by antisera to the p23,30 antigen. *Proc. Natl. Acad. Sci. U.S.A.* **74**: 711.

Giphart, M. J., J. F. Kaufman, A. Fuks, D. Albrechtsen, B. G. Solheim, J. W. Bruning, and J. L. Strominger. 1977. HLA-D associated alloantisera react with molecules similar to Ia antigens. *Proc. Natl. Acad. Sci. U.S.A.* **74**: 3533.

Humphreys, R. E., J. M. McCune, L. Chess, H. C. Herrmann, D. J. Malenka, D. L. Mann, P. Parham, S. F. Schlossman, and J. L. Strominger. 1976. Isolation and immunologic characterization of a human, B-lymphocyte-specific, cell surface antigen. *J. Exp. Med.* **144**: 98.

Parham, P., B. N. Alpert, H. T. Orr, and J. L. Strominger. 1977. Carbohydrate moiety of HLA antigens: Antigenic properties and amino acid sequences around the site of glycosylation. *J. Biol. Chem.* **252**: 7555.

Robb, R. J., D. L. Mann, and J. L. Strominger. 1976. Rapid purification of detergent-solubilized HLA antigen by affinity chromatography employing anti-β_2-microglobulin serum. *J. Biol. Chem.* **251**: 5427.

Robb, R. J., C. Terhorst, and J. L. Strominger. 1978. Sequence of the COOH-terminal hydrophilic region of histocompatibility antigens HLA-A2 and HLA-B7. *J. Biol. Chem.* (In press.)

Sallan, S. E., L. Chess, E. Frei, C. O'Brien, D. G. Nathan, J. L. Strominger, and S. F. Schlossman. 1978. Cell surface differentiation antigens: Prognostic implications in childhood acute lymphoblastoid leukemia. *Blood* (In press.)

Schlossman, S. F., L. Chess, R. E. Humphreys, and J. L. Strominger. 1976. Distribution of Ia-like molecules on the surface of normal and leukemic human cells. *Proc. Natl. Acad. Sci. U.S.A.* **73**: 1288.

Schwartz, B. D., A. M. Kask, and E. M. Shevach. 1977. The guinea pig MHC: Functional significance and structural characterization. *Cold Spring Harbor Symp. Quant. Biol.* **41**: 397.

Silver, J., W. A. Russell, B. L. Reis, and J. A. Frelinger. 1977. Chemical characterization of murine Ia alloantigens determined by the I-E/I-C subregions of the H-2 complex. *Proc. Natl. Acad. Sci. U.S.A.* **74**: 5131.

Springer, T. A., and J. L. Strominger. 1976. Detergent-soluble HLA antigens contain a hydrophilic region at the COOH-terminus and a penultimate hydrophobic region. *Proc. Natl. Acad. Sci. U.S.A.* **73**: 2481.

Springer, T. A., J. F. Kaufman, C. Terhorst, and J. L. Strominger. 1977a. Purification and structural characterisation of human *HLA*-linked B-cell antigens. *Nature* **268**: 213.

Springer, T. A., D. L. Mann, A. L. DeFranco, and J. L. Strominger. 1977b. Detergent solubilization, purification and separation of specificities of HLA antigens from a cultured human lymphoblastoid line, RPMI 4265. *J. Biol. Chem.* **252:** 4682.

Springer, T. A., J. F. Kaufman, L. A. Siddoway, M. Giphart, D. L. Mann, C. Terhorst, and J. L. Strominger. 1977c. Chemical and immunological characterization of HL-A–linked B-lymphocyte alloantigens. *Cold Spring Harbor Symp. Quant. Biol.* **41:** 387.

Strominger, J. L., D. L. Mann, P. Parham, R. Robb, T. Springer, and C. Terhorst. 1977. Structure of HL-A A and B antigens isolated from cultured human lymphocytes. *Cold Spring Harbor Symp. Quant. Biol.* **41:** 323.

Terhorst, C., P. Parham, D. L. Mann, and J. L. Strominger. 1976. Structure of HLA antigens: Amino-acid and carbohydrate compositions and NH_2-terminal sequences of four antigen preparations. *Proc. Natl. Acad. Sci. U.S.A.* **73:** 910.

Terhorst, C., R. Robb, C. Jones, and J. L. Strominger. 1977. Further structural studies of the heavy chain of HLA antigens and its similarity to immunoglobulins. *Proc. Natl. Acad. Sci. U.S.A.* **74:** 4002.

Utility of B- and T-Cell-Specific Antisera in the Classification of Human Leukemias

S. E. Sallan, L. Chess, E. Frei III, C. O'Brien, D. G. Nathan, J. L. Strominger, and S. F. Schlossman

Sidney Farber Cancer Institute and Children's Hospital Medical Center
Harvard Medical School, Boston, Massachusetts 02115

Several years ago, antiserum was prepared against a complex of two polypeptides of 23,000 and 30,000 daltons (p23,30) that had been isolated from a human B-lymphoblastoid cell line during purification of human histocompatibility antigens (Strominger et al. 1975; Humphreys et al. 1976). This antiserum was found to recognize an antigenic determinant localized to B cells and to a subset of null cells (which are surface IgG negative); it was not present on T cells. In culture, p23,30-positive null cells could differentiate to become surface IgG positive (Chess et al. 1976); they were presumed to represent a relatively immature cell in the B-cell lineage. The anti-p23,30 serum inhibited the mixed lymphocyte culture (MLC) (Humphreys et al. 1976), and p23,30 had other properties which suggested that it was the human analogue of the murine Ia antigen. The p23,30 complex appeared to function by triggering of B cells to become antibody-forming cells. In an in vitro system developed for the generation of plaque-forming cells from human B cells in vitro (Friedman et al. 1976), addition of the anti-p23,30 serum totally abolished the plaque-forming cell response (Friedman et al. 1977). Human myeloma cells are more differentiated cells in the B-cell lineage, i.e., they secrete immunoglobulin. Myeloma cells from several patients and those from a human myeloma cell line (Simpson) were lacking the p23,30 antigen, i.e., this antigen appeared to be a marker for an intermediate stage of B-cell differentiation. When the peripheral B cells differentiated further to become antibody-forming cells (plasma cells), this antigen disappeared.

The anti-p23,30 serum has had considerable utility in the classification of acute lymphoblastic leukemia (ALL). Several groups had used formation of E rosettes to distinguish T- and B-cell leukemias (Brown et al. 1974; Sen and Borella 1975; Tsukimoto et al. 1976). However, rosetting is not a very precise diagnostic procedure. Initially, in 24 cases of childhood ALL examined, 20 were found to be anti-p23,30 positive, i.e., they were either null or B-cell leukemias (Schlossman et al. 1976). Four cases were negative, and one of the four was a patient (BK) whose leukemic cells had been used previously for preparation of a rabbit antiserum. This antiserum was absorbed with a syngeneic B-cell line developed from the same patient during remission. The other three anti-p23,30 negative patients were also found to be anti-BK positive. Therefore, the 24 patients could be separated into those who were anti-p23,30 positive and those who were anti-BK positive. The BK antigen was also found on human thymocytes but not on peripheral human T cells. Since this distribution of the BK antigen, including its presence on leukemic cells, is the same as that of the murine thymus-leukemia (TL) antigen (Boyse and Old 1969) the BK antigen is now referred to as HTL and the antiserum as anti-HTL.

The survival of patients with childhood ALL could be related to classification of cell type by antigen in a larger series of patients seen at the Children's Hospital and the Sidney Farber Cancer Institute (Fig. 1). The disease-free survival of 76 children with Ia-positive ALL at our clinic was about 80% at 4 years, with a median follow-up of 28 months. We attained these results using the combined chemotherapy regimen shown in Figure 2. By contrast, the Ia-negative (p23,30-negative), HTL-positive cases had an extremely poor prognosis. The median duration of complete remission was 15 months, and all HTL-positive patients eventually relapsed on this regimen.

T-cell ALL patients have the ability to form E rosettes, whereas B- or null-cell patients do not. This fact has been used as a basis by which these two forms of childhood ALL may be distinguished. The availability of anti-p23,30 and anti-HTL sera has made possible a quick and accurate differentiation of these forms of ALL.

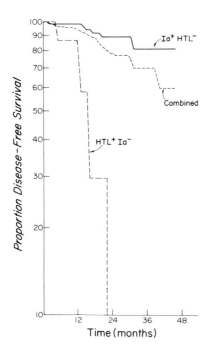

Figure 1
Childhood ALL. Disease-free survival by immunologic marker.

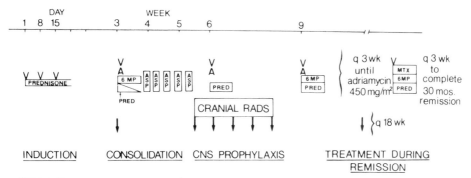

Figure 2
Schema for chemotherapy in childhood ALL. V, vincristine; VA, vincristine adriamycin; Pred, prednisone; mtx, methotrexate; 6MP, 6-mercaptopurine; Asp, asparciginase.

When patients are initially diagnosed, the leukemic cells are treated with one and then the other antiserum, and then with fluorescein-isothiocyanate–conjugated goat antirabbit IgG. The antigen content of the leukemic cells is then examined with the fluorescence-activated cell sorter (Figs. 3 and 4). Not all cases have fit into the simple pattern. A few cases have been found in which the cells expressed both antigens, and some rare cases have been seen in which the cells expressed neither of the two antigens. A high percentage of cases of CLL are both p23,30 and surface-IgG positive.

The importance of the immunologic classification of ALL is twofold. First, if a child has a p23,30-positive leukemia, the parents may be told that the chances of that child's survival are extremely good; i.e., that 80% of children with that disease are surviving at four years with the present chemotherapeutic regimen. On the other hand, if the child has T-cell ALL, then a very unfavorable prognosis must be given. From a prognostic standpoint, therefore, differentiation of the two forms of leukemia is extremely important. Second, the present form of chemotherapy for ALL has evolved as a method for treating the majority of children with the disease, i.e., those with B-cell or the null-cell ALL. In mouse models (the AKR leukemia being a model of T-cell leukemia and the L1210 leukemia being a model of B-cell leukemia), the drugs to which these two diseases respond maximally are quite different (Frei et al. 1974). At the present time, new therapeutic regimens based on those drugs that are effective in the murine T-cell-leukemia model are being developed for trial in childhood T-cell ALL. Our ability to distinguish these two forms of leukemia may make it possible for us to obtain therapeutic advances in the treatment of the T-cell disease.

Acknowledgments

This work was supported in part by research grants from the National Institutes of Health: CA 19589, CA 18662, AI 09576, AI 10736, and AI 12069.

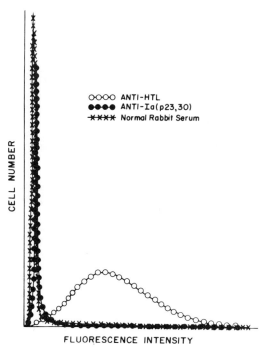

Figure 3
Acute lymphoblastic leukemia. Binding of anti-HTL and anti-Ia sera on T-cell leukemic lymphoblasts.

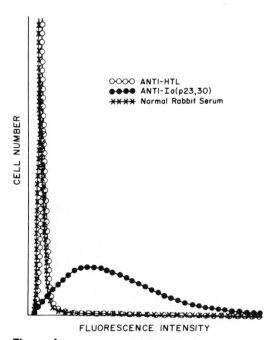

Figure 4
Acute lymphoblastic leukemia. Binding of anti-HTL and anti-Ia sera on B-cell or null-cell leukemic lymphoblasts.

REFERENCES

Boyse, E. A., and L. J. Old. 1969. Some aspects of normal and abnormal cell surface genetics. *Annu. Rev. Genet.* **3**: 269.

Brown, G., M. F. Greaves, T. A. Lister, N. Rapson, and M. Papamichael. 1974. Expression of human T and B lymphocyte cell-surface markers on leukaemic cells. *Lancet* **2** (7883): 753.

Chess, L., R. Evans, R. E. Humphreys, J. L. Strominger, and S. F. Schlossman. 1976. Inhibition of antibody-dependent cellular cytotoxicity and immunoglobulin synthesis by an antiserum prepared against a human B-cell Ia-like molecule. *J. Exp. Med.* **144**: 113.

Frei, E. III, F. M. Schabel, Jr., and A. Goldin. 1974. Comparative chemotherapy of AKR lymphoma and human hematological neoplasia. *Cancer Res.* **34**: 184.

Friedman, S. M., J. M. Breard, and L. Chess. 1976. Triggering of human peripheral blood B cells: Polyclonal induction of modulation of an in vitro PFC response. *J. Immunol.* **117**: 2021.

Friedman, S. M., J. M. Breard, R. E. Humphreys, J. L. Strominger, S. F. Schlossman, and L. Chess. 1977. Inhibition of proliferative and plaque-forming cell responses by human bone-marrow-derived lymphocytes from peripheral blood by antisera to the p23,30 antigen. *Proc. Natl. Acad. Sci. U.S.A.* **74**: 711.

Humphreys, R. E., J. M. McCune, L. Chess, H. C. Herrmann, D. J. Malenka, D. L. Mann, P. Parham, S. F. Schlossman, and J. L. Strominger. 1976. Isolation and immunologic characterization of a human, B-lymphocyte-specific, cell surface antigen. *J. Exp. Med.* **144**: 98.

Schlossman, S. F., L. Chess, R. E. Humphreys, and J. L. Strominger. 1976. Distribution of Ia-like molecules on the surface of normal and leukemic human cells. *Proc. Natl. Acad. Sci. U.S.A.* **73**: 1288.

Sen, L., and L. Borella. 1975. Clinical importance of lymphoblasts with T markers in childhood acute leukemia. *N. Engl. J. Med.* **292**: 828.

Strominger, J. L., L. Chess, H. C. Herrmann, R. E. Humphreys, D. L. Mann, J. M. McCune, P. Parham, R. Robb, T. A. Springer and C. Terhorst. 1975. Isolation of histocompatibility antigens and of several B-cell-specific proteins from cultured human lymphocytes. In *Histocompatibility testing* (ed. F. Kissmeyer-Nielson), p. 719. Munksgaard, Copenhagen.

Tsukimoto, I., K. Wong, and B. C. Lampkin. 1976. Surface markers and prognostic factors in acute lymphoblastic leukemia. *N. Engl. J. Med.* **294**: 245.

B-Lymphocyte Development and Growth Regulation

F. Melchers

Basel Institute for Immunology, Basel, Switzerland

Bone-marrow-derived (B) lymphocytes react to antigen by clonal growth and by increased production and secretion of hapten-specific immunoglobulin (Ig). B cells must express two types of reactivities, thus probably two types of receptors, to become reactive to antigen. One type of reactivity is the capacity to recognize haptenic determinants of antigen. This is done by expressing hapten-specific Ig molecules in the surface membrane of B cells. The hapten-binding capacity of B cells is distributed clonally among all B cells. Different B cells bind different haptens; one B cell expresses only one set of variable (V) regions of Ig molecules, which defines the given hapten-binding property of that B cell. However, the constant regions of these different hapten-recognizing Ig molecules on different B cells are the same for large sets of B cells. Thus, a large part of all small, resting B cells expresses μ-heavy chains in the surface membrane. The other type of reactivity of B cells is the capacity to be stimulated by mitogens. Mitogens activate a large part of all B cells to growth and to secretion of Ig molecules in a polyclonal fashion, stimulating the whole repertoire of B cells that express different sets of V regions with different hapten-binding properties, within the set of mitogen-reactive B cells. Mitogens do so presumably by binding to structurally unknown mitogen receptors. These mitogen receptors must be distinct from Ig molecules. It has been proposed that antigens must have mitogenic properties to be able to stimulate B cells to growth and Ig secretion. They must possess haptenic determinants that bind to Ig molecules and mitogenic structures that bind to mitogen receptors. Stimulation of B cells by antigen is initiated by the binding of haptenic determinants to surface-membrane-bound Ig molecules. This binding, however, is regarded as insufficient for triggering. Induction of B cells to growth and Ig secretion requires the subsequent interaction of mitogens, inherent on or bound to antigen, with their corresponding mitogen receptors. Hapten-specific Ig molecules are regarded as devices to concentrate sufficient quantities of hapten-mitogen complexes to the "right" B cells. T-cell-independent antigens may have inherent mitogenic properties. Major external B-cell mitogens, on which haptens will become T-cell independent, are lipopolysaccharides (LPS), lipoprotein (LPP), and Nocardia (NOC) mitogen. T-cell-dependent antigens may receive the mitogen from thymus-derived (T) lymphocytes to present it to the B cells. The origin, production, and structure of such T-cell factors is not very well understood at present. (The reader will find literature pertinent to this introduction in Melchers et al. 1977 and in Andersson et al. 1977d.)

In vitro single-cell cultures have been developed that let precursor B cells from fetal liver or bone marrow develop into antigen-reactive B cells, i.e., Ig-positive and mitogen-reactive mature B cells (Melchers 1977a,b). Precursor B cells thus can be defined as cells that do not yet possess expression of Ig molecules and/or of mitogen reactivities but that will acquire such expression. Mature B cells are identifiable by their capacity to be stimulated by mitogens such as LPS, LPP, or NOC, to polyclonal growth and to secretion of Ig molecules with all possible hapten-binding specificities within the given mitogen-reactive B-cell population. Development of precursor B cells to Ig-secreting cells occurs in two stages. The first stage is independent of externally added mitogen; Ig secretion is absent. The second requires the continued presence of mitogen; Ig secretion is present. Data concerning the frequencies of precursor B cells in fetal liver and in bone marrow, their time course of development to IgM, IgA, and IgG secretion, and their hapten-recognizing repertoire at different times of gestation and after birth have been published previously (Melchers et al. 1975a, 1977a; Phillips and Melchers 1976; Andersson et al. 1977a,b; Melchers 1977a,b). The reader is also referred to current summaries of work in other laboratories (Cooper et al. 1977; Owen et al. 1977; Phillips et al. 1978; Raff 1977). In this paper, an overview of our work and current understanding of B cell development is given. Furthermore, evidence is presented that precursor B cells divide in vitro during the first stage of development to LPS reactivity.

For the second stage—clonal growth of Ig-positive, mitogen-reactive, small B cells and their maturation to Ig secretion—in vitro cell cultures have been developed that let every mitogen-reactive B cell grow into a clone of IgM-, IgG-, or IgA-secreting cells (Metcalf et al. 1975; Melchers et al. 1975b; Andersson et al. 1977b; Lernhardt et al. 1977). Frequencies of mitogen-reactive B cells in neonatal and adult lymphoid tissues have been determined (Andersson et al. 1977a). Different B-cell subpopulations may possess different mitogen reactivities (Gronowicz and Coutinho 1974; Melchers et al. 1975a). The ability to stimulate in vitro every growth-inducible B cell to Ig secretion has made possible the determination of the repertoire of hapten-binding, idiotype-producing, mitogen-reactive B cells for a number of haptens and idiotypes (Andersson et al. 1977b; Eichmann et al. 1977; W. Gerhard and F. Melchers, in prep.). B cells, in conclusion, regulate their reactions to antigens by two repertoires of receptors: hapten-recognizing Ig molecules and mitogen-recognizing mitogen receptors.

Kinetics of growth and of maturation to Ig secretion have been analyzed in serum-free cultures (Melchers and Andersson 1974a,b,c), in mass cultures that allow extensive proliferation (Melchers et al. 1975b, Andersson et al. 1977d), and in cultures that allow every growth-inducible B cell to grow (Andersson et al. 1977c). Our previous analyses have made us conclude that the choice of a B cell between growth and maturation may not be irreversible but, rather, reversible during different stages of the cell cycle. In mass cultures, an increase in the rate of synthesis and secretion, and, therefore, of maturation, has been observed during increased length of growth, and, thus, during succeeding cell cycles (Melchers and Andersson 1974a). This may be the reason why B-cell clones cannot be grown indefinitely (Melchers et al. 1975b). Comparison between B-cell growth and maturation in mass and in single-cell cultures presented in this paper indicate that B cells in both types of cultures are equally limited in growth. They differ, however, in their capacity to mature to IgM secretion.

Functional correlations exist between the hapten-recognizing Ig molecules in the surface membrane of B cells and the growth- and Ig-secretion-inducing mitogen receptors. Thus, exposure of small, resting B cells from adult spleen to anti-Ig antibodies prevents maturation to Ig secretion. However, growth of these B cells is not inhibited when the cells are subsequently stimulated by a B-cell mitogen (Andersson et al. 1974a; Kearney et al. 1976). On the other hand, B cells, immature in either ontogeny or embryonic development, are inhibited by anti-Ig antibodies both in their mitogen-induced maturation to Ig-secretion and in their growth (Melchers et al. 1975a; Kearney et al. 1976). This has led to the suggestion that Ig molecules in the surface membranes of B cells are not mere concentrating devices for hapten-mitogen complexes (Coutinho and Möller 1974); they also form a functional, and possibly structural, complex with mitogen receptors. It has been proposed that the Ig-molecules modulate the functional conformation of the mitogen receptors and, thereby, modulate the signals given by mitogens to the B cells (Andersson et al. 1974a; Andersson and Melchers 1976).

MATERIALS AND METHODS
Mice

C_3H/Tif/BOM mice were obtained from Gr. Bomholtgaard (Ry, Denmark). C57Bl/6J and C57Bl/6J × DBA/2J F1 mice and timed pregnancies were obtained from the Institut für Biologisch-Medizinische Forschung A.G., Füllinsdorf, Switzerland.

Cells

Spleen cells, bone-marrow cells, fetal liver cells, and thymus cells were prepared as described previously (Andersson et al. 1977a; Melchers 1977a). Small cells from spleen and bone marrow were obtained by 1 g velocity sedimentation (Miller and Phillips 1969).

Media

RPMI-1640 medium was obtained in liquid from Microbiological Associates. Glutamine (2 mM), penicillin (5000 IU/ml), streptomycin (5000 IU/ml), HEPES buffer (10 mM, pH 7.3), β-mercaptoethanol (5×10^{-5} M), and fetal calf serum (10%) were added. For fetal liver and bone-marrow cell cultures, batch U951801 was used, and for spleen cell cultures, batch K255701D was used; both of these came from Gibco Laboratories. Cells were cultured in various Falcon plastic tubes or dishes and in LINBRO Dispo trays (model FB 16–27, TC, LINBRO Scientific).

Mitogens

LPS-s (S495/188049) was prepared for us by Drs. C. Galanos and O. Lüderitz, Max-Planck Institut für Immunbiologie, Freiburg, Germany. It was routinely used at 50 μg/ml.

Measurement of B-Cell Growth and Maturation to Ig-Secreting Cells

Cell growth was followed by counts of live cells in a Bürker hemocytometer under the microscope with the trypan-blue exclusion test. Enumerations of the number of IgM-secreting cells were done with the hemolytic plaque test; for this, protein A-coupled sheep red cells and rabbit anti-mouse IgM antibodies (anti-MOPC 104E myeloma 19-S IgM) as described previously (Gronowicz et al. 1976) were used.

Antisera for Anti-Ig Antibody Inhibition of B Cells

Sheep (anti-mouse Ig) (κ, γ) (a gift of Dr. A. Kelus, Basel Institute for Immunology) was complement inactivated and adsorbed to Sepharose 4B-coupled mouse Ig (a mixture of myeloma proteins purified from sera of MOPC 104E 19-S IgM (λ, μ), MOPC 21 IgG$_1$ (γ_1/κ) and MOPC 315 (α/λ) myeloma tumor-bearing mice. All myeloma tumors were kindly given to us by Dr. M. Potter, National Cancer Institute, National Institutes of Health, Bethesda, Maryland. The adsorbed antibodies were eluted from the immunosorbent as described (Andersson et al. 1974a), adjusted to protein concentrations between 0.5 and 5 mg/ml, and used at a final dilution of 10^{-2} in culture. Treatment of mouse lymphocytes with these antibodies was done as previously described (Andersson et al. 1974).

RESULTS AND DISCUSSION

The Development of Mitogen-Reactive B Cells from Precursor B Cells of Fetal Liver in Vitro

Development of B cells that are mitogen reactive and that synthesize Ig has been studied mainly in fetal liver (Melchers et al. 1975a; Phillips and Melchers 1976; Melchers 1977a,b; Melchers et al. 1977). The earliest cell synthesizing Ig in the mouse could be detected by biosynthetic incorporation of leucine and by lactoperoxidase-catalyzed radioiodination of fetal liver cells, followed by serological precipitation and identification as 7- 8-S IgM by gel electrophoresis from day 10 to 12 of gestation onward. These cells, large in size, may be identical to the precursor B cells that have been identified by adoptive transfer experiments to be precursors of antigen-reactive cells. From day 15 of gestation onward, small precursor B cells appear in fetal liver. They also synthesize 7- 8-S IgM but release that IgM at different rates of turnover. In vitro cell cultures let large-sized precursor B cells of day-14 fetal liver cells disappear, whereas small-sized precursor B cells appear after 3 days of culture. This may argue for a precursor-product relationship of large- and small-sized precursor B cells.

Single-cell suspensions of fetal liver cells, of as early as day 12 or 13 of gestation, develop LPS-reactive B cells in vitro within the same time as they do in vivo. All fetal liver cells acquire LPS reactivity in vitro at the equivalent of birth, no matter at which time of gestation they have been put in culture. This development of LPS reactivity is independent of the presence or absence of the mitogen LPS. When fetal liver cells have become LPS reactive in vivo at birth or in vitro at the equivalent of birth, they are then stimulated by LPS to growth and to maturation into clones of IgM-, IgG-, and IgA-secreting cells. The development of these Ig-secreting cells is mitogen dependent; in the absence of LPS, only 2–5% of all secreting cells develop. This background development of secreting cells is due probably to B-cell mitogens present in the fetal calf serum used to maintain the in vitro cultures. Peak responses of IgM- and IgA-secreting cells in culture occur at the equivalent of day 5 after birth; those of IgG-secreting cells, at day 7 after birth.

The efficiency with which fetal liver cells develop into LPS-reactive B cells is greatly increased, particularly at lower cell densities in culture, by the addition of thymus cells as "filler" or "feeder" cells (Andersson et al. 1977c). This has facilitated an analysis of the frequencies of precursor B cells developing to LPS reactivity and then to IgM-secreting cells in culture by limiting dilution analyses of fetal liver cells from day 13 to 19 of gestation and of liver cells until day 7 after birth. Table 1 shows that precursor B cells increase from 1 in 10^6 fetal liver cells at day 13 of gestation to 1 in 30 cells at birth (day 19), and decrease after birth to less than 1 in 10^5 liver cells at day 7 of neonatal life.

It remained remarkable that the magnitude, and not the development in time, of LPS-reactivity, measured as total IgM-secreting cells, was different in vivo and in vitro (Table 1). This could be explained in two ways: either the precursors of LPS-reactive B cells could divide only in vivo and not in vitro, or the precursors continued to migrate into fetal liver with continuing gestation. In the first case, the precursor potential of fetal liver would be unchanged from day 13 to day 19 of gestation, whereas, in the second case, the magnitude of LPS reactivity would increase through immigration of precursor cells from other, unknown sites in the body. Therefore, I attempted to analyze whether precursor B cells, before becoming LPS-reactive, would divide in culture.

The first indication that precursor B cells were dividing in in vitro cultures came from determinations of the sizes of IgM-secreting cell clones developing at the equivalent of day 5 after birth in culture in the presence of LPS, when precursors were limited to one per culture. Table 2 summarizes such analyses of cell cultures of day-15 and of day-19 (birth) fetal liver cells. Analyses were done at a fetal-liver-cell concentration that limited the

Table 1
Frequencies of Precursors of LPS-Reactive B Cells and Total IgM-Secreting Cell Responses of Fetal Liver Cells

Time in development	Frequencies of precursor B cells[a] in total fetal liver cells[b]	Total IgM-secreting cell response[c] $\times 10^{-2}/4 \times 10^6$ cultured cells
Prenatal		
day 13	1 in 10^6	5
day 15	1 in 10^5	25
day 17	1 in 2×10^3	810
day 19 (birth)	1 in 3×10^1	5300
Postnatal		
day 2	1 in 3×10^2	2400
day 4	1 in 3×10^3	205
day 7	1 in $\times 10^5$	20

[a] Frequencies were determined by limiting dilution analyses using Poisson's distribution (Melchers 1977b).
[b] Cells of C57Bl/6J × DBA/2J F1 embryos or neonatal mice were put in suspension culture with thymus cells at different times during prenatal and postnatal life.
[c] IgM-secreting cells, measured as plaque-forming cells (see Materials and Methods) each day from the equivalent of day 1 to day 10 after birth in culture, were added to yield the total response.

positive cultures (yielding IgM-secreting cells) to between 20–30% of cell cultures; according to Poisson's distribution, less than one precursor B cell was plated per average culture. A linear degression of the logarithm of the fraction of nonresponding cultures with decreasing fetal liver cell concentrations in culture was observed (not shown; see Melchers 1977b), which suggested that precursor B cells, yielding LPS-reactive B cells that develop IgM-secreting cells, were limiting in the cultures. The results in Table 2 show an almost 10-fold difference in the average clone size of IgM-secreting cells developing from single precursors of day-15 and of day-19 fetal liver cells, day-15 fetal liver cells yielding the larger clone sizes. Since it has been shown that LPS-reactivity is acquired at day 19, these results could have meant that LPS-reactive cells of day 15, from day 19 on, divide more rapidly than day-19 fetal liver cells. However, it was more likely to assume that LPS-reactive cells, developing in culture either from day 15 or from day 19, divide and mature to IgM-secreting cells at an equal rate. This would mean that precursor B cells, i.e., those from day 15 of gestation, before they become LPS reactive, could divide in culture. In the first instance, with no division of precursors in culture, the frequency of precursors would not increase between the equivalent of day 15 and day 19 of gestation in culture; whereas in the second, with precursors dividing in culture, it would.

This hypothesis was tested by culturing day-15 fetal liver cells until the equivalent of day 19 in mass culture and at the same fetal-liver-cell concentration that has been shown to limit the precursor B cells to around one (i.e., ~ 10^5 fetal liver cells/ml; see Tables 1 and 2 and Melchers 1977b). At the equivalent of day 19 of gestation, just as the cells were expected to gain

Table 2
Clone Sizes of LPS-Stimulated B-Cell Cultures Developing into IgM-Secreting Cells of Fetal Liver Cells Put in Culture at Day 15 or 19 of Gestation in Concentrations That Limit the Number of Precursor B Cells to Near One

Source of fetal liver cells	Number of IgM-secreting cells per individual culture[a]	Average number of IgM-secreting cells/culture	Positive cultures (%)
Day 15 of gestation[b]	0,0,510,836,0,258,0,0,0,0, 0,756,0,436,0,0,0,0,0,0, 212,76,0,0,0,460,918,0, 157,0	154	33
Day 19 of gestation[c]	0,0,0,0,46,0,57,0,32,0,0, 29,0,0,0,0,62,0,0,151,61, 34,0,0,42,21,0,39,0,0	19	36

[a] Cells were assayed in 30 individual cultures at the equivalent of day 5 after birth in culture.
[b] 10^5 C57Bl/6J × DBA/2J F1 fetal liver cells per culture were cultured in the presence of LPS (50 µg/ml) and syngeneic mouse thymus cells (3 × 10^6 cells/ml).
[c] Twenty liver cells were cultured per culture as in (b).

LPS reactivity, the frequency of cells showing LPS reactivity by the development of clones of IgM-secreting cells was determined in these cultures by limiting dilution analysis. The results in Figure 1 show that the frequency of LPS-reactive cells had increased in culture by over a factor of 10, indicating that the average precursor B cell of day-15 fetal liver had divided in culture more than three times until the equivalent of day 19. Furthermore, the clone sizes of individual LPS-reactive cells, plated at limiting dilutions, were now comparable in size to those of liver cells that had been put in culture at day 19 (birth) (Table 3, see also Table 2).

I conclude that precursor B cells of fetal liver can divide in vitro under our culture conditions, that this division is independent of exogenously added mitogen (LPS), and that these dividing precursors do not secrete Ig to the extent that it could be monitored by a plaque. Inasmuch as precursor B cells increase 3×10^3-fold in vivo between day 15 and day 19 (Table 1), but only 10-fold in vitro (Figure 1), we may still argue either that the precursor potential of fetal liver is constant between day 15 and day 19 of gestation or that our culture conditions did not allow maximally possible proliferation of precursors. It is also reasonable to assume that the 300-fold difference in the in vivo and in vitro precursor potential of day-15 and day-19 fetal liver cells came about by an increase of precursor B cells by immigration into fetal liver from other, unknown sites in the body.

Growth and Maturation to Ig Secretion of Mitogen-Reactive B Cells

At the stage of differentiation at which B cells become surface Ig positive and mitogen sensitive, i.e., antigen reactive, they fall into a resting state in which they do not synthesize DNA, do not divide, and appear small in size, until mitogen or antigen activates them to division and Ig secretion.

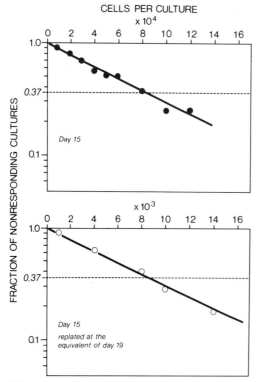

Figure 1
Titration of precursors of LPS-reactive cells yielded IgM-secreting cells at the equivalent of day 5 after birth in cultures of fetal liver cells of day 15 of gestation that had been put into culture either at limiting dilutions immediately (top) or after a mass culture period (at 10^5 cells/ml in the presence of syngeneic thymus cells) for 4 days until the equivalent of day 19 of gestation (bottom). Cell numbers per culture on top refer to day-15 fetal liver cells that had been put into culture originally at day 15 of gestation. Cultures contained 3×10^6 syngeneic C57Bl/6J × DBA/2J F1 thymus cells and 50 µg LPS/ml. The dotted lines indicate the number of cells with which 37% of all cultures did not yield a response, i.e., those which, according to Poisson's distribution, contained one precursor. Cell numbers on the bottom refer to day-15 fetal liver cells that were kept in mass cultures until the equivalent of day 19, then were replated in the presence of new thymus filler cells.

Structure, synthesis, and turnover of IgM in these resting, small B cells has been studied extensively (Melchers and Andersson 1973; Andersson et al. 1974b; 1974), whereas practically nothing is known of the structure, location, or synthesis of mitogen receptors such as LPS or LPP receptors. Activation of resting B cells by antigens or mitogens leads to proliferation and to maturation to Ig secretion. The biochemical changes occurring in B cells after mitogenic stimulation, particularly those connected with IgM synthesis and secretion, have been studied in detail (Melchers and Andersson 1973, 1974a).

Two findings appear particularly worth mentioning when regulation of

Table 3
Clone Sizes of LPS-Stimulated B-Cell Cultures Developing into IgM-Secreting Cells of Fetal Liver Cells Put in Mass Culture at Day 15 of Gestation and Replated at Equivalent of Day 19 in Concentrations That Limit the Number of LPS-Reactive B Cells to Near One

Number of IgM-secreting cells per individual culture	Average number of IgM-secreting cells per culture	Positive cultures (%)
26,0,0,67,71,0,31,0,0,0,0,0, 0,35,0,18,46,0,0,0,0,88,0, 27,0,33,32,51,0,0,	17.5	39

Cells were plated at 10^4 fetal liver cells of day 15 of gestation per culture. They were replated at the equivalent of day 19 in concentrations limiting the number of LPS-reactive B cells to near one. 10^5 C57Bl/6J × DBA/2J F1 fetal liver cells/ml were cultured in the presence of thymus cells (syngeneic, 3×10^6/ml) until the equivalent of day 19. Presence or absence of LPS (50 µg/ml) did not change the results.

growth and maturation to Ig secretion is considered within the cell cycle. First, one of the earliest changes detected in small B cells after mitogenic stimulation, i.e., within the first 4 hours, is the induction of secretion of IgM; this is characterized by a different size, by carbohydrate composition, and by the turnover rate of the IgM molecules found released from the cells (Melchers and Andersson 1974b). RNA-synthesis-dependent components of IgM synthesis are stabilized; IgM synthesis itself is reprogramed from a synthesis of a membrane-bound, hapten-receptor-type IgM to a synthesis of actively secreted IgM. The onset of DNA synthesis, detected by thymidine uptake, is only observed from 16 to 18 hours after activation (Melchers and Andersson 1974c). Maturation to Ig secretion, therefore, appears to precede DNA synthesis (and division?) within the cell cycle and can also occur when RNA and DNA synthesis have been inhibited (Melchers and Andersson 1974b, Andersson and Melchers 1974).

The second observation relevant to growth regulation within the cell cycle of B cells is that the ratios of the rates of synthesis or of secretion of IgM over those of all proteins made in the cell increase with increasing time of stimulation. Proteins other than IgM are made and secreted by B cells after mitogenic activation, but IgM synthesis and secretion increase selectively over synthesis and secretion of other proteins in B cells (Melchers and Andersson 1974a). Seven different malignant B-cell lines, lymphomas and myelomas all of which produce IgM, have been analyzed for the ratio of their rates of synthesis and secretion of IgM over total protein. These ratios, as well as the morphological appearance of the B cells, characterize these cells as transformed counterparts of normal B cells stimulated for various periods of time (Andersson et al. 1974c; Bergman et al. 1977).

Two alternate models of growth and regulation within the cell cycle could explain these findings. One presumes a dividing, nonsecreting, i.e., growing, nonmaturing "stem" cell which, at each division, gives probabilities of producing either two "stem" cells: one stem cell and one mature,

Ig-secreting, nondividing cell; or two such mature cells. Normal, mitogen stimulated B-cell clones would have probabilities to yield mature, nondividing cells, which would appear to increase with each additional division. Transformed cells, myelomas producing IgM, would appear to have a fixed probability at each division, characteristically different for each malignant line, to yield dividing "stem" cells or nondividing mature cells. This model of an irreversible choice of a B cell either to divide or to mature (and thereby to cease to divide) is contrasted by the other model in which the choice of a B cell is reversible and depends on the stage of the cell cycle in which the cell finds itself. Since mitogen-stimulated B cells selectively increase in mass-culture IgM synthesis and secretion over the synthesis and secretion of all other proteins of the cell upon prolonged growth, this would mean that synthesis and secretion of IgM are increased either in its rate (i.e., probably the number of mRNA molecules on polyribosomes making it) during a fixed time of the cell cycle or that, at a constant rate of synthesis, a longer period of the cell cycle is devoted to IgM synthesis and secretion (i.e., probably a constant number of mRNA molecules on polyribosomes being read more often).

LPS-stimulated B cells have been shown to divide every 18 hours. We have concluded that, within the first 126 hours of growth, every B cell in the clone divides and, at least between 96 and 126 hours of growth, every dividing B cell in this clone secretes sufficient IgM to let it be detected as a hemolytic plaque (Andersson et al. 1977c). More recent experiments (presented in Fig. 2b) show that IgM-secreting plaque-forming cells (PFC) can be detected in single B-cell cultures from as early as 54 hours of stimulation onward and that, again, all dividing B cells in this earlier time period of growth secrete sufficient IgM to form a hemolytic plaque. Because the growth curves of single B-cell clones, measured either by counts of cells or by numbers of IgM-secreting PFC, are superimposable, I conclude again that all B cells in a clone divide and that every dividing cell secretes sufficient IgM to form a hemolytic plaque. The clonal analysis of B-cell growth clearly argues for a reversible choice of a growing B cell for maturation to IgM secretion. It appears, therefore, that B cells can balance within the cell cycle between growth and maturation to IgM secretion.

The problem of a possible increase in the rate of IgM synthesis and secretion during succeeding cell cycles of a growing clone of B cells remains unanswered by these analyses. In fact, the data in Figure 2 reveal a striking difference between mass cultures and single-cell cultures of LPS-reactive B cells. Whereas in single B-cell clones between 54 and 120 hours of growth, every growing B cell secretes sufficient IgM to form a hemolytic plaque, this is not so in mass cultures. At 54 hours of growth, not more than 1% of all growing cells secrete enough IgM to form a plaque. The proportion of IgM-secreting PFC within growing cells increases with time of stimulation (as does the rate of IgM synthesis and secretion as determined per average cell by biosynthetic techniques [F. Melchers, unpubl.]); however, it never exceeds 20% of all growing B cells (Fig. 2a). In mass cultures 15–20% of all B cells initiate growth; whereas in single-cell cultures, 33% do (legend to Fig. 2). The difference in growth initiation, therefore,

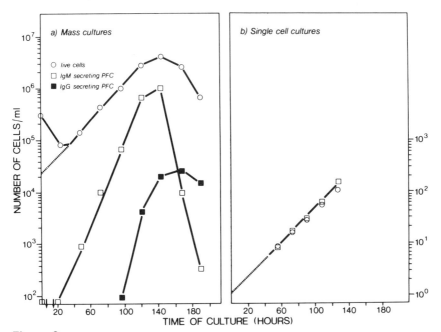

Figure 2
Growth and maturation of LPS-reactive small B cells measured (o——o) as numbers of live cells, (□——□) as IgM-secreting PFC, or (■——■) as IgG-secreting PFC. (a) Mass cultures of small lymphocytes, purified from large lymphocytes of $C_3H/Tif/BOM$ spleen by 1 g velocity sedimentation (Miller and Phillips 1969), were grown at 3×10^5 cells/ml in the presence of 50 μg LPS/ml but in the absence of thymus filler cells, as described by Melchers et al. (1975b). Not all cells initiated growth, as seen in the initial drop of the number of live cells in the culture. Extrapolation of the exponential part of the growth curve between 48 and 120 hours of stimulation (dotted line) shows that approximately 2.4×10^4 out of 3×10^5 spleen cells, i.e., approximately 15–20% of all B cells, initiated growth; (b) cultures emerging from one LPS-reactive B cell were originated either as colonies (clones) in semisolid agar (Metcalf et al. 1975) in the presence of LPS (50 μg/ml) but in the absence of thymus cells at a small spleen cell concentration of 3×10^4/ml (Andersson et al. 1977d). Thirty to 50 individual colonies, were counted every 18 hours from 54 to 120 hours of growth. (o——o) Averages of these determinations. For the determination of IgM-secreting PFC, small $C_3H/Tif/BOM$ spleen cells were diluted in suspension cultures containing 50 μg LPS and 3×10^6 syngeneic thymus cells/ml to approximately one LPS-reactive B cell per culture. At an average of six spleen cells per culture, 37% of all cultures did not yield an IgM-secreting PFC response, indicating, according to Poisson's distribution, that one LPS-reactive B cell was present in six spleen cells, i.e., that 33% of all B cells initiated growth and maturation to IgM secretion. From 54 hours of growth onward, 150 cultures were scored every 18 hours for IgM-secreting PFC. A discontinuous distribution of PFC per culture was obtained, as expected from our earlier results (Andersson et al. 1977c). (□——□) Numbers of IgM-secreting PFC, representing those of peak II emerging from one LPS-reactive B cell (see Andersson et al. 1977c).

is only in the range of a factor of 2 in the two types of cultures, which argues that maturation of IgM secretion is deficient in mass cultures.

This deficiency may be due to a quantitative difference in the numbers of IgM molecules secreted in a given time by B cells growing in these two types of culture. Thymus filler cells, present in large numbers in single-cell cultures and possibly in limiting numbers in mass cultures, may well supply maturation-supporting factors for which B cells are auxotrophic and which increase the rate of IgM synthesis and secretion. It is possible that all B cells growing in mass cultures also secrete IgM, most of them, however, in insufficient quantities to effect a plaque. At present, we do not know how many IgM molecules must be secreted in a given period of time by a B cell to let that B cell form a plaque (Gronowicz et al. 1976). As long as we have no means of determining the rate of IgM synthesis and secretion of single B cells or of cultures containing 5–1000 secreting B cells, or at least as long as we cannot distinguish between low- and high-rate secretion in single B cells, the problem of an increase in the rate of synthesis and secretion of IgM and, therefore, of increased maturation during the succeeding cell cycles of a growing clone of B cells cannot be probed experimentally.

A qualitative difference in the conditions of B cells for growth and maturation in single cell compared to mass cultures also should not be excluded. Different B-cell clones, growing together in culture and each secreting Ig molecules with different antigen-binding specificities and different idiotypes, may influence each other. Their products, Ig molecules, may inhibit one another's maturation to Ig secretion. (See also next section.) Such inhibitory interactions between different B-lymphocyte clones could be idiotypic-antiidiotypic in nature (Cosenza and Köhler 1972; Hart et al. 1972; Eichmann 1974) and may constitute in vitro effects of networks of lymphocytes (Jerne 1974). Of course, single clones of B cells would be devoid of such interactions; few B-cell clones together would have a diminished probability for such interactions.

The Role of Surface-Membrane-Bound Ig in the Regulation of Growth and Maturation of B Cells

Since it became evident that antigen selects its specific lymphocytes to initiate an antigen-specific immune response, it had been assumed that the binding of optimal concentrations of the haptenic determinant of an antigen to lymphocytes would alone be sufficient to trigger lymphocytes. Too little or too much antigen binding would render lymphocytes tolerant. The structure of Ig or its location in the surface membrane of B cells was expected to change upon antigen binding. This simple concept of Ig molecules being the growth- and maturation-initiating receptor of B cells became untenable when it became known that B cells, for many antigens, need the cooperation of T cells to initiate an immune response. Consequently, this lead to more complicated hypotheses as to how antigen would trigger or render lymphocytes tolerant (see *Transplantation Reviews,* vol. 23, 1975). One hypothesis proposed that binding by hapten to Ig alone, mediating signal one, would lead to unreactivity; whereas the binding of the cooperating structures from T cells, thought to be antigen-specific as well, would give signal two to the

B cells. Signals one and two together would then initiate an immune response (Bretscher and Cohn 1970). The discovery of B-cell- and T-cell-specific mitogens, polyclonal activators of lymphocytes (Andersson et al. 1972), made it quite clear, however, that lymphocytes could be activated to growth and maturation by circumvention of the binding step to Ig, because mitogens bind to mitogen receptors which appear structurally different from Ig (Nilsson et al. 1973; Watson and Riblet 1975). Furthermore, antigens could be rendered T-cell independent by being coupled to B-cell mitogens (Coutinho and Möller 1974). This led to the proposal that Ig molecules on the surface of B cells were mere concentrating devices for haptens and, by binding haptens, had no effect on B cells, either in their reactions to growth or to maturation.

The reactions of anti-Ig antibodies with B cells, however, pointed to a more delicate role for Ig molecules in the surface membrane of B cells, a role beyond that of a mere antigen-concentrating function. Anti-Ig antibodies alone do not stimulate B cells, either to growth or to maturation (Andersson et al. 1974a; Kearney et al. 1976), although other laboratories occasionally have found such stimulation (summarized by Melchers and Pernis 1977). Subsequent stimulation of anti-Ig-treated B cells to Ig secretion, however, was inhibited. This was shown both for mitogenic stimulation—that inhibited by anti-μ heavy-chain and anti-κ light-chain antibodies (Andersson et al. 1974a; Kearney et al. 1976), and for antigenic stimulation—that inhibited by relevant antiidiotypic antibodies (Cosenza and Köhler 1972; Hart et al. 1972). Inhibition of mitogenic stimulation to Ig secretion by anti-Ig antibodies suggested that a complex of structures on the surface of B cells is involved in regulation of growth and maturation, which consists of Ig molecules and mitogen receptors (Andersson and Melchers 1977). The occupancy of Ig molecules can modulate reactions that are induced by the binding of mitogen to mitogen receptors.

It was also recognized in these early experiments that thymidine uptake and, therefore, probably growth were not inhibited in anti-Ig-treated, mitogen-stimulated B cells (Andersson et al. 1974a; Kearney et al. 1976). The interesting possibility was suggested from these experiments that the modulating effects of Ig molecules were influencing a balance of growth and maturation within the cell cycle.

The experimental evidence was not as clear, inasmuch as different B-cell subpopulations at different stages of their differentiation showed different effects to inhibition of mitogenic stimulation by anti-Ig antibodies (Melchers et al. 1975a; Kearney et al. 1976, 1978). Earlier stages of B-cell differentiation appeared inhibited both for maturation to Ig secretion and for thymidine uptake.

I investigated the inhibition of LPS-induced activation by anti-κ light-chain antibodies in in vitro suspension cultures that allowed extended growth of lymphocytes (Melchers et al. 1975b; Andersson et al. 1977d). Small lymphocytes, from adult spleen as a source of mature B cells and from bone marrow as a source of earlier stages of B-cell differentiation, were enriched by velocity sedimentation, treated with anti-Ig antibodies, and then stimulated with LPS. Growth was measured by counts of cells and maturation by determination of the number of plaque-forming cells. Results in

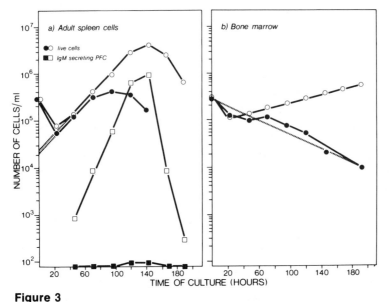

Figure 3
Growth and maturation of LPS-reactive, small B cells that were purified from large cells by 1 g velocity sedimentation (Miller and Phillips 1969) Cells were pretreated with the sheep (antimouse Ig) (κ, γ) antibodies at a final dilution of 10^{-2} and under conditions described by Andersson et al. (1974a). LPS was thereafter added to a final concentration of 50 μg/ml. Control data in the absence of anti-Ig antibodies are those also shown in Figure 2. (*a*) Adult speen cells from 6- to 8-week-old C_3H/Tif/BOM mice; (*b*) bone-marrow cells from 6- to 8-week-old C_3H/Tif/BOM mice. (o———o) Live B cells in absence of anti-Ig antibodies; (●———●) live B cells in presence of anti-Ig antibodies; (□———□) IgM-secreting PFC in absence of anti-Ig antibodies; (■———■) IgM-secreting PFC in presence of anti-Ig antibodies.

Figure 3 show that anti-Ig antibodies inhibited both maturation and growth of LPS-stimulated bone-marrow cells but inhibited only maturation of stimulated spleen cells. Approximately the same number of splenic B cells initiated growth with LPS, i.e., between 15% and 20% of all B-cells (see also Fig. 2), whether or not they had been previously treated with anti-Ig antibodies. It does suggest, but not prove, that the same splenic B cells grow in the presence and absence of anti-Ig antibodies. In the presence of anti-Ig antibodies, however, LPS-stimulated B cells cease to grow earlier (after approximately 90 hours, i.e., after five divisions). These differences are not understood at present.

The apparent difference in reactivity to anti-Ig antibodies and the subsequent mitogenic stimulation of more immature (bone marrow) and of mature (adult spleen) B cells may suggest that the proposed complexes of Ig molecules and mitogen receptors exist, in qualitative or in quantitative terms, in different conformations in these two types of B cells. A much simpler explanation should also be kept in mind. LPS-reactive B cells represent a much smaller proportion of all cells in bone marrow than they do in spleen (Andersson et al. 1977a). Other LPS-reactive cells which are not B cells may reside in bone marrow. It may be difficult, therefore, to distinguish, by numbers of cells or by thymidine uptake, B cells from a majority of non B cells in bone marrow.

Inhibition of maturation to Ig secretion but not of growth by anti-Ig antibodies of mature splenic B cells stimulated by LPS (as shown in Fig. 3) again suggests that a balance between growth and maturation exists within the cell cycle of growing B cells. The occupancy of surface Ig, in this case by anti-Ig antibodies, modulates this balance and suppresses maturation.

CONCLUSIONS AND OUTLOOK

Even when the nucleotide sequences of immunoglobin genes have all been determined (Tonegawa et al. 1978), it is highly unlikely that it will then become apparent from the amino-acid sequences, deduced from these nucleotide sequences, which "antigens" fit to the Ig molecules that are carried in the germ line. This information, however, is of prime interest for an understanding of the selective pressures on which variability in Ig molecules and thereby in antigen recognition evolves during ontogeny (Cohn 1968; Jerne 1971). Single-cell suspension cultures of precursors of Ig-producing B lymphocytes may allow us in the future to clone cells that express one combination of a heavy-chain and a light-chain gene carried in the germ line, to isolate these Ig molecules, and to study their binding specificities. Furthermore, such cultures may be the way to follow variation during ontogeny in vitro and to apply defined selective pressures to the evolving lymphocytes. Of course, these cultures of precursor cells (see also Dexter et al., this volume; Le Douarin, this volume; and Phillips, this volume) are aimed at identification of the growth- and differentiation-regulating substances (mitogens) effective at these early stages of lymphocyte differentiation.

Two molecules are required for the reception of antigen by B cells to grow and mature; these molecules are Ig and mitogen receptors. Reactions following the initial interaction of hapten-mitogen complexes with these receptor complexes may be expected to be general for all cells that are activated from a resting state to growth and differentiation; such reactions will not be lymphocyte specific. Lymphocytes, because of their apparent heterogeneity in antigen-binding properties, mitogen reactivities, and states of differentiation in practically every lymphoid organ, appear experimentally to constitute a very difficult system for study of these general reactions to the inside of the cell following receptor interactions in the surface membrane. The immediate future seems to lie more in the identification and structural characterization of the components of the antigen-recognizing receptor complex that regulates growth and maturation specifically for lymphocytes. In particular, we need to know the nature of the receptors for T-cell help and the nature of the factors effecting this help. Occupance of Ig on B cells, either within the hapten-binding sites by antigen or near the binding site by antiidiotypic (anti-Ig) antibodies, appears either to allow or to suppress mitogen-activated maturation to Ig secretion. These modulating effects of Ig molecules should be the consequence of two different conformations in which Ig can occur and be fixed.

I am inclined to think that mitogen receptors are lymphocyte specific. This may not be the case, however, for all receptors, since the mutation affecting LPS-reactivity in mice (Nilsson et al. 1973; Watson and Riblet 1975) is pleiotropic, not restricted to LPS-nonreactivity of lymphocytes only. In other cells, these receptors may function as receptors for other cellular reactions induced by LPS.

How T cells are induced to grow and mature is understood much less well, partly because the receptors for antigen are still unknown. There is evidence that T cells carry idiotypes (Binz and Wigzell 1977; Krawinkel et al. 1977) and, therefore, possess hapten-binding structures on their surfaces just as B cells do. In analogy to B cells, one may expect that T cells also need a complex of molecules to recognize antigen: a hapten-binding, idiotype-bearing molecule and a mitogen-receptor. Concanavalin A or phytohemagglutinin, exogenous mitogens for T cells (Andersson et al. 1972), certainly can polyclonally activate T cells, circumventing the hapten-binding step. Induction of T cells to immune function is often linked to genes within the major histocompatibility complex. One is tempted to speculate that genes within the major histocompatibility complex code for endogenous T-cell mitogens and/or for receptors that recognize these mitogens on T cells.

The apparent balance between growth and maturation within the cell cycle of growing, normal B cells is not a fixed one but drifts toward further maturation. This drifting eventually leads to the termination of growth through maturation, which evidently is the fate of most antigen-driven lymphocyte clones. Anti-Ig antibodies suppress maturation but let the clone expand. Anti-Ig antibodies, in particular antiidiotypic antibodies may be instrumental, therefore, in creating memory (Eichmann and Rajewsky 1975; Eichmann et al. 1977). The balance between growth and maturation appears fixed in states that are characteristic for each of a series of malignant lymphomas and myelomas (Andersson et al. 1974c; Bergman et al. 1977). The comparison of these malignant B cells with normal, mitogen-stimulated B cells in their kinetics of growth and maturation within the cell cycle may give us clues as to how transformed B cells differ from normal ones.

REFERENCES

Andersson, J., and F. Melchers. 1974. Maturation of mitogen-activated bone marrow-derived lymphocytes in the absence of proliferation. *Eur. J. Immunol.* **4**: 533.

———. 1977. Mitogen stimulation of B-lymphocytes. A mitogen receptor complex which influences reactions leading to proliferation and differentiation. In *Dynamic aspects of cell surface organization* (eds. G. Poste and G. L. Nicholson) vol. 12. North Holland Publishing Company, Amsterdam. (In press.)

Andersson, J., O. Sjöberg, and G. Möller. 1972. Mitogens as probes for immunocyte activation and cellular cooperation. *Transplant. Rev.* **11**: 131.

Andersson, J., W. W. Bullock, and F. Melchers. 1974a. Inhibition of mitogenic stimulation of mouse lymphocytes by anti-mouse immunoglobulin antibodies. I. Mode of action. *Eur. J. Immunol.* **4**: 715.

Andersson, J., A. Coutinho, and F. Melchers. 1977a. Frequencies of mitogen-reactive B cells in the mouse. I. Distribution in different lymphoid organs from different inbred strains of different mice at different ages. *J. Exp. Med.* **145:** 1511.

———. 1977b. Frequencies of mitogen-reactive B cells in the mouse. II. Frequencies of B cells producing antibodies which lyse sheep or horse erythrocytes, and trinitrophenylated or nitroiodophenylated sheep erythrocytes. *J. Exp. Med.* **145:** 1520.

Andersson, J., L. Lafleur, and F. Melchers. 1974b. IgM in bone marrow-derived lymphocytes. Synthesis, surface deposition, turnover and carbohydrate composition in unstimulated mouse B cells. *Eur. J. Immunol.* **4:** 170.

Andersson, J., A. Coutinho, W. Lernhardt, and F. Melchers. 1977c. Clonal growth and maturation to immunoglobulin secretion in vitro of every growth-inducible B lymphocyte. *Cell* **10:** 27.

Andersson, J., A. Coutinho, F. Melchers, and T. Watanabe. 1977d. Growth and maturation of single clones of normal murine T and B lymphocytes in vitro. *Cold Spring Harbor Symp. Quant. Biol.* **41:** 227.

Andersson, J., J. Buxbaum, R. Citronbaum, S. Douglas, L. Forni, F. Melchers, B. Pernis, and D. Scott. 1974c. IgM-producing tumors in the Balb/c mouse: A model for B-cell maturation. *J. Exp. Med.* **140:** 742.

Bergman, Y., J. Haimovich, and F. Melchers. 1977. An IgM-producing tumor with biochemical characteristics of a small B lymphocyte. *Eur. J. Immunol.* **7:** 574.

Binz, H., and H. Wigzell. 1977. Antigen-binding, idiotypic receptors from T lymphocytes: An analysis of their biochemistry, genetics, and use as immunogens to produce specific immune tolerance. *Cold Spring Harbor Symp. Quant. Biol.* **41:** 275.

Bretscher, P., and M. Cohn. 1970. A theory of self-nonself discrimination. *Science* **169:** 1042.

Cohn, M. 1968. The molecular biology of expectation. In *Rutgers symposium on nucleic acids in immunology* (eds. O. J. Piescia and W. Braun), p. 671. Springer-Verlag, Berlin and New York.

Cooper, M. D., J. F. Kearney, P. M. Lydyard, C. E. Grossi, and A. R. Lawton. 1977. Studies of generation of B-cell diversity in mouse, man and chicken. *Cold Spring Harbor Symp. Quant. Biol.* **41:** 139.

Cosenza, H., and H. Köhler. 1972. Specific suppression of the antibody response by antibodies to receptors. *Proc. Natl. Acad. Sci. U.S.A.* **69:** 2701.

Coutinho, A., and G. Möller. 1974. Immune activation of cells: Evidence for "one nonspecific triggering signal" not delivered by the Ig-receptors. *Scand. J. Immunol.* **3:** 133.

Eichmann, K. 1974. Idiotype suppression. I. Influence of the dose and of the effector functions of anti-idiotypic antibody on the production of an idiotype. *Eur. J. Immunol.* **4:** 296.

Eichmann, K., and K. Rajewsky. 1975. Induction of T and B cell immunity by anti-idiotypic antibody. *Eur. J. Immunol.* **5:** 661.

Eichmann, K., A. Coutinho, and F. Melchers. 1977. Absolute frequencies of LPS-reactive B-cells producing A5A-idiotype in unprimed, strep. A-carbohydrate-primed, anti-A5A-idiotype-sensitized and anti-A5A-idiotype-suppressed A/J-mice. *J. Exp. Med.* **146:** 1436.

Gronowicz, E., and A. Coutinho. 1974. Selective triggering of B cell subpopulations by mitogens. *Eur. J. Immunol.* **4:** 771.

Gronowicz, E., A. Coutinho, and F. Melchers. 1976. A plaque assay for all cells secreting Ig of a given type or class. *Eur. J. Immunol.* **6:** 588.

Hart, D. A., A.-L. Wang, L. L. Pawlak, and A. Nisonoff. 1972. Suppression of idiotypic specificities in adult mice by administration of antiidiotypic antibody. *J. Exp. Med.* **135:** 1293.

Jerne, N. K. 1971. The somatic generation of immune recognition. *Eur. J. Immunol.* **1:** 1.

———. 1974. Towards a network theory of the immune system. *Ann. Immunol.* (Inst. Past.) **125C:** 373.

Kearney, J. F., M. D. Cooper, and A. R. Lawton. 1976. B-lymphocyte differentiation induced by lipopolysaccharide. III. Suppression of B cell maturation by anti-mouse immunoglobulin antibodies. *J. Immunol.* **116:** 1664.

Kearney, J. F., J. Klein, D. E. Bockman, M. D. Cooper, and A. L. Lawton. 1978. B-cell differentiation induced by lipopolysaccharide. V. Suppression of plasma cell maturation by anti μ: Mode of action and characteristics of suppressed cells. *J. Immunol.* (In press.)

Krawinkel, U., M. Cramer, C. Berek, G. Hämmerling, S. J. Black, K. Rajewsky, and K. Eichmann. 1977. On the structure of the T-cell receptor for antigen. *Cold Spring Harbor Symp. Quant. Biol.* **41:** 285.

Lernhardt, W., J. Andersson, A. Coutinho, and F. Melchers. 1977. Cloning of murine transformed cell lines in suspension culture with efficiencies near 100%. *Exp. Cell Res.* **111:** 309.

Melchers, F. 1977a. B lymphocyte development in fetal liver. I. Development of reactivities to B cell mitogens *in vivo* and *in vitro*. *Eur. J. Immunol.* **7:** 476.

———. 1977b. B lymphocyte development in fetal liver. II. Frequencies of precursor B-cells during gestation. *Eur. J. Immunol.* **7:** 482.

Melchers, F., and J. Andersson. 1973. Synthesis, surface deposition and secretion of immunoglobulin M in bone marrow-derived lymphocytes before and after mitogenic stimulation. *Transplant. Rev.* **14:** 76.

———. 1974a. IgM in bone marrow-derived lymphocytes. Changes in synthesis, turnover and secretion, and in number of molecules on the surface of B cells after mitogenic stimulation. *Eur. J. Immunol.* **4:** 181.

———. 1974b. Early changes in immunoglobulin M synthesis after mitogenic stimulation of bone marrow derived lymphocytes. *Biochemistry* **13:** 4645.

———. 1974c. The kinetics of proliferation and maturation of mitogen-activated bone marrow-derived lymphocytes. *Eur. J. Immunol.* **4:** 687.

Melchers, F., and B. Pernis. 1977. Membranes and signalling. Report on Workshop 17. In *Immune system: Genetics and regulation* (eds. E. E. Sercarz, L. A. Herzenberg, and C. F. Fox), p. 747. Academic Press, New York.

Melchers, F., J. Andersson, and R. A. Phillips. 1977. Ontogeny of murine B lymphocytes: Development of Ig synthesis and of reactivities to mitogens and to anti-Ig-antibodies. *Cold Spring Harbor Symp. Quant. Biol.* **41:** 147.

Melchers, F., H. von Boehmer, and R. A. Phillips. 1975a. B-lymphocyte subpopulations in the mouse. Organ distribution and ontogeny of immunoglobulin-synthesizing and of mitogen-reactive cells. *Transplant. Rev.* **25:** 26.

Melchers, F., A. Coutinho, G. Heimrich, and J. Andersson. 1975b. Continuous growth of mitogen-reactive B-lymphocytes. *Scand. J. Immunol.* **4:** 853.

Metcalf, D., N. L. Warner, G. J. V. Nossal, J. F. A. P. Miller, K. Shortman, and E. Rabellino. 1975. Growth of B lymphocyte colonies *in vitro* from mouse lymphoid organs. *Nature* **255:** 630.

Miller, R. G., and R. A. Phillips. 1969. Separation of cells by velocity sedimentation. *J. Cell. Physiol.* **73:** 191.

Nilsson, B. S., B. M. Sultzer, and W. W. Bullock. 1973. PPD tuberculin induces immunoglobin production in normal mouse spleen cells. *J. Exp. Med.* **137:** 127.

Owen, J. J. T., R. K. Jordan, J. H. Robinson, U. Singh, and H. N. A. Willcox. 1977. In vitro studies on the generation of lymphocyte diversity. *Cold Spring Harbor Symp. Quant. Biol.* **41:** 129.

Phillips, R. A., and F. Melchers. 1976. Appearance of functional lymphocytes in fetal liver. *J. Immunol.* **117:** 1099.

Phillips, R. A., F. Melchers, and R. G. Miller. 1978. Stem cells and the ontogeny of B-lymphocytes. *Prog. Immunol.* **3:** (In press.)

Raff, M. C. 1977. Development and modulation of B lymphocytes: Studies on newly formed B cells and their putative precursors in the hemopoietic tissues of mice. *Cold Spring Harbor Symp. Quant. Biol.* **41:** 159.

Tonegawa, S., C. Brack, N. Hozumi, and V. Pirrotta. 1978. Organization of immunoglobulin genes. *Cold Spring Harbor Symp. Quant. Biol.* **42:** 921.

Watson, J., and R. Riblet. 1975. Genetic control of responses to bacterial lipopolysaccharides in mice. II. A gene that influences a membrane component involved in the activation of bone marrow-derived lymphocytes by lipopolysaccharides. *J. Immunol.* **114:** 1462.

Murine IgD and B-Lymphocyte Differentiation

J. W. Uhr, J. C. Cambier, F. S. Ligler, J. Kettman, and E. S. Vitetta

Department of Microbiology, University of Texas Southwestern Medical School
Dallas, Texas 75235

I. Zan-Bar and S. Strober

Department of Medicine, Stanford University School of Medicine
Stanford, California 94305

Murine IgD is of particular interest because it is the first immunoglobulin (Ig) to be described that exists solely in association with B lymphocytes. It presumably acts as an antigen-specific receptor that regulates the behavior of cells that will eventually synthesize and secrete other classes of Ig (isotypes).

Murine IgD was discovered in the course of experiments in which the lactoperoxidase technique was used for iodination of murine splenocytes (Melcher et al. 1974) and the reduced anti-Ig immunoprecipitates of NP40 lysates were analyzed by sodium-dodecyl-sulfate–polyacrylamide-gel electrophoresis (SDS-PAGE). Of the two heavy chain peaks observed, one peak co-electrophorsed with μ chain, whereas the other had a slightly faster mobility (Fig. 1). In its native state, the Ig class that gave rise to this latter peak was associated with light chain, but the Ig did not have the antigenicity of the known classes of mouse-serum Ig (IgG, IgA, and IgM). Additional biochemical and biological findings suggested that this new class of murine Ig was the counterpart of human IgD (Abney and Parkhouse 1974; Melcher et al. 1974).

We (Vitetta and Uhr 1975) therefore proposed a model that placed IgD into a scheme of B-cell differentiation and speculated about its possible function on the surface of B lymphocytes (Fig. 2). There are three major features to the model. (1) IgD is acquired after IgM in antigen-independent B-cell differentiation. Thus, cells bearing only IgM (μ^+) acquire IgD, thereby expressing two isotypes ($\mu^+\delta^+$) presumably with the same specificity for antigen. Some of the $\mu^+\delta^+$ cells then lose surface IgM to become cells that bear IgD only (δ^+). (2) It was proposed that cells that bear only IgM are tolerized upon interaction with antigen, whereas those that have acquired IgD are "triggered" by interaction with antigen. Thus, the acquisition of IgD changes a cell from one that is easily tolerized to one that is readily triggered. We therefore suggested that IgM plays a critical role in the development of immunologic tolerance to self-macromolecules. (3) We

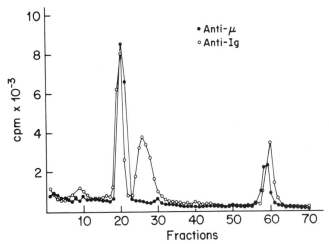

Figure 1
Cell-surface IgM and IgD on splenocytes of 25-day-old germ-free mice. Separate aliquots of the lysate from radioiodinated cells were immunoprecipitated either with anti-μ or anti-Ig. The precipitates were dissolved, reduced, and electrophoresed on 7.5% SDS-acrylamide gels for 17 hours (Vitetta et al. 1975).

judged the evidence claiming IgG as receptor to be unconvincing and suggested that cells bearing only IgD act as memory cells for the secondary IgG response.

We have tested the model in the following ways: separation of subsets of B cells based on the cell-surface isotype and characterization of their immune function, and determination of the roles of IgM and IgD in induction of immunologic tolerance and in triggering.

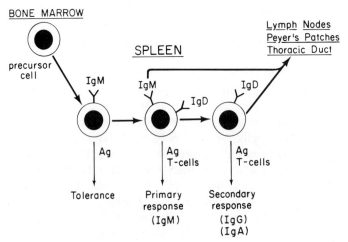

Figure 2
Model of B cell differentiation and function (Vitetta and Uhr 1975).

Function of Subsets of Cells Positively Selected by Virtue of Cell-Surface Isotype

To study the role of δ^+ cells in the immune response, we had to produce an anti-δ serum. Using a modification of the procedure described by Abney et al. (1976), we prepared a rabbit serum against mouse δ chain (Zan-Bar et al. 1977a). This serum was shown to be monospecific by the following criteria: (1) absence of heavy-chain peaks other than the δ peak on SDS-PAGE using immunoprecipitates from iodinated splenocytes; (2) immunofluorescent staining of appropriate numbers of adult spleen and lymph node cells; absence of staining of thymus and neonatal spleen; (3) independent capping of IgD and IgM on the same cell; and (4) absence of staining of adult splenocytes after treatment with either anti-κ or allotypic anti-δ (Goding et al. 1976) under capping conditions. This latter criterion is a critical one because it virtually excludes the possibility that the heterologous anti-δ contains other antibody specificities to unknown antigens that appear on mature B cells.

Using anti-δ sera that satisfy these criteria, Zan-Bar et al. (1977a, b) stained cells by an indirect fluorescence assay, then sorted positive and negative cells on the fluorescence-activated cell sorter (FACS). The major features of the protocol were: (1) the use of spleen cells either from virgin or primed animals for assessment of the cell populations responsible both for the primary and secondary antibody responses, and (2) the evaluation of specific antibody in the serum after adoptive transfer of sorted cells. Figure 3 shows the protocol for the evaluation of primed B cells.

The results described in Table 1 and Figure 4 demonstrate that γ^+ cells elaborate only IgG responses, whereas μ^+ and δ^+ cells elaborate both IgM and IgG responses. The early antibody response restored by purified populations of μ^+ or δ^+ cells obtained from the FACS was predominantly IgM whether primed or virgin B cells (data not shown) were selected. This finding indicates that the major precursor for the IgM response is a $\mu^+\delta^+$ cell. Coffman and Cohn (1977) have reached similar conclusions regarding the nature of the precursors of the primary IgM and secondary IgG responses from the results they obtained by using radioactive antigen-induced suicide coupled with prior treatment (protection) with antibody specific for various isotypes.

To determine the function of the cells bearing only IgM, IgD, or IgG, we needed a protocol involving two immunofluorescent stainings (I. Zan-Bar et al., in prep.). To prepare cells that bear predominantly IgM (μ_p^+), γ^+ and δ^+ spleen cells were removed by sorting on the FACS. Residual cells were used as the source of μ_p^+ cells. A similar procedure was employed to prepare δ_p^+ and γ_p^+ cells. The cells isolated in this protocol stain positively for only one isotype at the discriminator settings chosen to correspond to detection by the naked eye. We emphasize that such cells may bear trace amounts of other isotypes; hence, we have used the term isotype-predominant as exemplified by μ_p^+, etc.

The results of adoptive transfer experiments in which these cells were used are also summarized in Figure 4 (I. Zan-Bar 1978). μ_p^+ cells generate all responses, presumably by differentiating in the adoptively transferred host. In contrast, both δ_p^+ and γ_p^+ cells generate only IgG antibody.

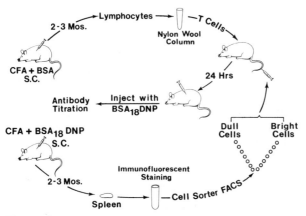

Figure 3
Experimental protocol used to assess the function of cells prepared on the cell sorter. Spleen cells from DNP-bovine serum albumin (BSA) primed mice were stained for surface IgM, IgG, or IgD. Bright and dull cells were separated on the FACS. The separated cell populations were injected with purified nylon wool T cells from BSA-primed mice into irradiated mice. The recipients were injected intraperitoneally with BSA-DNP in saline and the serum antibody was assayed. Two sorting experiments were carried out for each Ig isotype. In each experiment, a group of mice received a given dose of bright or dull cells (Zan-Bar et al. 1977a).

Table 1
Class of Anti-BSA Antibody Produced by Subpopulations of B Cells Selected on the FACS

		Percentage of response	
Antibody response	Donor cell isotype	IgM	IgG
Primary	μ^+	87	13
	μ_p^+	87	13
	δ^+	87	13
	δ_p^+	0	0
	γ^+	0	100
	γ_p^+	0	100
Secondary	μ^+	98	2
	μ_p^+	90	10
	δ^+	92	8
	δ_p^+	5	95
	γ^+	0	100
	γ_p^+	0	100

Data from Zan-Bar et al. (1977 a, b).

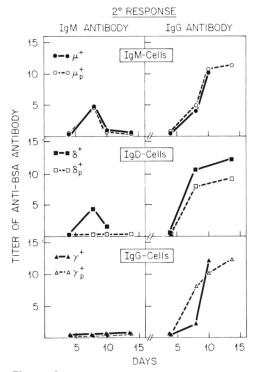

Figure 4
(*A*) IgM and (*B*) IgG anti-BSA responses elaborated by sorted cells. Splenocytes from mice primed previously with DNP-BSA were sorted for the isotypes indicated, adoptively transferred into irradiated syngeneic recipients, and the animals challenged with antigen as described in Figure 3. The mean antibody titer (log$_2$ anti-BSA) has been plotted for sera obtained from each animal 3–14 days after antigenic challenge (I. Zan-Bar et al., in prep.).

These studies, therefore, confirm the existence of the δ_p^+ memory cell as a precursor for the IgG secondary response and indicate that when the $\mu^+\delta^+$ cells becomes a δ_p^+ cell, that cell can no longer give rise to IgM secretion. The relationship between the γ_p^+ cell and the δ_p^+ cell is not known. The simplest relationship, consistent with our data, is that δ_p^+ is a precursor of γ_p^+. If so, there could be an intermediate cell with the phenotype $\delta^+\gamma^+$. The finding by Pernis (1975) that in vivo administration of anti-δ can markedly stimulate IgG synthesis in monkeys is consistent with the above observations.

INDUCTION OF TOLERANCE IN SPLENOCYTES FROM ADULT AND NEONATAL MICE

As discussed above, we hypothesized that acquisition of IgD on a μ^+ cell rendered it resistant to induction of tolerance. Because neonatal splenocytes lack IgD, it could be predicted that they would be readily tolerized in contrast to splenocytes from adult mice.

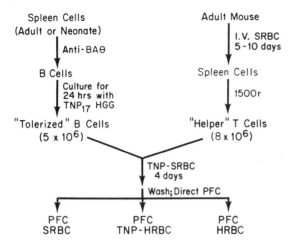

Figure 5
Protocol for the induction and assay of tolerance in murine splenic B cells. HRBC, horse red blood cells. Note that in some experiments cells were cultured with TNP-*Brucella* instead of TNP-SRBC (not shown) (Cambier et al. 1976).

This prediction was studied by an in vitro tolerance system shown in Figure 5. In this system, B-cell function is measured since excess T help has been provided, macrophages have been shown to be in excess, and a role for suppressor T cells has been excluded (Cambier et al. 1977a). Adult mouse spleen cells are exposed to trinitrophenyl $(TNP)_{17}HgG$, a tolerogen, for 24 hours before being immunized with TNP-sheep red blood cells (SRBC) in Mishell-Dutton cultures.

As shown in Figure 6A, antigen-specific suppression of the TNP response is observed in cell populations from adult mice previously exposed to $TNP_{17}HgG$ at concentrations greater than 1 μg/ml. TNP-SRBC is a thymic dependent (TD) antigen. However, in parallel cultures immunized with TNP-*Brucella*, a thymic-independent (TI) antigen, antigen-specific suppression of the TNP response was observed at much lower tolerogen doses, with 50% suppression occurring in populations previously exposed to approximately 0.01 μg $TNP_{17}HgG$.

When the identical protocol was used with spleen cells from 6- 7-day-old neonates, different results were observed (Fig. 6B). In cultures immunized either with TNP-SRBC or TNP-*Brucella*, suppression was seen at tolerogen concentrations approximately equal to those causing suppression of adult cells responsive to TNP-*Brucella* and less than 0.1% of those which suppress adult cells responsive to TNP-SRBC.

The findings indicate that neonatal B cells are far more susceptible than are adult B cells to induction of tolerance as determined by challenge with TD antigens. These findings suggest that clonal deletion of B cells occurs in early life and is a major mechanism for development of self-tolerance (Nossal and Pike 1975).

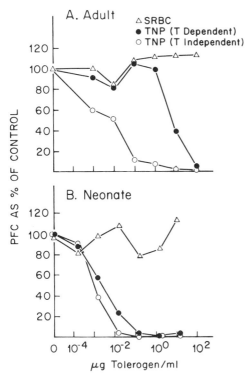

Figure 6
Effect of the dose of tolerogen on the induction of tolerance in (*A*) adult and (*B*) neonatal splenic B cells responsive to TD (TNP-SRBC) and TI (TNP-*Brucella*) forms of the TNP determinant (Cambier et al. 1977b).

The results also indicate that, whereas neonatal B cells responsive to TD antigens are hypersusceptible to tolerance induction as compared to adult B cells, B cells both from adults and neonates that are reactive to a TI form of the antigen display the same level of susceptibility. Moreover, this level of susceptibility is similar to that of neonatal responders to a TD antigen, i.e., the precursor cells are markedly susceptible to induction of tolerance. The results are consistent with earlier observations of Siskind et al. (1963) and Howard and Hale (1976); these workers, using pneumococcal polysaccharides, found that newborn mice are no more susceptible than are adults to tolerance induction. Pneumococcal polysaccharides are now known to be TI antigens.

The difference in susceptibility to tolerance induction between cells responsive to TD and TI forms of the antigen in adults can be explained by postulating that the two forms of antigen stimulate different B-cell populations: the TD antigen stimulates $\mu^+\delta^+$ and the TI antigen, μ_p^+. This concept is supported by the demonstration (Gorczynski and Feldman 1975) that the precursors stimulated by TI antigens to produce IgM plaque-forming cells (PFC) are, on the average, larger than IgM PFC-precursors stimulated by TD antigens. In addition, TI and TD responses to a given antigen have been shown to be additive (Jennings and Rittenberg 1976).

EFFECT OF PAPAIN ON TOLERANCE SUSCEPTIBILITY AND THE RELATIONSHIP TO REMOVAL OF SURFACE IgD

The preceding data identified a marked difference in tolerance susceptibility between adult and neonatal cells but provided no direct evidence that the difference was due to receptor IgD present on adult but not on neonatal cells. Prior studies had established that treatment of murine splenocytes with an appropriate concentration of papain results in cleavage of cell surface IgD with little or no removal of cell surface IgM (Cambier et al. 1977c) or of four other surface molecules. This marked susceptibility of IgD to proteolysis was exploited in experiments which involved treatment of splenocytes with papain and evaluation of the extent of removal of membrane IgD. The susceptibility of such treated cells to tolerance induction was then determined.

Adult splenocytes were treated with 0–100 μg/ml papain, washed, and

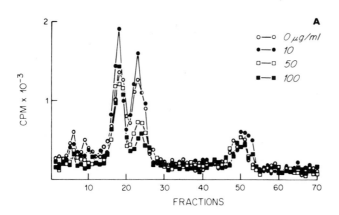

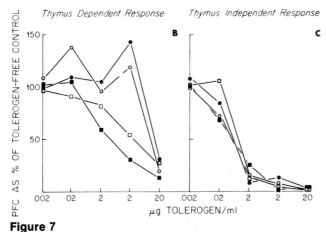

Figure 7
(*A*) SDS-PAGE of cell surface Ig from murine splenocytes treated with varying concentrations of papain. (*B*) Dose responses to tolerogen of papain-treated murine splenocytes responsive to TD immunogen. (*C*) Dose responses to tolerogen of papain treated murine splenocytes responsive to TI immunogen. Control responses were similar with or without prior papain treatment (Cambier et al. 1977c).

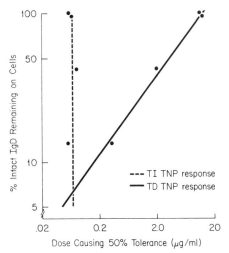

Figure 8
Relationship of the percentage of IgD remaining on murine splenocytes after papain treatment to (●) the tolerogen dose required to suppress plaque responses 50%. (——) TD response regression line, r = 0.98. (-----) TI regression line, r = 0.66 (Cambier et al. 1977c).

either radioiodinated or cultured for 24 hours with varying doses of tolerogen ($TNP_{17}HgG$). After cultivation, cells were washed and further incubated with TNP either on SRBC or *Brucella*. In the former case, SRBC-primed irradiated splenocytes were added as a source of helper T cells. As increasing amounts of δ were cleaved by papain (Fig. 7A), tolerance susceptibility of cells responsive to TNP on the TD carrier increased proportionately (Fig. 7B). Susceptibility to tolerance of cells responsive to TNP on the TI carrier remained at a constant high level regardless of the amount of IgD that had been cleaved (Fig. 7C).

Figure 8 summarizes five experiments. When the data are plotted on a log-log scale, there appears to be a direct relationship between the percentage of intact IgD remaining on the cells and the amount of tolerogen required to suppress the response to TNP-SRBC (TD) antigen by 50% (T_{50}). As described above, removal of IgD has no effect on susceptibility to induction of tolerance of TI antigen responsive cells.

These results suggest that acquisition of IgD confers resistance to induction of tolerance upon TD antigen-reactive cells but not TI antigen-reactive cells.

EFFECT OF ANTI-IG REAGENTS ON THE INDUCTION OF IMMUNE TOLERANCE OF SPLENOCYTES FROM ADULT MICE

Although the preceding experiments provided strong evidence implicating IgD as being responsible for the difference in tolerance susceptibility between adult and neonatal cells, it could be argued that papain was removing another molecule highly susceptible to proteolysis and important in tolerance induction. Therefore, we removed the IgD by capping with specific anti-δ and assessed the effect on tolerance induction (Vitetta et al. 1977).

The optimal conditions for modulating surface IgM and IgD on B cells from mouse spleens were determined by immunofluorescence studies. Figure 9 shows a representative experiment designed to determine the susceptibility of splenic B cells to tolerance induction after they have been treated with effective concentrations of antiserum. In Figure 9A, only anti-δ increased the susceptibility of the TD responders to tolerance induction. Addition of IgM and IgG to the anti-δ serum did not abrogate this effect. In contrast, anti-δ had no effect on the tolerance susceptibility of TI responders (Fig. 9B). The dose-response curve of TD responsive precursors to tolerogen after treatment with anti-δ resembled the dose-response curve for untreated TI responders.

Treatment with anti-μ markedly decreased the tolerance susceptibility of TI responders and also decreased slightly the tolerance susceptibility of TD responders (as judged by the 50% tolerance level).

The present studies confirm the earlier results in which papain was used (Cambier et al. 1977c); these data showed that removal of IgD increases susceptibility of treated TD responders but has no effect on TI responders. In addition, it was shown that treatment with anti-μ does not similarly increase the tolerance susceptibility of TD responders. This is a critical point because it could be argued that diminishing the concentration of either isotype would result in an increased susceptibility to tolerance induction. The present studies indicate that different reactivities are conferred on a $\mu^+\delta^+$ cell by IgM and IgD with regard to tolerance induction, i.e., these two isotypes can be responsible for conveying different signals to the same cell.

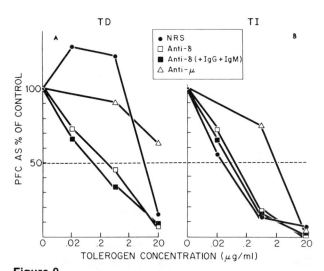

Figure 9
The effect of removing IgM or IgD receptors on the ability to tolerize splenic B cells responding to (A) TD and (B) TI forms of TNP immunogen. The tolerogen was $TNP_{17}HgG$. The control SRBC-PFC were unaffected when the dose of tolerogen was varied. Each point represents the average of the responses of duplicate cultures (Vitetta et al. 1977).

The absence of a similar effect of anti-δ on TI responders argues that they lack IgD or that, if it is present, it does not determine tolerance susceptibility.

It was also observed that treatment of B cells with anti-μ decreased the susceptibility of TI responders to tolerance induction. This finding implies that interaction of tolerogen with IgM receptors is a necessary event for tolerance induction of TI responders and that removing such receptors by capping with anti-μ prevents this interaction. It is provocative that the capping induced with anti-μ antibody itself did not give a tolerogenic signal.

EFFECT OF ANTI-IG REAGENTS ON THE IMMUNE RESPONSIVENESS OF ADULT SPLENOCYTES

In the final series of experiments, we studied the effect of anti-δ on immune responsiveness (Cambier et al. 1978). Splenocytes were treated with anti-μ, anti-δ, or anti-Ig followed by goat antirabbit Ig (GARIg) in amounts that reduced by 75–95% the number of cells bearing detectable quantities of the respective isotype(s).

Increasing concentrations (1–1000 μg/ml) of each Ig preparation were then added to cultures containing cells capped with the same antibody and either the TD or TI antigen (Fig. 10). The primary TD responses to TNP and SRBC were virtually abolished by anti-μ, anti-δ, or anti-Ig at Ig concentrations of 100–1000 μg/ml. The TI response to TNP, however, was blocked by anti-μ and anti-Ig at Ig concentrations greater than 100 μg/ml but was unaffected by anti-δ, even at 1000 μg Ig. These results indicate that both IgM and IgD receptors are required for activation of responsive cells to the TD antigen TNP-SRBC and that IgM but not IgD is required for a response to TNP on the TI carrier *Brucella*.

To show that the inhibition of the TD and TI response seen in Figure 10 was due solely to specific anti-Ig activity in the reagents, we added a mixture of IgM and IgG to the cultures that contained 400 μg/ml specific antibody. The inhibition of the TD and TI response by anti-μ or anti-Ig was abolished, whereas the inhibition of the TD response by anti-δ was unaffected. These results not only eliminate the possibility that the inhibition with anti-μ or anti-Ig was a nonspecific effect of putative antimembrane antibodies in the Ig preparations but also assure that the inhibition of the TD response by anti-δ was not due to contamination of the anti-δ preparation by anti-μ or anti-light-chain antibodies.

The results of the triggering studies contrast with the results of the tolerance studies which indicate that IgD plays a critical role in resistance in tolerance induction. In the tolerance studies, treatment with anti-μ or anti-δ revealed that these two isotype receptors give different signals to the cell: the IgD on the $\mu^+\delta^+$ cell appears unable to give a tolerance signal, whereas the IgM can give such a signal. In the triggering studies, no differences were observed in the two signals given by IgM and IgD; that is, both are necessary for triggering. The possibility has not been excluded that, for responses to TD antigens, IgD gives the triggering signal but IgM is essential to amplify this signal to threshold levels.

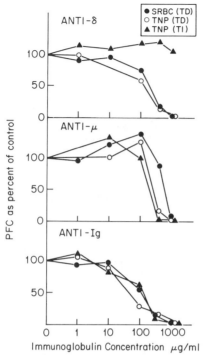

Figure 10
Effect of antibody concentration on the inhibition of in vitro IgM responses to TD and TI antigens. Cells were treated with the indicated antiserum and GARIg under capping conditions and cultured with an Ig fraction of the same antiserum used for capping (Cambier et al. 1978).

CONCLUDING DISCUSSION

In sum, the preceding data have tested several major features of the original model of IgD in B-cell differentiation and function. It is clear that membrane IgD develops late in antigen-dependent B-cell differentiation; the sequence appears to be μ^+, $\mu^+\delta^+$, and δ_p^+. The IgM receptor alone plays a critical role in the response to TI antigens, and the IgM and IgD receptors both are involved in the response to TD antigens. In contrast to the original model, γ^+ cells are present and are involved in producing only IgG responses to TD antigens. What is known about the relationships among subsets, in particular with respect to immunologic memory, is highly tentative and will certainly require frequent revisions as additional data are obtained.

Probably the most striking findings to arise from the present experiments are those that indicate that membrane IgD plays a critical role in the signal discrimination between tolerance induction and triggering of $\mu^+\delta^+$ cells. This observation represents an important clue into the mechanisms underlying development of tolerance to self-proteins. The finding that both isotypes are needed to trigger a $\mu^+\delta^+$ cell suggests that three signals may be needed for stimulation, since T help is essential. These data, therefore, present new insights into how ligands can deliver positive and negative signals to B lymphocytes and represent systems that are amenable to further analysis at the molecular level.

Acknowledgments

We thank Mr. F. Assisi, Ms. M. Bagby, Mr. Y. Chinn, Ms. S. Diase, Ms. C. Dos, Mr. M. Knapp, Mr. S. Lin, Ms. M. Neale, and Ms. G. Sloane for expert technical assistance. We are indebted to Dr. L. A. Herzenberg for his advice concerning the FACS and his gift of reagents. We thank Ms. J. Hahn for her excellent secretarial assistance.

This work was supported by National Institutes of Health Grants AI-11851, AI-12789, and AI-10293 and by American Cancer Society Grant IM-63. I.Z.-B. is a fellow of the Arthritis Foundation. J.C.C. is supported by NIH Postdoctoral Grant AI-05021. S.S. is an Investigator of the Howard Hughes Medical Institute.

REFERENCES

Abney, E. R., and R. M. E. Parkhouse. 1974. Candidate for immunoglobulin D present on murine B lymphocytes. *Nature* 525: 600.

Abney, E. R., I. R. Hunter, and R. M. E. Parkhouse. 1976. Preparation and characterisation of an antiserum to the mouse candidate for immunoglobulin D. *Nature* 259: 404.

Cambier, J. C., J. R. Kettman, E. S. Vitetta, and J. W. Uhr. 1976. Differential susceptibility of neonatal and adult murine spleen cells to in vitro induction of B-cell tolerance. *J. Exp. Med.* 144: 293.

Cambier, J. C., J. W. Uhr, J. R. Kettman, and E. S. Vitetta. 1977a. B cell tolerance. I. Analysis of hapten-specific unresponsiveness induced *in vitro* in adult and neonatal murine spleen cell populations. *J. Immunol.* 119: 2054.

Cambier, J. C., E. S. Vitetta, J. W. Uhr, and J. R. Kettman. 1977b. B-cell tolerance. II. Trinitrophenyl human gamma globulin-induced tolerance in adult and neonatal murine B cells responsive to thymus-dependent and independent forms of the same hapten. *J. Exp. Med.* 145: 778.

Cambier, J. C., F. S. Ligler, J. W. Uhr, J. R. Kettman, and E. S. Vitetta. 1978. Blocking of primary *in vitro* antibody responses to thymus-independent and thymus dependent antigens with antiserum specific for IgM or IgD. *Proc. Natl. Acad. Sci. U.S.A.* 75: 432.

Cambier, J. C., E. S. Vitetta, J. R. Kettman, G. M. Wetzel, and J. W. Uhr. 1977c. B-cell tolerance. III. Effect of papain-mediated cleavage of cell surface IgD on tolerance susceptibility of murine B cells. *J. Exp. Med.* 146: 107.

Coffman, R. L., and M. Cohn. 1977. The class of surface immunoglobulin on virgin and memory B lymphocytes. *J. Immunol.* 118: 1806.

Goding, J. W., G. W. Warr, and N. L. Warner. 1976. Genetic polymorphism of IgD-like cell surface immunoglobulin in the mouse. *Proc. Natl. Acad. Sci. U.S.A.* 73: 1305.

Gorczynski, R. M., and M. Feldmann. 1975. B cell heterogeneity—difference in the size of B lymphocytes responding to T dependent and T independent antigens. *Cell. Immunol.* 18: 88.

Howard, J. G., and C. Hale. 1976. Lack of neonatal susceptibility to induction of tolerance by polysaccharide antigens. *Eur. J. Immunol.* 6: 486.

Jennings, J. J., and M. B. Rittenberg. 1976. Evidence for separate subpopulations of B cells responding to T-independent and T-dependent immunogens. *J. Immunol.* **117**: 1749.

Melcher, U., E. S. Vitetta, M. McWilliams, M. E. Lamm, J. M. Phillips-Quagliata, and J. W. Uhr. 1974. Cell surface immunoglobulin. X. Identification of an IgD-like molecule on the surface of murine splenocytes. *J. Exp. Med.* **140**: 1427.

Nossal, G. J. V., and B. L. Pike. 1975. Evidence for the clonal abortion theory of B-lymphocyte tolerance. *J. Exp. Med.* **141**: 904.

Pernis, B. 1975. The effect of anti-IgD antiserum on antibody production in rhesus monkeys. In *Membrane receptors of lymphocytes* (eds. M. Seligmann et al.), p. 25. North-Holland Publishing Co., Amsterdam.

Siskind, G. W., P. Y. Paterson, and L. Thomas. 1963. Induction of unresponsiveness and immunity in newborn and adult mice with pneumococcal polysaccharide. *J. Immunol.* **90**: 929.

Vitetta, E. S., and J. W. Uhr. 1975. Immunoglobulin-receptors revisited. *Science* **189**: 964.

Vitetta, E. S., J. C. Cambier, F. S. Ligler, J. R. Kettman, and J. W. Uhr. 1977. B cell tolerance. IV. Differential role of surface IgM and IgD in determining tolerance susceptibility of murine B cells. *J. Exp. Med.* **146**: 1804.

Vitetta, E. S., U. Melcher, M. McWilliams, M. Lamm, J. Phillips-Quagliata, and J. W. Uhr. 1975. Cell surface immunoglobulin. XI. The appearance of an IgD-like molecule on murine lymphoid cells during ontogeny. *J. Exp. Med.* **141**: 206.

Zan-Bar, I., S. Strober, and E. S. Vitetta. 1977a. The relationship between surface immunoglobulin isotype and immune function of murine B lymphocytes. I. Surface immunoglobulin isotypes on primed B cells in the spleen. *J. Exp. Med.* **145**: 1188.

Zan-Bar, I., E. S. Vitetta, and S. Strober. 1977b. The relationship between surface immunoglobulin isotype and immune function of murine B lymphocytes. II. Surface immunoglobulin isotypes on unprimed B cells in the spleen. *J. Exp. Med.* **145**: 1206.

Zan-Bar, I., E. S. Vitetta, F. Assisi, and S. Strober. 1978. The relationship between surface immunoglobin isotype and immune function of murine B lymphocytes. III. Expression of a single predominant isotype on primed and unprimed B cells. *J. Exp. Med.* (In press.)

Differentiation of Precursor Cytotoxic T Lymphocytes Following Alloantigenic Stimulation

G. Sundharadas, M. L. Sopori, C. E. Hayes, P. R. Narayanan, B. J. Alter, M. L. Bach, and F. H. Bach

University of Wisconsin, Immunobiology Research Center and
Departments of Medical Microbiology, Pediatrics, Medical Genetics, and Surgery
Madison, Wisconsin 53706

Cytotoxic T lymphocytes (T_c) develop in a mixed leukocyte culture (MLC) in vitro when the stimulating (sensitizing) and responder cells differ for certain antigens of the major histocompatibility complex (Bach et al. 1976). The antigens that are recognized by precursor T_c and that are used eventually as the target antigens when the active T_c lyse target cells are referred to as cytotoxic-defined (CD) antigens (Bach et al. 1977).

Although there is some evidence in the literature that might suggest that CD antigens alone can activate precursor T_c and lead to the development of a highly significant cytotoxic response in many, if not most, situations, a maximal cytotoxic response is generated when precursor T_c are stimulated not only with the allogeneic CD determinants but also by the "help" that they receive from what appears to be a functionally different subset of T cells, the helper T (T_h) cells. The T_h cells in turn are activated by LD or LD-like antigens of the major histocompatibility complex (Bach et al. 1976).

Following the model introduced by Bretscher and Cohn (1970), one refers to the signal received by the precursor T_c when it recognizes an allogeneic CD antigen as signal 1 and to the signal received in the form of help to the precursor or developing T_c as signal 2 (Bach et al. 1976). It is our purpose in this paper to analyze further requirements for the various signals that lead to an optimal cytotoxic response. Because data relating to the generation of suppressor T (T_s) lymphocytes have been published recently by Sondel et al. (1977), we shall comment on possible relationships between the development of T_c and suppressor T lymphocytes.

METHODS

The methods used in this study are primarily those of the mixed-leukocyte-culture assay and the cell-mediated lympholysis assay, which have been described elsewhere (Peck and Bach 1975). In brief, allogeneic cells of two mouse strains differing either by the entire H-2 complex or by only certain regions thereof were mixed in an MLC and allowed to incubate from 4 to 5 days. At the end of this time, one could measure either the proliferative response (commonly referred to as the MLC response) by studying the incorporation of tritiated thymidine ($[^3H]$-TdR) into the responding cells or the cytotoxic capability of the cells present in the MLC on that day. Cytotoxic activity was measured in the cell-mediated lympholysis assay in which the effector cells that had been generated in the MLC were tested for their ability to lyse the sodium chromate $[^{51}Cr]$-labeled target cells carrying antigens to which the T_c had been sensitized. Results of the cell-mediated lympholysis assay were expressed as percentage of cytotoxicity.

The following mouse strains were used in this study. The letters in parentheses refer to the genotype derivation of the different regions of the H-2, including K, I, S, and D: B10.A (kkdd), B10.A(1R) (kkdb), B10.T(6R) (qqqd), C57BL/6 (bbbb), C57BL/10 (bbbb), B10.G (qqqq), B10.D2 (dddd), B10.S (ssss), and AQR (qkdd).

RESULTS

Expansion of T_c Development by a Factor Present in Supernatants Derived from MLC

As demonstrated by Altman and Cohen (1975), Plate (1976), and Sopori et al. (1977), a supernatant taken from a mixed leukocyte culture contained a "factor" that could expand a weak, ongoing, cytotoxic response. Shown in Table 1 are the results of one experiment in which this phenomenon was demonstrated. The responding strain, B10.A, differed from the stimulating or sensitizing strain donor, B10.A(1R), by the H-2 D region. In this combination, a weak but significant level of T_c was usually generated. Although the highest level of cytotoxicity generated in this particular experiment was $7.1 \pm 4.4\%$ at an effector to target ratio of 20:1, somewhat higher levels of cytotoxicity have usually been observed in other experiments with this

Table 1
Effect of MLC Supernatant on the Generation of T_c

Test system	Effector:target ratio	Percentage cytotoxicity ± S.D. on B10.A (1R) targets		
		None	B10.A + B10.A$_x$ supernatant[a]	B10.T(6R) + C57BL/6$_x$ supernatant[a]
B10.A + B10.A(1R)$_x$ (H-2 D difference)	5:1	2.5 ± 2.2	4.3 ± 1.1	14.5 ± 4.9
	10:1	5.8 ± 1.9	6.0 ± 3.3	26.7 ± 3.3
	20:1	7.1 ± 4.4	9.8 ± 3.2	31.6 ± 7.5

[a] Supernatant was obtained from a day-7 MLC and added to the test combination at the start of the culture at a concentration of 25%.

Table 2
Specificity of T_c Generated in the Presence of MLC Supernatant

Supernatant	Percentage cytotoxicity on the targets		
	B10.A	B10.A (1R)	B10.G
− MLC	1.9 ± 1.3	16.5 ± 0.1	2.2 ± 2.1
+ MLC	3.0 ± 2.0	42.4 ± 2.1	5.6 ± 3.6

Test combination: B10.A + B10.A(1R)$_x$ (H-2 D difference). MLC supernatant from B10.T(6R) + C57BL/6$_x$ combination obtained on day 7. Other details as in Table 1.

combination. Also shown in Table 1 are the results of an addition of the supernatant from an MLC in which responding and stimulating cells were syngeneic, i.e., where no recognition of alloantigens had taken place, and the results of experiments in which the supernatant from an allogeneic MLC (B10.T(6R) + C57BL/6$_x$) was added to the test system.

The results obtained are representative of experiments of this sort in that the supernatant of the allogeneic MLC resulted in very significant expansion of the level of cytotoxic response directed at the B10.A(1R) target cells at all three effector-to-target ratios tested, whereas the supernatant from the syngeneic MLC had no effect. It should be noted that similar results could be obtained with supernatants obtained from primary MLC that were collected on day 5 to day 7 and from supernatants from secondary MLC that were collected on day 1 or 2 following rechallenge.

The data presented in Table 2 simply demonstrate that the increased response obtained by including a supernatant in the cytotoxicity-generating MLC was specific. B10.A sensitized to B10.A(1R) showed specificity of killing when tested against B10.A, B10.A(1R), or B10.G target cells both in the absence and presence of the MLC supernatant.

Initiation and Expansion of T_c Development

The data presented up to this point demonstrate that an MLC supernatant contained a "factor" that could expand a weak but ongoing cytotoxic response. Another question to be asked is the following. What were the requirements for the initiation of the cytotoxic response? One experimental system available to answer such a question is that which uses, in the T_c-generating MLC, sensitizing cells that have been treated with UV light rather than with X-irradiation or mitomycin C. UV-light treatment abolished the functional expression of the LD antigens (i.e., there was neither an MLC proliferative nor a T-helper cell response to UV-light-treated stimulating cells), while still allowing those UV-light-treated cells to express their CD antigens (Lindahl-Kiessling and Safwenberg 1971). That the CD antigens were expressed can be demonstrated in a three-cell experiment in which a helper stimulus was provided by an LD-different cell (Bach et al. 1976).

As demonstrated in Table 3, stimulation of cells of strain B10.A with B10.A(1R) X-irradiated stimulating cells resulted in 11.4% cytotoxicity

Table 3
Effect of MLC Supernatant in the Presence of UV-Light-Treated Stimulating Cells

		Percentage cytotoxicity	
Test system	H-2 difference	− MLC supernatant	+ MLC supernatant[a]
B10.A + B10.A(1R)$_X$	D	11.4 ± 2.1	35.7 ± 3.2
B10.A + B10.A(1R)$_{UV}$	D	0.2 ± 6.3	−0.9 ± 3.5
AQR + B10.A$_X$	K	8.4 ± 3.0	22.2 ± 4.7
AQR + B10.A$_{UV}$	K	−1.1 ± 1.0	2.4 ± 3.0
B10.D2 + C57BL/6$_X$	entire H-2	60.2 ± 4.7	87.1 ± 6.3
B10.D2 + C57BL/6$_{UV}$	entire H-2	1.7 ± 3.2	−4.3 ± 3.8

[a] MLC supernatant was derived from the combination B10.D2 + C57BL/6$_X$. Other details as in Table 1.

in the absence of an MLC supernatant and 35.7% cytotoxicity in the presence of the supernatant. Alternatively, sensitization of B10.A with B10.A(1R) UV-light-treated cells resulted in no cytotoxic response against the specific target either in the absence or presence of the MLC supernatant. Stimulation with these UV-light-treated cells did not result in a significant proliferative response (data not shown). Similar findings, based on a different strain combination that differed by the H-2 K region, demonstrated the inability of UV-light-treated cells to elicit a response by themselves or in the presence of an MLC supernatant. Table 3 also shows a third strain combination that differed for the entire H-2 complex. In these cases the MLC supernatant caused a highly significant expansion of the cytotoxic response generated against the D, the K, or the entire H-2 complex when X-irradiated stimulating cells were used, but it had no significant effect when UV-light-treated stimulating cells were used.

One possible conclusion to be drawn from these data is that the requirements, in terms of signals to the precursor T_c, for the initiation of a cytotoxic response were different from those for expansion of a weak but ongoing response. It is possible, of course, that the same requirements existed for initiation versus expansion but that simply more of the same kind of helper signal (provided by the MLC supernatant factor) was needed for initiation than for expansion. One might argue against such a hypothesis on the basis of the very marked effect that the MLC supernatant had on expanding a relatively weak ongoing response and its complete lack of any effect in the presence of UV-light-treated stimulating cells, but this argument is not a critical one that would rule out such a possibility.

We approached the question of whether the signals required by a precursor T_c for the initiation of the response could be mediated by a soluble factor, assuming that the simple addition of the MLC supernatant on day 0 of the T_c-generating MLC, for instance, might have been the wrong time to add that factor. Thus, the following system was devised in which there might be a continuous supply of needed factors to the precursor CTL. We referred to the system as the agar system; the experimental setup using this model has been described in greater detail elsewhere (Sopori et al.

1977). An MLC was placed in the bottom of a petri dish in which the responding and stimulating cells either were different at certain regions of H-2 or were syngeneic with one another. Placed on top of this MLC was a layer of hard agar of 3 to 3.5 mm thickness. The test MLC combination, in which the development of T_c was assayed, was placed on top of the agar. By using this experimental system, one can study the effect of any factor produced by the MLC under the agar layer on the development of T_c in the combination above the agar layer.

Results of one experiment in which the agar system was used are shown in Table 4. The test combination, on top of the agar, involved B10.D2-responding cells that either were cultured alone or were stimulated with X-irradiated or UV-light-treated sensitizing cells of C57BL/6 (different from B10.D2 by the entire H-2 complex). In line 1, C57BL/6 cells were placed under the agar to assay whether the allogeneic cells used as stimulating cells in other lines of the table could exert this stimulatory effect through the agar. The absence of cytotoxicity against C57BL/6 suggested that the stimulating "antigens" could not get through the agar to the responding cells on top of the agar. In line 2, an allogeneic mixture was present under the agar to test whether any factor produced by an allogeneic mixture on the bottom would lead to activation of the B10.D2 cells on top and thus lead to the development of T_c in the absence of stimulating cells on top of the agar. Once again no significant level of cytotoxicity resulted.

Line 3 demonstrated that if the B10.D2-responding cells on top of the agar were "stimulated" with UV-light-treated allogeneic cells, no measurable level of cytotoxicity was generated if the allogeneic MLC under the agar also used UV-light-treated cells as the "sensitizing" cells. This is to be contrasted with the results shown in line 4 where the presence of a normal allogeneic MLC (in which X-irradiated cells were used as the stimulating cells) under the agar did allow the development of a specific cytotoxic response by the MLC on top of the agar, where UV-light-treated cells were used as the sensitizing cells. These results strongly suggested that the precursor T_c must have received other signals, in addition to the CD antigen stimulus (signal 1), for the initiation and expansion of the development of T_c to have occurred. These results also suggested that a normal MLC under the agar layer was capable of producing soluble factors which could diffuse through the agar and provide the additional second signal needed for development of T_c from precursor T_c.

Table 4
Initiation of T_c Generation by Cell-Free Factor in the Agar System

Test combination (top of agar)	Cells under the agar layer	Percentage cytotoxicity ± S.D. on C57BL/6 target cells
B10.D2	C57BL/6$_X$	−6.2 ± 3.8
B10.D2	B10.D2 + C57BL/6$_X$	0.7 ± 5.2
B10.D2 + C57BL/6$_{UV}$	B10.D2 + C57BL/6$_{UV}$	−3.3 ± 3.4
B10.D2 + C57BL/6$_{UV}$	B10.D2 + C57BL/6$_X$	20.3 ± 4.0
B10.D2 + C57BL/6$_X$	B10.D2 + C57BL/6$_X$	48.5 ± 5.8

The data presented up to this point based on use of MLC supernatants as a factor that was added directly to a cytotoxicity-generating MLC or on use of the agar system suggested that the signals required to initiate as well as to expand the developing cytotoxic reaction could be of a soluble nature; however, the data left open the question of whether the signals required for initiation differed from those required for expansion.

Effects of poly A:U and LPS on the development of T_c

We used another approach to obtain further information regarding this point. Two substances known to act as adjuvants in antibody responses, poly A:poly U (poly A:U) and lipopolysaccharide (LPS) were employed (Hamaoka and Katz 1973; Armerding and Katz 1974).

The results shown in Table 5 are representative of those obtained with these two adjuvants in terms of expanding a weak cytotoxic response. In this particular instance, cytotoxicity was generated against the H-2 D-region-associated CD antigens of B10.A; in the absence either of poly A:U or LPS, low-level cytotoxicity was generated against B10.A and B10.T(6R), both of which shared the sensitizing CD antigens. In the presence of either poly A:U or LPS, the level of cytotoxicity was markedly enhanced and the resulting high level of cytotoxicity was antigen specific.

Shown in Table 6 are results demonstrating that the action of poly A:U requires the presence of adherent cells, whereas the action of LPS does not. We have also tested the effects of poly A:U and LPS in the presence of UV-light-treated sensitizing cells and compared the results with those obtained when X-irradiated stimulating cells were used. In the example shown in Table 7, strain AQR was sensitized to B10.A, i.e., to K-region-associated CD antigens. In this combination, weak but significant cytotoxicity was generated against B10.A target cells when X-irradiated allogeneic sensitizing cells were used in the absence of poly A:U and LPS. Sensitization with UV-light-treated stimulating cells alone did not lead to any development of cytotoxicity, nor did the addition of poly A:U to such a culture enhance the level of cytotoxicity seen. On the other hand, the addition of LPS to a culture in which UV-light-treated sensitizing cells were used resulted in the development of a highly specific cytotoxic response.

These results suggested that LPS and poly A:U enhanced cytotoxic responses by different mechanisms. Poly A:U may have been able only to

Table 5
Effect of Poly A:U and LPS on the Generation of T_c

Additions	Percentage cytotoxicity ± S.D. on the targets			
	B10.A (kkdd)	B10.T(6R) (qqqd)	B10.S (ssss)	C57 BL/10 (bbbb)
None	15.0 ± 6.8	12.5 ± 5.0	−8.3 ± 7.1	1.7 ± 5.7
Poly A:U, 20 µg/ml	44.2 ± 2.6	42.2 ± 1.0	−8.7 ± 5.3	−8.9 ± 9.6
LPS, 5 µg/ml	65.1 ± 4.7	66.9 ± 4.7	12.6 ± 3.7	−5.0 ± 3.9

Test MLC combination: B10.A(1R) + B10.A$_X$ (H-2 D difference).

Table 6
Effect of Adherent Cell Removal on the Action of Poly A:U and LPS

	Percentage cytotoxicity ± S.D. on B10.A targets	
Additions	+ adherent cells	− adherent cells
None	9.3 ± 4.3	1.3 ± 2.6
Poly A:U, 20 µg/ml	33.1 ± 1.8	3.0 ± 2.6
LPS, 5 µg/ml	61.9 ± 3.0	55.9 ± 3.9

Test MLC combination: B10.A(1R) + B10.A$_X$ (H-2 D difference).

Table 7
Effect of Poly A:U and LPS in the Presence of UV-Light-Treated Stimulating Cells

		Percentage cytotoxicity ± S.D. on the targets	
Test combination	Additions	B10.A (kkdd)	B10.T(6R) (qqqd)
AQR + AQR$_X$	none	−4.7 ± 2.0	−3.1 ± 2.9
AQR + B10.A$_X$	none	5.6 ± 1.0	−1.8 ± 1.6
AQR + B10.A$_{UV}$	none	−8.5 ± 3.0	−2.3 ± 3.0
AQR + B10.A$_{UV}$	poly A:U, 20 µg/ml	−3.6 ± 3.8	−1.7 ± 1.5
AQR + B10.A$_{UV}$	LPS, 5 µg/ml	23.3 ± 3.1	−2.3 ± 1.5

amplify an ongoing cytotoxic response, whereas LPS may have had the capacity to promote the initiation of a cytotoxic response as well as to amplify the response. Since both the agents were potent at expanding an ongoing response, the inability of poly A:U to help initiate a response supported the notion that the signals for initiation and expansion of the cytotoxic response may have been different. That these two agents exerted their effect at two different steps in the development of T_c was also supported by the finding that, under the in vitro conditions used, poly A:U required the presence of adherent cells for its action, whereas LPS did not.

DISCUSSION

The data presented in this paper represent an extension of earlier results that suggested that the optimal development of T_c in a number of genetic situations requires not only the presentation of the target CD antigens to the precursor T_c but also the provision of help to the developing T_c (Bach et al. 1976). The several lines of evidence discussed above now raise the possibility that the requirements for help with respect to the initiation of the response may be different from those for the expansion of the response. Although we cannot critically rule out the idea that this difference is based on simply quantitative requirements for the same helper signal, several lines of evidence would tend to speak against such an interpretation.

Although the data presented in this paper deal with the development of T_c as they respond to CD antigens and receive signal 2 — presumably from responsive helper T lymphocytes that respond to allogeneic LD antigens or from a factor presumably produced by those helper T lymphocytes, there is still a question with regard to the generation of suppressor cells in a mixed leukocyte culture.

Studies of Schendel and Bach (1975) and Sondel et al. (1975) first suggested that the presence in a mixed leukocyte culture of one set of allogeneic stimuli may "preempt" the response of those same responding cells to different alloantigens at a later time. It was the work of Sondel and his collaborators in our laboratory that demonstrated that this preemption phenomenon is in all likelihood based on the development in an MLC of suppressor cells (Sondel et al. 1977). Perhaps most important from the point of view of the model presented here in terms of activating T_c, Sondel et al. were able to show that heat-treated stimulating cells (which appear to behave in a manner identical to UV-light-treated stimulating cells in that they do not by themselves elicit a proliferative MLC response nor a cytotoxic response) do stimulate the generation of suppressor cells. Although it is possible that the antigenic determinants encoded by the major histocompatibility complex genes that evoke suppressor cell generation are different from those that are presented as CD antigens to precursor T_c, the simplest working hypothesis at present is that the CD antigens on the heat-treated or UV-light-treated stimulating cells evoke the generation of the suppressor cells, and this generation occurs when help is not present.

The model that evolved from these studies represents a modification of that which we presented previously (Bach et al. 1976); it is shown in Figure 1. First, one would have to ask whether the signal to the developing T_c (referred to in Fig. 1 as the "poised" T_c) might be different from the signal required at an earlier or later stage. Is the signal for initiation dif-

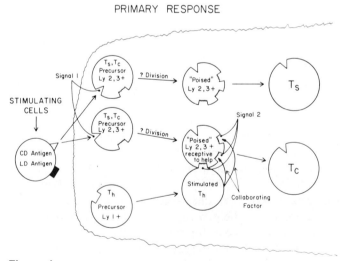

Figure 1
A model illustrating the signals involved in the development of T_c in a primary response. This represents a modification of the model presented previously (Bach et al. 1976).

ferent from that required for expansion of the response or does the signal for initiation include both the factor that expands the response plus something else? Alternatively, as we have stressed several times above, could the signals be the same but the quantitative requirements for initiation versus expansion be different? In addition, since suppressor lymphocytes can be generated in the absence of help, the question is raised as to whether the presentation of the CD antigens as signal 1 alone to the Ly 2, 3$^+$ precursor cell, or whether the presentation of the CD antigens in the presence of a factor other than the helper factor (which is yet to be defined) leads to the generation of the suppressor T lymphocyte. Whether the balance between suppressor and cytotoxic cell generation in some way involves the level of help that is given as well as the timing of that help remains the basis of current experiments.

Acknowledgments

This work was supported in part by National Institutes of Health Grants CA-16836, AI-11576, AI-08439 and by the National Foundation–March of Dimes Grants CRBS 246 and 6-76-213. This is paper no. 141 from the Immunobiology Research Center and paper no. 2189 from the Laboratory of Genetics, The University of Wisconsin, Madison.

REFERENCES

Altman, A., and I. R. Cohen. 1975. Cell-free media of mixed lymphocyte cultures augmenting sensitization *in vitro* of mouse T lymphocytes against allogenic fibroblasts. *Eur. J. Immunol.* **5:** 437.

Armerding, D., and D. H. Katz. 1974. Activation of T and B lymphocytes in vitro. I. Regulatory influence of bacterial lipopolysaccharide (LPS) on specific T-cell helper function. *J. Exp. Med.* **139:** 24.

Bach, F. H., M. L. Bach, and P. M. Sondel. 1976. Differential function of major histocompatibility complex antigens in T-lymphocyte activation. *Nature* **259:** 273.

Bach, F. H., M. L. Bach, O. J. Kuperman, H. W. Sollinger, and P. M. Sondel. 1977. Cellular recognition of major histocompatibility complex antigens. *Cold Spring Harbor Symp. Quant. Biol.* **41:** 429.

Bretscher, P., and M. Cohn. 1970. A theory of self-nonself discrimination. *Science* **169:** 1042.

Hamaoka, T., and D. H. Katz. 1973. Mechanism of adjuvant activity of poly A:U on antibody responses to hapten-carrier conjugates. *Cell Immunol.* **7:** 246.

Lindahl-Kiessling, K., and J. Safwenberg. 1971. Mechanism of stimulation in the mixed leukocyte culture. In *Proceedings of the 6th Leukocyte Culture Conference* (ed. R. Schwarz), p. 623, Academic Press, New York.

Peck, A. B., and F. H. Bach. 1975. Mouse cell-mediated lympholysis assay in serum-free and mouse serum supplemented media: Culture conditions and genetic factors. *Scand. J. Immunol.* **4:** 53.

Plate, J. D. 1976. Soluble factors substitute for T T cell collaboration in generation of T-killer lymphocytes. *Nature* **260:** 329.

Schendel, D. J., and F. H. Bach. 1975. H-2 and non-H-2 determinants in the genetic control of cell-mediated lympholysis. *Eur. J. Immunol.* **5:** 880.

Sondel, P. M., M. W. Jacobson, and F. H. Bach. 1975. Pre-emption of human cell-mediated lympholysis by a suppressive mechanism activated in mixed lymphocyte cultures. *J. Exp. Med.* **142**: 1606.

———. 1977. Selective activation of human suppressor cells by a nonproliferative stimulus. *Eur. J. Immunol.* **7**: 38.

Sopori, M., B. J. Alter, F. H. Bach, and M. L. Bach. 1977. Cell-free factor substitutes for "signal 2" in generating cytotoxic reaction. *Eur. J. Immunol.* **7**: 823.